The Land of the Solstices

Myth, geography and astronomy in ancient Greece

TOMISLAV BILIĆ

BAR INTERNATIONAL SERIES 3039 | 2021

BAR PUBLISHING

Published in 2021 by
BAR Publishing, Oxford

BAR International Series 3039

The Land of the Solstices

ISBN 978 1 4073 5862 8 paperback
ISBN 978 1 4073 5863 5 e-format

DOI https://doi.org/10.30861/9781407358628

A catalogue record for this book is available from the British Library

COVER IMAGE *Helios, Nux and Eos on a black-figure lekythos by the Sappho Painter (The Metropolitan Museum, accession number 41.162.29). Open Access, CCO 1.0..*

BAR
PUBLISHING

BAR titles are available from:

BAR Publishing
122 Banbury Rd, Oxford, OX2 7BP, UK
EMAIL info@barpublishing.com
PHONE +44 (0)1865 310431
FAX +44 (0)1865 316916
www.barpublishing.com

Of related interest

Beyond Paradigms in Cultural Astronomy
Proceedings of the 27th SEAC conference held together with the EAA
Edited by A. César González-García, Roslyn Frank, Lionel D. Sims, Michael A. Rappenglück, Georg Zotti, Juan A. Belmonte, Ivan Šprajc
Oxford, BAR Publishing, 2021

BAR International Series **3033**

Astronomy and Power: How Worlds Are Structured
Proceedings of the SEAC 2010 Conference
Edited by Michael A. Rappenglück, Barbara Rappenglück, Nicholas Campion, Fabio Silva
Oxford, BAR Publishing, 2016

BAR International Series **2794**

SEAC 2011 Stars and Stones: Voyages in Archaeoastronomy and Cultural Astronomy
Proceedings of the SEAC 2011 conference
Edited by F. Pimenta, N. Ribeiro, F. Silva, N. Campion, A Joaquinito and L. Tirapicos
Oxford, BAR Publishing, 2015

BAR International Series **2720**

Lieux de culte et parcours cérémoniels dans les fêtes des vingtaines à Mexico - Tenochtitlan
Elena Mazzetto
Oxford, BAR Publishing, 2014

BAR International Series **2661**

Water Management: The Use of Stars in Oman
Harriet Nash
Oxford, BAR Publishing, 2011

BAR International Series **2237**

Sense and Nonsense in Homer
A consideration of the inconsistencies and incoherencies in the texts of the Iliad and the Odyssey
John Wilson
Oxford, BAR Publishing, 2000

BAR International Series **839**

Acknowledgements

The writing of this book was a rather solitary endeavour, so the list of acknowledgements will be relatively short.

The kernel of the present text, but not its development, i.e., neither the methodology I have devised and used throughout, nor either the overall or detailed argumentation, is my Ph.D. thesis at the Zagreb Faculty of Humanities and Social Sciences. It was inspired by the lectures of my thesis supervisor Aleksandar Durman, as well as those of Marina Milićević Bradač, also of the Zagreb Faculty, and an examiner on my thesis committee. I am grateful to both for their support and guidance, both during my graduate and post-graduate studies. Also, my gratitude extends to Ivor Janković of the Institute for Anthropological Research in Zagreb, another examiner on the committee and a scholar well-versed in all kinds of methodological questions.

Further, I would like to thank my home institution, the Archaeological Museum in Zagreb, more precisely, its successive directors during the last decade, for allowing me the freedom to pursue my professional interests.

I must also emphasize the unconditional support of Ivan Mirnik, my senior colleague at the department, who always acted as, consciously or not, a true pillar of sanity, of unmatched poise and confidence. I would also like to thank the many colleagues with whom I have discussed (whether directly or online) various subjects treated in the book, as well as the librarians at the Archaeological Museum in Zagreb, the National and University Library in Zagreb, the Library of the Faculty of Humanities and Social Sciences in Zagreb and the British Library, for all their help in obtaining the often difficult to get books and papers. Also, my thanks extend to numerous colleagues who were so kind to send me their published works, either in electronic form or as hard copies, often answering a request of someone they have never heard of.

Last, but not least, I would like to thank the editorial team at BAR Publishing, especially Jacqueline Senior and Ruth Fisher, for accepting this manuscript, as well as for their helpful insights and a collegial and constructive atmosphere in seeing the manuscript through the production process. I am also grateful to two anonymous readers for the press for their insightful comments. Likewise, since earlier versions of some segments of this book were previously published in scholarly journals, I would like to thank the editors of these journals and, especially, the readers who reviewed these papers for their comments and suggestions, which have contributed significantly to improve these papers, and thus also the present book.

On a more personal note, I must thank my family, my wife Ivana and our two sons Tristan and Nikola, for their patience for the long hours spent apart because of my silly and seemingly never-ending task.

Finally, there is one person that has always absolutely and unconditionally supported my pursuits, never for a moment doubting their feasibility or usefulness: my mother, Nada, to whom I would like to dedicate this book.

Contents

List of figures

Abstract

Following a recent upsurge of interest in ancient geography and astronomy, together with the ever-present scholarly fascination with myth, this book offers a fresh study of what is commonly but erroneously known as 'solar myth'. This subject-matter is at present at the margins of scholarly interest, mainly due to the now outdated comprehensive theories of myth that used solar phenomena as an interpretative key by which they proposed to explain the greatest number of, if not all, traditional narratives conveniently called myths by modern scholars. In comparison with these dated approaches, this book offers a more rigorous methodology and more selective interpretation applicable to a group of particular myths, those referencing solar phenomena. The class of 'solar' myths discussed in this book is thus formed out of traditional narratives—and iconographic designs on the objects of material culture—that either explicitly include references to solar movement or the recognition of such references does not require strained interpretations. The book approaches its subject-matter from a cross-disciplinary perspective, in that it broaches the subject of ancient myth alongside ancient geography and astronomy, with the particular geographical and astronomical topics studied here being subsumed under the term 'meteorology' by the ancient Greeks. These considerations also prompted a discussion of the relation of myth and science in general, which is considered here from the perspective of the modern history of science.

The key hermeneutical tool used in this book for assessing traditional tales belonging to the corpus of stories usually called Greek myth, as well as from other comparable corpora—including material evidence from prehistory—and to offer a new interpretation of some of these stories, is the recognition of some myths as models accounting for natural phenomena in traditional societies. By traditional societies I understand those devoid of precise professional terminology and experts that would use it in discussing natural phenomena. More precisely, following some other researchers in choosing only one class of Greek myths and interpreting them in terms of solar movement, I have joined the consensus of scholars who almost unanimously discard comprehensive theories of myth and approach its study in an eclectic fashion (with an important caveat that some strands in this eclecticism are more privileged than others). These proper 'solar' myths (in the meaning of narrative models referencing or describing and accounting for solar movement) are not particularly prominent neither in the corpus of Greek myths nor elsewhere. Even so, a number of stories indeed incorporate references either to diurnal or annual solar movement and related phenomena in their narrative, which are as a rule rather explicit, or else do not require strained interpretations. These stories, as well as the iconographic designs that were derived from them, deserve an adequate interpretation, and one is offered in this book.

Another important hermeneutic tool provided by the recent—and some not so recent—developments in the history of science and, more broadly, in analyses of different cognitive strategies applied by humanity when dealing with its environment, is the recognition of the role of anthropomorphism in accounting for phenomena in objective reality. With respect to the former, it is the endorsement of myth as a legitimate explanatory tool in accounting for the phenomena, together with a neo-Tylorian recognition of two different ways in accounting for causality—by personalistic forces in traditional societies, and in terms of impersonal forces in non-traditional ones—that offer a strong basis for interpreting the group of myths discussed in this study as referring to or being the accounts of, i.e., descriptions and explanations of, natural phenomena. With respect to anthropomorphism, it is now recognized as a general cognitive strategy shared by myth and science and an almost universal explanatory method, which provides an opportunity to re-examine Greek and non-Greek myths that refer to solar movement from this perspective. Greek material was not previously discussed in this context, or it was touched upon only tangentially, but Greek tradition offers a large body of evidence for both myth and, more generally, anthropomorphism as an explanatory method accounting for, among other things, natural phenomena. This book focuses on one, albeit prominent, process in nature, i.e., the diurnal and annual solar movement, referred to and accounted for in terms of myth and the reflections of this practice in—specifically—Greek astronomy and geography proper.

The understanding of some myths as narrative models representing a specific aspect of physical reality, applicable on the class of stories discussed throughout the book, i.e., those referencing the solar movement, approaches some recent developments in the study of human cognition and the role myth played in its development. Once again, this understanding of myth was never before, to the best of my knowledge, applied onto specifically Greek myths, or to stories involving a reference to solar movement in general. The status of 'solar mythology' in modern scholarly studies is, perhaps at this point understandably, extremely unenviable, and my interpretation of this class of myths is an attempt to validate its presence—on a significantly reduced scale in comparison to earlier comprehensive theories of myth—in the study of, in the first place, Greek myths. An important hermeneutic tool in this respect is the recognition that myths should be studied in their ethnographic context. Throughout this book myths are studied alongside contemporary or later astronomical and geographical developments, thus showing their complementarity but also the continuity between these two explanatory methods.

1

Introduction

The early 21st century, in the wake of the developments in the second half of the 20th, can be confidently defined as a period of eclecticism in its approach to myth; the same applies to the apparently polyparadigmatic field of study of religion in antiquity, which naturally includes the study of Greek myth.[1] However, some approaches are unsurprisingly favoured on account of others—those more in accordance with current scholarly sensibilities—while others are considered outdated or superseded by more appropriate ones. Thus, even in eclecticism and polyparadigmaticism there is a relatively clearly defined centre and margin. This seems the only way in which research can properly function and is thus hardly controversial; but one should always be aware of the commitment—for various reasons—to a dominant paradigm and its ideological background as important factors in streamlining current research. Concomitant to this, one should also be aware of the often-ideological background in discarding earlier scholarship, occasionally based on inadequate research and hastily accepted by the majority of scholars in order to 'move on' unhindered by the often silly—in hindsight—and methodologically flawed approaches of the past. Once again, this seems to be the way in which research regularly develops, but it does not seem as non-controversial as the preceding point on the centre-margin relation in polyparadigmaticism. In itself, this does not entail that the current approaches are in some way flawed; but it does suggest that their attention is perhaps too heavily focused on some facets of evidence on account of others and that some lines of evidence are considered irrelevant on incompletely justified grounds. It seems that precisely those categories of evidence that were central, or at least very important, to earlier approaches, are considered peripheral to those that succeeded them and now dominate the field. This does not evaluate properly their relevance (neither did their central place in earlier paradigms), but more correctly measures the interest of current scholarship and the preoccupations of researchers working under a dominant paradigm, or rather the central core of a polyparadigmatic system. This has an unwanted consequence of leaving the body of evidence—and its interpretation—central to a once dominant paradigm in the state of research appropriate for the time,[2] which puts it in an unfair position in comparison with the evidence appropriated by current research. The subject of solar movement, especially its annual aspect, as envisaged in myths, is a prominent example of such practice. Once both an attractive field of study and an important interpretative tool, it is now considered a subject almost unworthy of serious scholarly engagement. In addition, another reason for its peripheral status in current research is its technical nature, which seems to have restrained modern scholars from engaging with this subject. Finally, with the move from substantive to socio-cultural framework of interpretation of religion,[3] the 'physical' content of myth was overshadowed by other concerns. These circumstances leave the subject of solar movement mostly in the hands of amateur researchers,[4] which, in its turn, in a vicious circle even more downgrades the potential attractiveness of the subject for the researchers working under the current polyparadigmaticism, i.e., it relegates it to a distant margin of scholarly interest frequently tainted by 'unscientific' and 'unserious' feelings it evokes.

It seems inevitable that the once discarded interpretative approach is marginalised together with the body of evidence it once held central for its paradigm. The treatment of (especially annual) solar movement in myths is a particularly illuminating example of such practice. It was once the main pillar of the solar-myth (or nature-myth in a wider sense) school that dominated the 19th-c. discussions on the subject, which culminated in W. H. Roscher's work, especially his famous *Mythologisches Lexicon*.[5] Despite the heavily ideologised dismissal of the solar/nature-myth paradigm,[6] as well as the inadequacies of its rival and conqueror, the anthropological-fertility paradigm, the eccentricities of the former frequently did not require an especially meticulous criticism in order to be refuted. But since the threat in the form of solar/nature-mythology no longer lurks in the background (it was already long-dead when R. Dorson in 1955 wrote an article titled 'The Eclipse of Solar Mythology'), it seems an opportune moment to attempt to reassess some of its subject matter, emancipated from the interpretative method that was convincingly proven to be inherently unusable.[7]

1 On the concept of paradigm in the humanities see Ober 1989: 136–37 and Versnel 1994: 11–14, 86 (= 1990a: 65). On the eclectic approach to the study of myth see immediately below.

2 Cf. Goodison 1989: xiii, citing M. L. West's inaugural lecture of 1975; Hämeen-Anttila 2006: 6.

3 Versnel 1994: 7–11. For the ideological background of the denouncement of intellectualist explanations of religion by modern scholars see Horton 1968: 629–32 and Versnel 1994: 8 n. 13.

4 Such as Bailey 1997; 1998.

5 Versnel 1985–86: 134–35 = 1994: 289–92; Konaris 2010: 484–87; 2016: 123–30.

6 See Dorson 1955; Arvidsson 2006: 125–31, 168–73, 176; Konaris 2010; 2011; 2016.

7 For a consensually negative assessment of the work of the leading exponent of solar-myth theory, F. M. Müller, see Dorson 1955: 394–405; de Vries 1984[1967]: 36–40; Kirk 1974: 43–44; Rogerson 1974: 34–36, 40–43, 56, 176; Ackerman 1975: 116–17; Dowden 1992: 18–19; Davidson 1993: 146–48; Graf 1993: 25–26; Turner 1993: 346–47; Lincoln 1999: 67; Von Hendy 2001: 78–83; Segal 2004: 20; Csapo 2005: 19–30; Arvidsson 2006: 73, 76–83, 87–90, 127, 130–31, 311; Calame 2009a: 23–24; Konaris 2016: 106–23.

In general, I will attempt such a reconsideration of some of the subject matter treated by the solar/nature-mythology, although my emphasis will not be on the actual source-material gauged by the 19th-c. scholarship faithful to this paradigm, since most of it was constructed by the proponents of the paradigm themselves by strained or completely implausible interpretations. Thus, this essay is intended as a reintroduction of the study of 'solar' *material* engaged by myth into mainstream scholarship, but without the package of discarded and untenable interpretative approaches. In doing this I will decidedly focus on Greek myth, but will occasionally incorporate material from other ethnographic contexts, in the first place the Mesopotamian and, in a lesser measure, Egyptian. At the same time, I will not offer a monolithic theory of myth, but only an interpretation of a provisionally yet clearly defined group of traditional narratives in their respective ethnographic contexts—in the first place a group of Greek narratives and their particular cultural setting.

More specifically, I will try to provide an analysis of the presence of the results of the observation of diurnal and annual solar movement, as well as other related phenomena (such as the fixed, i.e., geographical, arctic circle),[8] in certain narratives, in the first place those that are customarily classified as myths. This presence manifests itself in two different ways: as incorporation of motifs drawn directly from the observation and/or interpretation of solar movement (representing the main body of evidence) and as actual narratives built upon solar movement (represented by a modest number of stories). In addition, I will study the perpetuation of this presence in those texts usually understood as closer to the modern scientific standard.

With respect to the latter category (the narratives built upon solar movement), the spiral of scientific development[9] has now, I believe, reached the point when some of the central preoccupations of 'solar mythology', devoid of its extremist and extravagant interpretative environment, can be reintroduced into discourse on ancient beliefs, certainly not as forcefully as when it dominated the field, but only as a useful hermeneutic approach in certain rigorously selected cases.[10] An obvious and very pertinent case in point is the anthropomorphisation of celestial bodies, in the first place the sun, and the explanation of their movements and related phenomena in terms of human behaviour. What I am arguing for here is simply an acknowledgement of the fact that the interpretation of some mythic motifs, episodes or narratives in terms of solar movement can be justified if applied selectively and judiciously. Indeed, this type of interpretation is widely if implicitly accepted in the treatment of such straightforward myths as that of Helios' journey in his chariot or cup, but the hermeneutic principles involved in such an interpretation are rarely followed out to their logical end.

Without taking into consideration the now often neglected substantive facet of mythic motifs, narratives and clusters of stories, notably those referencing solar movement discussed in the main part of this book, the fertility, shamanistic,[11] ethological or sociobiological,[12] myth-ritual[13] or initiatory[14] paradigms fail to provide a framework for persuasively explaining, for example, the Laestrygonian episode from the *Odyssey*, which will be treated in chapter 2 of this study, even though I readily accept that some of these interpretive approaches can cogently explain at least some portions of the traditions discussed here (including the Laestrygonian episode).

The substantive facet of the myths that will be studied in this essay, i.e., their subject matter, will in the course of time become the object of study of the newly-emerged disciplines of geography and astronomy. Considering the widely diverging understandings of basic terminology and the concepts behind it, it seems fair to say that deciding whether there is any relation between myth on one side and geography and astronomy on the other (the two disciplines which jointly treated the problem of solar movement) in the first place depends on the definition of myth one accepts, as well as on the definition of science.[15] The aspect of this relation in which I am interested pertains to the potentially shared subject matter, more precisely, the common segments of the respective objects of study shared by these two intellectual strategies, but also, in a lesser measure, to their potentially corresponding approaches and goals, in the first place to the question whether myth can have and was intended to have a descriptive and explanatory function (as science certainly by definition did and still does). In this context it is important to emphasise that I

[8] The geographical (fixed) arctic circle is at a latitude corresponding to a polar distance equal to the obliquity of the ecliptic (i.e., at ca. 66°30'), while the arctic circle in Greek sense (i. e. an always-visible circle) varies with latitude.

[9] Ackerknecht 1954: 124–25, cited in Versnel 1990a: 66–67 = 1994: 87–88.

[10] Compare Goodison 1989 for a similar approach in a compatible field of study and Fowler 1993: 41 for a plea for reintroduction of myth-and-ritual paradigm—not in its original form, but naturally updated to modern standards—in the study of myth.

[11] On which see the refutation in Bremmer 2002: 27–40, 145–51; 2016; 2018 and Zhmud 2012: 207–12.

[12] This approach is mainly championed by W. Burkert, on whose work see Versnel 1990a: 61–62, 64–65 = 1994: 77–79, 83–84, Csapo 2005: 163–80 and Iles Johnston 2018: 50–52; on the ideological background of the 'biological' in 'sociobiological' see Csapo 2005: 162, 172 (on the 'social' see below).

[13] Convincingly criticized by Fontenrose 1971; cf. Iles Johnston 2018: 34–55, 57–63.

[14] For the rise of the initiation paradigm and its supplanting of the earlier fertility paradigm see Graf 2003: 5–6, 19–20 (cf. Versnel 1990a: 50 = 1994: 59 and Fowler 2013: 174); for its ideological roots see Graf 2003: 6, 8, 19–20 and Lincoln 2003, esp. pp. 249–50; for its diminishing relevance see Graf 2003: 20 and Lincoln 2003: 250; for an analysis of its approach to myth from an emic and reflective perspective see Graf 2003: 15–19.

[15] The Greeks of the early period would have used the term 'meteorology' for the discipline that treated solar movement; I will not use this term, since it would be confusing to modern readers, who expect something else of this discipline. At the same time, when I use terms such as 'astronomy' and 'geography' in discussing pre-Hipparchean material, I am aware of the anachronistic character of such practice. When used in this context, the terms such as 'astronomical concepts' or 'geographical tenets' must be understood in a qualified sense, but I believe it would be pedantic to qualify them on every occasion.

am going to treat ancient scientific and narrative accounts of solar phenomena concurrently not with respect to their accuracy, i.e., their correspondence with objective reality, but as attempts at description and explanation (with differing accents of these respective approaches on these two activities) of the specific phenomena of physical reality. At the same time, I will put an accent on narrative accounts, given that the scientific ones are easily understandable (with the exception of some controversial points, which will be treated in detail), since they are given in an idiom almost identical to the one used in modern western scientific culture, its direct descendant.

Naturally, the observation and interpretation of solar movement was always recognised as an important subject in the study of ancient science, but it was rarely associated in any profitable way with the widespread solar elements in ancient myth and cult, except perhaps in Egypt, where the solar element in myth and religion is difficult to ignore. However, the interconnectedness of these two traditions—mythic and 'rational'—with respect to the sun and solar phenomena was never truly discarded: Copernicus' in the long run successful reintroduction of the sun in the centre of universe was, in a way, its final offshoot.[16]

It is well out of scope—and contrary to its fundamental postulates—of this study to attempt to define myth in general; but I will try to bring into sharper focus those pertinent characteristics of some narratives usually labelled myths and their relation with—specifically—geography and astronomy that will allow a coherent interpretation of a group of mythic motifs and myths in terms of the subject matter shared by the two cognitive approaches conveniently labelled 'mythic' and 'rational'. At the same time, I am not privileging the group of myths that in some way treat the solar movement as representing 'real' myths, as opposed to others that do not fall under this provisional definition, nor do I find them paradigmatic for myths in general; in this way I believe I have succeeded to circumvent the circularity often met in discussions on myth.[17] In this sense, it will be argued that the term 'mythic', with reference to traditional accounts of solar movement in general and solstices in particular discussed in the main part of the book, could be considered as designating a discourse that treated identical referents with the discourse styled 'scientific', with the main difference (for the purpose of this study) between the two in the use of their characteristic ways or idioms in describing and explaining them.

1.1. Selective interpretation of myth

In comparison to the 'standard' comprehensive theories of myth, a more selective interpretation applicable to a group of particular traditional accounts, those referencing solar phenomena, is offered here. This approach is in this respect similar to Dowden's, whose initiatory theory explains, as he himself is happy to recognise, only one class of Greek myths.[18] Similarly, Fowler interprets a group of myths (Cephalus and Procris, Cephalus and Eos, Philonis, Ixion, Golden Fleece) in terms of weather and seasonal rituals, in the first place rain magic, without applying this interpretative approach on the entire corpus of Greek myth.[19] With respect to ancient precedents to this approach, all ancient hermeneutic traditions selectively focused on those particular myths that were relevant, or seemed to be relevant, to their particular interpretative strategy.[20]

The interpretative approach advanced here proposes to explain another class: narratives referencing either diurnal or annual solar movement, as well as certain related phenomena. The myths involved either explicitly include references to solar movement or their recognition does not require strained interpretations. This is a simple and I believe noncontroversial general principle that does not involve many theoretical and methodological subtleties, even though the recognition of references to solar movement in some cases will require a more detailed explanation. At the same time, the principle is rigorous up to a point that allows a profitable—or any type of—discussion on the subject, with the alternative being the perpetuation of ignoring this type of testimonies. It will be shown that the corpus of narratives built upon solar movement is very modest, but that there is a large body of myths that incorporate motifs drawn directly from the observation and/or interpretation of solar movement. However, the modest number of narratives actually built upon solar movement, to paraphrase Parker, is not what matters, but their very existence needs to be accounted for, and this will be attempted here.[21] Both these categories of material testify to the importance of the observation and interpretation of solar movement in various mythic traditions, in the first place Greek, while the interest of myth-makers in these phenomena also speaks something of the nature of the respective mythic traditions as an appropriate medium for discussing this type of phenomena. Finally, I will not discuss in any detail the unambiguous and noncontroversial 'standard' account of the diurnal solar movement in Greek tradition, which is accounted for by the well-known model of chariot-ride, including its best known iteration in the Phaethon myth, except where it has some repercussions on other less-known pieces of

[16] Cf. Yates 1964: 153–55; Kahn 2001: 161; Gee 2013: 180–83. The importance and centrality of the sun—if in a different way—was already heavily emphasized in scholastic cosmology (Grant 1994: 226–27, 231–33, 235, 311, 452–54).

[17] Csapo 2005: 2; Hawes 2014: 73.

[18] Dowden 2011b: 494.

[19] Fowler 1993: 33–34, 36–40; 2013: 184–85, 461–62. Parker (2011: 25–26) distinguishes between different classes of myths dealing specifically with gods.

[20] Hawes 2014: 35, cf. 6, 11, 13, 24, 79–80, 89–91, 114–16, 194, 201, 205, 212, 225. In a somewhat similar fashion Plato offered an astronomical explanation of some myths (*Polit.* 268e–274e, *Tim.* 22a–23c), although elsewhere he discarded the physical explanation of myth in general (*Phaedr.* 229c–230a).

[21] Parker 2011: 79 (cf. Konaris 2016: 271) argued that it is not the modest scope of the worship of natural phenomena in Greek cults practice that matters, but that its very existence should be explained.

evidence. In-depth discussions of this familiar model are readily available in a number of studies focusing on early Greek astronomy or myth.[22] Yet this mythic model must always be borne in mind as a paradigmatic example of the Greek practice of accounting for solar phenomena in anthropomorphic terms. [23]

The selective approaches conform to the dominant scholarly paradigm on the subject, which strongly discards the notion of monolithic theories of myth.[24] In general, one could argue for the move from dominantly dogmatic to dominantly sceptic approaches to myth as a characteristic of modern—i.e., late 20th and early 21st-c.—studies of myth. Somewhat surprisingly, this process resulted in a creation of a 'hermeneutic environment' analogous to the one characteristic of the approach of Greeks themselves to their myths.[25] This position entails, even if its proponents do not always state it explicitly, the acceptance and implementation of different interpretative strategies in myth analysis.

1.2. Ethnographic context

This polyparadigmaticism is consonant with the emphasis on myth's dependence upon and relation to its cultural setting, entailed in the recognition of the so-called ethnographic context, which is a rather commonsensical concept that offers more possibilities for understanding and interpreting myths than many of the 'monolithic' theories. In its turn, the concept of ethnographic context nicely illustrates the complementarity—although not an identity—of myth and scientific accounts. This interpretative tool will be used, often implicitly, throughout this study, giving support to a concurrent treatment of narrative and descriptive accounts of phenomena of physical reality.

Ethnographic context was already evoked by Lévi-Strauss, even though he used it sparingly in his work. He insisted that myths should be interpreted by reference 'to the ethnography of the societies in which they originate', analysing their different aspects, such as technology, economy, political and familial structures, systems of representation, aesthetic expression, ritual, religion, etc.[26] The concept was further elaborated by Detienne, who introduced the notion of a specifically Greek ethnographic context, which is composed of both 'economic, technical or religious facts' and also 'different branches of knowledge—botany, medicine, the study of religious festivals' etc.[27] It is thus extremely heterogeneous, and must be studied by an analysis of various complementary sources—including both myths and scientific writings.[28] It is possible to recognize these notions characteristic for Detienne's method in the studies of other adherents of the Paris School (Vernant, Vidal-Naquet, etc.), who in their discussions of Greek myth focus on the ethnographic context of ancient Greece, consisting of a wide range of cultural phenomena.[29] Mythic narratives are thus studied in their cultural context that encompasses different branches of knowledge, including the ones that would later be called respectively mythic and scientific.

The notion of ethnographic context is, however, often left implicit in modern classical scholarship. Specifically with respect to the subject of this book, the interrelation of various intellectual and other cultural phenomena is implicitly acknowledged, without recourse to the concept of ethnographic context, when ancient literary tradition is accepted as a relevant source for the study of ancient geography. This remains the case even though the perception of geographical space of the authors working in this tradition is recognised as more subjective in comparison with a more 'scientific' outlook of some ancient geographers, such as Eratosthenes.[30] Certainly, myths—together with poetry in which they often appear—are acknowledged as an important source in the study of ancient geography,[31] while, in its turn, ancient geographical tradition shows a remarkably 'literary' (i.e., descriptive and narrative) character.[32] A consideration of both corpora (if one wishes to retain the traditional division between them)

[22] E. g. Dicks 1970: 31–34; Gantz 1993: 30–34.

[23] As opposed to Eos' chariot (*Od.* 23.243–46), Helios' chariot only appears in a post-Homeric tradition (Gantz 1993: 30). For its earliest appearances (7th–6th c.) see Hes. fr. 311 M-W (Gantz 1993: 31, 33 is undecided on which part—if any—of Hyginus' report came from the Hesiodic author); [Hes.] fr. 390 M-W *ap.* Σ A. R. 3.309–13b (p. 229 Wendel) (Malkin 1998: 188–89 believes that this fragment could actually be Hesiodic); Mimner. fr. 12.1–3, 9–11 West; *hDem.* 88–89 (late 7th–early 6th c., Faulkner 2011: 10); *Titanomachia* fr. 7 West (late 7th to late 6th c., West 2002: 109, 129–30); *hHer.* 68–69 (second half of the 6th c., Vergados 2013: 130, 145–47). Earliest iconographical attestations: *LIMC* V.1.1008, 1015 s.v. *Helios* 1 (670–660 BC: Theran neck-amphora), *Helios* 2, 95, 97 (510–500 BC: Attic lekythoi), *Helios* 101 (late 6th c.: Attic lekythos).

[24] Kirk 1970; 1972a: 76; 1972b: 84; 1984[1973]: 54–55, 58–61; 1974 (cf. des Bouvrie 2002: 22); Honko 1984[1972], 46–47; Rogerson 1974: 174; Burkert 1985: 9; Wyatt 2005[1987]: 28–29; Versnel 1988: 121 = 1994: 89; 1990a: 27, 66 = 1994: 19, 86–87; 1990b: 29; 1994: 7, 11, 13; Edmunds 1990: 17; Dowden 1992: 16, 24–25; Davidson 1993: 151, 160; Patton & Doniger 1996: 2–3, 21; Lightfoot 1999: 231; Von Hendy 2001: 251, 277; Csapo 2005: 7–8, cf. 290–91; Bremmer 2011: 539 (conditionally); Parker 2011: 22, 25; Dignas & Audley-Miller 2018: viii; Iles Johnston 2018: 6–7.

[25] Hawes 2014: 10–13, 24–25, 35, 93, 103–15, 120, 152 (citation from p. 25); on Plutarch in this context see Hardie 1992: 4760.

[26] Lévi-Strauss 1976: 65; cf. Segal 1980b: 32–33; Champagne 1992.

[27] Detienne 1994: 130; for ethnographic context cf. 1981: 98–99, 106, 108; 1994: 143; 1999: 140 = 2009: 36–37.

[28] Detienne 1994: 131, cf. 1981: 107. For Detienne's method cf. Champagne 1992 (esp. chapter 6); Calame 2011: 512–13.

[29] Champagne 1992 (who refers to the exponents of this approach as 'contextualists' or the 'contextualist school'); Von Hendy 2001: 271–72. For the notion of (neo-structuralist) 'school' cf. Buxton 1981: x; Versnel 1990a: 27 with n. 7 on pp. 68–69 = 1994: 18 with n. 7; 1990b: 28–29; Champagne 1992.

[30] Meyer 2001: 225–27, cf. Lightfoot 2014: 8.

[31] See Dueck 2012: 20, 24, 27, 51 for the importance of myths in the study of ancient geography and pp. 20–35 for geography in poetry and myth. For geography in poetry see also Nicolet 1991: 8, Skinner 2012: 121–22, 124, Lightfoot 2014: 8–9, Dan 2017: 172, 189–91 and Kaplan 2018: 196–201 and for fiction in general Clarke 1999: 23–25 and Roller 2018: 325–26. For astronomy in poetry, including 'philosophical', see Kidd 1997: 12–13 and Van Noorden 2015: 173 with n. 30; for cosmology, see Hardie 1986: 5–11, 22.

[32] Van Paassen 1957: 56–57 (on Eudoxus' geographical work) and *passim*; Romm 1992: 3–5, 7; Clarke 1999: 23–25; Meyer 2001: 225–27; Lightfoot 2014: 8. For 'descriptive geography' see in general Dueck 2012: 20–67, cf. Poiss 2014: 70.

seems as a prudent methodological approach to the study of a concept that appears, if with a different emphasis and from a different point of view, both in myths and in non-mythic texts.

1.3. Inclusive definitions of science

Several modern definitions of science are sufficiently broad to encompass traditional disciplines that are excluded if the criteria for delineation of modern science are applied. On the other hand, the definitions reached by judging ancient science by the standards of modern seem too restrictive and are useful for distinguishing 'modern science' in antiquity, but not 'science in antiquity'.[33] They exclude much of material that—according to a recent nuanced classification of ancient geographical knowledge—belongs to the parts of the discipline that can be styled intuitive/naïve and scholarly/canonical, encompassing only the scientific/fully reasoned category,[34] or that—according to a similar classification of ancient astronomical knowledge—belongs to the parts of the discipline that can be styled metaphysical and cosmological or speculative astronomy, encompassing only the mathematical astronomy, itself developing from the scientific or descriptive astronomy.[35]

These nuanced classifications offer some useful distinctions which suggest that not everything deserving of the name geography or astronomy should be judged by the criteria of mathematical, or even descriptive, science. At the same time, the place of myth—specifically those myths that refer to solar movement in some way—in these classifications is not specified. Instinctively, it would probably be placed in the 'lowest' respective categories, those farthest from the mathematical part of the disciplines. The inclusion of this type of myths in a discussion of ancient geography or astronomy seems justified since empirical observations were in different ways involved in the creation of mythic accounts referencing solar phenomena, while anthropomorphic causality and narrative accounts of phenomena were certainly helpful in their understanding, thus satisfying at least the minimal criteria for inclusion under the terms 'geography' and 'astronomy', however qualified.

Another favourable view of the scientific nature of early astronomy finds in it a number of traits of modern science: 'systematic and selective observation, and the collection, coding, and eventually the visual storage of data; the analysis of stored data for regularities and cohesive structure; and the formulation of predictions on the basis of these regularities'.[36] The materialisation of these procedures can be recognised in Neolithic, Chalcolithic and Bronze Age 'megalithic observatories', interpreted as symbolic analog models of time and space used for marking, measuring, predicting and verifying periodical events, including the annual solar movement.[37] What is missing in this discussion is a clear understanding that any 'theory' (a term used by Donald) that stood behind these models must have been expressed in terms of anthropomorphic causality, i.e., myth. Another interpretation of early buildings, this time solstitial orientations of various Egyptian temples dedicated to the solar deity, as analog devices, explicitly recognizes that the Egyptian knowledge of astronomy was expressed in stories involving gods, i.e., in mythic discourse.[38]

These inclusive definitions of science allow the comparison of mythic cosmologies with later attempts at studying similar phenomena. Specifically, the acknowledgement of myths—narrative accounts of phenomena in terms of anthropomorphic agents—as speculative explanations based upon, in equal measure, intuitive ideas and empirical observations is of special importance in terms of my discussion in general.[39] One could thus attempt to assess traditional prehistoric observations and interpretations of solstices (as an important concept in the context of solar movement), dating perhaps as early as the Upper Palaeolithic period,[40] by using a set of epistemological and methodological criteria for defining science, such as empiricism, that is, information obtained from direct observation of phenomena as foundation of all knowledge, theory as a result of the study of these phenomena, and prediction of phenomena on the basis of theory.[41] With respect to Donald's example ('observatories'), the observation of solar movement can be recognised in the orientations of certain structures towards the rising and setting points of the sun at solstices. These orientations could have been implemented in the layout of these architectural features only by the application of the data gathered through the observations of the annual solar movement. But the observations were interpreted in terms of regularities—the recognition of the cyclic nature of the annual solar movement—resulting in the manifest ability to predict the sun's return to its extreme rising and setting points, without which the orientations would be pointless. In addition, the recognition of regularities in the sun's behaviour must have been verbalised, i.e., described and explained, in terms of anthropomorphic causality in the form of stories. One such description and speculative explanation of the annual solar movement based upon

[33] E.g., Neugebauer 1945: 7; Clagett 1955/1957: 4, cf. Lloyd 1970: 1; Graham 2013: 256. For a criticism of such approach see Bowen 2018: 293–94.

[34] Dan, Geus & Guckelsberger 2014: 19–21, 26–31.

[35] Couprie 2011: xxviii–xxxii. For a similarly sensitive and inclusive approach to ancient ethnographic knowledge, although with no attempt at classification, see Skinner 2012, *passim* (but especially pp. 7–8, 14–17, 43–44, 49–50, 59, 133, 233–36, 241–43, 255).

[36] Donald 1991: 339.

[37] Donald 1991: 335, 337–39, cf. 340.

[38] Lettvin 1980: 134–35. Lettvin's inferences with respect to the orientations of Egyptian temples are indeed supported by later studies (Shaltout & Belmonte 2005; 2006; Shaltout, Belmonte & Fekri 2007; 2008; Belmonte, Fekri, Abdel-Hadi, Shaltout & González García 2010). For the Egyptian use of metaphors rather than technical vocabulary in describing celestial phenomena and for the overlap in astronomical and religious use of certain terms see Ross 2020: 191 with n. 3.

[39] Cf. Popper 1982: iii.165; Gattei 2009: 33, 102 n. 4.

[40] Ruggles 2005: 318, 386; Hayden & Villeneuve 2011: 332, 347, 350–51.

[41] Donald 1991: 339; Rochberg 2004: xv.

intuitive ideas—but also empirical observations—is the story of Apollo's seasonal voyage, which can be interpreted as a narrative model that represents a specific aspect of physical reality in the context of early Greek society (chapter 7). Similar narrative models can be recognised in Homer's account of the land of Laestrygonians (chapter 2) and the island of Syriê (chapter 5), while some resonances of this approach can be found in Pytheas' much later account of the island of Thule, even though the latter is certainly based on solid scientific theory and a corpus of observational data.

This takes us back to the notion of ethnographic context, which seems an appropriate tool for approaching evidence belonging to different categories of texts without prejudicing the intentions of their authors solely on account of their literary classification (i.e., an epic or lyric poem, a mythic or mythographic narrative, an astronomical handbook, a periplus or periegesis, etc.). Therefore, a reference to solar movement in a dominantly literary work—such as a retelling of a myth—should be treated as an expression of familiarity with the concept, even though there is a good chance that this expression would be more idiosyncratic (or idiosyncratic in another way) than an account of solar movement in a manifestly descriptive and explanatory treatise.

1.4. Myth and physical phenomena

The appropriation of the results of the observation of solar movement by the myths discussed in this book, in the first place by the limited number of myths built upon solar movement, and their idiosyncratic treatment of these pieces of information can be explained by two fundamental characteristics of myth in general: its narrative structure—myth being a 'syntagmatic chain of "motifemes"'—and its referencing of phenomena of common reality.[42] The first of these characteristics explains the emphatic anthropomorphisation of these phenomena, which include seasonal processes and cosmology,[43] while the second acknowledges myth's descriptive and explanatory function.[44] It is pertinent for my subject that Burkert's definition applies to such phenomena as solar movement in general and solstices in particular, which might have been treated in story-form, perhaps only descriptively, but possibly with an attempt at causal explanation in an appropriate idiom. This approach to myth is indeed another comprehensive explanation, but on a more basic level, which only tangentially touches upon the contents of myth. It rather focuses on its form and gives it the widest possible referent, conditioned only by the latter's social importance (admittedly, the category of social importance is here defined in somewhat circular terms). Burkert's definition of myth can thus be understood as an umbrella-term able to encompass different hermeneutical approaches to particular classes of motifs, episodes, stories and clusters of narratives.

In this context, a number of scholars adhere to the view that myth can describe and, perhaps, account for physical phenomena in its specific idiom characterised by anthropomorphisation and narrativisation (without claiming that it is its sole or even dominant purpose), and this view will also be supported by the present study. Earlier scholars who believed that myth accounts for physical phenomena in anthropomorphic terms usually did so in the form of comprehensive theories that were easily falsified by providing examples of myths blatantly inexplicable in these terms. This line of interpretation can be traced from the Sophist and Presocratics to the modern period and it was more vulnerable the more all-embracing it professed to be.[45] But—as I already noticed—the application of an iteration of this hermeneutical method, one formulated upon a firm set of arguments, to a circumscribed group of narratives selected by rigorous criteria, will prove successful in interpreting a number of instances otherwise insufficiently explained and at the same time it will provide theoretical support for these instances when its pertinence was 'instinctively' recognised.

1.5. Myth and ancient science

Some nuanced iterations of this hermeneutical method already appeared in the studies that pay more attention to details and contexts, and some of these more recent interpretative strategies will be employed and further developed in this study, especially those that problematize the relation of myth to science. In general, the most important postulate of the currently standard view on this relation seems to be that any unverified projection of modern classifications onto ancient material is anachronistic and that modern criteria should not be applied to ancient systems of thought.[46] Ancient traditions employed their own idioms in approaching physical reality, with anthropomorphisation and narrativisation being the most important characteristics

[42] Burkert 1988: 11. For the elaboration of this definition see Bremmer 1988; it is discussed by Von Hendy 2001: 269, 277 and Calame 2009a: 1, 152. For the definition of myth as 'applied narrative' see further Burkert 1979a: 23 (with Iles Johnston 2018: 1–2); 1979b: 29; 1982 (cited in Calame 2003: 4). Elsewhere, Burkert (1979a: 57, cf. des Bouvrie 2002: 23, 30; 1985: 120) refers to myth's 'suspended reference', or to its function 'to illuminate reality'.

[43] Burkert 1979a: 23. For Burkert's notion of phenomena of collective importance see further des Bouvrie 2002: 24.

[44] For myth as an explanatory mechanism see Popper 1980[4][1959]/2005: 127; Kirk 1972b: 87–88; Celoria 1992: 24, cf. 31; Brisson 1998: 7, 9; Naddaf 1998: xlii n. 25, cf. 2005: 2, 4, 37–38; Calame 1996b: 23 (further cited in des Bouvrie 2002: 15–16, 25, 47); 1999: 140.

[45] An example is the intellectualistic theory of religion, for which see Ross 1971: 105, 115; Ackerman 1975: 117–19, 124, 134; Segal 1980a: 173–74; 1980b: 4–6, 19, 26, 43; 1999: 7, 12–13, 16–18, 20–21, 39, 41, 69, 80–81, 84, 136–37, 143, 147, 149; 2004: 14–17, 19, 24, 30–31, 63, 67; Guthrie 1993: 21, 26. The proponents of the so-called myth-ritual theory upheld similar beliefs (Ackerman 1975: 124; Segal 1980a: 173–75; 1980b: 38–40, 43; 1999: 3, 41, 80, 134; 2004: 24, 27, 67).

[46] Blumenberg 1985: 12, 48 (cf. Segal 1999: 149–50); Finkelberg 1986: 322–23, 325, 327–28, 332–35; Allen 1988: ix, cf. 1988: 56; Lloyd 1990: 7, cf. 1989: 176; Kingsley 1995: 80–81, 87, 90; Calame 1996a: 22, 46 = 2003: 9–10, 27–28 (cf. Buxton 1999: 10); 1999: 121; Morgan 2000: 44–45; Betegh 2004: 178–79, 221, 284; Rochberg 2004: xii, xv; 2010: xxi.

of traditional idioms. But anthropomorphic causality was programmatically excluded from serious consideration by the Greek philosophers' creation of the explicit categories of metaphor and myth, on one side, and the concept of 'naturalness' of phenomena on the other.[47] More generally, the conflicting categories such as *logos*, 'rational', 'philosophical', 'scientific' and 'literary' on one hand, and 'myth' (*muthos*), 'fictitious', 'magic' and 'metaphorical' on the other, were introduced and utilised in polemical texts precisely in order to differentiate the novel from the traditional 'style of inquiry'.[48]

This creation of categories and their hierarchical arrangement by an apparently objective criterion of assessment of their truth-value or explanatory potential is similar to the phenomenon of successive paradigm shifts, or changes in interpretative frameworks, in science. Indeed, the entire so-called *muthos-logos* transition can be understood precisely as a paradigm shift, rather than a substantial transformation of intellectual activities.[49] This also applies to dichotomies contrasting science, rationality, etc. with myth, superstition and other similar notions, in human intellectual activities in general.[50] The intellectual transformation that occurred in Greece should rather be contextualised in its political, social and historical environment, in the first place the creation of polis and adjacent phenomena, such as public debate.[51]

The understanding of the so-called *muthos-logos* transition as a paradigm shift is supported by and at the same time explains the existence of shared subject matter between these two approaches. The continuities between mythic and philosophical interests in similar phenomena are particularly evident in scientific theories aimed at accounting for natural phenomena and in philosophical cosmologies,[52] with the result in 'science [becoming]... an alternative... to some myths, or some aspects of some myths... it is scientific cosmogony, and scientific astronomy, and scientific meteorology... that are the successors, or at least the rivals, of mythical "cosmology", "astronomy", and "meteorology"'.[53] The paradigm-shift model emphasizes the shared subject matter but at the same time also highlights the diachronic nature of the process. However, in order to explain the juxtaposition or even intertwining of these two approaches, a rather long period of transition in which this mixing was possible must be envisaged. This is actually quite unprofitable and uninformative, since this period would, in Greek case, encompass the entire antiquity. It is more in accordance with the available evidence to note the appearance of a paradigm shift, but also to recognize the parallel existence of a previously dominant mythic paradigm with the now dominant philosophical-scientific one within the same ethnographic context. At the same time, the latter should not be understood as a monolithic construct, but rather as a continuum stretching from the fairly inclusive to extremely critical approaches, as noted above.

1.6. Anthropomorphisation and narrativisation

The presence of anthropomorphic deities is among the most typical features of myth. To a modern reader (including many scholars), this characteristic seems utterly opposed to scientific method. The deities' human form, however, appears to be not an end in itself, but could represent a base for accounting for physical phenomena in terms of human behaviour, such as in the case of Greek Helios or Mesopotamian deities in omen text, in which celestial phenomena were described and explained in terms of behaviour of anthropomorphic gods.[54] Another example, hitherto hardly acknowledged, is recognised in the concept of 'Hyperborean' Apollo as articulated in myth and cult, which, it will be argued, represents an analogous account in anthropomorphic terms referring to a phenomenon—in this case the annual movement of the sun with an emphasis on solstices—in terms of gods, that is, in mythic terms.[55] This method is not as arbitrary and unprofitable as it might seem at first. Anthropomorphism in both religion and science is an almost universal explanatory method, a strategy of hypothesising about the surrounding world and attempting a plausible interpretation of it.[56] It has its place in scientific interpretations of nature when understood as a model-building on the basis of metaphor and analogy. It is, moreover, 'uniquely intelligible' and accounts for a large number of phenomena with avoiding multiplication of hypotheses.[57] In effect, the traditional method of

[47] Detienne 1986: 102; Lloyd 1989: 210; 1991: 418, 422, 431–32.

[48] Lloyd 1990: 10, 15, cf. 46, 105; 1989: 209–11. See in more detail Hack 1939: 152; Jaeger 1946: i.43; 1947: ii.9–10, 211; Lloyd 1966: 404–405; 1989: 101–102, 172, 209–12; 1990: 7–8, 10, 23, 34, 45–46, 67, 97; 2003: 101–103, cf. 108; Detienne 1986: 107–109 (discussing Cassirer's notions); Mansfeld 1990[1986]: 43, 52–53; Lincoln 1996: 11; 1999: x, 18 (= 1997: 363), 43, 155–56 (= 2002: 225–26); Morgan 2000: 2–4, 7, 10, 17, 23, 24, 29–31, 34, 40, 45, 46, 53, 84; Struck 2004: 23–24, 50, 64, 67–70, 155, 170.

[49] Cf. Popper 1980⁴[1959]/2005: 126–28; Segal 2004: 33–34. For criticism of the *muthos*-to-*logos* model see Segal 1980b: 3, 44 n. 11; Vernant 1982: 103–104; 1983: 371–72, 400; Blumenberg 1985: 27; Buxton 1999: 1; Lincoln 1999: 3 = 1997: 341, 209, cf. 1996: 2; Morgan 2000: 33; Clay 2007: 210; Calame 2009a: 3; 2011: 521; Pirenne-Delforge 2009: 39; Fowler 2011: 46, 48–49; Hawes 2014: 18.

[50] Hack 1939: 38–39 (cf. Segal 1980b: 44 n. 11); Kirk 1974: 286, cf. 1984[1973]: 58–59; Lloyd 1975: 199–200 = 1991: 146; Rochberg 2010: 408.

[51] See especially Vlastos 1947: 173, 175; 1953: 361–63; 1955: 75–76; Lloyd 1972: 385, 394 = 1991: 131, 139; 1989: 78–81; 1990: 8, 96–97, 105, 124–25, 141; 1991: 124–25; Vernant 1982: 10–11, 51, 39, 51–52, 59, 81, 107–108, 122, 124–32; 1983: 215–16, 237, 249, 380–81, 387–90, 396–97, 404; 1990: 92–93, 96, 99 (cf. von Reden 1999: 65; Hahn 2001: 19, 21–26, 28–35, 53; Seaford 2004: 176, 182–88, 201–202, 207); Zhmud 2012: 254-255 (citing Zaicev).

[52] Cornford 1912: ix, 6, 11, 19, 42–43, 61; Durkheim 1995[1912]: 431; Nilsson 1949: 186; Onians 1951: 248 n. 3; Morrison 1955: 62; Kahn 1960: 158; Lloyd 1966: 50, 193, 195, 199–200, 208; 1967: 32; 1970: 9; Vernant 1990: 255; Naddaf 2005: 2, 7–8, 64.

[53] Mansfeld 1990[1985]: 13–14.

[54] Havelock 1963: 168–71, cf. 180; 1966a: 47–50, 54; 1978: 50; Rochberg 2004: 175. In chapter 13 and Appendix 1 it will be shown how the diurnal and annual solar movement were conceptualised and described in Mesopotamia in terms of an anthropomorphic deity. For Greek myths accounting for both the annual and diurnal solar movement in terms of an anthropomorphic deity see chs. 5–7.

[55] See chapter 7.

[56] Guthrie 1993: 3–4, 31–32, 36, 38, 62–64, 82–83, 89–90, 102–103, 176, 197 (cf. Segal 2004: 33).

[57] Guthrie 1993: 189.

accounting for causality by personalistic forces represents only an idiomatic difference with respect to the scientific rendering of causality in terms of impersonal forces.[58] This conclusion respects the differences between the two intellectual approaches to reality, but at the same time treats them as less essential than usually thought.

The very fact that solar movement was accounted for in terms of anthropomorphic causality conditioned that myth's syntagmatic structure was configured by a typical action pattern extracted from human experience.[59] In general, the gods' actions were readily modelled on human behavioural patterns, as clearly seen in the Apollo myth with its voyages, chariot driving (cf. Helios), feasts etc., relating the phenomena of objective reality accounted for by myth to the social world of humans.

The account of a natural phenomenon was thus given in the form of a narrative, another fundamental characteristic of myth that seems incongruent with scientific method.[60] The use of stories to describe and account for natural phenomena could be explained on a general level as a consequence of the linear nature of language, which prescribes linear narrative as a dominant form for the description and explanation of physical reality.[61] Myth—a syntagmatic chain of actions and events—is in its turn characterised by the logic of *emplotment*, i.e., creating a 'story' out of 'dry' data.[62] Yet the information gathered by the observation of the diurnal or annual solar movement had to be processed in some way before it could be used in stories that described and explained these events and processes in nature. It is precisely at this point that anthropomorphism, as an all-round cognitive strategy, made the data intelligible and suitable for further processing. However, it is almost impossible to outline a clear boundary between anthropomorphisation and narrativisation, and it seems that the latter was a close companion of the former from the very start of the process of making sense of the observational data in traditional societies.[63]

1.7. Observational data in myths

It transpires that observational data can be present in myths in two basic forms. But while the recognition of the presence of technical information, including the data on both diurnal and annual solar movement, in mythic accounts is hardly controversial, acknowledging the possibility of rendering such information in terms of anthropomorphic causality by way of stories seems more contentious. At the same time, it could be easily supported in Greek context by the existence of an undoubtedly anthropomorphic deity such as Helios (I will also argue in chapter 7 that the same applies to some myths and cultic realities involving Apollo), and this example alone necessitates a plausible explanation for the existence of an anthropomorphised celestial body and the fact that solar phenomena were accounted for in terms of his behaviour in Greek ethnographic context. The explanation of this phenomenon is offered by the recognition of anthropomorphism as an accessible cognitive strategy in dealing with natural phenomena. Even though the anthropomorphisation of solar phenomena in Greek myths is readily recognised by modern scholars, it is often left unexplained and glossed over hastily. My study aims to rectify this unfortunate practice by offering a firm hermeneutical framework for evaluating such mythic renderings of natural phenomena.

If some myths are renderings of observational data, this entails that this information was indeed deemed important enough for a society to preserve and transmit, but stories were built around or upon them because of the profitable explanatory potential of anthropomorphism, which was in its turn elaborated in sequential narrative. Otherwise—and more often—the results of solar observations were simply incorporated in myths or stories as information worth communicating to posterity, without actual narratives being built upon them.

A coherent and intelligible cognitive strategy thus emerges from these considerations, based upon anthropomorphic causality and narrativisation and manifesting itself in mythic renderings of natural phenomena, i.e., events, processes, sequences, and periodicities in nature. This strategy is different from the 'analytic' or 'logico-scientific' approach, characteristic of science, but this differentiation, as already noted, is not as definitive as it appears, i.e., it does not exclude the former from the involvement with natural phenomena.[64] Both approaches could operate in a single ethnographic context simultaneously, either juxtaposed to or intertwined with each other, as concrete examples of both Mesopotamia and Greece plainly show.

[58] Horton 1967a: 50, 55, 64–66, 69–70; 1967b: 164; 1968: 632–33 (cf. Ross 1971: 111; Guthrie 1993: 35, 188; Segal 2004: 31).

[59] Frazer 1913: 88; Boas 1921: 230–31, 234–35; 1938: 614–17, 619, 622–24; Havelock 1963: 187; Lloyd 1966: 192–93, 207–209; Meletinsky 2000[1976]: 59; Burkert 1979a: 57 (cf. des Bouvrie 2002: 23, 30); 1985: 120; Blumenberg 1985: 273 (cf. Von Hendy 2001: 323, 326).

[60] Havelock 1963: 166–67, 170–71, 173–74, 180, 218, 234, 236, cf. 42–43, 85 n. 18; 1966a: 48–50, 63; 1978: 42–43, 50, 92, 106, 114–15, 122, 183, 220–21, 332; 1983: 13, 21, 24 (cf. Adkins 1983: 208–10, with criticism on pp. 210–11; Detienne 1992: 11; Halverson 1992: 150–51; Naddaf 1998: xv); Marshack 1972: 133, 197, 279, 283, 316, 330. For criticism of Havelock's theses, none of which affects the rendering of phenomena in story-form, see Halverson 1992.

[61] Burkert 1979a: 23.

[62] Calame 1999: 140–41. For a narrative representation of a particular *geographical* space in genealogical myth see Calame 2009a: 120–22. Thalmann 2011: 24 similarly argues that Apollonius' *Argonautica* is a narrative representation of space encompassing the Greco-Roman *oikoumenê*. See also Hawes 2017: 6 and Clarke 2017: 16, 21 for similar interactions of myth and geography (cf. Thalmann 2011: 40).

[63] An alternative explanation asserts that the presence of anthropomorphic (divine) causality in myth is a consequence of its being basically a story (a sequential or syntagmatic narrative, a causally connected series of diachronical events), communicating an event or process unfolding in time, which requires that it is expressed in terms of anthropomorphised characters (Marshack 1972: 119, 132–33, 283). This hypothesis connects anthropomorphism and narrativisation in a causal relation, although it seems to be putting the cart before the horse. The reasons for provisionally favouring the former as primary are its explanatory potential and the fact it is a necessary condition for the creation of a narrative.

[64] Havelock 1963: 298, cf. 1966b: 68–69; Bruner 1986: 11–14, 42–43, 51–52, 88; Donald 1991: 256–58, 273; Wyatt 2005[2001]: 160–61, 166, 174, 184 n. 48; Rochberg 2004: 40.

This model accounting for the appearance of some myths allows the interpretation of the class of myths referencing either diurnal or annual solar movement and related phenomena in terms of their attempt at describing and explaining these processes, whether they only incorporated motifs drawn from the observation and/or interpretation of solar movement or were actually structured analogously to the phenomena themselves. It is not necessary to unconditionally accept this reconstruction of myth-creation as a general principle in order to acknowledge its applicability to the motifs, narratives and groups of stories referencing solar movement that will be studied in this essay. Indeed, it is obvious that this interpretative strategy cannot explain large portions of the vast body of narratives we customarily classify as myths. But in the case that neither this narrower application is accepted, one must still find a method of explaining the unequivocal presence of the results of solar observation in some myths, which is adequately and economically accounted for by this model; I am not aware of the existence of such an explanation. The—surely unsatisfactory—alternative is to continue ignoring them or treat them as 'primitive' nature-worship.

1.8. Mythic models

The recognition of the appropriation of the results of solar observation in myth, as already noted, does not entail its correspondence to science in all or even many aspects of their respective approaches to reality. But one especially profitable parallel was implicitly utilised in the preceding discussion: that of model and model-building. Indeed, model-building in terms of anthropomorphic causality rendered in story-form seems to be the best definition of the category of narratives studied in this book, i.e., those, however modest in number, specifically built on an analogy with solar movement (those myths that simply incorporate a reference to solar movement in their narrative can be explained in terms of any other hermeneutical method, but still display an awareness of and interest in the results of solar observations). Students of nature, ancient and modern, regularly engage in model-building to represent specific aspects of the surrounding world, in this way describing and explaining empirical phenomena on the base of analogy,[65] and this practice became commonplace in Presocratic, especially Milesian, cosmogonies and cosmologies, in particular with reference to 'meteorological' phenomena.[66] These models do not have to fully correspond to a physical reality in every aspect, but certain models may represent more aspects of the real world or fit some of them more accurately.[67] Mythic discourse also encompasses explanatory models representing specific aspects of physical reality comparable to scientific models and having a similar cognitive function.[68] Parallel to scientific models, myth can not only reflect reality, but also 'refract' it by selectively emphasising certain aspects of the phenomenon it treats and by simplifying it to a level considered most productive for its purposes.[69] Its notorious fictionality should not be taken as a serious hindrance for its use in constructing models, since philosophy of science now decidedly recognises the use of fiction in scientific model-building as well as the epistemological value of such models.[70] However, in light of the unacceptability of monolithic theories of myth an all-encompassing notion of mythic models cannot be accepted, and it seems imprudent to insist on an almost complete correspondence between mythic and scientific models. At the same time, it could be granted that some myths were intended to account for natural phenomena in this way, namely those that have as their referents phenomena and processes that will later be studied by the disciplines of meteorology (in early Greek terms), astronomy and geography. The term mythic model is thus an acceptable and profitable designation for narrative accounts of physical phenomena expressed in terms of anthropomorphic causality, and I will consequently use it throughout the book when referring to mythic accounts of solar phenomena.

As already noted, typical action patterns extracted from human experience are the basic analogues in models built upon anthropomorphic causality, but all models, mythic or otherwise, are naturally based on analogies and metaphors.[71] Indeed, the cognitive processes behind metaphor, analogy and model on one side and scientific (theoretical) model-building and theory on the other are all based on the premise of explaining something in our experience in terms of something else, and any assertion of identity is in some measure metaphoric, emphasising the similarities and minimising the differences, with the explicit category of metaphor designating a specific interval it bridges.[72] In response to new experiences, humans in general use comparisons with some previously known facts, but while traditional societies use explanatory analogies derived from social contexts, resulting in a personalistic causality, modern societies use those derived

[65] Sambursky 1960: 14; Black 1962: 229–31, 233; Hesse 1966, *passim*; Barbour 1976: 6, 30, 32–33, 37–38; Giere 1999: 94; Bailer-Jones 2002: 108, 124.

[66] Sambursky 1960: 40, cf. v, 134, 242; Guthrie 1965: ii.299–300; Lloyd 1966: 232–71, 319, 321–22, 324–25, 336, 417; Furley 1987: 21–22; Guthrie 1993: 55, 153–54; Algra 1999: 49; Naddaf 2005: 73.

[67] Sambursky 1960: 14; Black 1962: 39, 41–42, 44–46, 220–21, 238; Ross 1971: 111; Barbour 1976: 6; Giere 1999: 92–93; Roby 2014: 158, 161, 167, 173–74, 177.

[68] Horton 1962: 212–13, 216–18; 1964: 97–99 (cf. Ross 1971: 111, with criticism on pp. 111–15); Ramsey 1964 (cited in Barbour 1976: 60); Barbour 1976: 6–7, 16, 27, 49, 51, 68–69; Meletinsky 2000[1976]: 155; Donald 1991: 213–15, 259, 267. For mythic models accounting for cosmogony in Hesiod's *Theogony* see Lloyd 1966: 205; Kirk 1974: 296297; Most 1999: 344.

[69] Horton 1964: 100–101 (cf. Ross 1971: 111, with criticism on pp. 111–12); Buxton 1992: 7; 1994: 87–88, cf. Bremmer 2008[2003]: 60.

[70] Roby 2014: 157–58, 173–74, 177–78.

[71] Black 1962: 236–39; Hesse 1966, *passim* (cf. Barbour 1976: 43 and Bailer-Jones 2002: 109, 113, 117, 119); Barbour 1976: 43; Bailer-Jones 2002: 114, 117–19, 124; Mourelatos 2008: 38. For the virtual identity of analogy and model see Black 1962: 232; Hesse 1966, *passim*; Lloyd 1966: 319, 321–22, 324–25, 336, 417; Barnes 1982: 40; Pender 2003: 64, 72 with n. 44; Roby 2014: 160. In general, the opinion that denies any role to figurative discourse in scientific treatment of natural phenomena is now superseded (Hesse 1988: 3; Bailer-Jones 2002: 115; Rochberg 2004: 177).

[72] Black 1962: 229–31, 233, 238; Guthrie 1993: 47, 98, citing Oatley 1978; Lakoff & Johnson 2003[1980], *passim*.

from the world of inanimate things.[73] Once again, this is an important difference, but one that implies the shared subject matter of these two cognitive and hermeneutic strategies.

1.9. Conclusion

Precisely what novelties in the analysis of literary testimonies are offered by this study of a specific category of narratives—those referencing in some way solar movement—within the larger corpus of what is usually styled myth? In the first place, the focus on a rather circumscribed subject matter—and a highly specific one—allows some measure of (self-)control in applying the hermeneutic approach that allows myth to encompass (but not be reduced to) narrative models of phenomena in physical reality, characterised by anthropomorphic causality, to a category of texts studied here. Secondly, this approach offers a theoretical background for both the introduction of narrative accounts in the study of how the ancients understood natural phenomena and for the interpretation of certain motifs used in ancient narratives or the narratives themselves in terms of natural phenomena as their referents. These non-scientific texts can thus be acknowledged as accounts of natural phenomena in terms of anthropomorphic causality. In this way, a strong substantiation is provided for recent analyses—in principle, rather than in details—that either study certain large-scale mythic texts as descriptions of cosmological voyages[74] or interpret them in terms of mythic geography or cosmography.[75] This interpretation makes the application of analytical methods which would, following different assumptions, be reserved only for the analyses of scientific texts (natural phenomena unquestionably being their proper subject-matter), methodologically justifiable in the interpretation of such literary works, or the appropriate sections of these works. This on the other hand means that a complete ethnographic context of an idea, i.e., all its iterations, should be studied in order to obtain a comprehensive picture that a certain tradition entertained of some subject, such as diurnal or annual solar movement or solstices. The context must be studied by an analysis of various complementary sources, both mythic and non-mythic. In implementing this approach, I am following a well-trodden path, but in addition I have attempted to substantiate and elaborate it in this introductory chapter, with a special focus on the main subject of this book, i.e., solar phenomena.

Alongside focusing on the study of the category of texts referencing the ‘meteorological’ concepts of diurnal and annual solar movement as they appear in Greek tradition, I will also offer a conspectus of analogous ideas in several traditions in contact with Greek civilisation. The latter was manifestly in a multi-faceted contact with Mesopotamian tradition,[76] perhaps also—less clearly—Egyptian.[77] It appears that Greek, Mesopotamian and (with some reservations) Egyptian traditions accounted for the solar movement in general and solstices in particular in a corresponding fashion (using anthropomorphic causality as a primary tool) and incorporated these phenomena similarly into their respective mythic discourses. Indeed, the Egyptian evidence from the New Kingdom onwards (perhaps even earlier) approached the Greek and Mesopotamian manner in describing and accounting for solar movement, even though an unequivocal evidence of the Egyptian understanding of solstices, as opposed to the solar movement in general, is lacking.

A similar incorporation of solar phenomena into mythic discourse cannot be argued for prehistoric traditions, since they left us no literary testimonies. However, as noted above, it is a well-established fact that the solstices were recognised in some prehistoric traditions.[78] The knowledge of this concept must have been incorporated into their respective cosmological systems and one could speculate that it was accounted for in some way, most probably in local mythic idioms. Material evidence suggests that these strategies could not have been very different from their Greek or Mesopotamian counterparts, although the example of Egypt recommends more restrained conclusions on the question of similarities between various systems.

All these traditions observed the annual solar movement and recognised the solstices, incorporating this knowledge into their respective worldviews. When they undertook to describe, communicate and account for these phenomena, they did so in the form of stories with anthropomorphic agents. This practice was used until new impersonal models of accounting for phenomena were introduced, but even then, the traditional idiom was used simultaneously with the new one—in the case of the Greco-Roman civilisation for almost the entire duration of antiquity. But even in modern science anthropomorphism, metaphor, analogy

[73] Boas 1921: 203–204; Abrams 1953: 31–32 (cf. Black 1962: 240); Lloyd 1966: 175, 179, 209, 304–305, 356–57, 359, 380, 415; Horton 1964: 98–99; 1967a: 64–66, 69; 1968: 632–33 (cf. Ross 1971: 111; Guthrie 1993: 34, 71; Segal 2004: 31–32).

[74] For example, Marinatos 2001, cf. Käppel 2001: 16, 19, 21 Abb. 1 and Beaulieu 2016: 53 for the *Odyssey*; Endsjø 1997: 374–75, 377, 380, Stephens 2003: 18, 178, 217–37, Noegel 2004: 129–30, 132, 135–36 and Beaulieu 2016: 84–85 for the *Argonautica*.

[75] Nakassis 2004.

[76] For the influence from Mesopotamian ideas (sometimes mediated by the Levantine or Anatolian civilisations) on Greece see Burkert 1988; 1992; 2004; West 1995: 212–17; 1997; 2014: 31–32, 97, 126; 2018; Penglase 1994; S. Morris 1997; Wyatt 2005: 102–24, 189–237; Bremmer 2008; Rutherford 2009; 2018; Louden 2011; López-Ruiz 2010; 2014; Haubold 2013 (a different focus); Bachvarova 2016; Currie 2016: 160–222; Scurlock 2018; Rochberg 2020.

[77] For a review of Egyptian influence on Greek cosmological notions see West 1971: 36, 47, 62, 92; Naddaf 2005: 80, 82, 100–101, 107, 191 n. 24.

[78] Appart from the Upper Palaeolithic observations of solstices noted above, one could also emphasize the examples of early 5th-millenium BC Goseck (Bertemes, Schlosser 2004: 50–51; Bertemes 2008: 41) and Ippesheim enclosures (Schier 2008: 52–54), various British and Irish prehistoric sites, such as Stonehenge, Durrington Walls, Newgrange, Maes Howe, Brainport Bay, Drombeg and Dorset Cursus (Ruggles 1999: 12–13, 17, 19, 29–32, 35, 37–38, 40–41, 100, 127, 129, 136–39, 240 n. 93; 2005: 48–49, 125, 135, 237, 405–409; Pearson *et al.* 2007: 630, 633), as well as Maltese temples, such as that at Mnajdra (Hoskin 2001: 30–31).

and model-building continue to be important tools—if not nearly as decisive and dominant as before—for accessing physical reality.

Thus, humans have used the traditional narrative idiom based on anthropomorphic causality in order to account for the diurnal and annual solar movement from the Upper Palaeolithic until fairly recently. Their accounts deserve an unbiased study, relieved of the burden this particular subject matter carries on account of the 19th-c. (and some 20th-c.) interpretative strategies that focused their extravagant, excessive and now outdated theories precisely upon this body of material. This inquiry is an attempt to reintroduce the study of solar movement in myth-interpretation, i.e., to transfer it from the margin of a polyparadigmatic field it now occupies somewhat nearer to its centre, without reintroducing the untenable principles of the solar-myth school. Two distinct, yet interrelated, corpora of references to solar movement will be studied concurrently: the motifs drawn directly from the observation and/or interpretation of solar movement incorporated in mythic narratives and stories structured as anthropomorphic analogues to the phenomena themselves. The criteria for inclusion in this particular class of myths are straightforward: the stories should either explicitly include a reference to solar movement or the recognition of such a reference does not require strained readings. The interpretation of this class of myths represents another attempt at approaching the heterogeneous body of traditions usually classified as myths selectively, in line with the current polyparadigmatic hermeneutic environment.

Part One

Annual solar movement

2

The Laestrygonians and the geographical arctic circle

ἑξῆμαρ μὲν ὁμῶς πλέομεν νύκτας τε καὶ ἦμαρ,
ἑβδομάτῃ δ' ἱκόμεσθα Λάμου αἰπὺ πτολίεθρον,
Τηλέπυλον Λαιστρυγονίην, ὅθι ποιμένα ποιμὴν
ἠπύει εἰσελάων, ὁ δέ τ' ἐξελάων ὑπακούει.
ἔνθα κ' ἄυπνος ἀνὴρ δοιοὺς ἐξήρατο μισθούς,
τὸν μὲν βουκολέων, τὸν δ' ἄργυφα μῆλα νομεύων·
ἐγγὺς γὰρ νυκτός τε καὶ ἤματός εἰσι κέλευθοι.

So for six days we sailed, night and day alike, and on the seventh we came to the lofty citadel of Lamus, to Telepylus of the Laestrygonians, where herdsman calls to herdsman as he drives in his flock, and the other answers as he drives his out. There a man who never slept could have earned a double wage, one by herding cattle, and one by pasturing white sheep, for the paths of the night and the day are close together.

Odyssey 10.80–86, translation by Murray revised by Dimock in *LCL* 104 (1995)

The first mythic episode that will be studied in the light of the tenets expounded in the introductory essay will be the story of Odysseus' encounter with the Laestrygonians reported in the tenth book of the *Odyssey*, which was very early recognised as containing some 'secondary' meteorological (in Greek sense) reference outside the narrative itself. This episode is certainly not structured on an analogy with solar movement, but it equally certainly contains a reference to its annual aspect.

The episode is one of those more familiar to a modern reader—probably due to its violent character and the attractive folkloric conception of men-eating giants. The story indeed represents a familiar concept of a hero escaping a cannibal-giant after his dealings with ogre's womenfolk.[1] Yet the folklore elements are not as pronounced here as in similar episodes, especially, for example, that involving the Cyclopes. Furthermore, a number of details make the *Odyssey*-version, which otherwise performed various functions within the narrative,[2] both a poet's reworking of a familiar folkloric theme and a narrative that clearly involves a reference to the annual solar movement and certain related phenomena, such as the sun's varying visibility at different latitudes in different parts of the year and, specifically, the so-called midnight-sun phenomenon. This fact, which will be argued for below and is accepted by a majority of scholars, allows me to include this mythic episode in the category of myths that refer to solar movement and to interpret it in these terms.

[1] Page 1973: 27–29.

[2] Cf. Heubeck 1989: 47 and de Jong 2001: 253–54.

As fifth in line of Odysseus' adventures, it follows the hero's visit to Aeolus' island and precedes the adventures he encountered on Circe's Aeaea. The Laestrygonians are in fact responsible for the destruction of a major part of Odysseus' fleet returning from the Trojan War. The Ithacans were more or less untouched before this episode, although they had suffered certain losses in previous encounters. From this point on Odysseus' fleet numbered one ship only, the hero's, which will carry him on until destroyed in a storm provoked by the impious slaughter of Helios' cattle on Thrinacia.

Can any geographical information be extracted from Homer's narrative, in light of the fact discussed in the introduction that ancient literary tradition, including poetry and myths, is recognised as a relevant source for the study of ancient geography?[3] And how precise is this information? As gathered from Homer's narrative, the land of Laestrygonians lies a six days' sail northwest (perhaps north or west) from the island of Aeolus,[4] which itself lies a nine days' sail west of Ithaca.[5] If we accept the identification of Ithaca with the modern homonymous island (or a nearby location), and the identification of Aeolus' Island with the islands in the Tyrrhenian Sea,[6] we may conclude that in geographical terms the man-eating giants were placed somewhere in the western Mediterranean. This was obviously the opinion of ancient scholars, who placed them on Sicily[7] or—in a later tradition—on the border between Latium and Campania.[8] Those modern authors who differed from the ancient interpretations in general opted for the region of the

[3] Nicolet 1991: 8; Endsjø 1997: 373–74, 379–82, 384–85; Clarke 1999: 23–25; Meyer 2001: 225–27; Thalmann 2011: 8–11, 13, 171; Dueck 2012: 20–35, 51; Klooster 2012: 55, 57–59, 65, 75–76; Lightfoot 2014: 8–9; Dan 2017: 172, 189–91. I will use the name Homer for the author(s) of the *Iliad* and *Odyssey* throughout the book, without prejudicing the question of authorship of these poems. I think it would be overly pedantic to qualify the use of this name on every occasion, but I am not committed to any particular position with respect to this issue.

[4] *Od.* 10.80–81. Cf. Wolf 2009: 23, 46 (with Fig. 5 on p. 24).

[5] *Od.* 10.25–30.

[6] As early as Thuc. 3.88.1. The identification was undisturbed by the fact that the island was floating, *Od.* 10.3.

[7] Hes. fr. 150 M-W in P.Oxy. 1358 fr. 2 col. i.25–26 (cf. Ballabriga 1986: 23, 139–41); Thuc. 6.2.1 (from Antiochus of Syracuse, Fowler 1996: 77; 2013: 634); cf. Theopomp. *FGrHist* 115F225a; Lycoph. 662–65, 956; Str. 1.2.9; *HN* 3.89, 7.9; Sil. Ital. *Pun.* 14.33–34, 125–26.

[8] Cic. *Att.* 2.13.2; Hor. *Carm.* 3.17.1–6, cf. 3.16.34; *HN* 3.59; Stat. *Silv.* 1.3.83–84; Sil. Ital. *Pun.* 7.276–77, 410, 8.529–30; cf. Ov. *Fas.* 4.69–70, *Met.* 14.233.

Thyrrhenian Sea[9] or, more adventurously, Balaclava on the Crimea[10] or a Norwegian fjord.[11]

The level of precision of Homer's geographical information is rather disappointing—it is only with later authors that concrete geographical toponyms (other than Ithaca) are given in this context. It thus seems highly adventurous to attempt to identify Homer's locations with toponyms we can securely localise on a modern geographical map. This seems to undermine my claim that there are easily recognizable references to the body of knowledge belonging to modern disciplines of geography and astronomy in this Homeric passage. However, there is another way of looking at this problem.

2.1. Interpreting the 'meteorological' facet of the Lastrygonian episode

Both ancient and modern localisations in general ignored Homer's description of 'meteorological' conditions current in the land of Laestrygonians, much as Homer himself ignores them in his narrative,[12] and focused on the physical description of the harbour, occasionally mentioning the poet's knowledge of short winter nights characteristic for the northern latitudes.[13] Indeed, we may go so far to conclude that if vv. 82–86, the basis for all interpretations that recognize a reference to the far north in this episode, were to be excluded from the narrative, the story would function without any noticeable change. On the other hand, while physical description of the harbour is crucial for the plot at this juncture, neither the fact that the 'meteorological' description is redundant nor the context in which the episode appears, which is in large part defined by this description, should be disregarded.

'Meteorological' information, however redundant for the plot, is obviously here because it traditionally belongs to an encounter with the 'Laestrygonians'; there is no other plausible explanation for it. Further, there is no doubt that the setting of this episode is cosmological, and that it does not belong to the realm of terrestrial geography as expounded in *periploi* and *periegeses*;[14] but it is specifically the annual solar movement that is referenced here, as will be shown presently.

The main pieces of information providing the indication for the region's location are the peculiar meteorological characteristics of the land of Laestrygonians. The lines 82–86 clearly pronounce the most important extra-narrative feature of the Laestrygonian country, analysed in various ways by both ancient scholars and modern investigators, who have often come to different conclusions in their interpretations of its geographical reference.[15] Here Homer describes the following unique feature of this land: a shepherd returning from the pasture with his flock in the evening, immediately before or after the sunset, meets a herdsman driving his cattle out to pasture. Yet he is not going on some kind of night-pasture, because here 'the paths of night and day are close together'; he is simply going on a day-pasture, immediately before or after the sunrise. Two herdsmen hail each other as they pass by, and the poet pragmatically concludes that one would be able to earn double wages here, if only he could do without sleep. Therefore, in the land of Laestrygonians there is almost no night; here 'the days are so long that evening and morning twilight run into each other'.[16]

This 'meteorological' fact will allow us to determine the exact cosmological position of the land of Laestrygonians, i.e., its location with respect to the understanding of the structure of the world as a whole. This location was determined with respect to the annual solar movement and the phenomena it causes with respect to latitude and season.

Homer's prosaic shepherds and herdsmen are probably rationalisations derived from the same mythic or cosmological concept that is found in Hesiod's description of how Night and Day meet and pass by each other at a brazen threshold.[17] Moreover, their cattle has to be in some way connected with Helios, in the same or similar way in which his holy herds on Thrinacia and Erytheia are associated with the sun god.[18] Helios has his cattle at two opposite points on the earth, both where he rises and where he sets.[19] The latter could be those on Erytheia, while the former could be those on Thrinacia, not far from Circe's Aeaea, associated with sunrise.[20] The exchange of night and day (Night and Day are obviously Hesiod's personifications of natural phenomena conditioned by the

[9] Butler 1897: 184–86; Bérard 1929: iv.224–80; Carpenter 1946: 107–108, 110; 1973: 46; Bradford 1963; Geisthövel 2008: 72–73, 76–80; Wolf 2009: 47–52, 142, 212.

[10] Dubois 1843: 111–14 and after him West 2005a: 49–52, who also lists earlier references for this theory in n. 46; Bury 1906: 88; Bachvarova 2016: 102.

[11] Cornelius 1925: 344 (cf. Meuli 1931: 540.11–12).

[12] Cf. West 2014: 206.

[13] E.g., West 1997: 407; 2005a: 52 with n. 47; 2014: 86, 206–207. Cf. Harrison 1882: 60; Wilamowitz-Moellendorff 1884: 168; Gisinger 1924: 534.15–22; Cornelius 1925: 344; Dreyer 1953: 20–21; Butterworth 1970: 179; Knight 1995: 148; Käppel 2001: 17.

[14] For some recent interpretations of the nature of Odysseus' voyage as cosmological in character (interpreted in terms of mythic geography) see Marinatos 2001; Käppel 2001: 16, 19, 21 Abb. 1; Nakassis 2004; Beaulieu 2016: 53; Fowler 2017: 249–50.

[15] See for example Σ *Od.* 10.82, 85–86 (ii.452–54 Dindorf); Nitzsch 1840: 101–105; Packard 1874: 33–34, 36–37, 39–40; Merry *et al.* 1886: 406–407; Dörpfeld 1925: 252–56; Pocock 1958; 1968; Carpenter 1946: 107; Vos 1963: 18, 22–23, 25–26; Woodbury 1966: 611–12, 616; Page 1973: 42, 122 n. 26; Heubeck 1989: 48; Burgess 1999: 198–99; 2016: §§12–13; Marinatos 2001: 396; Scodel 2003; Nakassis 2004: 224; West 2005a: 47; Louden 2011: 160; Fowler 2017: 247–48.

[16] West 2005a: 47.

[17] *Th.* 748–50; Frame 1978: 62–63; Heubeck 1989: 48; Scodel 2003: 84–85.

[18] *Od.* 11.107–109, 12.127–37, 261–64 (Thrinacia); [Apollod.] *Bibl.* 1.6.1; Σ Pi. *I.* 6.47b, iii.254–55 Drachmann (Erytheia) (Gantz 1993: 419–20).

[19] *Od.* 12.379–82; Nakassis 2004: 226–27.

[20] *Od.* 12.3–4. These were later located on Sicily (Panyas. fr. 16 West; Nymph. *FGrHist* 572F3; Philosteph. fr. 15 *FHG*; Sen. *QN* 3.26.7; *HN* 2.220; Appian. *BC* 5.12.116; Polyan. *FGrHist* 639F7; Ael. *VH* 10.8; Σ A. R. 4.965, p. 299 Wendel). For parallels in other traditions see Marinatos 2001: 389 (Mesopotamia) and Sick 2004: 457–59 (Vedic India).

diurnal solar movement) and the presence of solar cattle at Laestrygonia clearly point to solar movement as the reference of this description, and the latter has contributed to the search for the land of Laestrygonians in the farthest west, at the extreme western point in the solar course.[21] Similarly, the understanding of v. 86 as meaning that the paths of day and night are not near to each other in the land of Laestrygonians, but rather that these paths are near the land itself,[22] clearly places their land either in the extreme west or east.[23] But Homer's vagueness with respect to geography aside, his account should on principle be treated charitably as coherent, rational or at least commonsensical.[24] An interpretation that presupposes that Homer believed in an eastward movement of the sun across the sky, which is a necessary consequence of accepting the land of Laestrygonians in the farthest west, even though it recognises solar movement as the referent of this description, does not seem very convincing in this light (the other, eastern localisation, is less implausible, but the shortness of night is equally unconvincingly explained). That does not mean it is completely out of the question, but that a plausible interpretation preserving the text's coherence, if one is available, should be preferred.

2.2. Crates' interpretation of the Laestrygonian passage

It seems that precisely such an explanation was offered by the 2nd-c. grammarian and critic Crates of Mallos.[25] His interpretation, according to which Homer was speaking of the area with almost continuous summer solstice daylight and, thus, where the sun's setting point lays near its rising point, keeps solar movement as the referent of Homer's description, but focuses on its annual, rather than diurnal, aspect. Although Crates' approach to Homer and his interpretative method were highly idiosyncratic, this particular interpretation conforms to the Homeric narrative much better than many of the modern interpretations of the episode, and thus seems the most plausible reading. Moreover, his localisation is actually in general implicitly accepted by a number of modern scholars who place the land of Laestrygonians in the far north, whether they would like to be placed in the same footnote with the scholar from Mallos or not.[26] Crates is regularly accused of reading Stoic or, rather, his own principles into Homeric poems, understanding the latter as cosmic allegories.[27] However, no recognizable Stoic principles are present in his interpretation of Homer's Laestrygonians, and he certainly did not understand Homer's verses as allegorical in any way. Homer is here simply describing an existing meteorological feature of a land, and this is precisely the way in which Crates understood him. It is possible that this particular non-allegorical line of interpretation goes back at least to none other than Pytheas of Massalia, who located Thule, an island at the latitude of the geographical arctic circle, a six-days' sail to the north of another island, Britain.[28] In this way he combined his geographical explorations with what seems to be an intriguing example of Homeric scholarship.[29] Thus a long tradition of understanding the Laestrygonian episode as including a reference to the annual solar movement, the 24-hour night at the summer solstice at a latitude that would later be defined as that of the geographical arctic circle, and the recognition of the sun's rising and setting points at solstices is manifested from the late Classical or early Hellenistic period onwards. It could be dismissed as a later scientific interpretation of a 'mythic' description, but the presence of solar motifs is in need of explanation in terms of their referents. This particular explanation satisfactorily accounts for their presence without using allegoresis or any other invasive method of interpretation and seems acceptable under modern standards of scrutiny. Indeed, it obviously seemed acceptable to a number of respectable scholars, who represent a majority opinion, even though many of them use a less technical vocabulary in comparison with Pytheas and Crates in describing identical phenomena and offering identical conclusions. Perhaps this reflects their reluctance to interpret a literary or mythic text in terms of information from the domain of physical phenomena apparently drawn from observations, whether directly or at secondhand.

---

[21] Packard 1874: 36–37, recounting Lauer's (1851) hypothesis; Packard 1874: 40; Merry *et al*. 1886: 407.

[22] Packard 1874: 33–34, 39–40; Merry *et al.* 1886: 406; Carpenter 1946: 107; Vos 1963: 19, 22; Woodbury 1966: 611–12; Page 1973: 42, n. 26 on p. 122; Heubeck 1989: 48; Fowler 2017: 248; cf. also Scodel 2003.

[23] Marinatos 2001: 396 (east or west); Meuli 1921: 54, Vos 1963: 23, 25–26, 29, Woodbury 1966: 612, 616, Heubeck 1989: 48, Nakassis 2004: 224 and Louden 2011: 160 (east).

[24] Page 1973: 44; Lloyd 1990: 18; Buxton 2009: 58–61, 250–51; Bilić 2016a: 216.

[25] For Crates and his geographical theories see especially Kroll 1922; Wright 1925: 18–19; Mette 1936; Uhden 1936: 106–15; Thomson 1948: 202–203; Abel 1974: 1051–55; Quinn 1982: 97–108; Aujac 1987c: 162–64; Giampaglia 1998; Asmis 1991; Romm 1992: 129–31, 179–80, 188–89; Porter 1992; 2003; Broggiato 2001: xiii–lxix; Stallard 2010a: 44–48; 2010b: 17–29; 2016: 3–7; Bilić 2012a. The following discussion is a modified version of Bilić 2012a: 302–13; some of it also summarily appears in Bilić 2017: 5–6.

[26] Harrison 1882: 60; Wilamowitz-Moellendorff 1884: 168; Gisinger 1924: 534.15–22; Knight 1995: 148; Ballabriga 1998: 122–24, 137–39; Käppel 2001: 17; West 1997: 407; 2005a: 52 (with n. 47 for earlier references); 2014: 86, 119–20, 206–207; Irby 2012: 86; Cerri 2014: 168.

[27] Philod. P.Hercul. 1676 fr. 2 = fr. 99 Broggiato; Gemin. *Elem. Astron.* 16.27 = fr. 37 Broggiato; Heracl. *All.* 27.2 = fr. 3 Broggiato; Asmis 1991: 141; Porter 1992: 72; cf. Graf 1993: 194, 198.

[28] For Thule at the latitude of the geographical arctic circle see Pyth. fr. 8c Bianchetti = fr. 6 Roseman = fr. 6c Mette = Eratosth. fr. 34 Roller; cf. fr. 12a Bianchetti = T 27 Roseman = fr. 14 Mette, also Cleom. *De motu circ.* 1.4.222 Todd; see also Pyth. fr. 14 Bianchetti = T 10 Roseman = fr. 6d Mette and Crates fr. 37b Mette (Mette 1936: 268.5–7); for its location with respect to Britain see Pyth. fr. 8a Bianchetti = fr. 2 Roseman = fr. 6a Mette *ap*. Eratosth. fr. 35 Roller (IIC2 Berger); fr. 9a Bianchetti = T 18a Roseman = fr. 13a Mette; cf. Tim. *FGrHist* 566F74 = Pyth. fr. 8f Bianchetti = T 23 Roseman = fr. 11b Mette (Müllenhoff 1870: 1.385, 472; Thomson 1948: 146 n. 1; Cary & Warmington 1963: 256 n. 47; Bianchetti 1998: 147, 173; Roller 2006: 72, 73 n. 139). See also the interpolation in Solinus, Mommsen 1895: 219, cf. xci (Tierney 1967: 116; Hawkes 1977: 34 with n. 75, 35).

[29] Bilić 2012a: 319–22; 2016a: 221–23. In particular, Pytheas might have been working with the Massaliote edition of Homeric poems, which might have been 'published' already in his time (West 2001: 67–68 with n. 69).

This interpretation of the episode as referring to the (in later terms) latitude of the geographical arctic circle defined in terms of the annual solar movement raises a number of questions related to the understanding and development of this concept in Greek ethnographic context. Whoever was the author of the tradition behind Homer's description—reflected in Hesiod's portrayal of the house of Night accounting for the diurnal solar movement—must have been aware of the concept, even if in vaguest terms. It seems impossible to know whether it was Homer or some of his predecessors who applied the model accounting for the diurnal solar movement onto the annual, but this is a first occurrence of a phenomenon that will be met regularly in the continuation of this study: the interchangeability of motifs used in building models accounting for respectively diurnal and annual solar movement. This practice was probably conditioned by the similarity and interdependence of these two phenomena and a limited number of motifs—derived from appropriate typical action patterns extracted from human experience—from which mythic models accounting for both aspects of solar movement could be built.

2.3. Crates' interpretation and arctic circle

It seems that ancient Greeks had two different understandings of the annual solar motion in its relation to latitude and arctic circle. Here I am using the ambiguous term 'arctic circle' intentionally, reflecting the original Greek practice of referring by this expression to both the fixed geographical arctic circle and the variable latitude-dependent circles of permanent visibility. This double understanding, apparently not reflecting a diachronic development, is nicely illustrated in Crates' reported opinions on the subject. I will thus structure my discussion of these two understandings upon the testimonies on his position, which will allow me to discuss some earlier evidence for the understanding of these phenomena in Greek thought across various literary genres, but with a clear focus on scientific tradition. In addition, Crates' opinions on the (fixed) arctic circle are expressed in a literary genre that bridges different disciplines as we define them today—in the first-place literary criticism and scientific astronomy and geography. It is thus especially relevant to a number of subjects treated in this book.

In general, according to one group of sources, Crates certainly understood the phenomenon of the change in duration of the longest day depending on latitude, and presumably was familiar with the concept of the fixed or geographical arctic circle; according to another, he did not quite understand the difference between the latitude-dependant always visible and the fixed arctic circle.

Thus according to fr. 50 Broggiato he compared the main feature of the land of Laestrygonians with Aratus' description of the constellation Draco,[30] placing their territory 'about (*peri*) the head of Draco'.[31] There are two possible explanations for this comparison. According to one, he understood Eudoxus'/Aratus' description of the head of the figure's movement ('go[ing] around near the place where the limits of setting and rising mingle together') as referring to the asterism's motion, i.e., to the fact that it was the last circumpolar asterism for the latitude of Athens.[32] This is probably the most conventional and conservative interpretation of 'the limits of setting and rising'. In this case, we should understand Crates' περὶ τὴν κεφαλὴν τοῦ δράκοντος, literally 'about the head of Draco', as 'about the area (directly) below the head of Draco', that is, the area 'where the head of Draco is at the zenith'.[33] The head of Draco was at the zenith for a latitude much to the south of the fixed arctic circle, but very near the always visible circle of Rhodes (54°), used by the majority of ancient astronomers as *the* arctic circle. Therefore, it would seem that Crates, associating them with the constellation Draco, located the Laestrygonians with reference to the always visible circle of Rhodes taken as the arctic circle. Perhaps another level of interpretation of this practice can be added in terms of interaction of rational and non-rational approaches to phenomena. Neugebauer thus claims that the origin of the use of the latitude of 36° should not be recognised as a reference to the latitude of Rhodes, but rather associated with 'a cosmologic doctrine of "pythagorean" flavor',[34] dividing a quadrant of 15 parts (90°) of a circle of 60 parts (360°), as determined by the obliquity of the ecliptic (c. 24°, thus 4 parts), by an arctic circle 6 parts (36°) removed from the pole. Thus a 'neat numerical pattern' of 4-5-6 parts was obtained,[35] resulting in what was later recognized as the Rhodes-determined always visible circle. Whether Crates consciously used this tradition is another question, but it would certainly strengthen his cosmological argument.[36] Nevertheless, it appears that an important 'scientific' concept was perhaps derived from cosmological speculation (since Neugebauer was extremely critical towards any manifestation of ancient science lacking mathematical background, his suggestion should be treated as coming from an extreme

[30] *Phaen.* 61–62; Mair & Mair 1921: 382; Gain 1976: 83; Kidd 1997: 77; Evans & Berggren 2006: 5; Broggiato *BNJ* 2113F20. Cf. the Latin translations of Cic. *ND* 2.108 (*Aratea* fr. 10 Soubiran); German. *Arat.* 60–62; Avien. *Arat.* 164–68; Hyg. *Poet. astr.* 4.3 (Crates fr. 37i Mette).

[31] Σ HQ *Od.* 10.86 (ii.453 Dindorf) (Page 1973: 43; Broggiato *BNJ* 2113F20) = fr. 37e Mette; cf. Σ HV *Od.* 10.86 (ii.454 Dindorf) = fr. 37f Mette, Eustath. *Od.* 10.86, i.369 = fr. 37g Mette, and Σ Q and MDKVUA *Phaen.* 62 = fr. 37h Mette.

[32] Translation from Broggiato *BNJ* 2113F20. Aratus' conception of Draco's movements is certainly derived from Eudoxus, as confirmed by Hipparchus (Hipparch. 1.4.7–8, pp. 32.25–34.21 Manitius = Eudoxus fr. 16 Lasserre = Attal. fr. 5 Maass). For the discussion of Eudoxus' authorship of the quotations in Hipparchus and the chronological priority of the texts used by Hipparchus over Aratus see Dekker 2013: 2, 41–43. An asterism or a star that in its lower culmination—i.e., in its transit over the meridian near the horizon in its circling around the north celestial pole—merely touches the horizon and stays visible during its entire trajectory, never dipping below the horizon.

[33] Thus also Page 1973: 43. Mette (1936: 86–87, 91–92) shared this opinion, but he further believed that Crates had associated this region with the one in which the longest day is of 23 hours.

[34] Neugebauer 1972: 247.

[35] Neugebauer 1972: 247, cf. 1975: ii.733

[36] For Crates' (fr. 34ab Mette) division of the sphere into five or six zones according to this pattern (6-5-4) see Mette 1936: 66–67 and Uhden 1936: 108.

sceptic), which conforms to Popper's understanding of pre-scientific theories closely resembling scientific approach with a potential for transforming into proper scientific theories;[37] this would be an example of precisely such an occurrence with important repercussions for further development of geography and astronomy. Indeed, a number of respectable ancient scientists, including Eudoxus, Aristotle and Eratosthenes, adhered to the Rhodian standard, placing the arctic circle and the northern limit of the northern inhabitable zone at 54°.[38]

But it seems that this standard was abandoned rather early by some prominent scholars, perhaps dominantly under the influence of Pytheas of Massalia and his observational and/or calculated data.[39] In its turn, this development will lead us back to an alternative understanding of the annual solar motion in its relation to latitude and arctic circle discernible in Crates' exegesis of the Laestrygonian epizode. Thus, Eratosthenes placed Thule 46300 stadia north of the equator, corresponding to a latitude of 66°8'34".[40] How are we to reconcile this concept with Eratosthenes' clear adherence to the 'Rhodian' standard expressed elsewhere and the limit of the northern habitable zone at 54°? It is certain that he accepted Pytheas' data on the location of Thule with his unambiguous reference to the island's latitude, but it is uncertain whether he committed to the opinion according to which the island was inhabited or habitable. It is obvious that he accepted a schematic division of the earth onto 5/6 zones with the boundary of the arctic/northern temperate zone at 54°, but at the same time, confronted with Pytheas' evidence, also used a system that incorporated Thule in the discussion on the extent of the *oikoumenê*. It seems that he was the first scholar to define the northern limit of the *oikoumenê* as represented by the geographical arctic circle, even though in what is left of his writings he nowhere explicitly defined the latter, nor used the annual solar movement to define it (as did Pytheas). In his wake, Hipparchus and Ptolemy also deviated from the Rhodian standard and there is no suggestion that they used it in their calculations.[41] Their conceptions of the northern limit of the *oikoumenê* approach the latitude corresponding to the fixed arctic circle, of which both were well aware.

It was only Posidonius who clearly defined the arctic circle as the region 'where the tropics serve for arctic circles' calling the inhabitants of the polar zone *periskioi*, after the fact that for them the shadow travels all round in 24 hours during the summer season; it does so on the arctic circle only at the summer solstice.[42] This description places the arctic circle at 66°,[43] exactly where Pytheas placed his Thule. Yet Posidonius' definition is an abstraction derived from Pytheas' testimonies, whose very terminology for describing the situation obtaining at Thule he actually uses in his more general description. Pytheas himself, as we have seen, embedded his own observations and calculations within a Homeric framework, i.e., the latter's account of the 'midnight sun' phenomenon at the latitude of the geographical arctic circle on the summer solstice. Indeed, as we will see later, Pytheas conceptualisation of solar phenomena was frequently influenced by mythic models, and he sometimes expounded it in traditional terms, collapsing the boundary between mythic and rational understanding of phenomena.[44]

The development of the notion of the fixed arctic circle in Greek thought was thus far from linear. Already in Homer's time—presumably even earlier, since this part of his account shows a clear traditional nature—the Greeks were aware of the 'midnight sun' phenomenon at some far northern latitude. Pytheas was clearly familiar with this knowledge and utilized its articulation in his account of the phenomena at the latitude of the fixed arctic circle. In the meantime, a cosmological notion with strong numerological characteristics dominated the field, placing the arctic circle at the latitude of 54°. The coexistence of these two traditions is evident in Eratosthenes' oeuvre, but it was subsequently abandoned, with a culmination in Posidonius' clear definition of the geographical arctic circle based on the annual movement of the sun.

This summary analysis of the development of the understanding of the fixed arctic circle brings us back to Crates, whose position was apparently similar to Eratosthenes', since there are reliable and convincing pieces of information that suggest that there existed a strain in Cratetean thought concerning the problem of the arctic circle differing from the 'Rhodian' standard. Thus, still according to fr. 50 adduced above, Crates compared the meteorological conditions characteristic of the land of Laestrygonians with those obtaining in the far north—more precisely, in the area with the longest day on the summer solstice of 23 hours, and a meagre one hour of night, where, 'the setting of the sun is near to its rising',[45] and where, therefore, those that could stay awake could have earned double wages.[46] The land of Laestrygonians was also placed on the latitude with the longest solstitial day of

[37] Popper 1980⁴[1959]/2005: 278; 1982: iii.161; Gattei 2009: 34–35, 52.

[38] For Eudoxus see Dicks 1960: 24–25 and Stevens 1980: 270; Arist. *Mete.* 362b3 (cf. Kidd 1988: ii.742, 744–45; Bianchetti 1998: 163, 185, 194–95); Eratosth. IIB19 Berger/M1 Roller *ap.* Gemin. *Elem. Astron.* 16.6–9, IIB20 *ap.* Achill. *Isag.* 29, IIB24/M2 *ap.* Macrob. *Somn.* 2.6.3–5, IIB25 *ap.* Anon. *Geogr. Expos.* 1.2.

[39] Bianchetti (1998: 43, 153) believes that it was Pytheas who was the first to recognize the difference between the always-visible and fixed arctic circle.

[40] Diller 1934: 264; cf. Bunbury 1883: i.664; Uhden 1936: 100; Dilke 1985: 33–34; Shcheglov 2003–2007: 173; Marx 2014: 200–201. Alternatively, 46800 stadia = 66°51'24" (Dicks 1960: 153) or 'little over 64°' (actually, close to 65°) (Bianchetti 1998: 150)

[41] Hipparch. fr. 61 Dicks with Pyth. fr. 11 Bianchetti = fr. 6b Mette = T 5 Roseman (61°2'); see the discussions in Dicks 1960: 185; Roseman 1994: 43; Bianchetti 1998: 181. Ptol. *Geog.* 1.7.1 (Marinus of Tyre *FGrHist* 2114F1), 1.20.8, 23.1 (= 1.23.22 Nobbe), 24.5, 2.3.14 (62°40'–63°15'), 3.5.1 Müller, 6.16.1, 7.5.12, 16, 6.7–8, 8.3.3 Nobbe; cf. *Alm.* 2.6, p. i.1.114.9–11 Heiberg; Anon. *Summaria ratio geographiae in sphaera intelligendae* 2.9, 15 (*GGM* ii.490, 493).

[42] Fr. 49 E-K = 13 Theiler = *FGrHist* 87F28.

[43] Cf. Ptol. *Syntax.* 2.6, p. 114.21–115.7 Heiberg, who places his 33rd parallel, 'where the longest day is 24 equatorial hours', at 66°08'40".

[44] Cf. Beaulieu 2016: 7–8

[45] Heath 1991[1932]: 133; cf. Evans & Berggren 2006: 163.

[46] Gemin. *Elem. Astron.* 6.10–12, p. 72.2–20.

23 hours by a scholiast on the *Odyssey*,[47] while Priscianus Lydus may have compared a 23-hour longest day in the north with the situation obtaining at Thule.[48] Furthermore, according to Hyginus, the inhabitants of the region 'under the head of Draco' have a night of less than a third of an hour (presumably on the summer solstice), which further associates the region with Homer's Laestrygonians.[49] On the other hand, a number of sources mention a 20-hour longest day on Thule,[50] exactly as in Ptolemy, where the island, with the longest day of 20 hours, is placed at 63° north.[51] Consequently, the Laestrygonians would be placed somewhat to the south of the fixed arctic circle.[52] According to Ptolemy, however, the summer solstice day of 23 hours is located at 66° northern latitude,[53] which would place the Laestrygonians effectively on the fixed arctic circle and the northern limit of the *oikoumenê* according to Eratosthenes.

Both the discussion in Geminus and that in Eratosthenes (who heavily drew on Pytheas' data) associate some of these concepts with the testimonies of Pytheas, who wrote on some northern areas where 'the night becomes very short ... so that, a little while after setting, the sun rises straightaway'.[54] The similarity of Pytheas' description to Homer's was probably the reason why Crates and Geminus believed that it referred to the conditions imagined by the poet. After his mention of Pytheas, Geminus continues his paraphrase of Crates: 'For indeed, since around these places[55] the longest day is 23 equinoctial hours, the night lacks only one hour of being shortened to nothing, so that the setting [point] draws near the rising [point and is separated from it only] by the very short arc of the summer tropic', glossing (still following Crates?) *Od.* 10.86 as 'the setting [point] lies near the rising point'.[56] Here 'setting' and 'rising' naturally refer to the sun.

This 'solar' explanation of Homer's text certainly seems plausible enough, lacking any allegoresis and only recognising a secondary reference to a meteorological fact incorporated into the story. But the main objection to Crates' exegesis of this Homeric passage is that he seems to have believed that Aratus' verses and Eudoxus' concept behind them—to which he compared Homer's description of the Laestrygonians—also refer to the risings and settings of the *sun*, to which Homer's description certainly applies, while it is reasonable to assume that both the Hellenistic poet and the astronomer-mathematician were speaking instead of the risings and settings of the *constellation Draco*, more precisely, the asterism's head ('go[ing] around near the place where the limits of setting and rising mingle together').[57] These limits of rising and setting would in Crates' interpretation refer to the northernmost rising and setting point of the sun at the summer solstice at the latitude near the geographical arctic circle. In any event, these testimonies clearly show that Crates was familiar with the concept of arctic circle significantly differing from the 'Rhodian' standard.

At the same time, a 'solar' interpretation of Eudoxus'/Aratus' description seems to have been accepted by a number of scholars in antiquity. They interpreted 'rising' and 'setting' as referring to the eastern and western semicircle of the horizon or the section of the sun's path from the eastern horizon to the meridian and from the meridian to the western horizon, respectively, with the head of Draco touching upon the meridian's northern intersection with the horizon.[58] Strabo, Achilles, Leontius, Hippolytus and the scholiasts (perhaps also Posidonius) are here referring to the sun's diurnal path, which is only a conjecture on their part.[59] At the same time, Kidd attempted to interpret Posidonius' suggestion for 'improving' Crates' emendation of Homer's description of the Aethiopians' dwellings by taking it as referring to the sun's annual path.[60] According to this explanation, 'setting' would mean the sun's path from the summer to the winter tropic, while 'rising' would mean the sun's path from the winter to the summer tropic; 'rising' and 'setting' would thus 'meet' at the tropics.

Even though it seems unlikely that Crates would have interpreted Eudoxus'/Aratus' description as applying

---

[47] Σ P *Od.* 10.86 (ii.454 Dindorf) = fr. 37d Mette.

[48] *Solutiones ad Chosroen* 4, p. 67.9–15 Bywater = fr. 37b Mette. Mette (1936: 85 with n. 1), however, claims that all he said was that in certain areas the longest day lasts for 23 hours, while on Thule there is a 24-hour day, since it is on the fixed arctic circle; Mette discards as an interpolation the notion of a 5- or 6-day continuous daytime in the area where the longest day is of 23 hours

[49] *Poet. astr.* 4.3.3 = fr. 37i Mette.

[50] Σ MDΔKVUA *Phaen.* 62 = fr. 37h Mette (without Mette's emendation (1936: 87 with n. 1, 271)); Σ Dionys. Per. 582 (*GGM* ii.451); Steph. Byz. s.v. Θούλη; Eustath. Dionys. Per. 581 (*GGM* ii.329.30–33).

[51] *Alm.* 2.6 (i.1.114.9–11 Heiberg). Cf. *Geog.* 7.5.16, 8.3.3 Nobbe for the 20-hour day at Thule.

[52] Cf. Mette 1936: 86.

[53] *Alm.* 2.6, i.1.114.18–20 Heiberg. Ptolemy places the 24-hour day, thus the fixed arctic circle, at 66°08'40", ii.1.114.21–115.7 Heiberg.

[54] Fr. 13a Bianchetti = fr. 8 Roseman = fr. 9a Mette *ap.* Gemin. *Elem. Astron.* 6.9, p. 70.24–72.2 Manitius. Translation from Evans & Berggren 2006: 162; cf. Roseman 1994: 140 and Bianchetti 1998: 103. Pytheas specifically mentions the regions with the longest day of 21 or 22 hours. Compare also fr. 13b Bianchetti = fr. 9 Roseman = fr. 9b Mette *ap.* Cosm. Indic. *Christ. Top.* ii.80.6–9 Wolska-Conus.

[55] Geminus refers to the region where Crates placed Homer's Laestrygonians in 6.10 and 12, which he associated in 6.10 with northern regions described by Pytheas in 6.9.

[56] Crates fr. 50 (*BNJ* 2113F20) = Gemin. *Elem. Astron.* 6.11–12, 72.10–20 Manitius. Translation from Evans & Berggren 2006: 163; cf. Heath 1991[1932]: 133. Broggiato (2001: 217–18) attributes the notion exclusively to Crates.

[57] Translation from Broggiato *BNJ* 2113F20.

[58] Posidon. fr. 49 E-K = 13 Theiler (according to Mette 1936: 73–74); Str. 2.3.8, discussing Posidon. fr. 49 E-K = 13 Theiler (Kidd 1988: i.268–70); Achill. *Isag.* 35 (p. 71 Maass), cf. 22 (p. 52 Maass) (at both places—three times in total—Achilles replaces the last word in *Phaen.* 62 (ἀλλήλῃσιν) with ἠελίοιο, rendering Aratus' phrase as 'the limits of risings and settings *of the sun* blend'); Σ Q *Phaen.* 62; Σ MQDΔKVUAS *Phaen.* 62 (cf. Kidd 1997: 200); Leont. *De sphaer.* 4, p. 564 Maass; probably also Hipp. *Haer.* 4.47.4. Achilles and Hippolytus also adduce the possibility of a reference to the horizon that divides the world into two hemispheres, that above and that below the earth.

[59] Cf. Philo *De opific. Mundi* 9.33–35, who in a language similar to Aratus' speaks of the *akroi* of light and darkness (eventually, day and night) separated by the 'boundaries' (*horoi*) represented by morning and evening, referring clearly to the diurnal solar movement.

[60] *Od.* 1.24; Crates fr. 37 Broggiato, cf. fr. 34f Mette; Posid. fr. 49 E-K = 13 Theiler *ap.* Strabo 2.3.7–8; Kidd 1988: i.269–70.

to anything but Draco, this is what the scholia appear to claim explicitly to have been the case, and this is also what several other ancient scholars enumerated above believed (alternatively and not impossibly all these reports are actually derived from Crates'). In this case, we should understand Crates' περὶ τὴν κεφαλὴν τοῦ δράκοντος, literally 'about the head of Draco', as 'about the area (generally) below the head of Draco', i.e., 'in the far north beneath the circumpolar stars'. The most plausible interpretation—and the one supported by the least ambiguous testimonies—seems to be that Crates indeed associated the position of the land of Laestrygonians with the latitude of the fixed arctic circle, no matter whether he understood it as such or not. It seems unreasonable to believe that he understood the changeability of day lengths with respect to different latitudes,[61] but at the same time did not understand the difference between the fixed arctic circle and the latitude-dependent always-visible circles.[62] The introduction of Draco in the discussion was probably only secondary and was motivated by a cosmological doctrine demanding the neat division of the cosmic sphere. Indeed Crates only once explicitly explained 'the rising(s) and setting(s)' by the phenomena associated with the constellation Draco as a determinative of arctic circle. i.e., *its* risings and settings; it would be more correct to say that even here he simply juxtaposes the two accounts (that of Homer and that of Eudoxus/Aratus) but still distinguishes between the risings/setting of Draco in the latter and that of the sun in the former.[63] In every other instance he implicitly associated Eudoxus'/Aratus' 'rising(s) and setting(s)' with the annual solar motion, by simply adducing the Homeric passage alongside *Phaen.* 61–62.[64] Thus Crates associated solar risings and settings of a certain type (i.e., the northernmost annual rising and setting point of the sun) with the fixed arctic circle and the limit of the sun's annual northern passage on c. 66° northern latitude; in this he was apparently followed by Posidonius (*via* his student and Posidonius' teacher Panaetius?).[65] His exegesis of the Homeric passage on the Laestrygonians most probably followed that of Pytheas, but with the (unnecessary) addition of the constellation Draco.

2.4. The limits of the annual solar movement

The notion of the limits of the annual solar movement was hardly a novelty in Pytheas' time, even if we leave the passage in the *Odyssey* out of consideration. A short examination of the development of this concept in Greek thought will help to contextualise both the Homeric passage and the later tradition in the wake of Pytheas' work. The sun's 'proper limits' were already mentioned by Heraclitus of Ephesus,[66] which is certainly an allusion to the solstices, whether in their temporal or spatial aspect (or both—see chapter 5).[67] In a related fragment he described the Bear as 'forming the limits of morning and evening',[68] which seems to suggest that for Heraclitus arctic circle (I am intentionally using this ambiguous phrase) was defined by the Ursa Major constellation. It is possible that he associated the limit of the sun's annual movement in the north with (in later terms) the latitude of the geographical arctic circle, which can be supported by his 'limits of morning and evening' exemplified by the Bear, the 'limits' here in the meaning Crates and others assigned to 'the limits' in Eudoxus'/Aratus' description of Draco (i.e., the northernmost rising and setting point of the sun at the summer solstice at the latitude near the geographical arctic circle).[69] He might have conceptualized the idea by the northernmost solar azimuths at the summer solstice or the maximum solar declination at the same date of the year (see chapter 5); in this case, 'the Bear' would be signifying 'far north' and/or the maximum declination of the sun relative to the Ursa Major constellation and not the always-visible circle for the latitude of Greece.[70] If Heraclitus used 'the Bear' in the latter meaning (the one understood by Strabo), he would adhere—intentionally or otherwise—to the doctrine associating the turning of the sun with the latitude of 54° north, one defined by the always-visible circle for the latitude of Greece. It appears that Aristotle did something similar, associating the northern uninhabitable zone with the Bear[71] and at the same time defining the boundary of the northern temperate zone with the always-visible

[61] Which he certainly did, as is clear from Geminus' report (*pace* Abel 1974: 1052.3–1053.14).

[62] A provisional argument for Crates' familiarity with the concept of the fixed arctic circle can be extrapolated from a map dating from late antiquity discussed in Bilić 2012a.

[63] Σ Q *Phaen.* 62 (p. 98 Martin). Crates' actual explanation of the association of Aratus and Homer is rather unconvincing (cf. Σ MDΔKVUA *Phaen.* 62, p. 100 Martin).

[64] Σ HQ *Od.* 10.86; Σ HV *Od.* 10.86; Eust. *Od.* 10.86, i.369; Σ MDKVUA *Phaen.* 62; cf. Gemin. *Elem. Astron.* 6.10–12. Similar associations (with reference to the *diurnal* solar movement) of Posidonius, Strabo, Achilles, Leontius, Hippolytus and some of the scholiasts on Aratus referred to above could have originated with Crates.

[65] Panaetius a student of Crates: Crates T21 Broggiato (= *BNJ* 2113T21) = Panaet. fr. 5 Alesse with Broggiato 2001: xvii, 137; Posidonius a student of Panaetius: Posid. T1a, 9, 10, fr. 41c E-K = T1a, 7, 8, fr. 432 Theiler with Kid 1988: ii.12–13.

[66] P.Derv. 4.8: τοὺ[ς ὅρους] – [ὅ]ρους ἑ[αυτοῦ] or ε[ἱμαρμένους] (Lebedev 1989: 39, 46–47; Schönbeck 1993: 8, 17–20; Janko 2002: 8–9; Kouremenos *et. al.* 2006: 68–69, 130); in a corresponding passage Plutarch (*De. exil.* 11.604a = 22B94 D-K) has the term *metra* (in the meaning of 'limits', cf. Seaford 2004: 286), while in *De Is. et Os.* 48.370D he retains *horous*.

[67] Lebedev 1989: 43; cf. Kirk 1962: 285; Kahn 1979: 109, 156, 160, 199. Cf. Diogenes of Apollonia (64B3 D-K *ap.* Simpl. *in Phys.* ix.152.13–14 Diels), who also mentions the *metra* of winter and summer as well as night and day.

[68] 22B120 D-K. *Terma* (pl. *termata*) has a double meaning: 'turning post', 'turning point' (which is the only meaning present in the *Iliad*) but also 'boundary', 'endpoint', *terminus* (Kahn 1979: 51, 161; Purves 2010: 56 with n. 88).

[69] Marcovich (2001: 337–38) suggests that Heraclitus had the phenomenon of the midnight sun in mind when he described the limits beyond which there is no distinction between the rising and setting of the sun in 22B120 D-K. West (1971: 157–58) also associates the fragment with the limits of the sun's annual movement.

[70] *Pace* Strabo, who here believes that Heraclitus employed the term 'the Bear' precisely for the 'arctic (always visible) circle'.

[71] *Mete.* 362b9, cf. fr. 248 Rose, p. 195.5–6 = fr. 695, p. 749B.37–38 Gigon = *BNJ* 646F1.6; cf. Aeschin. *In Ctes.* 165, where 'the Bear' is associated with the extent of the *oikoumenê*.

circle, presumably that for the latitude of Greece.[72] Thus the Bear might have designated the limit of the temperate zone as well as the always-visible circle for Aristotle, but its relation to the sun's annual movement is not specified. In any case, this limit would be placed at 54° north, following the 'Rhodian' standard or '*teoria aristotelica*'. Furthermore, Hecataeus of Abdera's (perhaps a generation later than Pytheas) northern island of Helixoia suggests both an association with Helikê, that is, Ursa Major, and the noun *helix* or the verb *helissô*, 'spiral / to turn round or about', perhaps designating the sun's annual turning round a solstitial turning-post associated with Ursa Major, but the latitude at which this is supposed to happen—if this interpretation corresponds to Hecataeus' intentions—is not specified.[73] Much later, Ptolemy explicitly defined the limit of the northern temperate zone with the Bears (at 54° or c. 66°?), a passage in Dionysius Periegetes and two passages in Festus Avienius (one a translation from the former) suggest the connection of Ursa Major with the limits of annual solar movement, and it is possible that Nonnus' northern turning-point beside which the Bears move also defines the northern limit of the sun's annual passage.[74] Since Dionysius Periegetes and Festus Avienius explicitly associated the Bear and the limit of sun's annual movement with the 24-hour day at Thule, it is certain that they at least had the latitude of the geographical arctic circle in mind. Thus, it seems clear that two approaches to the question of the latitude of 'the' arctic circle were current in pre-Posidonian Greece, both exemplified in Crates' explication of the Laestrygonian epizode: one defined it with respect to the latitude of Greece and the other with respect to the annual solar movement, both occasionally using the northern circumpolar constellations of Draco and Ursa Major in their respective accounts of the phenomenon.

In light of this discussion, it seems possible that Eudoxus did originally refer to the risings and settings of the sun in his description of the movement of Draco's head, but associated them with the latitude of 54° north, the latitude corresponding to the Draco-determined always visible circle, in which he was followed by Crates. Eudoxus was indeed interested in the observation of the lower culmination of Draco's head,[75] and he used it together with the front feet of UMa in determining the always-visible circle.[76] This supports the fact noted above that Eudoxus was apparently unaware of the geographical arctic circle, but rather calculated the always-visible circle with reference to the latitude of Rhodes, i.e., used the 'Rhodian' standard.

On the other hand, it appears that Homer addressed the multifaceted concept of solstices in the Laestrygonian episode in reference to the increasing length of summer days in high latitudes and the almost 24-hour daylight on the summer solstice near the geographical arctic circle, reapplying a narrative model hitherto used in Greek tradition to account for the diurnal solar movement (as exemplified in Hesiod's account of the day/night exchange), while Heraclitus, on the other hand, focused on the association of the maximum declination of the sun relative to the circumpolar constellation of Ursa Major, as later did Pytheas (perhaps at the same time noticing the extreme points of the sun's risings and settings on the horizon). Eudoxus then introduced another circumpolar constellation into the concept, Draco, perhaps at the same time introducing the 'Rhodian' standard for the determination of the 'latitude' of arctic circle (if the latter was not introduced even earlier by the 'Pythagoreans', or perhaps by Anaximander in his parallelogram of the *oikoumenê*, for which see chapter 5), and it was Crates who finally merged all these models into one, albeit somewhat arbitrarily and incoherently.[77] This judgment, however, needs to be taken with caution, since the testimonies for Crates' exegesis are both incomplete and of questionable merit.

2.5. Arctic circle in epic poetry

Other than the Laestrygonian episode, there are additional testimonies for the concept of arctic circle in the epic poetry: a clear reference to the always-visible circle in Homer and a tentative one in Hesiod. The Homeric reference is found in a passage that appears both in the *Iliad* and *Odyssey*; together with the description of Telepylus and the Ortygia/

[72] *Mete.* 362b3, supported by b9–12. In [Arist.] *Pr.* 942a4 the 'regions of the Bear' are said to be 'outside the solstice' (942a1), that is, presumably the limit of the temperate zone in the Aristotelian sense. In *Mete.* 350b6–7 Aristotle placed the Rhipaean Mountains 'under the Bear' (cf. Hippocr. *Aer.* 19 = Aristeas *BNJ* 35F8 for the region 'under the Bears' associated with the Rhipaean Mountains), which is understood by Kiessling (1914: 850.56–58, 851.20–24, 852.22–34) as referring to the fixed arctic circle.

[73] *BNJ* 264F11a = 73B1 D-K; Macurdy 1920: 140–41; Krappe 1947: 223; Thomson 1948: 403; Geus 2000: 72 (from Ephor. *BNJ* 70F41); Winiarczyk 2011: 55 n. 66; Lang, *Commentary* on *BNJ* 264F11a; Gagné 2021: 340 (unfortunately, this study appeared too late to be included in the present book to the extent it merits). Hecataeus' Hyperborean (is)land and its association with solstices are perhaps paralleled in *HN* 4.89, where *cardines mundi* are placed in the land of the Hyperboreans, together with limits (*extremi*) of the movements of the stars. Ferrari (2008: 145–46 with n. 64) believes that these *cardines* represent the solstice(s), as *cardine Phoebus* certainly does in Avienius (*Arat.* 653–54), with *anni cardo* designating the summer solstice at *HN* 18.264 and *cardines temporum* designating the solstices and equinoxes at *HN* 18.218, 220, 222, but it is more probable that Pliny had polar axis in mind here (cf. Beaulieu 2016: 48).

[74] Ptol *Tetr.* 2.2.6; Dionys. Perieg. 580–86, Lightfoot 2014: 228–29, 394–95; Avien. *Or. Mar.* 649–50, Murphy 1977: 42–43, Antonelli 2009; *Descr. orb. terr.* 761–63, *GGM* ii.184; Nonn. *Dion.* 38.406–407, 25.398 (perhaps referring to the constellation's lower culmination, rather than the solstice), cf. 1.454 (cf. 6.236–37), 38.284–85.

[75] Fr. 16 Lasserre with Arat. *Phaen.* 61–62. It appears that the Athenians already in the second half of the 5th c. BC may have attached some importance to the observation of the lower culmination of Draco's head scraping along the horizon (Boutsikas 2007: 143; cf. Boutsikas 2007: 137–43; 2020: 119–26, 131, 150, 152). Indeed, as it will be shown in chapter 5, the Athenians observed the annual movement of the sun on the horizon at least as early as the second half of the 5th c. BC.

[76] Fr. 64a Lasserre.

[77] Since Aristotle had chosen Corona Borealis as the circumpolar constellation defining the arctic circle, as well as taking the zenith position of the asterism rather than its lower culmination into consideration (Arist. *Meteo.* 362b9–12; cf. Str. 2.2.2), it seems possible that Eudoxus was the first author to have used the lower culmination of Draco and UMa—in the case of the latter he might have been preceded by Heraclitus—as the determinative of the arctic circle according to the 'Rhodian' standard, at least insofar the scientific tradition is concerned.

Syriê complex mentioned in book 15 of the *Odyssey*,[78] it represents a sum total of possible references to arctic circle in Homeric poetry. Since these are scattered and incidental remarks, I will avoid any attempt to reconstruct Homer's coherent 'doctrine' on the concept of arctic circle based on them. Rather, I will try to interpret them in their own right (as I have already done with the Laestrygonian episode), notwithstanding either their haphazardness or (potential) incoherence.[79] Indeed, the only thing that the author of the verses probably referring to the always-visible circle found both in the *Iliad* and *Odyssey* had to be aware of in order to define his concept of arctic circle is the diurnal movement of circumpolar stars, with no need of any theoretical background to account for this phenomenon.[80] He seems to have expressed this awareness when he described the constellation of the (Great) Bear as 'she alone is never plunged in the Ocean'.[81] This could be a description of an arctic circle defined by some of the stars in the Ursa Major constellation, but since UMa is certainly not the only circumpolar asterism, the passage was already in antiquity understood as requiring further explanation. The issue revolves about the reading of the word οἴη in Homer's text. Aristotle and Strabo believed he had the whole 'arctic circle' in mind,[82] and the latter thought Heraclitus was of the same opinion,[83] in which they were all probably right, since this is by far the most reasonable interpretation of Homer's description.[84] Crates shared this opinion, although he expressed it somewhat differently.[85] At the same time, while it is not very likely that the claim found in a late source reflects Hesiod's actual opinion on the Bear never touching the waters of the Ocean,[86] it would not be surprising if he did mention the concept somewhere, perhaps in the *Astronomy*, where he indeed mentioned the constellation of Ursa Major (fr. 163 M-W).[87] If we look for analogous concepts in contemporary traditions,[88] it is possible that a text dating from Sennacherib's reign suggests the circumpolarity of the constellation.[89] But there is no need to postulate an outside influence for such a simple observation. The circumpolarity of the Bear(s) at various latitudes was the subject of numerous discussions during antiquity,[90] often in pair with Canopus, its southern counterpart.[91]

What Homer certainly had in mind is to describe 'the Bear' as a circumpolar constellation in an idiom he was familiar with. But if we take the expression *verbatim*, there begin to appear certain problems, since Homer's statement can in no way refer to the whole constellation of Ursa Major, at least the constellation as understood by later authors. We are thus forced to conclude that Homer had in mind only the seven most conspicuous stars of the Bear α–η[92] (although the southernmost of those (η) was circumpolar almost to Syene) when he used it to designate the always-visible circle.

After Homer, the most important asterism for determining the arctic circle for the latitude of Greece seems to have been Draco, closely followed by (again) Ursa Major. The former was already known to Hesiod, who mentioned it in his *Astronomy*, although without drawing attention to its circumpolarity.[93] Elsewhere, in another context, he described a serpent or Ophis as the guardian of the apples of the Hesperides.[94] This beast was always awake,[95] guarding the tree without ceasing,[96] and was also immortal.[97] The serpent's/dragon's epithets imply the circumpolar nature of the constellation and indeed the 'dragon' that guarded the apples of the Hesperides was subsequently catasterised as the constellation Draco.[98] Its counterpart, another 'never-sleeping' deathless dragon,

[78] I will not discuss the 'turnings of the sun' at Syriê here but will provide a full-scale analysis of this line from the *Odyssey* in chapter 5.

[79] Skinner 1969: 7, 12.

[80] Circumpolar stars were, for example, well-known in Egypt and played an important role in eschatology of the Pyramid Texts (Faulkner 1966: 155–57; but today the 'Imperishable Stars' are believed to represent all the stars north of the zodiac circle, Spalinger 2020: 267, not only the circumpolar stars). Thus no theoretical background in modern sense was necessary in order to conceptualise the phenomenon.

[81] *Il.* 18.487–89 = *Od.* 5.273–75; cf. Mus. *Her.* 213–14.

[82] Arist. *Poet.* 1461a20–21 (cf. Lucas 1968: 242); Str. 1.1.6.

[83] 22B120 D-K; Str. 1.1.6 (Kirk 1962: 289; Marcovich 2001: 337).

[84] Cf. Berger 1903: 79 and Burnet 1920: 135 n. 5 on Strabo's interpretation of Heraclitus.

[85] Fr. 27 Broggiato = Broggiato *BNJ* 2113F11a; cf. Broggiato *BNJ* 2113F11b, d. He either attempted a change of the word's gender, modifying Homer's 'the Bear' to 'the arctic (circle)' (Dicks 1970: 49), removed the adverb 'solely' (Maas 1892: 189–90; Broggiato 2001: 190) or kept the adverb but believed it referred to the position of the Bear with respect to Orion (Giampaglia 1998: 507–10; Broggiato *BNJ* 2113 *Commentary* to F11).

[86] Fr. dub. 354 M-W *ap.* [Lact. Plac.] *Narr.* 2.6, p. 639 Magnus.

[87] Merkelbach&West 1967: 79, 150 suggest that this fragment might belong to the *Astronomy*—thus Diels-Kranz, who print it as 4B6, as well as West 1985: 91–92 and Henrichs 1988: 262—although they themselves print it under the *Catalogue of Women*. A *communis opinio* of previous scholarship was that the catasterism does not belong to the Hesiodic corpus (Henrichs 1988: 261 with n. 75, 264).

[88] For the question of the dates of Homeric texts see e.g., Janko 1982: 228–31; 1998: 1; Powell 1991: 189–221, 232 n. 31; 1997: 3–4, 23–24, 31; 2004: x, 25, 32, 38, 40–42, 50–51, 60, 165–66, 168, 171; West 1995; 2014: 1, 22, 27, 37–38, 40–41, 43; Malkin 1998: 45, 110, *Appendix*; Dowden 2004: 194 n. 30; 2011a: 48; Fowler 2004a: 224–25; 2004b: 376; 2013: 125–26; Frame 2009: 3, 577–78, 603, 610, 618, 620; Burgess 2015: 93–97; Manoledakis 2015: 305, 307; Bachvarova 2016: 396–400 (an overview of earlier discussions). The dates range from c. 800–750 (Powell; Malkin argues for the 9th c. but allows a date after c. 750) to after 630 BC (West). Hesiod is a near contemporary, e.g., Janko 1982: 228–31 (700–650 BC).

[89] Thompson 1940: 90, 18; Koch-Westenholz 1995: 154.

[90] Nearch. *FGrHist* 133F16 (cf. 133F1); Onesicr. *FGrHist* 134F28 – Juba *FGrHist* 275F28, 134F10; Orthag. *FGrHist* 713F2; Baeton *FGrHist* 119F4; Megasth. *FGrHist* 715F7a, cf. F7b and F4; Deimach. *FGrHist* 716F3; Eratosth. IIIA9, fr. 67 Roller, IIIA10, fr. 68 Roller (Berger 1880: 180–81; Dicks 1970: 127–28); Hipparch. fr. 17 Dicks (Dicks 1970: 127; Shcheglov 2005: 365), fr. 43 Dicks = Eratosth. fr. 57 Roller, fr. 45 Dicks, fr. 47 Dicks = Eratosth. fr. 59 Roller, etc.

[91] Arist. *De caelo* 198a3–6 (perhaps); Eudox. frs. 64ab, 74 Lasserre *ap.* Hipparch. 1.11.1–8; Posid. fr. 205 E-K = 74 Theiler *ap.* Procl. *In Tim.* iii.125.5–17 Diehl, etc.

[92] Cf. Hipparch.1.5.6, p. 46.7–9 Manitius.

[93] Fr. 293 M-W.

[94] *Th.* 334–35.

[95] Panyas. fr. 15 West; Ov. *Met.* 9.190; Hyg. *Astron.* 2.6, etc.

[96] Agroit. *FGrHist* 762F3b.

[97] [Apollod.] *Bibl.* 2.5.11. Cf. Ogden 2013: 37.

[98] [Eratosth.] *Cat.* 3, 4; Hyg. *Astron.* 2.3, 6; Σ S, MKVUA Arat. 45, Q Arat. 46, Q, S Arat. 69; Σ B, S, G Germ. *Arat.*, pp. 60–61, 118–19 Breysig; cf. an illustration in a 12th c. MS of Germanicus, Cod. Matrit. A16, f. 56r (Thiele 1898: 145 Fig. 62; Cook 1940: iii.1.489 Fig. 318) and another in the late 8th c. astronomical compendium Cod. Col. Eccl. Metrop. LXXXIIIII (Thiele 1898: 159 Fig. 69). Cf. Ogden 2013: 38.

the guardian of the Golden Fleece in Colchis,[99] was also explicitly identified with the constellation Draco.[100] It is not clear whether the concepts of serpents/dragons as ever-watchful guardians and, more specifically, the two never-sleeping guardians of valuable treasures at the opposite extremities of the world, can in any way be associated with the circumpolar nature of the Draco constellation. The majority opinion would at present incline towards the negative answer, allowing the identification of the constellation with the mythic figure only in the Hellenistic period. Supported by the lack of explicit evidence for Hesiod's identification of the two, this conclusion seems the more prudent one. However, the prevailing view that the catasterising tradition only started off in the Hellenistic period is now challenged.[101] More specifically, the tradition is already explicitly present in the Hesiodic *Astronomy* and in the cyclic poems.[102] It is thus not at all impossible that the identification of the always-visible serpentine constellation Draco with the always-awake serpentine guardian of fabulous treasures was already present in early Greek tradition, as exemplified by the Hesiodic corpus. At this point, however, this cannot be supported by clear—as opposed to circumstantial—evidence.

2.6. Laestrygonia, the sun and the Otherworld

The meteorological phenomenon characteristic for Laestrygonia is by far the most important information for determining the cosmological position of this region, but it is not the only one. However, the two supporting pieces of information which will be discussed presently would be clearly insufficient to argue for a cosmological location of the land without the meteorological description. In this particular context the interpretation of this additional evidence seems plausible, but even if it seems insufficiently so, this will not undermine the strength of the conclusions drawn from the interpretation of the meteorological description of the land.

Thus, the name of the Laestrygonian city, *Têle-pylos*,[103] could represent another indication of the solstitial nature of the land of Laestrygonians in Homer's narrative. It is usually translated as 'distant gate' or something similar,[104] which seems more plausible than the alternatives.[105] It is thus associated with some gates, most probably located far away from the geographical focus of Homeric epics—the Aegean. The form of the name is unique in Greek nomenclature,[106] which suggests a programmatic use by the poet, or, rather, by the long epic tradition behind this episode.[107] The 'Distant Gates' (for which further see chs. 2, 7 and 8), can indeed be coherently incorporated into the cosmological frame of the narrative's secondary reference and was convincingly compared to the gate-mountain Māšu in the *Gilgameš Epic*,[108] which undoubtedly represents a solar portal,[109] most likely designating—among other things—a solstitial point in the annual solar orbit.[110] The undoubtedly seasonal and solar character of the land of Laestrygonians strongly suggests that Telepylus indeed represents a solstitial 'portal', the northernmost point and limit of the annual solar voyage.

Furthermore, the queen of Laestrygonians is characterised in a way suggestive of her otherworldly nature,[111] Telepylus itself can be understood as 'Pylos… an entrance to the Beyond',[112] while the entire Laestrygonian episode can be understood in terms of solar eschatology.[113] It will be clear from the following chapters that the association of solar movement with the Otherworld was widespread, but at this point it is sufficient to note that it was predominantly diurnal solar movement that was involved in such type of speculation (as in the second *nekuia* in the *Odyssey* or in Egypt during the New Kingdom period), even though some eschatological models were also built upon the annual. Since the eschatological features of Laestrygonia are not particularly pronounced in Homer's narrative, I will offer an extended discussion of this subject at a more appropriate place. With respect to the Laestrygonian episode, the queen's character, together with the gates she is apparently guarding (perhaps even the entire episode as such), could thus substantiate the cosmological

[99] A. R. 2.405–407, 1208–1209; 4.129; Dionys. Skytobr. *FGrHist* 32F14; [Apollod.] *Bibl.* 1.9.16; Ov. *Met.* 7.36, 149, 212–13; *Her.* 6.13, 12.49, 60, 101, etc. Cf. Ogden 2013: 58, 60–63.

[100] Val. Flacc. *Argon.* 8.56–61. For Draco as a circumpolar constellation see *Argon.* 2.64–65.

[101] E.g., Dowden 1992: 11 and his *Commentary: Synthesis* on *BNJ* 468; Dekker 2013: 2; Bilić 2017: 14. Boutsikas' study presupposes several early catasterisms as part of Archaic and Classical-period ritual complexes (e.g., Boutsikas 2020: 116–32, 162–69).

[102] *Astronomy*: Sale 1962: 133 n. 18, 138–40; 1965: 15; Dowden 1992: 11; Gantz 1993: 213, 218, 271–72, 726; Fowler 2013: 293 n. 110; cyclic poems: Gantz 1993: 213, 215; West 2013: 209–11 (fr. 14a* of the *Little Iliad*). For a presupposed catesterism behind Alcman's *Partheneion* see Dale 2011: 30–31; Boutsikas 2020: 133, 152; for the 5th-c. catasterisms in Athenian tragedians see Ogden 2013: 125, 127, 164–65.

[103] *Od.* 10.81–82. On this point cf. Bilić 2017: 5.

[104] Woodbury 1966: 612; Frame 1978: 60; West 1997: 406–407; 2005a: 62; Käppel 2001: 18; Nakassis 2004: 225; Marinatos 2001: 403. For this question cf. Merry *et al.* 1886: 405; Meuli 1931: 537.14–23; Heubeck 1989: 48; Scodel 2003: 86; Fowler 2017: 248.

[105] Nitzsch 1840: 100 *ad* 10.81–82; Packard 1874: 32–33; Merry *et al.* 1886: 405; Page 1973: 34–37, who gives the alternative explanations of Homeric scholia, Σ BQTV *Od.* 10.82 (ii.452 Dindorf); Carpenter 1946: 107; Coxon 2009: 275.

[106] Page 1973: 34–35.

[107] Page 1973: 37–38; for the Laestrygonian episode originally part of a pre-Homeric *Argonautica* see Meuli 1921: 54, 89–91; 1931: 539.36–49; Page 1973: 38–39; Heubeck 1989: 49; West 2005a: 52–53; 2014: 85, 119–20, 206–207; Fowler 2013: 219; Bachvarova 2016: 102–103 and the following chapter.

[108] West 2005a: 62; cf. West 1997: 406–407, after Germain 1954: 414–17, 421–22; Burgess 1999: 199.

[109] *EG* 9.37–45 (George 2003a: i.668–69; 2003b: 71).

[110] Huxley 2000: 126–28. I am here emphasizing this particular understanding of the cosmological role of mount Māšu, but in Mesopotamian tradition it is more often a feature of the models accounting for the diurnal solar movement. The already mentioned interchangeability of motifs used in building models accounting for respectively diurnal and annual solar movement predictably allowed it to be used in both types of models.

[111] *Od.* 10.113 (cf. *Il.* 5.47, 8.368; Hes. *Th.* 739); Frame 1978: 58. She was later identified with Lamia, both because of her name and nature (Σ UEAGP Theocr. *Id.* 15.40c (Wendel 1914: 309.11–13). The Italian city of Lamia, named after the homonymous Libyan queen (Σ Aristoph. *Pac.* 758d), is probably to be associated with the Laestrygonian city (Fontenrose 1974: 103)). Cf. Ogden 2013: 91 n. 114.

[112] Fowler 2017: 248.

[113] Frame 2009: 40–43 (referring to *Od.* 10.131–34).

localisation of her land proposed here by adding it subtle but distinctive eschatological features.

2.7. Conclusion

The land of Laestrygonians in the *Odyssey* appears to represent a narrative model describing the annual solar movement and its determining moments, the solstices or, more precisely, the summer solstice. The model accounts for this phenomenon in terms of a distant land in the far north at (in later terms) a latitude near the geographical arctic circle. This latitude is characterized by an almost non-existent night about the summer solstice, which manifests itself in the extremely short period between sunset and sunrise, with these points being very close to each other on the northern horizon (near the azimuth of 0°). It is probably closely related to and derived from the model accounting for the diurnal solar movement in which Day and Night as personified agents represented the homonymous meteorological phenomena. The knowledge Homer is presenting in the form of a narrative model does not entail any non-personalistic theoretical background to account for this process, but only an awareness of the connection of the midnight-sun phenomenon with the annual solar movement and the geographical position of the observer. This could have been derived from empirical observations (performed by sailors, merchants, etc.), but it was nevertheless incorporated into a mythic model delineating the causal relation of the annual solar movement and its determining points with the latitude-dependent changeable day lengths. In this way, it is a clear example of a myth referring to solar movement and belongs to a class of myths that are the subject of this book. Its ethnographic context is provided both by a similar mythic model for the diurnal solar movement and a scientific tradition that developed in its wake. This tradition includes Pytheas' testimony and Posidonius' dependence upon the latter for his definition of the geographical arctic circle, as well as a parallel but distinct concept possibly originating in Pythagorean circles and focusing on the always-visible circle for the latitude of Greece. Crates of Mallos was apparently a focal point of such considerations, combining in his Homeric exegesis mythic tradition with modern astronomical and geographical insights. In this practice he was perhaps partially inspired by Pytheas, while Posidonius, two generations removed from Crates, showed much less interest in this line of interpretation.

Finally, in this chapter two important features of myths referencing solar movement were recognized: the first is the interchangeability of motifs used in building models of the diurnal and annual solar movement, the second is the presence of eschatological considerations in such myths. Both these characteristics will be further explored in the following chapters.

3

The Bear Mountain

It was emphasised in the introduction that analogies and models were employed, if in different measure and with significantly different types of analogues (i.e., personalistic in the former and non-personalistic in the latter), in both 'mythic' and 'rational' discourse to account for specific phenomena of physical reality. In addition, and as a consequence, the concept of solstices and related ideas, previous to the formation of a specialized technical non-personalistic language, must have been formulated in terms of mythic causality. Even later, when Presocratic philosophers formulated their 'scientific' (i.e., non-personalistic) concept of solstices, it was also expressed in a non-technical language, that is, once more, in terms of myth-like or concrete models. In the preceding chapter it was demonstrated how Homer's Laestrygonian episode from the *Odyssey* can be interpreted in terms of 'mythic geography/astronomy' (from the point of view of the early Greeks the best term would be 'mythic meteorology'), and, furthermore, how it can be discussed alongside later concepts of 'scientific' cosmology of the philosophers of nature such as Heraclitus and even with the scientific tradition proper, as exemplified in Pytheas, Eratosthenes and Posidonius, with Crates as a special and highly idiosyncratic case. It is now time to analyse another but closely related mythic model accounting for a similar nexus of phenomena, which will be approached through a study of Apollonius' Cyzicus episode from his *Argonautica*.[1] It will be shown that this model is actually a variant of and complementary to the one present in the *Odyssey* and that both probably derive from a closely related source or tradition. In this way the analysis presented in this chapter is complementary to and seamlessly segues from the study of the Laestrygonian episode in the previous chapter.

As opposed to the Laestrygonian episode from the *Odyssey*, which contains a clear solar reference, the Cyzicus episode can be interpreted in similar terms only through a comparison with the former, since it does not contain any clear reference to solar movement. Since this was my first and unconditional criterion for the inclusion of a motif, story or set of stories into the category of myths discussed in this book, this requires some explanation. Indeed, the episode's unequivocal connection and correspondence to the Laestrygonian episode unanimously accepted by scholars allows me to claim that a reference to solar movement could and should be recognised in Apollonius' account and its immediate and more remote predecessors, inspirations and sources. Since this correspondence is widely accepted and seems unambiguous, I feel free to take it as a starting-point in interpreting the episode from the *Argonautica*. At the same time, I will maintain—accepting the arguments of earlier scholars which form, it might be claimed, a *communis opinio*—that this episode, or some variant of this episode, was indeed part of a pre-Homeric Argonautic tradition, which was most probably the source for Homer's account as well and thus should be considered the earliest Greek testimony for this particular model of the annual solar movement.

3.1. The Cyzicus episode

On a general level, it seems clear that Apollonius' *Argonautica* has—or some of its episodes have—a recognisable cosmological setting, which in principle allows the drawing of comparisons between the two epics as a whole.[2] More specifically, the Cyzicus and Laestrygonian episodes are similarly structured and have a number of narrative traits in common, which allows for a parallel analysis on both the level of structure and detail. In this context, a specific topographical piece of information was often used as a starting-point in comparison between the two accounts:[3] a fountain *Artakiê* is found in the land of Laestrygonians,[4] while in Apollonius' *Argonautica* we find a description of an island or a peninsula in the Propontis with a fountain of precisely the same name.[5] The fountain was mentioned before Apollonius,[6] perhaps together with Argo's visit,[7] thus indicating that it was a traditional part of the episode, rather than the poet's invention. It could be suggested that either Callimachus or Apollonius discovered it in some local account of Argo's visit, antiquarian, mythographical or otherwise.

Apollonius called this island *Arktôn oros*, 'Mountain of Bears',[8] and mentioned two different peoples as its

[1] The following chapter is an abbreviated and slightly modified version of Bilić 2017. For the analysis of this episode in Apollonius see Jessen 1895: 757.39–758.43; Fitch 1912; Bacon 1925: 69–71; Levin 1971: 87–109; Clauss 1993: 148–75; Hunter 1993: 41–43; Knight 1995: 84–93, 147–52; Green 2007: 222–28; André 2010; Thalmann 2011: 91–100; Fowler 2013: 217–20.

[2] For a cosmological setting of Apollonius' *Argonautica*, especially the description of the return voyage of Argo, see for example Endsjø 1997: 374–75, 377, 380, Dräger 2001: 81–84 (criticized by Thalmann 2011: 136 n. 61), Stephens 2003: 18, 178, 217–37, Noegel 2004: 129–30, 132, 135–36 and Beaulieu 2016: 84–85 (cf. Lindsay 1965: 21–22). This is especially evident in A. R. 3.309–11 and 4.630. For the *Argonautica* in general as a 'mythic' adventure see e.g., Fontenrose 1974: 478–87 and Davies 1988: 282.

[3] Wilamowitz-Moellendorf 1884: 166–68; Heubeck 1989: 47–49; Clauss 1993: 159–61; Knight 1995: 147–52; Ballabriga 1998: 135; West 2005a: 48 (with n. 37 for earlier literature); 2014: 120, 207; André 2010; Louden 2011: 159.

[4] *Od.* 10.108.

[5] A. R. 1.936–39, 957.

[6] Alc. fr. 440 Lobel-Page; Callim. fr. 109 Pfeiffer = Σ A. R. 1.955–60c (West 2005a: 49 n. 40).

[7] Callim. frs. 108, 109 (*Dieg.* V.34–39) Pfeiffer with Σ A. R. 1.954 (cf. Pulbrook 1988: 57–58; Cameron 1995: 253–55).

[8] A. R. 1.941, cf. 1150.

inhabitants, one of those being monstrous six-handed 'Earth-born'.[9] While Jason and his companions were absent the Earth-born descended the mountain and blocked the entrance to the Chytus harbor, where Argo was anchored, with boulders. Heracles, who did not ascend the mountain, tried to repel the aggressors, who assaulted him by throwing giant rocks. In the meantime the rest of the Argonauts returned, and the Earth-born were annihilated.[10]

There is no recognisable solar reference in this account, but it is immediately obvious that the parallels between this episode in the *Argonautica* and the Laestrygonian episode in the *Odyssey*—where the reference is clear—are very pronounced. The two passages indeed have a number of narrative details in common: the fountain *Artakiê*, a narrow-entranced harbour, the ascent of part of the crew, gigantic invaders, descendants of Poseidon,[11] the attack on an anchored ship(s) with hurled rocks, and the leader absent from the battle. Moreover, the narrative structure of the episode is common to both, and it is hard to believe they are unrelated.[12] This is easily shown through a syntagmatic analysis of the structure of these two stories:

(1) The travellers arrive to a(n) (is)land inhabited by monstrous creatures, of which they are not aware upon their arrival; (2) They (recklessly) enter a narrow-entrance harbour and (3) are attacked by the monstrous natives,[13] who are allied by the harbour's configuration; (4) The outcome of the battle, although different in two cases, raises great grief for the travellers; (5) After the battle the travellers continue their voyage.

This analysis of action patterns (in Proppean/structuralist terms 'functions' or 'motifemes'), together with narrative details (i.e., the contents), clearly shows both the structural and narrative similarities between Homer's and Apollonius' account. It is almost certain that Apollonius followed Homer in his elaboration of this episode, but in addition used a pre-Homeric source also used by the author of the *Odyssey*.[14] Nonetheless, this does not explain why the action was transferred from 'distant gates' where 'the paths of night and day are close together' to a 'Bears' island' in the Propontis, in the process apparently losing all references to solar movement. Part of the answer may lie precisely in the secondary reference of the mythic narrative utilised by Apollonius, as exemplified when it appears in a similar form in the *Odyssey*, here with a clear reference to the annual solar movement.

Before proceeding to this discussion, one detail in Apollonius' toponymy should be analysed more thoroughly, since it will be important in the continuation of this study. It exemplifies how 'mountain(s)' and 'island(s)' are interchangeable features in the accounts of solar movement and represent paradigmatic shifts that do not change the meaning of the account of which they are part. Both Homer and Apollonius called the fountain *Artakiê*, but Apollonius' name for the island or peninsula on which it was situated is *Arktôn oros*, 'Mountain of Bears'. Subsequently, he seems to equate the mountain *Dindumon* with another named 'Mountain of Bears'[15]—unless he is still speaking of the homonymous *island*, which is unlikely—most probably situated on the island itself. As *di(n)dumos* was—at least by popular etymology—probably derived from 'two peaks',[16] or, according to Philostephanus, from *didumoi mastoi/mazoi*, 'twin breasts',[17] it is possible that one of those peaks was called *Arktôn oros*. In any case, it seems that Apollonius uses the same name both for the mountain—or one of its peaks—and the island.

3.2. Celestial bears at the solstice island

We should now turn to the discussion of the secondary reference of Apollonius' myth and its connection with the corresponding myth of Laestryognian Telepylus, which will help us in determining why this particular narrative was applied to the island of Cyzicus.

The association of the island from the *Argonautica* with bears is unmistakable.[18] But if Apollonius' narrative stands in any connection with Homer's—and that it does seems clear from the preceding discussion—then the possibility that *celestial* Bears are involved deserves to be investigated, given that it was shown that Homer's narrative refers to the latitude of the geographical arctic circle, even though it was unknown to Homer as such. As discussed in chapter 2, in their turn Ursa Major and other

[9] A. R. 1.942–46, 989–91.

[10] A. R. 1.985–1011. Variants of the episode were known to Herodorus *BNJ* 31F7 (Clauss 1993: 149–50; Knight 1995: 149; West 2005a: 49; Fowler 2013: 219); Deilochus *BNJ* 471F7 (Fitch 1912: 50–55; Meuli 1921: 90; Levin 1971: 99 n. 3; Clauss 1993: 149; West 2005a: 49; Green 2007: 223; Fowler 2013: 218); Ephor. *BNJ* 70F61; Hyg. *Fab.* 16; Conon 41 *FGrHist* 26F1; Val. Flacc. 2.657, 659, 3.45, 126, 221; [Apollod.] *Bibl.* 1.9.18; cf. Agath. *BNJ* 472F2.

[11] This is not explicitly stated or implied in Homer's narrative, however, but it was already known to the Hesiodic author (Hes. fr. 150.26–27 M-W); cf. Meuli 1931: 537.61–67; Pease 1943: 72, 80.

[12] For this opinion see esp. Meuli 1921: 54, 89–91.

[13] Lycoph. 662–63 mentions how Heracles slew a major part of the Laestrygonians in an unelaborated episode. Tzetz. (Σ) Lycoph. 662 (ii.220 Scheer) explains how this incident occurred when they attempted to rob him of the cattle of Geryon (Holzinger 1895: 270; Mair & Mair 1921: 549), and D. S. 4.24.1 mentions how he passed through the plain of Leontini (the commonest localisation of Telepylus in antiquity) with the herds of Geryon. It seems that this encounter with the Laestrygonians, although similar to the one he had with the Earth-born in the *Argonautica*, belongs to a different context.

[14] The very presence of important cosmological elements in Apollonius' version (i.e., the Ursae constellations), which naturally fit into the model expounded in the *Odyssey* and will be discussed below, prove that Apollonius could not have depended solely on Homer's rendering of the episode.

[15] A. R. 1.1150.

[16] Σ A. R. 1.985 (p. 87.2–3 Wendel); *EM* 276.35–36 *s.v.* Δίνδυμον; Hasluck 1910: 22–23, 214; Mooney 1912: 131; Pliny (*HN* 5.142) indeed calls the mountain *Didymus* and Silius Italicus (17.20) mentions *gemino Dindyma monte*. See also Lucian. *Astrol.* 23 (the authenticity of the work is defended by Lightfoot 2003: 191–96); Σ Clem. Alex. *Protrept.* 34.18, p. 310.29–31 Stählin & Treu; *EM* 276.36–43 s.v. Δίνδυμον.

[17] Philosteph. fr. 2 *FHG ap.* Σ A. R. 1.985 (pp. 86.22–87.1 Wendel). The potential significance of this name will be discussed later.

[18] Cf. on the island's name Σ A. R. 1.936–49a (p. 81 Wendel); perhaps also Σ Parthen. fr. 3.3–4 Lightfoot (Lightfoot 1999: 142), but the evidence here is slight.

circumpolar constellations (most notably Draco) were used to define the always-visible circle for the latitude of Greece, while specifically Ursa Major's association with the solstices, attested from Heraclitus onwards, made the constellation a determinative for either the geographical or astronomical (always-visible) arctic circle and, at the same time, the extent of the habitable temperate zone.

It could be argued that the island (at first mythic/cosmological, later located in the Propontis) was indeed named after the constellation(s).[19] Further, the scholiast on Apollonius claims that the island of *Arktôn oros* got his name after the fact that Zeus' nurses were here transformed into bears.[20] Zeus' nurses were Helice and Cynosura, in their turn catasterised into the constellations of Ursa Major and Minor.[21] Aglaosthenes, if he is really the author of the narrative given in the scholion on the *Odyssey*, adds more information to the story. Zeus, fleeing Cronus, had transformed himself into a serpent; later, the nymphs Helice and Cynosura, previously transformed into bears, were catasterised by the storm god into celestial Ursae. Similarly, 'Epimenides' claimed that Zeus' serpentine form was catasterised into the constellation Draco, along with his nurses who were transformed into celestial Bears.[22] In this way a coherent story can be reconstructed from ancient authors' and scholars' rather ingenious interpretations: Zeus' nurses were transformed into bears at the island of *Arktôn oros* and subsequently catasterised, with either of these two incidents responsible for the island's name. If we were unfamiliar with the Laestrygonian episode, this series of explanations could safely be labelled aetiological, i.e., *post factum* explanations of existing reality considered characteristic—together with catasterising tradition—of the Hellenistic period. But the structural and narrative similarities between the two episodes make this explanation insufficient. Something in the Cyzicus episode, especially if it is derived—as it almost certainly is—from a pre-Homeric *Argonautica*, must refer to the crucial meteorological characteristic of Laestrygonia, and the only (or almost only, see below) feature that could possibly refer to the set of phenomena referenced in the latter is precisely the one reconstructed from the ancient tradition, i.e., the association with the Ursa Major constellation in its role as a determinative of arctic circle, both geographical and latitude-dependent.

How old is this catasterising tradition? I have already casted serious doubts (chapter 2) onto the dominant view that the catasterising tradition only started off in the Hellenistic period and argued that it is already present in the Hesiodic *Astronomy*. Indeed, Jacoby was following this dominant view in his highly influential comments on the date of the catasterising tradition in his notes on 'Epimenides', and the association established between Zeus' nurses transformed into bears on the island of Cyzicus and celestial Bears was consequently dated to the Hellenistic period.[23]

But contrary to the prevailing opinion Chronus was already in the 5th c. rather explicitly associated with the constellation Draco, mentioned alongside the Ursae.[24] Later, Draco was described in terms suggesting the 'all-seeing time',[25] while Orphic sources mention the dragon-like Heracles-Chronus.[26] Furthermore, he was, this time as Cronus, associated with the (north celestial) pole.[27] In this context the Pythagorean identification of the Ursae with the hands of Rhea, the consort of Cronus,[28] as well as their name for the (northern celestial) pole—the seal of Rhea[29]—seems significant.[30] Thus, C(h)ronus was associated with Draco in the 5th c., and Helice and Cynosura must have been associated with the Ursae already at this time, perhaps as early as the 6th c.

The latter suggestion depends upon the dates of Aglaosthenes and 'Epimenides'. The former is dated to the late 4th–3rd c. and thus, belonging to Hellenistic period, of no particular relevance to this discussion.[31] The problem with the date of 'Epimenides', on the other hand, is rather complicated, since the name stands for both an author and a tradition attached to his name, with various dates proposed for different works belonging to this tradition. Modern scholarship in general dates the works attributed to Epimenides from the late 6th to mid-4th c. BC.[32] More specifically, Jacoby ascribed the story of catasterism of the Bears to (in his opinion late) anonymous *Cretica*, but still regarded Epimenides' authorship of this fragment possible.[33] On the other hand, Dowden explicitly ascribes the fragments to Epimenides' verse poem (in this he follows Diels-Kranz), which he dates to a wide period of 550–300 BC, but surely before the prose version he dates

[19] Σ A. R. 1.936–49a (p. 81 Wendel); cf. Nic. *Alex.* 7 with Σ *Alex.* 7a, c.

[20] Σ A.R. 1.936–49a (p. 81 Wendel); Hirschfeld 1895: 1172.52–53; Gundel 1912: 2860.53–57; Robertson 1996: 267; Blakely *Commentary* to *BNJ* 31F7.

[21] Arat. 30–37 and Σ Arat. 46, perhaps after Epimen. 3B22–23 D-K = *PEGr* fr. 36 F, 49 F or Anon. *FGrHist* 468F3a, 5; Aglaosth. *BNJ* 499F1 *ap.* [Eratosth.] *Cat.* 2, Hyg. *Astron.* 2.2, Σ Germ. *Arat.* pp. 59, 115 Breysig; Σ Q *Od.* 5.272 = *Proleg. in Arat.* 22 and *Exc. var. de phaen. Arat.* 5a Martin, from Aglaosthenes?; D. S. 4.80.1–2; Hyg. *Astron.* 2.2; Manil. *Astron.* 2.30; Germ. *Arat.* 32–33, 39, 714; Serv. Dan. *Georg.* 1.246 (Gundel 1912: 2859.64–2860.46; Robertson 1996: 267). Helice was sometimes considered identical to Callisto (Theocr. *Id.* 1.123–26 (cf. Gow 1952: ii.26–27); Lact. Plac. *Theb.* 7.414–15, pp. 480–81 Sweeney; Serv. Dan. *Georg.* 1.246), herself catasterised into Ursa Major (Hes. fr. 163 M-W = 4B6 D-K, cf. fr. dub. 354 M-W; Sale 1962: 122–23, 140; also, *BNJ* 468F3b with Dowden's commentary).

[22] *BNJ* 468F5 = *PEGr* fr. 36 F.

[23] Cf. Robertson 1996: 267–68.

[24] Crit. *Pirith. TrGrF* 43F3.5 = 88F18 D-K = Eur. fr. 594 Nauck (cf. Beck 1977–78: 115). For the ancient identification of Cronus with Chronus see Pherecyd. frs. 14, 60, 65, 66 Schibli; Epicur. *ap.* Epiph. *Adversus haereses* 1.8 = p. 589.11–21 Diels; Orph. fr. 56 Kern = *PEGr* fr. 120I F (actual identification not included by Bernabé), 203I F, 207I F (West 1963: 162; 1994: 289–90, 307; Schibli 1990: 17 n. 8, 9, 39, 79, 91, 97, 102–103, 130, 135–39; Kirk *et al.* 1983: 56–57, 67; Graf 1993: 185–86).

[25] Hipp. *Ref.* 4.47.2 (cf. Beck 1977–78: 115).

[26] Frs. 54 (Hieron. *BNJ* 787F3), 57 Kern = *PEGr* fr. 76I–II F, cf. 76III–IV F, 78 F.

[27] Procl. *Resp.* ii.213.4 (Molina Moreno 2001: 60).

[28] Arist. fr. 196 Rose = fr. 159 Gigon = 58C2 D-K. Incidentally, Bremmer (1999: 75) interprets this information as Pythagorean rationalization of traditional myth.

[29] Arist. fr. 204 Rose = fr. 179 Gigon.

[30] Burkert 1972: 171.

[31] Müller in *BNJ* 499, *Biographical Essay*.

[32] Toye, *BNJ* 457, *Biographical Essay*.

[33] *FGrHist Comm.* iii.365; he dated the work to c. 100 BC (iii.342, 349).

to shortly before the mid-4th c.[34] Fowler similarly accepts the presence of catasterisms in Epimenides' original work, but attributed it to a prose pseudepigrapha paraphrasing or deriving from it.[35] Finally, Ogden categorically attributed the catasterisms to the 6th-c. Epimenides.[36] Thus it seems possible, if not probable, that the catasterism of the Bears was indeed present in Epimenides' original work. There is nothing in this concept that conflicts with its early formation in the period when Hesiodic *Astronomy*, the cyclic poems and Alcman's *Partheneion* were current (see chapter 2), thus in the 6th c. BC, at the higher end of Epimenides' proposed dates.

This particular catasterising tradition could thus be associated with the notion of a mythic solstice island defining either geographical or astronomical (always-visible) arctic circle, concomitant to the role of the constellation Draco in determining the always visible circle for the latitude of Greece combined with the similar function of Ursa Major. As we have seen, several authors, both philosophers/scientists and poets associated the Bear with the solstices, thus making the constellation a determinative for the fixed arctic circle and, at the same time, the extent of the habitable temperate zone. These stories must at first have been referring to a mythic island, i.e., a cosmological location, and only afterwards were applied to the island of Cyzicus. Otherwise, the application would remain completely arbitrary.

Another connection of *Arktôn oros* with celestial Bear is found in Nicander's *Alexipharmaca*. He described the island of Cyzicus as Ἄρκτον ὑπ' ὀμφαλόεσσαν.[37] This is explained by the scholiasts by the fact that omphalos represents the northern celestial pole around which the Bear turns.[38] Although Nicander's reference could be associated with Cyzicus' *Arktôn oros*,[39] the appearance of a 'Bear' having an omphalos above Cyzicus is strange. If Nicander's unusual construction does refer to the celestial Bear, as it seems probable,[40] then we have here a curious pairing of the constellation with the island that, as we know, received its name after bears (or Bears). 'Bear' in association with 'omphalos' certainly suggests that the poet was aware of the account according to which the island got his name after the fact that Zeus' nurses were here transformed into *celestial* bears.

[34] Commentary to *BNJ* 468F3a, b, 5 and *Commentary: Synthesis*. Toye, *BNJ* 457 *Biographical Essay*, dates the poem(s) of Epimenides to the late 6th–second half of 5th c. BC. Bernabé 2007: 137, 156–57 includes the fragments in the *Cretica* (fr. 36) and *Theogony* (fr. 49), respectively, but without committing to their nature.

[35] Fowler 2013: 396 on Epimen. fr. 2 Fowler (= 3B24 D-K, *PEGr* fr. 37 F, *BNJ* 457F18), cf. 649–52.

[36] Ogden 2013: 164.

[37] *Alex.* 7.

[38] Σ *Alex.* 7a, c (on Hellenistic scholarship on Nicander, eventually distilled into the preserved scholia, see briefly Cameron 1995: 192–93).

[39] Thus Gow 1951: 106; Gow & Scholfield 1953: 190.

[40] Nicander was said to have been an astronomer in an unreliable tradition, recorded in *Arati Vita* I, p. 8.29–31 Martin, *Vita* II, p. 12.1–3 Martin and *Vita* IV, p. 21.5–6 Martin (Cameron 1995: 195–96, 199).

If the information obtained through the preceding analysis is compared with the interpretation of Hecataeus' Helixoia as the 'island of the solstice', discussed in the previous chapter, the compatibility of the Cyzicus episode to the Laestrygonian in terms of their referents in physical reality becomes immediately apparent. Hecataeus of Abdera described an island in the north associated with an Apollo with emphasized solar (calendrical) characteristics.[41] He called this island Helixoia, probably in association with the constellation of Ursa Major, Helice.[42] Both Helice and, consequently, Helixoia, derive their name from the noun *helix* or the verb *helissô*, 'spiral / to turn round or about'.[43] Thus Helixoia meant to Hecataeus 'the island of turning', that is, 'the island of the solstice', and he definitely connected both the island and the concept with the movement of the UMa constellation.

3.3. A pre-Homeric Argonautica

The Laestrygonian episode, unconnected to UMa as it stands, is well supplemented by Apollonius' narrative, while, at the same time, it itself provides for it a solar connection the latter lacks, and together they form a coherent whole. The island of solstice, called the Island of Bears, is inhabited by monstrous Laestrygonians/Earth-born, has a 'city' named Distant Gates, and a harbour with an extremely narrow entrance. This island—known as '*Mountain* of Bears'—could be imagined as a mountain barrier over which the sun cannot pass,[44] and a twin mountain—whether corresponding to mount Māšu in Mesopotamian tradition[45] or bearing the name Di(n) dymum[46]—fits perfectly into this scheme.

[41] *BNJ* 264F7. On Hecataeus' account see a fuller discussion in chapter 6.

[42] *BNJ* 264F11a; cf. Hdn. Gr. iii.1.281 Lentz; Krappe 1947: 223; cf. Macurdy 1920: 140–41; Thomson 1948: 403; Winiarczyk 2011: 55 n. 66; Gagné 2021: 340.

[43] For the etymology of Helice see Σ Arat. 35, 37, cf. Hipp. *Ref.* 4.48.8 (Gundel 1912: 2859.22–34); Chantraine 1970: ii.339 and Beekes 2010: i.411 s.v. ἕλιξ. For the UMa revolving (part. of *helissô*) about the pole see Johann. Gaz. *Ekphras.* 1.190. It has actually been argued that the very origin of the Bear constellations' names must be associated with the circling path of the celestial Bears (Huld 1999: 125–26).

[44] Kiessling (1914: 850.56–58, 851.20–24, 852.22–34) believed that when Aristotle placed the Rhipaean mountains 'under the Bear' (*Meteo.* 350b6–7) he had in mind the fixed arctic circle defined by this constellation, thus a kind of barrier for the sun in the north (cf. Hippocr. *Aer.* 19 = Aristeas *BNJ* 35F8 for the region 'under the Bears' associated with the Rhipaean Mountains).

[45] For Māšu = Twins see Wiggermann 1992: 180; Burgess 1999: 178 n. 26; George 2003a: ii.492, 669, 863, 865; Woods 2009: 191, 197; Marinatos & Wyatt 2011: 393. On the twin-peaked Māšu as the solstice mountain see chapter 2 and below chapter 13. Apollonius was undoubtedly familiar with the model involving the sun rising between two mountain-peaks in the context of diurnal solar movement (A. R. 3.161–63 (with an emendation to πόλον in 161), cf. Gillies 1928: 21, 38; Hunter 1993: 69; Campbell 1994: 141–42; Green 2007: 117, 258; Race 2009: 228–29).

[46] If the derivation of the name Dindymon from *didumoi mastoi/ mazoi* would not be only a fanciful etymology, it would make excellent sense for the name of the 'solstice mountain' corresponding to both the Egyptian model accounting for the diurnal solar movement (Appendix 2) and a later model possibly accounting for the annual, perhaps originating in Mesopotamia (chapter 13), both involving 'breasts'. The interchangeability of motifs used in building models accounting for both diurnal and annual solar movement was already discussed in chapter 2.

But why would the island of Cyzicus be chosen as a solstice island, since it has no connection whatsoever with the main meteorological feature of such a location?[47] In other words, why was this tangible island in particular associated with celestial Bears?

As already mentioned, many scholars believe that the Laestrygonian episode was originally part of a pre-Homeric *Argonautica*.[48] Due to its geographical setting, it seems plausible that the local pre-Apollonian Cyzicene tradition transferred the same episode to their own fountain Artace,[49] known at least from Alcaeus onwards under the Homeric name *Artakiê*.[50] This localization, however, must have followed the identification of Aea with Colchis,[51] which is firstly attested in the *Korinthiaka*,[52] a mid-6th c. epic poem attributed to Eumelus of Corinth.[53] In pre-Homeric Argonautic tradition Aea was, on the other hand, placed beyond the known world towards the east, on or off the hither shore of Oceanus,[54] and was in no way connected with Cyzicus. This Argonautic tradition was indeed concerned with the sun and its movement,[55] which, in the light of my earlier discussion, substantiates the probability that the episode initially belonged to it and was taken over by a pre-Apollonian but post-*Korinthiaca* Argonautic tradition and transferred to Cyzicus. In this transfer the mythic island had lost any connection with solstices or an obvious connection with celestial Bears, but the association with solstices and solar movement was kept in the Laestrygonian episode.

An alternative explanation of the association of Cyzicus with the solstice island from the early Argonautic tradition can be proposed, even though it seems much less likely. The area of Cyzicus was thus associated with the observations of circumpolar stars, which were probably related to the concepts of the always-visible and fixed arctic circles, as well as with astronomical observations that resulted in the determination of the ratio for the lengths of daylight and night at the summer solstice used in further astronomical and geographical calculations. We have already mentioned that Dicaearchus reported that the head of Draco was at the zenith for Lysimacheia, a town in Thrace opposite to, and at the same latitude as, the Cyzicus island (40°23'–40°31').[56] This report naturally comes from an actual observation. In addition, the ratio of 5:3 for the respective lengths of daylight and night at the summer solstice, used by both Eudoxus and Aratus, presupposes a latitude of 40°42',[57] while Leontius used the latitude of 41° in making his 'Aratean globe'.[58] Eudoxus' observations were certainly made at Cyzicus, where his school had been located at one point in his career, and the corresponding ratio was derived from these observations.[59] Thus a strong tradition of astronomical observations existed in this area, which was especially associated with the circumpolar constellation Draco and the length of daylight/night at solstices. These observations could have encouraged the association of these asterisms and the phenomena and concepts entailed by them with this particular locality. Perhaps it was this tradition that inspired the localisation of the Argonautic visit on Cyzicus, rather than the fountain's similar name, as claimed by West. Naturally, this association would then have to be post-Eudoxean, which does not seem very likely in light of such early testimonies as that of Deilochus and Herodorus that positively localised the episode on Cyzicus.[60]

3.4. Conclusion

A mythic island in the north, associated with solstices and either the always visible or fixed arctic circle is, thus, the cosmological location of both Homer's Laestrygonia and Apollonius' *Arktôn oros*, both probably deriving from a pre-Homeric *Argonautica*. This is also the myth's secondary reference to a phenomenon of common reality and collective importance the solstices as the determinative moments in the annual solar movement—articulated by an adaptation of a traditional tale.[61] But while the account from the *Argonautica* as we have it (i.e., the one by Apollonius with several pieces of information offered by his more or less immediate predecessors) lost the connection with solar movement and kept only an ambiguous reference to Ursa Major, the story in the *Odyssey*, while retaining a clear reference to the annual solar movement and the latitude-dependent phenomena that accompany it, lost the reference to circumpolar constellations which was apparently present in the original tradition. It can also be noted that the *Argonautica* account also lost the otherworldly features present in the Laestrygonian episode.

---

47 West 2005a: 49, 52.

48 Meuli 1921: 54, 89–91; 1931: 539.36–49; Page 1973: 38–39; Heubeck 1989: 49; West 2005a: 52–53; 2014: 85, 119–20, 206–207; Fowler 2013: 219; Bachvarova 2016: 102–103.

49 West 2005a: 53.

50 Cf. West 2005a: 49.

51 Heubeck 1989: 49.

52 Fr. 17 West.

53 West 2002: 109, 130–31; 2005a: 41; 2007b: 193; cf. Allen 1960: 80, Allen 1993: 90.

54 Mimnerm. fr. 11 and 11a West = fr. 11 Alen; cf. Allen 1993: 88, 90; Endsjø 1997: 376; Nesselrath 2005: 157; West 2005a: 41; 2007b: 193–95. For the implied location of Aea in the extreme east see also Hes. fr. 241 M-W (cf. Endsjø 1997: 377); Hecat. *FGrHist* 1F18a; Pi. *P.* 4.26, 251, cf. Σ A. R. 4.257–62b; Antim. fr. 76 Matthews.

55 Mimnerm. fr. 11a West (Helios reposes in a golden chamber); cf. fr. 12.5–10 with fr. 23 West = fr. 12 Allen (Helios sleeping in a hollow golden winged bed). See also Davies 1998: 280 n. 20. For the sun 'going to bed' in the far north see Pytheas fr. 13a Bianchetti = fr. 8 Roseman = fr. 9a Mette. These two testimonies are also associated by Ballabriga 1998: 127–28 and once again show the influence of mythic models on Pytheas, i.e., on his conceptualisation of solar phenomena in traditional terms (cf. Beaulieu 2016: 7–8).

56 *De motu circ.* 1.5.59 Todd.

57 Eudox. fr. 68 Lasserre; Arat. *Phaen.* 497–99 (cf. Hipparch. 1.3.5; Hipparch. 1.3.7, 9 = Attal. fr. 21 Maass; Gemin. 5.23–24; Cleom. 1.4.18–30 Todd; Leont. *De sphaer.* 4, 6, pp. 564, 567 Maass) (Dekker 2013: 29).

58 Leont. *De sphaer.* 4–6, pp. 564–66 Maass (Dekker 2013: 30, 32).

59 Huxley 1963a: 89–90; Zhmud 2006: 98–99, 284.

60 For the date of the former see Fowler 1996: 64, who accepts the dating before the Peloponnesian War; for the latter see Blakely *BNJ* 31 *Biographical Essay* (ca. 400 BC).

61 For the folk-lore ('traditional') elements in the Laestrygonian episode see Carpenter 1946: 110, 136–37; Page 1973: 25–48; Fontenrose 1974: 110–11, 533; West 2014: 207.

One could thus attempt to offer a tentative chronology of the development of various strands constituting this tradition. In the early pre-Homeric Argonautic tradition, the island of Bears with the fountain Artakiê was associated with the UMa constellation and with what would later be called the latitude of the geographical arctic circle, i.e., the area characterised by the midnight sun phenomenon. The author of the *Odyssey* took over this episode, together with the high-latitude solstice phenomenon and the fountain but left out the circumpolar constellation association. He also added elements of the already existing model accounting for the diurnal solar movement, perhaps also the association with the otherworld, although it is equally likely that either or both were already present in his source material. At the same time, later Argonautic tradition localised the island at Cyzicus (together with the fountain or precisely because of the existence of a similarly-named fountain on the island in the Propontis), which must have happened after the goal of the Argonautic voyage was set as Colchis in the recess of the Pontus, i.e., after the mid-6th c. (some traces of this tradition are visible in the works of Deilochus, Herodorus and Ephorus, which precludes the post-Eudoxean localisation). The particular catasterism of Zeus' nurses explained the association of initially mythic and later Propontine island with Bears (its earliest positive occurrence is associated with Epimenides after the mid-6th c.), and a version of the story, not necessary the one we are familiar with today, must have preceded the transfer to the Propontis. Alternatively, the catasterism was concocted following the localisation on Cyzicus in order to make sense of the otherwise inexplicable association of Propontis with celestial Bears, which was recorded in some other way in the original tradition (something like 'an island where the sun approaches the Bear', as evidenced in later tradition). Finally, this state of affairs was taken over and reworked is some way by Callimachus and in full extent by Apollonius.

These myths thus represent instantiations of a narrative model that accounts for the concept of solstices, critical points in the annual solar motion. They are not structured upon solar movement but incorporate an elaborate reference to its annual aspect in their narratives, showing a lively concern for this type of phenomena. These narratives belong to a class of myths that treat specific astronomical/geographical, i.e., meteorological phenomena and should be interpreted accordingly. They are certainly not unique examples of this type, and the existence of a class of such myths should be acknowledged. The inclusion of the Cyzicus episode in this class was justified by its unambiguous connection with the Laestrygonian episode, in which a solar reference is perfectly clear. The former adds an important element to the model recognisable in the latter, i.e., the circumpolar constellation(s) involved in the delineation of arctic circle(s), but also helps in contextualising the tradition in a mythic-cosmological environment that is not so easily recognisable from the account in the *Odyssey*, i.e., the affiliation of this mythic episode with the early Argonautic tradition, itself heavily involved with solar movement (see further chapter 11). These two contextual arguments (the close correspondence with the Laestrygonian episode and the affiliation with the early Argonautic tradition) justify the inclusion of the Cyzicus episode as we have it in the class of myths referencing solar movement.

4

Snatched away by the gust of wind

Any connection with the domain of eschatology was poorly attested in the Argonautic and Odyssean myth referencing the annual solar movement discussed in preceding two chapters. Quite the opposite, in several variants of mythic models of solar movement that can be recognized in a group of myths treating the Boreads' pursuit of Harpies to the Plotai/Strophades Islands, this domain features prominently. More specifically, this cluster of myths engages with certain eschatological tenets in the form of stories of mythic abductions, including those performed by the Harpies and the north wind-god Boreas. Since an intimate connection of eschatological tenets with solar movement—especially diurnal, but also annual—was widespread in ancient civilisations, as will be shown in the remainder of this book, its presence in this group of myths is not surprising. Indeed, it seems that a tangible analogical relation between the annual solar movement and human life-cycle, with special emphasis on eschatology and solstices, was recognized and utilized in antiquity.[1]

This particular facet of 'solar' myths will thus be treated somewhat more extensively in this chapter in the context of mythic abductions performed by different agents involved in this set of myths, as well as by other similar characters, such as the dawn goddess, who represent paradigmatic variants embedded in a relatively stable syntagmatic structure. Moreover, the stories treating these specific mythic abductions actually seem to be partially structured upon the phenomenon of solar movement, which was manifestly not the case with the previously discussed mythic cluster, and thus fully deserve to be included in the class of myths referencing solar movement. However, references to solar movement in this group of myths are less straightforward in comparison with the stories discussed in previous two chapters, so I will be forced to offer a strong argument for their presence in order for these stories to be accepted in the class of myths discussed in this book.

An important variant of this mythic model—or perhaps another model with a number of similarities with the island-barrier model found in this set of myths—involves a mountain barrier in the far north in place of an island. This model apparently accounted for specifically annual solar movement and is represented by a Greek concept—attested in similar variants in other traditions—of a vast northern mountain barrier. This barrier was often in its turn associated with several other concepts that can be interpreted in terms of the annual solar movement (for which see chs. 5 and 6), as well as with the Ursa Major constellation, blocking the sun in its annual voyage in the far north (some traces of this model were already recognised in the 'Mountain of Bears' in the Argonautic tradition). This mythic model accounted for the sun's observable halt when it reaches its maximum northern declination (the halt can be observed both at the horizon and at the meridian and can also be defined in relation to certain asterisms), and its return to the south. Alternatively, the turn could have been performed at some northern island characterised precisely by the sun's activity at the summer solstice, which represents a paradigmatic change resulting in a similar model (for this island, already encountered as Thule in chapter 2 and as the Island of Bears and Helixoia in chapter 3, and as we will shortly see, also instantiated in Plotai/Strophades, see further a detailed discussion in chapter 5). Presumably, the same or similar concept might have been present in other traditions, but only the model involving a northern barrier (as opposed to an island) can be recognised in Mesopotamia and Egypt.

4.1. The island of turning

I will first discuss the tradition involving an island as a turning-post, since it is intimately connected with the Harpies-Boreads cluster of myths and its eschatological resonances. This tradition revolves around a number of floating islands, one of which was already discussed in ch.1. There the location of Aeolus' island was briefly discussed with respect to the location of the land of Laestrygonians. Another, presumably early, identification of Aeolus' island can be associated with solar movement. Thus, Stesichorus mentions together the words 'Aeolian' and 'Strophades'.[2] The most plausible explanation for the juxtaposition of these two mythic locations is that Aeolus' floating island was identified with the Plôtai/Strophades, the floating islands of 'turning' that appear in the story of pursuit of Harpies. The 'Floating Islands'—Plôtai—are indeed sometimes identified by later authors with the Aeolian Islands in the Tyrrhenian Sea,[3] and the fact that

[1] Gee 2020: 125, 168–70, 186, 248. Gee's book as a whole discusses the incorporation of contemporary scientific tenets, in the first place cosmological, in eschatological discourse (see especially Gee 2020: 4–5, 7, 11, 131, 137, 151, 155, 190, 235, 275–76).

[2] 222B fr. 74 Campbell = fr. 258 Finglass. Garner (1993: 160 with n. 20) explains these fragmentary lines differently, dismissing a reference to the Strophades, while Davies & Finglass 2014: 559 do not accept the reading 'Aeolian'. The fragment is indeed problematic, but the later tradition of identification of these two localities suggests that their juxtaposition in Stesichorus is plausible.

[3] Dion. Perieg. 461–66, Lightfoot 2014: 222, 369–70; cf. Eustath. Dion. Perieg. 461, *GGM* ii.304–305 and paraphr. 450–54, i.380 Bernhardy. A. R. 3.42 describes Hephaestus' island as *plagktê*, and the island is naturally identified by the scholia with the Aeolian Islands (Σ A. R. 3.41–43a, p. 216 Wendel); the Florentine scholia (p. ii.532.12 Brunck & Schaefer), however, has πλωτή instead. Thus, again the Aeolian Islands are referred to as the Plôtai, even if by mistake.

the island of Aeolus was described as floating (*plôtos*) by Homer[4] was probably decisive for this identification. A reference to solar movement is absent from the poorly preserved papyrus fragment of Stesichorus, but it can be gauged from the fuller expositions of the Boreads' pursuit of Harpies myth.

According to Apollonius, who is apparently following Antimachus, the Boreads pursued the Harpies over the sea all the way to the remote Floating Islands (*Plôtai*).[5] These were sent to molest Phineas by Helios or 'the Titan Phaethôn'.[6] When they swore never to disturb him again, the sons of Boreas turned (*hupostrephô*) back towards their ship. On account of this event *Plôtai*, Floating Islands, had their name changed into Strophades, Islands of Turning.[7] This account is very ancient and one of its versions, including the name-change, was already known to the Hesiodic author,[8] who seems to have identified them with the Echinades.[9]

The name Strophades is thus derived from the verb *strephô*, which has two primary meanings that are of some interest for this discussion: (1) to turn about something, i.e., to reverse the direction of a movement, and (2) in certain contexts it can assume the meaning of 'circular motion', 'revolving', 'rotating'.

The second meaning is manifest when the word and its derivatives are used to describe the rotatory movements of circumpolar asterisms. It was used by Homer with reference to the Ursa Major, in what is perhaps a rendering of an Akkadian phrase,[10] as well as by the early tragedians.[11] But it could also be used to describe the turning-point in the asterism's revolution: its culmination, i.e., transit over the meridian.[12] Here it seems to combine the two meanings, 'revolving' and 'turning about'.

The term was also used to describe the revolving of the sphere of the fixed stars around the celestial pole or of the revolution of the world on its axis,[13] as well as to describe the apparent orbital revolution of the sun, i.e., its daily course in the sky.[14] Otherwise, it could be applied to the sun's turnings at the solstices.[15] This meaning of the term can be further recognized in the phrases describing the change in the direction of solar movement in association with the mythical feast of Atreus.[16] Thus in a cosmic context the phrase either designates a movement that involves a change in direction or a revolving motion around a pivot of some kind.

Consequently, the name Strophades can be rendered either as 'Revolving Islands' or the 'Islands of Turning'. Homer and Sophocles used the expression describing the continuous circular motion with specific reference to the Bear, and various other authors used it in similar contexts, while Hesiod, Antimachus and Apollonius, on the other hand (together with other authors), explained the name of the islands with a fact that there the sons of Boreas *turned back*, after arriving at the end of their pursuit, not associating in any way this episode with a circular motion.[17] It could thus be concluded that both meanings of the terms discussed in this section are appropriate in this cosmological context, and the mythic islands of Strophades could be understood as either 'Revolving Islands' or the 'Islands of Turning', with the latter obviously more appropriate in the particular narrative context. It thus seems that the stories of the Boreads' pursuit of the Harpies involved a cosmological location. The alternative would be that the myth, known already to Hesiod, is simply an aetiology for an already existing (yet nowhere attested) name, which seems less plausible. But there is as yet no apparent connection with the sun which would vindicate the inclusion of this set of myths into the class of narratives referencing solar movement respecting the criteria introduced in the early part of this study.

Vergil's rendering of this mythic episode, radically different from that of Hesiod and other authors, places it, however, in a different—if complementary—mythic context, one that finally includes a clear reference to the sun. He describes Aeneas' visit of the island of Harpies in unmistakable terms of Odysseus' visit of the island of Helios, thus in a way 'replacing' the story of Helios' cattle with that of the Strophades.[18] The fact that Virgil made

[4] *Od.* 10.3. For ancient interpretations of *plôtos nêsos* in this way see Cook 1940: iii.2.975.

[5] A. R. 2.285.

[6] Asclep. *FGrHist* 12F31 (Gantz 1993: 350, 352); Opp. *Cyn.* 2.617–20.

[7] Antim. fr. 71 Matthews *ap.* Σ A. R. 2.296–97ab, pp. 149.23–150.5 Wendel; A. R. 2.262–300 (partially cited at [Apollod.] *Bibl.* 1.9.21); cf. Hyg. *Fab.* 19; Σ A. R. 2.285, p. 149.12–14 Wendel. See also Matthews 1996: 216.

[8] Hes. fr. 156 M-W *ap.* Σ A. R. 2.296–97a, pp. 149–50 Wendel.

[9] Hes. fr. 155 M-W. The islands were normally placed in what is today called the Ionian Sea (Antimach. fr. 71 Matthews according to Nishimura-Jensen 2000: 296 n. 26, but more probably only the scholiast's opinion, Σ A. R. 2.296–97b, p. 150.12–13 Wendel; Str. 8.4.2; Verg. *Aen.* 3.209–18; Ov. *Met.* 13.709–10; Pomp. Mel. 2.7.110; Val. Flacc. 4.512–13; *HN* 4.55; Ptol. *Geog.* 3.16.239). The Aegean localisations (Σ A. R. 2.285, p. 149.14–15 Wendel (?) and Hyg. *Fab.* 14.18) are blunders.

[10] *Od.* 5.273–74 (= *Il.* 18.487–88); [mul]MAR.GÍD.DA *kal* MU DU-*az ma-a i-lam-ma-a*, 'The Wagon stands all year, namely, it circles round' (*Enuma Anu Enlil*, Tablet 50, Text III.28c; Reiner & Pingree 1981: 18, 42–43; Huxley 1997: 193; Hunger & Pingree 1999: 68).

[11] Soph. *Tr.* 130–31, fr. 432.11 Pearson. Eur. *Ion.* 1154 is similar. Later usage: Epic. *Ep.* 2.112 p. 52.12 Usener (circumpolar stars); Arat. *Phaen.* 43 (UMi), etc.

[12] Anacr. 33.1–2 *PMG* (according to Campbell 1988: 203 the Bear is here turning *by* the hand of Boötes); Theocr. *Id.* 24.11 (Gow 1952: ii.417–18).

[13] Soph. fr. 432.8 Pearson; Pl. *Tim.* 40B (cf. 39A); Arist. *Mot. An.* 699a29; Epic. *Ep.* 2.112, p. 52.14 Usener; Arat. *Phaen.* 498–99 (cf. Gemin. *Elem.* 5.24, pp. 50.16–52.2 Manitius), 512, etc. For the planets see Plut. *De gen. Socr.* 22.590E.

[14] Orph. fr. 236 Kern = *PEGr* fr. 539 F; Timon fr. 841.6 SH; Dion. Perieg. 584, Lightfoot 2014: 228, 394–95, cf. Eustath. Dion. Perieg. 581, *GGM* ii.329. Cf. Eur. TrGF fr. 1009 Kannicht for the moon, where the meaning is inconclusive.

[15] Aristid. *Aegypt.* 63 (p. 348.1–2 Jebb); Porph. *De antr. Nymph.* 23.14–16.

[16] Soph. fr. 738 Pearson; Eur. *El.* 739; Σ MTAB Eur. *Or.* 998; Tzetz. *Chil.* 1.18.451; cf. Σ TAB Eur. *Or.* 998.

[17] The Hesiodic author elsewhere described the pursuit (but not the moment of turning) in terms of circular motion (fr. 150 M-W P.Oxy. 1358 fr. 2 col. i.20, 28–29; cf. Gisinger 1929: 321; Romm 1992: 29).

[18] See esp. *Aen.* 3.220–21, 247–52 compared to *Od.* 12.260–419; cf. Macr. *Sat.* 5.2.14 and Miller 2009: 124–25.

Caeleno the prophet of Apollo—in a setting undoubtedly replicating the events that unfolded on the island sacred to the Sun—points to his identification of the two in this instance.[19] Thus solar elements at least in this version of the story of Harpies and the Strophades seem prominent, and we should notice the emphasis on the 'island of turning' throughout the tradition, which might in this context point to solar turnings, as well as the fact that it was the sun god who sent the Harpies upon Phineas in the first place. His creatures—a notion supported by Vergil's version—would obviously seek their master's islands of turning in order to save themselves from pursuers. Another reference to the Strophades as the islands of *solar* turning is recognised, but at the same time dismissed as irrelevant, by Eustathius. He compared the Strophades with Dionysius the Periegetes' description of the island of Taprobane above which Cancer revolves (*strophaligxi*), thus placed directly under the tropic.[20] However, it seems only natural to give the name 'Strophades', the Islands of Turning, to the island(s) under the tropic, and Eustathius was perhaps too rash in dismissing it.

Finally, in their capacity as floating islands the Strophades are in nature identical to another famous floating island, Delos,[21] in its turn also inevitably compared to Aeolus' island.[22] Delos was inextricably connected with Apollo and his birth; as will be shown in chapter 7, in the complex of stories revolving upon his connections with Hyperborea, to which his Delian birth belongs, Apollo should be understood as seasonal in character, i.e., as a divine agent whose behaviour is structured upon the annual solar movement.[23] This is thus another example of solar connections exhibited by floating islands and/or islands of turning in Greek tradition.

Thus, a number of elements point to the presence of solar references in the cluster of myths pivoting on the Islands of Turning. Admittedly, these are not as clear and unambiguous as the ones analysed in previous two chapters, but still allow a strong inference that they are indeed present. It is certainly not a one-to-one correspondence between the solar movement and the structure of the stories belonging to this mythic cluster that is argued for here. Rather, it is proposed that the concepts derived from solar observations were incorporated into these stories and that occasionally some episodes or parts of episodes were even structured on an analogy with solar phenomena—such as the pursuit and the act of turning. Other interpretative approaches can clarify further elements in the structure of these mythic narratives, as well as explain a number of narrative details inexplicable in terms of solar references; but the particular narrative structures and elements analysed here are best explained in precisely the terms outlined here.

4.2. The Harpies and eschatology

Another persuasive connection of the set of myths analysed so far in this chapter with solar movement can be recognised in the crucial involvement of the Harpies in these stories. These mythic creatures were hitherto mentioned only in passing, but an analysis of their appearance in myths outside of this particular mythic nexus reveals their connection with eschatological concepts of solar mythology. As already mentioned, this subject was hitherto advanced only incidentally in relation to the land of Laestrygonians, but it will be demonstrated that an intimate connection with solar movement is one of characteristic features of several ancient eschatological concepts (in this chapter, however, I will focus solely on Greek evidence).

At the same time, another phenomenon that was not discussed in previous chapters but is of some importance in understanding ancient models accounting for—in the first place, diurnal—solar movement will be introduced: an ambivalence in localisations of certain mythic regions and figures in the farthest west and in the underworld. Since both localisations are heavily reliant on different understandings of the diurnal solar movement, it seems fitting to start the discussion precisely with this subject. It is the dual localisation of the Harpies' abode that will introduce this subject into our discussion.

There existed a very early tradition that associated the Harpies with the farthest west,[24] and it seems reasonable to postulate a cosmological location of the Strophades, the islands of Helios and the Harpies, 'beyond the Ocean'. At the same time, another early tradition associated them with Hades.[25] This and other similar apparent inconsistencies (such as the dual localisations of certain mythic 'exiles' Cronus, Ophion, Typhon, Ogygus, Briareus, Cyclopes and Cadmus) can be plausibly explained in terms of different conceptualisations of the diurnal solar movement. Accepting and elaborating upon D. Nakassis' scheme, I have argued that the frequent duplication of certain localisations in classical myth, most often in the far west and below the earth, could be explained through two alternative ways in which the early Greeks conceptualised the solar phenomena of sunrise and sunset: according

[19] For Vergil's equation of Apollo with Sol see *Aen.* 6.18–19 (Miller 2009: 137), 8.720 (Miller 2009: 208–209, 249), 11.912–15 (Hardie 1986: 356 n. 54), 12.164, 176, 197–98 (cf. Hardie 1986: 322 n. 54 *pace* Fontenrose 1943 and Galinsky 1969: 454) and *Ecl.* 4.10 (cf. Nigidius Figulus *De diis* 4 *ap*. Serv. Dan. *Ecl.* 4.10) (Tarn 1932: 158; Alföldi 1997: 7; Miller 2009: 30, 259, 278).

[20] Dion. Perieg. 593–95 (Lightfoot 2014: 228); Eustath. Dion. Perieg. 591 (*GGM* ii.330).

[21] Pi. fr. 33d (88) Snell-Maehler (Str. 10.5.2; Σ HM *Od.* 10.3 (cf. *Anecd. Par.* iii.464.7 Cramer; Eustath. *Od.* 10.3, i.363; Plut. *De fac.* 6.923C; Sen. *QN* 6.26.2); Favorin. fr. 96.25.2 Barigazzi), *Pa.* 7b.49; Call. *hDel.* 35–54, 191–94, 213; a 3rd-c. embroidery going back to a Hellenistic painting (Furley & Bremer 2001: ii.138); Varro in Macr. *Sat.* 1.7.29; Verg. *Aen.* 3.73–77 (cf. Sen. *QN* 6.26.2), etc.

[22] Σ HM *Od.* 10.3; Eustath. *Od.* 10.3, i.363, DP 461, *GGM* ii.304.

[23] Olmsted 1994[2019]: 137; Bilić 2012b: 509–10, 515–19, 525–27; cf. Gernet 1981[1933]: 116; Parker 2005: 417–18 (hesitatingly).

[24] *Il.* 16.149–51; Hes. *Th.* 237, 265–67, 270–79; Epimen. fr. 48 F *PEGr* = 9 Fowler = 3B9 D-K = *BNJ* 457F6b (cf. 3B7 D-K = *BNJ* 457F6a); *Titanom.* fr. 9 West (late 7th c., West 2002: 109, 130; 2003: 30); Acus. fr. 10 Fowler = 9B5 D-K= *FGrHist* 2F10; Gantz 1993: 6, 18.

[25] Pherec. 7B5 D-K = fr. 83 Schibli; cf. Verg. *Aen.* 6.289; Sil. Ital. 13.599.

to the 'uni-polar' model the alternation between night and day occurs at the *axis mundi*, represented by and located in Hades/Tartarus, while according to the 'bi-polar' model it rises and sets on the eastern and western horizon, respectively. In both cases, mythic discourse was employed in the elaboration of these important natural phenomena, but two models accounting for the diurnal solar movement were developed, resulting in an apparent confusion when not understood as such.[26]

This dichotomy and its solar context can be exemplified in the semantic range of the word *zophos*. Rather than repeating the comprehensive analysis offered elsewhere,[27] I will focus here on this revealing term. The word is used several times in Homeric epics to designate the west in general.[28] The fact that it can be linked with the name of the west wind, Zephuros, agrees with this usage.[29] Most often, however, it is used in the context of the diurnal solar movement: in a description of a sunset, paired with the dawn, or opposed to the dawn, east and sunrise.[30] These testimonies undoubtedly and unsurprisingly testify to the dependence of the term 'west' upon the diurnal solar movement. However, at the same time the word was also used already in the Homeric epics to designate 'the nether darkness'.[31] It is in this 'Tartarean' sense that Hesiod described Chaos as *zophoeros*, 'assimilating' it to Tartarus.[32] Furthermore, the commonly used phrase *hupo zophon êeroenta* always refers to Tartarus,[33] Hades' realm,[34] or the world of the dead.[35] Here the solar element is less pronounced, but in light of the term's primary reference to sunset it seems clear that its application to a subterranean region must be derived from the latter's involvement in the sun's nocturnal passage, or perhaps the absence of its involvement.

These two meanings appear indistinguishable in several uses of the term, which further shows the compatibility of these two concepts in Greek tradition, the farthest west and the underworld, dependent upon solar movement. Thus Pindar used the word in describing the impossibility of navigating beyond the Pillars of Heracles 'towards *zophon*', while Pindaric scholia adds how beyond Erytheia there is 'chaos and *zophos*', which at the same time suggests the farthest west and an otherworldly setting.[36] At the same time, an association of *zophos* with night is also prominent, Hades' helmet being described as possessing the night's *zophos*,[37] while in Orphic and other sources the night itself is addressed as *zophera* or *zophoeidês*.[38] Night is naturally associated in Greek tradition both with the west and the otherworld, which will be further substantiated in chapter 10. Therefore, it is clear that the same word could be used both in describing the 'western' darkness and at the same time the 'underworld gloom'. In this way, the placing of the Harpies in the farthest west and at the same time in Hades ceases to be as outright inconsistent as it appeared at first glance and rather represents their involvement in two different models accounting for diurnal solar movement, one involving sunset in the far west and sunrise in the far east as decisive features in the alternation between night and day and the other locating the focal point of the alteration below the earth.

The Harpies' association with Hades—and no less their association with the farthest west—opens the subject of the association of their Islands of Turning with eschatological concepts of solar myths. This subject was competently treated by many authors, and it will be only summarily discussed here, due to its bearing on the localisation and nature of this particular mythic feature. It will be shown that mythic 'snatchings' represent a model accounting for death and the ensuing fate of human 'soul' involving (but not reduced to) the results of solar observations, i.e., the structuring of the former on an analogy with the latter. This model was created in order to explain a less-known experience in terms of a much-better known one, with the analogy based on a postulated correspondence between the sun which, as a divine being, lives continually and resurrects both daily and annually, and humans, who presumably experience a similar fate in some form appropriate to their ontological position.[39] This model naturally depends on a personalistic account of causality of periodical natural phenomena and processes, which is what makes it mythic. At the same time, it represents a rare example of a myth structured upon an analogy with solar movement—even though it should not be reduced to this analogy—and fully deserving to be included into a corpus of myths referencing solar phenomena.

The Harpies, whether written with an initial capital letter or not, or whether styled 'gusts of wind' (*thuella*),[40] were involved in early poetry in a series of 'snatchings' regularly described by the verbs *harpazô* or *anereipomai*,[41]

[26] Bilić 2013a with earlier literature (following Nakassis 2004; cf. Burgess 2016: §15; Fowler 2017: 247–48).

[27] Bilić 2013a.

[28] *Od.* 9.26; Strabo, following Crates (Str. 10.2.12 = Crates fr. 31c Mette, cf. Str. 1.2.20, 28 = Crates frs. 21d and 21b Mette; all under fr. 52 Broggiato) argued that Homer had north in mind here, as well as in *Od.* 10.190 and *Il.* 12.240; this was perhaps also the opinion of Dion. Perieg. 500 (Lightfoot 2014: 224, cf. 376 ad 500), but cf. Dion. Perieg. 421 (Lightfoot 2014: 220).

[29] SVF 2.697 = Aët 3.7.2 Mansfeld & Runia 2020; West 1997: 153.

[30] *Il.* 12.240; *Od.* 3.335, 10.190, 13.241; cf. A. R. 1.451; [Theocr.] *Id.* 25.85; Q. S. 8.489.

[31] *Il.* 15.191, 21.56, 23.51; *Od.* 11.57, 155, 20.356; Archil. fr. 24.15–18 West (West 1997: 500); Aesch. *Pers.* 839; Philisc. fr. 1.6 West (emended).

[32] Hes. *Th.* 814; Lincoln 2009: 384 n. 34.

[33] Hes. *Th.* 653, 658, 729; *hHerm.* 256–57.

[34] *Od.* 11.57, 155 (according to [Plut.] *De Hom.* 2.97 (2, lines 1014–16 Kindstrand; Keaney & Lamberton 1996: 163) Homer thus called the portion obtained by Hades); *hDem.* 80, 337, 402, 446, 464, 482; Minyas fr. 7* West = Hes. fr. 280 M-W; [Hes.] *Scut.* 227; Anon. 925e.9–10 PMG.

[35] *Il.* 21.56, 23.51.

[36] Pi. *N.* 4.69; Σ BCDEQ Pi. *O.* 3.79d (Brown 1968: 44, n. 20).

[37] [Hes.] *Scut.* 227.

[38] *Orph. fr. PEG* 97 T = 107 F I Bernabé; *Orph. Hymn.* 71.9, 78.4; Nican. *Alex.* 501; Anth. Pal. 14.72.2.

[39] Cf. Gee 2020: 125, 168–70, 186, 248 on the 'tropic' notion of existence.

[40] For *harpuiai = thuellai* see Hesych. α 7409.

[41] *Od.* 1.241, 4.727, 8.409, 14.371, 15.250, 20.66, 77 (cf. Q. S. 10.395, 428); Hes. *Th.* 990 (cf. [Hes.] fr. 375 M-W with supplement of Hartung and Urlichs); *hAphr.* 218; Ibyc. *PMG* 289; [Hes.] fr. 375 M-W; Eur. *Hipp.* 454; Xen. *Cyneg.* 1.6. For etymologies deriving *harpuiai* from the act of

sometimes with a goal at the 'mouth of the Ocean'[42] or 'the ends of the earth'.[43] It seems that there existed an eschatological concept involving the Harpies snatching—in the form of gusts of wind—the soul and taking it (westward) to the Ocean, where we can also expect to find their Island of Turning. In any case, this island is a cosmological location, while its specific position was in the main defined by the concepts derived from solar movement (for corroborating evidence see the remainder of this chapter).

4.3. Other mythic snatchings

The example of the abduction of Oreithyia by Boreas—occasionally a father of the Harpies[44]—is a similar case of a mythic snatching. All accounts agree she was snatched up by the deity and use similar vocabulary based on *harpazô* in describing the abduction.[45] This mythic story was recognised as a metaphor for or a model of death by Socrates, who reported the interpretation of the Boreas-Oreithiya myth as the girl's death caused by the blast of Boreas.[46] It seems that this interpretation was generally accepted in the 4th c. Thus, the representations of Boreas carrying off Oreithiya on certain 4th century Greek bronze hydria probably represent some sort of eschatological speculation, since for at least some of them a funerary context is confirmed.[47]

The myth of Sarpedon in the *Iliad*, corresponding structurally to Memnon's, Eos's son, myth in the *Aithiopis*,[48] and thus echoing the post-mortem destiny of this solar (in terms of his family connections) hero, can be adduced as further evidence in this context. He was once revived by the 'blast of Boreas'[49] and was 'carried away' (*pherô*) by Hypnos and Thanatos.[50] The element *sarp-* in his name might be connected with the verb-root *harp-*.[51] Furthermore, his name might be connected to the Sarpedonian rock to which Boreas carried Oreithyia,[52] and thus associated with this particular mythic nexus. The rock was localised somewhere in the north, perhaps—in terms of rationalising localisations—in Thrace.[53] It can thus be safely presumed that this story of a divine abduction by the north wind was at some point understood as a metaphor or model accounting for the concept of death in general and certain specific eschatological ideas connected to or structured on an analogy with solar phenomena in particular.

On a more general level, the 'snatching off' was used as a model accounting for death as such,[54] but also, more particularly, of premature death of the young. This is sometimes attributed to the Harpies themselves,[55] but more often to no particular agent,[56] or Fates[57] or Nymphs,[58] who thus took over the role usually performed by the Harpies, representing paradigmatic shifts in a relatively stable syntagmatic structure. These characters do not exhaust the agents involved as paradigmatic variants in such contexts; a similar function—described in similar vocabulary—was performed by Echidna,[59] Scylla,[60] the

snatching see A. R. 2.188–89 (Stern 2003: 77, 93); Plb. fr. 21 Büttner-Wobst (Walbank 1979: iii.746); Heracl. *De incred.* 8; Herodian. iii.1.281 Lentz; Hesych. α 7410, 7411; *EM* s.v. Ἅρπυς; Fulgent. *Mythol.* 1.9; *VM* 1.27, 2.13 [= 2.21 Kulcsar], 142 [= 165 Kulcsar], 2.271 Kulcsar, 3.5.5–6. For the connection of the noun *harpuia* to Homeric/Hesiodic *anereipomai* see Nagy 1973: 157–59; 1990: 242–45; cf. Rohde 1925: 56–58, 80 n. 4.

42 *Od.* 20.63–65, with additional verb *propherô* (for which cf. *Od.* 8.409 in a similar context); in *Il.* 2.302 and *Od.* 14.207 the Kêres carry away (*pherô*) the dead soul to Hades; for *pherô* in accounts of snatching see Hes. fr. 26.23 M-W, 43a.80 M-W; [Hes.] fr. 375 M-W; Sappho fr. 58 Voigt (P. Oxy. 1787 fr. 1.20; P. Köln 21351.18).

43 Sappho fr. 58 Voigt (P. Oxy. 1787 fr. 1.20; P. Köln 21351.18) (cf. Allen 1993: 55) on Tithonus. Elsewhere, the location of Tihonus' sojourn with Eos is an island in the eastern Ocean, the place of sunrise (Lycoph. 16–19 with paraph. 16, 18 (i.2–3 Scheer) and Tzetz. (Σ) 15, 17, 18 (ii.14–15, 18, 20 Scheer); *HN* 6.198 with Ephor. *BNJ* 70F172; Nonn. *D.* 16.45–46, perhaps also 33.183–84 and 38.287; Lightfoot 2014: 310 n. 25).

44 Pherec. 7B5 D-K = fr. 83 Schibli.

45 Pherec. *BNJ* 3F145; Simon. 534 *PMG*; Acus. *FGrHist* 2F30, 31; Soph. fr. 956 Pearson (most probably); Pl. *Phdr.* 229cd; Herag. *FGrHist* 486F3; Choer. fr. 7 *PEG* = Anon. *De Persia FGrHist* 696F34g; A. R. 1.214; Heracl. *De incred.* 28.

46 Pl. *Phdr.* 229cd.

47 Zuntz 1971: 356; Tzifopoulos 2011: 171; cf. Picard 1940: 73ff, cited by Richter 1946: 366. For the hydria in question see Richter 1946: 365, group V with Pl. XXVIII = *LIMC* III.1.138, 2.21–22 s.v. *Boreas* 69–73, VII.1.66–67, 2.58 s.v. *Oreithiya* 21–22.

48 Schein 1997: 356–57; Currie 2016: 64–67, 72, 89–90. I am not arguing for the primacy of either account; for the latest assessments of the subject see Currie 2016: 55–72, 100 and Davies 2016: 16–19.

49 *Il.* 5.697.

50 *Il.* 16.454, 671–72, 681–82.

51 Nagy 1990: 141–42; cf. Vermeule 1981[1979]: 169, 242 n. 36. When Nonius Marcellus quotes from Caecilius Statius' *Harpazomene* he either uses the form *Arpaiomene/Arpazone* or *Sarpazomene* (frs. 1–2, 4–5 Ribbeck = frs. 50–51, 54–56 Warmington *ap.* Non. *De Compediosa Doctrina* I–III (i.16.6, 186.4, 228.7, 295.1 Lindsay)).

52 Simon. 534 *PMG*; Pherec. *BNJ* 3F145; Soph. fr. 637 Radt; A. R. 1.216 (cf. Vermeule 1981[1979]: 169).

53 Simon. 534 *PMG*; Pherec. *BNJ* 3F145; A. R. 1.213, 216–17; Σ A. R. 1.216–17a, p. 27 Wendel; Σ Eur. *Rh.* 29. For Sarpedonian rock in Thrace and Cilicia see Hesych. σ 227; for the Sarpedonian rock Herodian. iii.1.27.2, 2.730.21, 914.10. For the Sarpedonian *cape* (Soph. fr. 46 Radt) in Thrace see Hdt. 7.58.2; perhaps Aesch. *Suppl.* 870 (Fowler 2013: 461 n. 38); Crates fr. 127 Broggiato = *BNJ* 2113F29; Str. 7 fr. 52 Meineke; Zenob. 5.86; Hesych. σ 230; Suda σ 145; Phot. *Lex.* p. 502.3.

54 Snatched away by Hades = death: Callim. *Ep.* 34.1207–1208 Gow-Page = 2.5–6 Pfeiffer; *Anth. Pal. append.* ii.268.1–2 Cougny (= Kaibel 570.1–2); *Anth. Pal.* 7.13.3, 186.6, 476.7–8; Σ A *Il.* 5.564; *Anth. Gr.* 7.603.1–2, 671.1–2 (Charon); Plut. *De sign. Socr.* 22.591C; arcosolium of Vibia (Casagrande-Kim 2012: 165–66 with Fig. 44 on p. 283); arcosolium of Pulina in the Hypogeum of the Octavii (Casagrande-Kim 2012: 239 with Fig. 139 on p. 340; Borg 2013: 65, 68 with Fig. 40 on p. 66). See also Hes. *Op.* 356 and Callim. fr. 229.2 Pfeiffer for *H(h)arpax*.

55 *Anth. Pal.* append. i.264.14 Cougny = Kaibel 1046.14 (Peek 1979: 80; Pomeroy 2007: 170); the four female-headed birds carrying children in their arms on the so-called Harpy Tomb at Xanthus (Harrison 1908a: 177 Fig. 20; Vermeule 1981[1979]: 170 and Fig. 21 on p. 171; cf. *LIMC* IV.1.446 s.v. *Harpyiai* 3); terracotta figure of ca. 525–500 BC depicting a winged female daemon with winged shoes carrying off a baby (Pfister 1940: 41); Etruscan hydria of ca. 500 BC (with Gorgon's head) (*LIMC* IV.1.340, 2.195 s.v. *Gorgones (in Etruria)* 117) (for Erinyes/Gorgons similar to the Harpies see Aesch. *Eum.* 48–51).

56 Soph. *El.* 848; Dion. Hal. 7.68.5; Plut. *Cons. Apoll.* 30.117BC, 18.111D, *De Alex. Magn. fort. aut virt.* 13.343F; Lucian. *De Luct* 13; *Iscr. di Cos* 427.8–10 = Paton & Hicks 1891: 209 no. 322.8–10.

57 *Anth. Pal.* append. ii.479.1 Cougny = Kaibel 324.1; cf. *Anth. Pal.* append. i.264.14 Cougny = Kaibel 1046.14 for the Harpies Klôthes).

58 Callim. *Ep.* 24.1–2; *Anth. Gr.* append. ii.268.8–10 Cougny = Kaibel 570.8–10 = *IGUR* III.1344.15–20; *Anth. Gr.* append. ii.271.1 Cougny = Kaibel 571.1; *CIL* VI.29195. In this case the abductee might actually be translated to some blissful place (cf. in mythic context Theocr. *Id.* 13.72; Ant. Lib. *Met.* 32.4). See Borgeaud 1988: 77, 105–107.

59 [Apollod.] *Bibl.* 2.1.2.

60 *Od.* 12.99–100; cf. Ath. 1.22.13B.

Theban Sphinx,[61] who showed some preference to the young,[62] Lamia, the famous snatcher of children,[63] and Poine or Ker[64] or even the Bacchants.[65] Vermuele aptly summarises this nexus of ideas:

> 'The change of experience which is mythologically marked by Sleep, Death and Eros seems to be more than a euphemism for the Greeks. It is figured in many ways: by immortal passion for mortals; sudden disappearances underground, into the mountains, overseas, up into the winds and clouds; a succession of rapes by Zeus, Poseidon, Hermes, Apollo, Boreas, Zephyros; and on the female side most conspicuously by Eos...'[66]

4.4. The snatchings in their solar context

Hitherto only slight indications of the solar traits present in this concept were documented, but there is more evidence which, in connection with the postulated solar character of the Harpies' Islands of Turning, suggests that at least some variants of the mythic concept of 'snatching off' as a model for death were structured on an analogy with solar movement. Thus, occasionally the sudden death of young men was attributed to either Sun or a solar Apollo.[67] Furthermore, another prominent agent of abduction is the goddess of dawn, as noted in the citation from Vermuele's study. Eos indeed carried off Clitus,[68] Cephalus[69] and Tithonus.[70] Among these abductees is also Orion, who was taken by Eos/Hemera as a young man,[71] and was killed at the island of Ortygia.[72] These testimonies do not prove that all 'snatchings' were structured on an analogy with solar movement. However, the ones performed specifically by Harpies and Eos (perhaps also Boreas) certainly involve a presence of the results of the observations of solar movement, while other agents—or some of them—could represent paradigmatic shifts in the syntagmatic structure of a model accounting for the 'change of experience' represented by death.

With the introduction of Ortygia, we have come full circle, decidedly connecting the 'snatchings' with the Island(s) of Turning—a plausible inference adduced earlier in this chapter, but with little direct evidence. This evidence will be provided here. In the first place, Eos' Ortygia surely corresponds to the location of Tithonus' sojourn with Eos—i.e., the destination of this particular 'snatching'—on an island in the eastern Ocean, the place of sunrise, at the ends of the earth. The cosmological location of Homer's Ortygia is furthermore inseparable from another island he adduces, Syriê, the island of the sun's turnings or solstices, which will be discussed *in extenso* in the following chapter.[73] Here it is enough to state that the Ortygia/Syriê complex is part of a mythic model accounting for the annual solar movement involving the island of the sun's turning at a cosmological location corresponding to (in later terms) the latitude of the geographical arctic circle (I ask for little patience here, since I cannot argue for this interpretation at the moment, as an entire following chapter will be devoted for this argument). Indeed, Eos seems to have carried off Cephalus precisely ἐν Συρίᾳ, while Memnon, her son, was also buried in Syria, thus once more connecting the Island(s) of Turning with solar eschatology.[74] Another connection of the same type might be recognizable in an emphasized paradisiacal character of the island Syriê, where gentle death is brought to mortals by Apollo's arrows.[75] Ortygia/Syriê, islands at the ends of the earth, the place where Tithonus, Cephalus and Orion were taken by the goddess of dawn, are thus part of a model involving the 'snatchings' of mythic figures, including Harpies and Eos as the abductors, structured on an analogy with solar movement and pivoting around the idea of a solar island of turning. This nexus of mythic motifs was rather explicitly set alongside the land of Laestrygonians (for whose solar resonances see chapter 2) by the Hesiodic author in a fragment dealing with the Boreads' pursuit of Harpies.[76] Since the nexus of myths analysed in this chapter contains clear solar resonances both on the level of structure and detail, Hesiod's juxtaposition of these two traditions points to an awareness of the cosmological character of both. In the fragment Hesiod enumerates 'the mountain of Atlas' and 'Aetna' (25), the toponyms that point to the

---

61 *Oedipodea* fr. 3 West = Pisand. *FGrHist* 16F10 (but see Davies 2015: 4); Aesch. *Sept.* 776; Eur. *Ph.* 1021; Asclep. *FGrHist* 12F7b; Hermocl. *Ithyph.* 27 Powell = Douris *FGrHist* 76F13.

62 Eur. *Ph.* 1026–27; the throne of the Olympian Zeus (Paus. 5.11.2); for the evidence in art see further Gantz 1993: 495–96; for similar activities of the monstrous fox see Palaeph. 5; [Apollod.] *Bibl.* 2.4.7; Ant. Lib. *Met.* 41.8.

63 Duris *FGrHist* 76F17 *ap.* Phot. *Lex.* 205.13–14 s.v. Λάμια, Suda λ 84 s.v. Λαμία, cf. Σ Aristoph. *Vesp.* 1035d; D. S. 20.41.3; Hor. *Ars P.* 340; Σ Ael. Arist. *Pan., ad* 102.5 Jebb (see extensively in Ogden 2013: 90–91). The Delphian Lamia/Sibaris did not show any such preference (Nic. *ap.* Ant. Lib. *Met.* 8), but Fontenrose (1974: 100) and Ogden (2013: 89) believe she must have.

64 Paus. 1.43.7; *Anth. Pal.* 7.154.3; unnamed in Callim. *Aet.* fr. 26.11–14 (from Agias/Dercylus fr. 8A Fowler) and Stat. *Theb.* 1.596–626 with [Lact. Plac.] *ad* 1.602–603, p. 74 Sweeney (Fontenrose 1974: 104, cf. Harrison 1908a: 213; Fowler 2013: 496; Ogden 2013: 87–88).

65 Eur. *Bacc.* 754.

66 Vermeule 1981[1979]: 162.

67 Str. 14.1.6, perhaps from Apollod. *FGrHist* 244F99b; D. L. 1.39 = *Anth. Pal.* 7.85.1–2 (Th 237 Wöhrle) (here actually an aged Thales is being snatched away by a Zeus Helios); Σ E *Od.* 5.1; *CIL* VI.29954; cf. Graf 2009: 12, 19, 27–28 on Apollo's role in this phenomenon.

68 *Od.* 15.250.

69 [Hes.] fr. 375 M-W; Eur. *Hipp.* 454; Xen. *Cyneg.* 1.6; [Apollod.] *Bibl.* 3.14.3; Ant. Lib. *Met.* 41.1.

70 *hAphr.* 218; Sappho fr. 58 Voigt (P. Oxy. 1787 fr. 1.20; P. Köln 21351.18); Ibyc. *PMG* 289; Eustath. *Il.* 6.347, ii.326, 11.1, iii.134. Cf. depictions of Eos/Thesan carrying off Cephalus, Tithonus or Memnon's body (*LIMC* III.1.794–96, 2.585–86 s.v. *Eos/Thesan* 24–25, 30, 36–44).

71 Heracl. *Hom. All.* 68.3; a young man's death in general is described as a snatching by Hemera (Heracl. *Hom. All.* 68.5; Buffière 1962: 72; Vermeule 1981[1979]: 163) or Eos (Σ E *Od.* 5.1, specifying that the death is sudden).

72 *Od.* 5.121–24 (the verb used is *haireô*, which should be compared with *Od.* 20.66, describing how the *thuellai* took away (*anaireô*) Pandareus' daughters); Σ PQT *Od.* 5.121 (Euphor. fr. 103 Powell); [Apollod.] *Bibl.* 1.4.4. For Cephalus as the constellation Orion see Fontenrose 1981: 99, 101–102, 104, 110 n. 31.

73 *Od.* 15.404.

74 [Apollod.] *Bibl.* 3.14.3; Simon. 539 *PMG.* For Eos as Memnon's mother see *Th.* 984–85 and *Aeth. Arg.* 2a West with [Apollod.] *Ep.* 5.3, cf. *Od.* 4.188.

75 *Od.* 15.405–11.

76 Hes. fr. 150.25–27 M-W. Cf. Gisinger 1929: 320, 325. I will henceforth use 'Hesiod' for 'the Hesiodic author'.

far west, but also—in the case of the latter—to Sicily. The beginning of the next verse is lost, but it continues with 'Ortygia' and 'the people sprung from Laestrygon' (26), Poseidon's son (27). It would seem that here Hesiod associated the abode of the Laestrygonians with Sicily,[77] yet the 'mountain of Atlas' and—no less—Ortygia need not have been so early (for Homer, at least) identified with any concrete geographical place. It seems that we can see at work in this fragment the clearly unfinished process of localisation of Homeric toponyms onto the map of western Mediterranean (see the next chapter). But it is certain that in v. 26 Ortygia and the Laestrygonians are associated with *each other*, and this association could well predate their respective identifications with Sicilian localities. At the same time, this would explain the subsequent placement of the Laestrygonians on Sicily (see chapter 2), due to their association with an Ortygia (in this case, the Sicilian one). But if Ortygia could be associated with the Island(s) of Turning, as argued here, and, taking the concept to another level, with the solstices (for an extensive discussion on this subject see the following chapter), this would add another argument to my 'solar' interpretation of Homer's description of the Laestrygonian land, inspired by Pytheas' and Crates' exegesis, with its almost perpetual daylight on the day of the summer solstice.

In this way it seems possible to equate the Strophades, the Harpies' islands (also the Floating Island(s)), with both the island of Helios' cattle and the Ortygia/Syriê complex. The latter also has some connection with floating, since Ortygia was the ancient name of Delos, the archetypal floating island, which was already itself compared both with the Strophades and Aeolus' island.[78] Naturally, the myths involving these locations show considerable differences between them both on the level of structure and detail. Certainly, there is no one-to-one correspondence between any of these myths and solar movement and various other influences were involved in their formation. Nevertheless, all seem to include some reference to solar movement and are thus justifiably included in the class of myths that share this basic characteristic. In addition, this nexus of myths seems to show an additional concern about eschatology, which can consequently also be interpreted in terms of solar resonances, more precisely, in terms of structuring the experience of death on an analogy with solar movement, as in the second *nekuia* in the *Odyssey* or in the Egyptian New Kingdom tradition. As opposed to the mythic complex discussed in first two chapters, the nexus engaged with in the present chapter thus concerns itself both with eschatology and with structuring narratives on an analogy with solar movement. These two features appear to be inextricably connected with each other in that it seems that eschatological considerations were of prime importance for drawing the results of solar observations into mythic narratives and structuring these narratives upon these data.

4.5. An alternative model—cosmological solstice mountain

In addition to the model involving a northern island of solar turning, Greek tradition shares with a number of other traditions a familiarity with another mythic concept that can be connected with both the 'snatchings' and 'turnings'. This concept, involving the shadowy Hyperboreans and the vast northern mountains associated with them, usually called the Rhipaeans, as well as the Ursae constellations, reflects another mythic model of solar movement, or rather a variant of the solstice-island model. In this particular model, the 'snatchings' are attested in the myth involving the northern mountain and Boreas' abduction of Oreithiya, while the 'turnings' are implicitly present in all accounts that involve the mountains as a barrier to the further progress of the sun. Together with Boreas, this mythic nexus is manifestly connected with the Boreads and the Hyperboreans, even though there is some doubt in the correctness of traditional etymology of the latter's name as derived from the north wind.[79] Even if there is some truth in this skepticism, all ancient authors discussed here presupposed it,[80] so its acceptance or dismissal is of little relevance. The association of both the north wind Boreas and Hyperboreans with the Rhipaeans is, on the other hand, uncontroversial. The Rhipaeans were indeed closely associated with the former[81] and the latter.[82] In all these testimonies it is in general impossible to distinguish the mythic person Boreas from the homonymous wind, which corresponds to my position regarding the relationship between meteorological phenomena and their elaboration in myths, where they were treated in terms of anthropomorphic agents and their actions.

The natural connection between Boreas and the Hyperboreans allowed some ancient authors to explicitly claim that Boreas' sons Zetes and Calais actually arrived from Hyperborea,[83] which might be taken as implied elsewhere. Indeed, the garden of Phoebus-Apollo, where Oreithyia was brought by Boreas, located at the ends (*eschata*) of the earth, could only refer to Hyperborea, particularly because the 'sources of night' are also to be

[77] Cf. Ballabriga 1986: 23, 139–41.

[78] For Delos as Ortygia see Pind. *Pa.* 7b.48 = fr. 52h *POxy* 2442, 841; Call. *Ap.* 58–59 with Σ 2.59; *Aet.* fr. 18.7 Pfeiffer (*POxy* 2167, fr. 3); A. R. 1.419, 537, 4.1704; Lycoph. 401–402 with paraph. (i.37 Scheer); Nicander *FGrHist* 271–72F5; Σ EV *Od.* 5.123; Σ BHQ *Od.* 15.404; Σ Pind. *Pyth.* hypoth. a, *Nem.* 1.4a, c; Σ Ap. Rhod. 1.308a (p. 35 Wendel), 419 (pp. 38–39 Wendel); Σ (Tzetz.) Lycoph. 401 (pp. ii.149–50 Scheer); Strab. 10.5.5; Verg. *Aen.* 3.124, etc.

[79] Ambühl in *BNP* s.v. *Hyperborei*.

[80] Cf. Gagné 2021: 52 n. 242.

[81] Dam. *FGrHist* 5F1 (cf. Bolton 1962: 39–40); [Hippocr.] *Aer.* 19; A. R. 4.286–87; Val. Flacc. *Argon.* 2.515–16; *HN* 4.88, cf. 7.10 and Solin. 15.20, etc.

[82] Hellan. *FGrHist* 4F187bc; Aesch. fr. 197 Nauck = 330 Mette; Dam. *FGrHist* 5F1 (cf. Bolton 1962: 39–40); Callim. *Aetia* fr. 186.8–9 Pfeiffer, cf. fr. 215 Schneider *ap.* Σ A. R. 4.282–91b, p. 280 Wendel; A. R. 4.286–87 with Σ A. R. 4.282–91b, p. 280 Wendel; Str. 7.3.1; Pomp. Mel. 3.5.36; *HN* 4.88–89, etc. Probably also Posidonius, who identified the Rhipaeans with the Alps (frs. 240ab E-K = 402 Theiler) and placed the Hyperboreans there (or perhaps at the Apennines) (fr. 270 E-K = 70 Theiler).

[83] Duris *FGrHist* 76F86; Phanod. *FGrHist* 397F1.

found there.[84] This makes Hyperborea an appropriate destination for 'snatchings' and thus part of the mythic nexus revolving about eschatological concepts involving a reference to the annual solar movement; indeed, 'sources of night' most probably refer to the long winter nights in the far north, i.e., they should be associated with the phenomena accompanying the annual solar movement. In his turn Apollo was elsewhere connected specifically with the Rhipaeans: Hecataeus of Abdera thus used the name Rhipaean for the mountains he associated with Apollo, swans and the priests of Boreas, while Apollodorus derived Apollo's epithet *Hekaergos* from ῥιπή and Phanodicus associated it with the Hyperborean Hecaerge.[85]

In addition to these 'snatchings', the concept of 'turnings' of the summer solstice sun associated with northern mountains is a well-attested feature in Greek tradition. The connections of the Rhipaeans with the annual solar movement were already noted in chapter 2, where it was maintained that references to the Rhipaeans in the second half of the 5th c. Hippocratean author of the tract *Airs, Waters, Places* and Aristotle indicate some connection with the northern limit of the sun's annual movement.[86] Their solar associations are not exhausted by these testimonies, nor with their apparent connection to the concept of solar eschatology discussed above, but are amply attested in a number of sources that more or less clearly use them as part of a model accounting for the sun's annual approach to and return from the far north. At the same time, these mountains were also involved in another model accounting for solar movement, this time diurnal, which is another manifestation of the already discussed phenomenon of interchangeability of motifs in building models accounting for respectively diurnal and annual solar movement (this particular instance of this practice will be further discussed in chapter 9).

Among the earliest sources mentioning the Rhipaean Mountains is Alcman's ambivalent phrase Ῥίπας ὄρος,[87] for which it is not clear whether it means 'the mountain of Rhipe', or simply 'the mountain of the stormy blasts' (ῥιπᾶς). In any case, it is more than probable that the name of the mountain itself is actually derived from its association with the stormy blasts (of the north wind),[88] which leaves the question purely academic. Alcman further described these mountains as 'breast of black night',[89] while Sophocles referred to them as 'night-wrapped Rhipaeans'.[90] The context in the latter clearly shows he located them in the far north,[91] which could also apply to the former. This insistence upon the association of the Rhipaeans with night could indicate a region not visited by the sun, north of the northern tropic,[92] such as that described by the Hippocratic author of the treatise *De victu*, who claims that the north wind comes from the region outside the *oikoumenê* which 'the sun does not approach'—precisely the region 'under the Bears' where the Rhipaean Mountains are located according to a near-contemporary author of another Hippocratic treatise *Airs*, or even the much earlier Aristeas, and Aristotle.[93] It thus appears that the tradition associating the Rhipaeans with the northern limit of the sun's annual movement is as early as the late 7th c. and was maintained throughout the Archaic and Classical period.[94] In the Hellenistic period similarly, according to one reading of a fragment of Pherenicus, the Hyperboreans lived beyond the *dromos* of the sun or Boreas.[95] 'Beyond the course of Boreas' is in line with the most popular etymology of their name,[96] but 'beyond the course of the sun' would associate them with the Rhipaean mountains as the limit of the sun's northern advance on the summer solstice.[97] They would thus live in a region 'outside the solstice', that is, beyond the *oikoumenê*. Further theory connecting the Hyperboreans with the annual solar movement is reported by Pliny, locating them *medios inter utrumque solem, antipodum occasus exorientemque nostrum*, 'midway between the two suns, the sunset of the antipodes and our sunrise'.[98] The two suns could refer to the summer and winter sun, which would place the Hyperboreans due east, since there the sun rises for us and at the same time sets for the antipodes. However, Pliny's rebuttal of this thesis—the fact that a vast expanse of sea intervenes between our world and the antipodean—does

[84] Soph. fr. 956 Pearson, Radt; Pearson 1917: iii.118; Dowden, Commentary on *BNJ* 35F9. Compare Aeschylus' Aethiopians, living 'by the sources of the sun' (*PV* 808–809), referring to the east or south-east.

[85] Hecat. *BNJ* 264F12 *ap.* Ael. *NA* 11.1, cf. 11.10; Apollod. *FGrHist* 244F97; Phanod. *FGrHist* 397F5.

[86] Kiessling 1914: 850.56–58, 851.20–24, 852.22–34. Since both the restoration and interpretation of Aesch. fr. 68 *TrGF*, Radt = fr. 102 Mette, where the Rhipaeans are apparently mentioned together with Helios or Hyperion, are extremely tentative, one cannot build a solid argument for Aeschylus' understanding of any connection of the mountains with the annual solar motion upon it. For the date of the Hippocratic treatise see Jouanna 1999: 375.

[87] Alcm. 90 *PMG*.

[88] Bolton 1962: 40. See already Serv. *Georg.* 3.382; Isid. *Etym.* 14.8.8; also, Gervase of Tilbury, *Otia imperialia* (ii.763 Leibnitz); cf. Kiessling 1914: 855.55–58. Compare Homeric ῥιπῆς βορέαο in *Il.* 15.171 and 19.358; cf. Callim. *Hymn.* 4.25–26; Dion. Perieg. 666 + 674, Lightfoot 2014: 232 (Bolton 1962: 40).

[89] Compare Philostephanus' etymology of mount *Dindumon* in the preceding chapter: 'twin breasts'.

[90] Soph. *OC* 1248.

[91] Soph. *OC* 1245–47 with Stenger in *BNP* s.v. *Rhipaia orê*. *Pace* Σ Soph. *OC* 1248 and its interpretation of Alcman. See also fr. 956 Pearson, Radt discussed in a previous paragraph.

[92] Cf. Bolton 1962: 40, who associates Alcman's and Sophocles' testimony with the annual solar movement.

[93] *BNJ* 35F8; Hippocr. *De victu* 2.38, *Aer.* 19 (Dowden in *BNJ* here apparently follows Bolton 1962: 41 in attributing *Aer.* 19 to Aristeas, cf. Fowler 2013: 606); Arist. *Mete.* 350b6–7. For the dates of Hippocratic works see Jouanna 1999: 375, 409. For a similar, but much later, concept of the zones outside the arctic circles defined by the absence of the sun see Heracl. *Quaest. Hom.* 51.2.

[94] For the early date of Aristeas see Bolton 1962: 4–7; Bridgman 2005: 31; Dowden *BNJ* 35 *Biographical Essay* (620–580 B.C.) (for a 6th-c. date see Bremmer 2016: 66–67; 2018: 107). For the date of Alcman (late 7th c.) see Calame in *BNP*.

[95] Fr. 671 SH; emendation ὑπέρ for the MS ὑπό by Drachmann i.113, following Voss; Ferrari 2008: 144 with n. 58 (sun); Bolton 1962: 22–23 (Boreas).

[96] On which see Geus 2000: 67 with n. 50.

[97] To the best of my knowledge, a *dromos* of Boreas is nowhere mentioned in Greek; for the sun's *dromos* see (only the earliest sources are listed) Emped. fr. dub. 154 D-K (Plut. *De esu carn.* 993E); Hippocr. *Flat.* 3; Pl. *Crat.* 397d, *Leg.* 7.821c (by implication); [Pl.] *Axioch.* 370b; Arist. fr. 11 Rose = fr. 948 Gigon *ap.* Sext. Emp. 9.27; Alex. Aet. fr. 1.3–4 Powell.

[98] *HN* 4.90, cf. Solin. 16.2.

not explain his previous statements. He could not have believed that there are actually two suns and that there is an intervening region between them anywhere on earth. Thus although Pliny locates the Hyperboreans with respect to the solar movement, his report is incomprehensible. Elsewhere he locates the Hyperboreans with respect to the azimuth points: they inhabit the region from due north (i.e., azimuth 0°) or the mid-point between the north and the summer solstice sunrise (ca. 74° observed from Italy),[99] or even the summer solstice sunrise itself (ca. 58° observed from Italy),[100] to cape Lytharmis in Celtica and the river Carambucis where the Ripaean (*sic*) mountains terminate.[101] This is a series of localisations that employs a frame of reference utilising solstice points to define the sections of the *oikoumenê*, which is at least as early as Ephorus and was possibly even known to Homer (Syriê) (see chapter 5).[102]

Alongside Aristeas, the Hippocratic treatise *Airs* and Aristotle, the Bears were again associated with the Rhipaeans in a fragment of *Pirithous*, where the two Ursae guard the pole of Atlas, here certainly referring to the north celestial pole.[103] The Bears are described as having ῥιπαί wings, which is a clear allusion to Rhipaean Mountains.

Furthermore, Herodotus, a contemporary of the author of the *Pirithous*, although not naming it, described a high mountain range in the north, beyond which certain people live who sleep for six months of the year, by which we should probably understand the Rhipaeans.[104] His description refers to the meteorological situation obtaining at the north pole and thus his 'beyond' should be understood as 'well beyond'. This tradition was usually associated with the Hyperboreans, who were often localised at the north pole and/or inhabiting the region with a six-month continuous daylight/night,[105] or with certain Celtic people.[106] The concept of continuous daylight or night in the far north can also be recognised in Xenophanes' writings, who was born shortly after Dowden's proposed dates for Aristeas and who died in Herodotus' early youth. He describes a month-long solar eclipse at the same time expounding a curious theory of different suns for different *klimata* and zones of the earth, which might refer to the meteorological situation obtaining at high northern latitudes.[107]

The early testimonies adduced in this section suggest that the post-Homeric Greeks were aware and gave some consideration to the phenomena caused by the annual solar movement in high northern latitudes. They described these phenomena and their causal relation with solar movement at least partly in terms of mythic discourse, which was used even by Aristotle and later authors, although perhaps more self-consciously with respect to its potential as technical vocabulary. As opposed to the set of myths treating 'snatchings', these sources do not refer to any recognisable eschatological tenet, solar or otherwise, but rather focus on the cosmological aspect of the tradition. Indeed, many of the authors writing on these subjects, such as Xenophanes and Aristotle, were not interested in such considerations, at least not in the context of their discussions of high latitudes or the annual solar movement, which might explain the lack of eschatological resonances in their accounts. At the same time, it could be claimed that the paradisiacal region of the Hyperboreans always connoted some eschatological ideas, or at least an otherworldly or marginal character of this location.

Following these early testimonies, there is little Greek evidence for the concept of a mountain range forming the northern limit of the annual solar movement and/or the southern boundary of the northern uninhabitable zone. It seems that this particular model accounting for the annual solar movement and interrelated phenomena was not particularly conspicuous in later tradition, even though it did resurface later in some medieval zonal diagrams[108] and writings,[109] and it also appears in Islamic mythic geography as the Qūfāyā mountains of Ibn Khaldūn,[110] probably also deriving from ancient Greek concepts.[111] Thus both mediaeval European and Islamic geography were familiar with the concept of cosmological Rhipaean Mountains associated with the latitude of the fixed arctic circle and the northern boundary of the northern temperate zone. It seems that this later tradition interpreted Aristotle in much the same way as Kiessling did, and that the tradition ultimately stemming from Aristeas thus survived

[99] *HN* 2.119.
[100] *HN* 2.125.
[101] *HN* 6.34.
[102] For Hecataeus' localisation of Hyperborea with respect to the annual solar movement see chapter 6.
[103] Crit. *Pirith. TrGF* 43F3.3–5 Snell, Kannicht = 88F18 D-K = Eur. fr. 594 Nauck²; West 1979: 145 n. 91. On the question of authorship see Gantz 1993: 293, 852 n. 26, Collard & Cropp 2009: 627–35 and Winiarczyk 2013: 47 n. 135. For *axis Atlanticus* associated with the Hyperboreans see Avien. *Or. mar.* 663–64. For Atlas among the Hyperboreans cf. [Apollod.] *Bibl.* 2.5.11 (cf. Tzetz. *Chil.* 2.36.357, 372, 377–78; Pediasimus *De Herc. laboribus* 11), most probably from Pherecydes, from whose narrative (*BNJ* 3F17) it is clear that his Atlas was not located in north-west Africa (Pearson 1917: iii.118; Jacoby *Comm.* i.394–95; van der Valk 1958: 123, 127 n. 89; Fowler 2013: 291–92, 296, 298–99, 606); for 'Hyperborean' Taygete, Atlas' daughter, see Pi. *O.* 3.13–16, 28–31 with Σ Pi. *O.* 3.53b–e; for a 'northern Atlas' cf. Dion. Hal. 1.61.1 (reading uncertain; Bolton 1962: 189 n. 16); Antioch. 5 in MS Vat. 1056, f. 28ᵛ (Boll 1903: 58, 260–61); also, Prisc. *Perieg.* 570–76, *GGM* ii.195 and Σ D. P. 560, *GGM* ii.451.
[104] Hdt. 4.23.2, 25.1; cf. Bolton 1962: 42; Bridgman 2005: 47–49.
[105] Str. 1.3.22; Pomp. Mel. 3.5.36; *HN* 4.89; Solin. 16.3; Claud. *In Ruf.* 2.240; Mart. Cap. 6.664.
[106] Steph. Byz. s.v. Γέρμαρα; Page 1973: 6. For identification of the Celts with the Hyperboreans see Heracl. Pont. fr. 102 Wehrli = fr. 49 Schütrumpf = *FGrHist* 840F23; Asclep. *FGrHist* 12F19 (Bridgman 2005: 68); Posid. fr. 270 E-K = 70 Theiler = *FGrHist* 87F103 (Krappe 1947: 223) with fr. 240a E-K = 402 Theiler = *FGrHist* 87F48.
[107] Xenoph. 21A41, 41a (Aët. 2.24.5, 8 Mansfeld & Runia 2009/2020); Ballabriga 1998: 125–27; Couprie 2018: 143–44, who further argues that the 'eclipse' here has a more general meaning of 'disappearance'.
[108] MS Walters 73, f° 7ʳ (Bober 1956–57: 94 Fig. 22; Edson 1997: 70 with Fig. 4.6); map from a MS of *Navigatio Sancti Brendani* (Scafi 2006: 168–69 with Fig. 7.8a; 2013: 84–85 with Fig. 53); Cotton MS Tiberius C.I., f° 11ᵛ and MS Oxford, St. John's College 17, f° 40ʳ (Bober 1956–57: 82 Fig. 17, 85 Fig. 18, with a mistake of the copyist).
[109] Gervase of Tilbury, *Otia imperialia*, ii.763 Leibnitz.
[110] Rosenthal 1967a: i.50, 55, 56, 96, 149–66; cf. Van Donzel & Schmidt 2010: 97 = 2009: 107.
[111] Tallgren-Tuulio 1936: 170, cited by Rosenthal 1967a: i.149 n. 167.

for a long time—interpreted in a more precise technical sense—and died out only stubbornly.

4.6. A reinterpretation of the northern mountains model

At the same time, it appears that another understanding—or a reinterpretation—of the model accounting for the annual solar movement by recourse to a turning-point in the far north appeared in some authors belonging to the tradition of Greek scientific geography. A vast northern mountain blocking the sun in the north thus represents a mythic model corresponding to a technical (or semi-technical) term and concept of *peirata gaiês*, in the meaning of the limits of the sun's annual movement and the limits of habitable zone(s).[112] The expression was used by Eratosthenes, who called the parallel that passes through the Cinnamon-producing country a (southern) *peras* of the *oikoumenê*;[113] Strabo also mentions its northern *peras*, the parallel through Ierne, a substitute for Pytheas' and Eratosthenes' Thule.[114] Marinus and Ptolemy placed Thule at the northern *peras* of the known world,[115] and the southern *peras* at the winter tropic.[116] Thule was, as we have seen in chapter 2, a cosmological location most frequently placed at the northern turning of the sun.[117] In this connection, it was occasionally located at the *perata* long before Ptolemy, perhaps even by Pytheas himself, who claimed he had visited the whole northern coast of Europe as far as the *kosmou peratôn*, which could only be Thule.[118] From Pytheas the term might have been taken over by Eratosthenes, and later by Strabo and others. Furthermore, Plutarch placed Britain at one of the *oikoumenês peratôn*, which is to be associated with the concept of Britain as a kind of Thule and/or its geographical proximity to Thule as noted by Pytheas,[119] while for Geminus the latitude where the summer tropic is entirely above the earth—which is how Pytheas defines the latitude of Thule—is a *peras*.[120] At the same time, the limits of the annual movement of the sun were also explicitly designated as *perata*. Thus Aristides described the *perata* of the sun's annual movement, as well as its northern and southern boundaries (*horous*),[121] the Stoics spoke of the northern and southern *perata* of the helix produced by the combination of the sun's diurnal and annual movement,[122] Ptolemy refers to the tropics as *perata*,[123] while Cosmas Indicopleustes uses the expression *boreion peratos kai notion* to designate the northern and southern limits in the annual motions of the sun.[124] This effectively equates the *perata* with *tropai*. As we shall see in the following chapter, the latter term is the standard designation of the limits of the sun's annual movement, both in a temporal and spatial sense, used most probably already by Homer, but certainly by Hesiod.

Thus a truly scientific model accounting for the annual solar movement and its critical points was constructed analogously to a mythic model involving a high northern mountain range. At the same time, it is conceivable that the term *perata* can actually be etymologically derived from or connected to mountains, i.e., most probably a world-encircling mountain. The examples are Hittite *peru(r)*, *perun-* ('rock'), Vedic *párvata-* ('mountain') and Avestan *pauruuatā-* ('mountain chain').[125] At the same time, the word *peirata* was associated with a concept of 'circumscribing band', 'belt, girdle, band', and the encircling Ocean, all derived from the original concept of 'bond'; both functions are indeed characteristic for the Ocean: it binds the earth by encircling it.[126] From a cosmological point of view these two proposed derivations are not mutually exclusive, both referring to a feature limiting the *oikoumenê*. Thus, the Rhipaeans could represent two analogous and interrelated cosmological features: the northern section of the mountain circumscribing the *oikoumenê*, and at the same time the northern limit of the annual solar motion. This concept was apparently utilized by ancient Greeks in their cosmological speculations, both in terms of mythic narratives and the theories of the philosophers of nature, but also by the authors adhering to the tradition of scientific geography.

4.7. A region outside the sun's course in non-Greek traditions

A region not visited by the sun, i.e., outside of the limits of the annual solar movement north of the summer tropic, is also described in other traditions. I do not wish to argue for a genetic connection between the Greek concept and the corresponding Mesopotamian (Neo Babylonian), Egyptian (New Kingdom) and Indian (Vedic) ones, or even for an influence of any of these traditions on any other. I merely wish to document the existence of parallel traditions or shared cross-cultural concepts in order to show that the notion was reasonably common, which is not particularly strange if the phenomena it accounted for are indeed identical. All the traditions are rather explicit on this point, since all discuss solar movement, which makes this conclusion commonsensical. In addition, the study of these parallel traditions can offer a valuable outside perspective on the Greek understanding of the concept

[112] The following is a modified version of a section of Bilić 2013a: 267–69.
[113] Eratosth. fr. 34 Roller; in the same fragment he seems to use the verb *peratoô* for the northern boundary of the *oikoumenê*, but the expression could equally be Strabo's.
[114] Str. 2.5.8, 34; Roller 2010: 167.
[115] Ptol. *Geog.* 1.7.1 (Marinus *FGrHist* 2114F1), 7.6.8.
[116] *Geog.* 1.7.2–3 (Marinus *FGrHist* 2114F1).
[117] Pytheas fr. fr. 8c Bianchetti = fr. 6 Roseman = fr. 6c Mette, etc.
[118] Pyth. fr. 21 Bianchetti = fr. 7a Mette = T8 Roseman = Eratosth. fr. 14 Roller.
[119] Plu. *De def. or.* 2.410A. For Pytheas, Thule and Britain see chapter 6.
[120] Gemin. 5.21. For Pytheas' formulation see fr. 8c Bianchetti = fr. 6 Roseman = fr. 6c Mette.
[121] Aristid. *Aegypt.* 36.38–39 (ii.275 Keil).
[122] Aët. 2.23.6 Mansfeld & Runia 2020 = 2.23.8 Mansfeld & Runia 2009 (not in SVF, cf. Mansfeld & Runia 2009: 554–55, 557–58; 2020: 1015, 1019).
[123] Ptol. *Syntax.* 1.12 (p. 67.22–68.1 Heiberg).
[124] Cosm. Indic. *Topogr. Christ.* 9.4 Wolska-Conus.
[125] Watkins 1995: 162, 530.
[126] Onians 1951: 314–17, 321, 332, 384 n. 1.

of solar movement, which is why its inclusion is well-merited.

Thus, on the Babylonian '*mappa mundi*' depicted on a clay tablet BM 92687 we can see a central continent encircled by *marratu*, a 'cosmic ocean'. Beyond the ocean seven or eight triangular areas identified as *nagû* are depicted. They probably represent islands beyond the ocean, as confirmed by the *Epic of Gilgameš* and Late Babylonian royal inscriptions,[127] or perhaps the mountains rising from the sea.[128] Similarly to Greek tradition, here two paradigmatic variants (islands/mountains) seem to converge or are not easily distinguished. The northern *nagû* is labelled on the map as 'a place where the sun is not seen', and a further fragment now attached to the description adds the label 'Great Wall' to it.[129] 'Great Wall of Heaven and Earth' is indeed placed at the ends of the earth in a Sargon Neo-Babylonian legend.[130] The absence of the sun in this *nagû* can be explained by the fact that the sun never passes to the portion of the sky north of the tropic of Cancer.[131] This excludes the notion of a completely sunless area, similar to Homer's description of the land of Cimmerians,[132] but takes into account the annual motion of the sun instead. Thus, it could be argued that this *nagû* describes an area whose southern limit is (in modern terms) at 66° north, while the Great Wall would be an appropriate barrier stopping the sun from going any further north.

Moreover, the *nagû* from the map are probably connected to a notion that *ziqpu*-stars rise and set into the cosmic ocean amidst *nagû*, because the seven *nagû* are mentioned in the *ziqpu*-star text AO 6478.1.[133] If the *nagû* are indeed to be understood as a cosmic mountain-barrier or any kind of marker used to denote the point of rising or setting for heavenly bodies, then the twin mountain Mašu, from which the sun rises and into which it sets (for which see chapter 13), could be convincingly conceived as part of this wall.

A parallel for these Greek and Mesopotamian concepts appears in Egyptian writings. The New Kingdom *Book of Nut* thus describes the region 'outside' the sky, inaccessible to the sun, engulfed in primeval darkness and primeval waters, with unknown boundaries, and annulled cardinal directions.[134] The text, however, is somewhat enigmatic, and the translations accordingly vary.[135] Text L of the *Book* describes a region beyond the sky in total darkness, where the sun does not appear or rise.[136] According to Text Dd, it apparently stretches from Nut's north-western side up to her north-eastern (or, possibly, south-eastern?) side,[137] with the Duat being located on her northern side.[138] This seems to describe a portion of the sky in which the sun does not rise, and in which a total darkness rules. However, it is not completely clear where this region is actually located, especially since the *Commentary* further adds that Nut represents the northern sky, with her head in the west and her hind parts in the east.[139] Nevertheless, it seems pretty clear it should be looked for to the north of the tropic of Cancer,[140] which is precisely where it is placed in the original text from Seti's tomb. It thus seems that the Egyptians of the New Kingdom recognised the northern solstitial boundary of the annual solar motion, associated it with the segment of the horizon bounded by the extreme solar azimuths from which the sun rises during the year and into which it sets, and believed that the region further north was completely devoid of sunlight.

Finally, Indian cosmology was also familiar with the concept of an unapproachable region beyond the sunrises and sunsets.[141] However, the only mountain in any way associated with the sun was Meru, which was described as blocking the sunlight and was located in the very centre of the world, with no apparent connection with the annual solar movement.[142]

Thus, a number of ancient cosmologies recognised a region in the far north, outside the sun's path, separated from the *oikoumenê* by some feature blocking the sun. The concept was treated differently by different traditions, but the underlying structure remained stable, conditioned as it was by the unchanging nature of the phenomenon it referred to. Thus, it cannot be claimed that the concept originated with any of these peoples (incidentally, the Mesopotamian example is probably chronologically the latest), but rather that the sun's annual path was studied independently by a number of peoples, and that each of them created a somewhat different mythic model accounting for it, with the emphasised similarities between

[127] *EG* 11.139; Horowitz 1988: 157; 1998a: 30–32.

[128] Flygare 2006: 15. Unger (1937: 6–7) believes that a fresco from Teleilat Ghassul (Jordan, mid-4th millennium BC) (Harley & Woodward 1987: Plate 1) also represents a 'cosmological map' depicting an earth encircled by an ocean beyond which the world-encompassing mountains and another, 'celestial', ocean are placed; beyond the latter eight 'islands' (*nagû*?) are located. The map—if it is a map in the first place—is extremely schematic and there is little contextual support for Unger's interpretation.

[129] Finkel 1995: 26–27. Horowitz (1998a: 38 n. 29) suggests that the description of the fifth *nagû* on the reverse might relate to this northern *nagû*, translating the l. 14' as 'its (the sun's) shine he does not see'.

[130] ll. 20, 30; Flygare 2006: 15; Horowitz 1997: 98; Westenholz 1997: 42–45.

[131] Kahn 1960: 83 n. 3; Dilke 1985: 13; Horowitz 1988: 158, 1998a: 33, after C. B. F. Walker.

[132] Cf. West 1997: 145, 426.

[133] Horowitz 1998a: 182–83; 1998b: 51; cf. Flygare 2006: 23, n. 40.

[134] Hornung 1999: 114. The core of this text might be as early as the Old Kingdom (Quack 2016: 233).

[135] Allen 1988; 2003; Clagett 1995.

[136] Allen 1988: 1, cf. 66 n. 25 (the commentary in P. Carlsberg, col. 1.6), 74 (the commentary in P. Carlsberg 2.27); 2003: 27; Clagett 1995: 374, cf. 371 (the commentary in P. Carlsberg, col. 1.6–7).

[137] The text in Seti's tomb: 'north-eastern' (Frankfort 1933: 73); P. Carlsberg, 4.29: 'south-eastern' (Allen 1988: 74).

[138] Clagett 1995: 383.

[139] P. Carlsberg, col. 1.1–2 (Clagett 1995: 371).

[140] Cf. Allen 1988: 5.

[141] *Ṛg Veda* 1.105.16 and 3.30.12 with Anghelina 2010: 69 n. 10; *Atharva Veda* 10.8.16 (Whitney 1905: 598); *Kaṭha Upanishad* 4.9 (Hume 1921: 355).

[142] *Viṣṇu Purāṇa* 2.8 (Wilson 1840: 219); Pingree 1978: 554. Later Meru was 'transferred' to the north pole (Pingree 1978: 555; 1981: 12; Minkowski 2010: 23; Plofker 2010: 38), with Dhurva, the north pole star, above it (Eliade 1958: 100, 375; Pingree 1978: 554–55; Mabbett 1983: 69; Schwartzberg 1992a: 359; Plofker 2010: 35).

models being conditioned by their shared referents. At the same time, one cannot exclude the possibility of mutual influences, interactions and transmissions of concepts, which might be evoked on a case-to-case basis to account for these similarities.

4.8. Conclusion

The widespread concept of a mountain barrier (or some other similar feature representing a paradigmatic alternative) in the far north thus represents a mythic model accounting for the annual solar motion, more precisely, for its northernmost section. In Greek tradition the barrier, as a rule referred to as the Rhipaean mountains, was associated with the constellation of Ursa Major (of which more will be said in the following chapter), the Hyperborean people and region (and through them with the complex of stories involving a solar or seasonal Apollo, for which see chapter 7), and the blasts of the north wind. This model was through mythic narratives of snatchings performed by the Harpies, Boreas and Eos further associated with certain concepts of solar eschatology involving the Island(s) of (solar) Turning. These Islands, where the Boreads, pursuing the Harpies, 'turned back', and where those that were snatched off might be carried, represent a paradigmatic variant of possible markers of the northernmost limit of the annual solar movement. These particular Islands can, in their turn, be understood as corresponding to the Ortygia/Syriê complex, itself belonging to the series of floating and/or revolving solar islands and, furthermore, already by the Hesiodic author closely associated with the Laestrygonians. Hesiod's knowledge of a detail involving the Laestrygonians' ancestry, not recorded by Homer but also known to Apollonius in a structurally parallel episode (see chapter 3), points to a pre-Homeric version of the story that most probably did not involve Sicily or the western Mediterranean in general.[143] Rather, this early version should be associated with the mythic model accounting for the annual solar motion, more precisely, to its northernmost section, that was used by Homer in the *Odyssey*, which most probably belonged to a pre-Homeric Argonautic tradition. Thus, several interconnected mythic models of the annual solar motion, some of those involving hitherto in this context unrecorded eschatological resonances, can be recognised in the stories involving the Harpies, mythic snatchings, floating or revolving solar islands of turning and the fabulous mountains and peoples in the far north. This mythic nexus can thus justifiably be classified to the category of myths referencing solar movement, even if its solar references—including the instances of episodes structured on an analogy with solar movement—are not always immediately apparent.

In this chapter I have had the chance to bring into sharper focus the eschatological facet of the myths involved with the solar movement. Crucially, it appears that eschatological considerations might have played a pivotal role in introducing the practice of structuring mythic narratives on an analogy with solar movement. This course of action consisted of incorporating the results of solar observations in myths not in the form of references to this phenomenon inserted into a narrative but by constructing narrative models based on an analogy with the phenomenon itself. The analogy between the sun(-god)'s diurnal and/or annual reappearance and the desired fate of the human soul appears to have been a powerful and inspiring one, utilised by various traditions in their eschatological speculations.

Another concept that was introduced in this chapter is the existence of two differing models accounting for the phenomena of sunrise and sunset. One of those, the uni-polar model, accounts for the alternation of night and day with recourse to the Tartarean *axis mundi*. The bi-polar model, on the other hand, appears more commonsensical and in a better accordance with the observed phenomena, recognising the sun' risings and settings on the eastern and western horizon, respectively. However, both models satisfactorily account for the phenomenon they intend to describe and thus explain. Where they differ is in focusing on different specific moments in the course of the diurnal solar movement, the bi-polar model concentrating upon the moments of sunrise and sunset and the uni-polar model focusing on the specific moment of the daylight/night alternation.

Finally, while in the first couple of chapters I have focused solely on the testimonies from Greek tradition, in this chapter I have added the complementary material from other civilisations. This allowed a comparison between similar attempts at dealing with corresponding natural phenomena. My discussion has shown the existence of shared cross-cultural traditions in approaching the phenomenon of the annual solar movement, which is in itself unsurprising, but which offers a valuable outside perspective on the Greek attempts at understanding the solar movement.

[143] See the next chapter for the tendency to localise the episodes from the *Odyssey* to this region.

5

The island of the sun's turning

νῆσός τις Συρίη κικλήσκεται, εἴ που ἀκούεις,
Ὀρτυγίης καθύπερθεν, ὅθι τροπαὶ ἠελίοιο...

There is an island called Syria, if perchance you have heard of it, above Ortygia, where are the turning places of the sun.

Odyssey 15.403–404, translation by Murray revised by Dimock in *LCL* 105 (1995)

All the concepts discussed in the preceding chapter—turnings, snatchings, wind-blasts, floating islands, mountain barriers—seem to belong to a single, yet complex and multifaceted mythic nexus or to a set of myths having as one of its referents the annual solar movement. The reference to this aspect of solar movement might be present in the form of a narrative structured on an analogy with the phenomenon, or in the form of an incorporation of the results of the observations of solar movement in a narrative otherwise independent of the phenomenon itself. More precisely, the concepts of turnings etc. seem to feature in different iterations of this mythic nexus on several interrelated levels: cosmological, eschatological, meteorological (in ancient Greek sense), geographical and astronomical. From the point of view of early Greek tradition, however, all iterations belong to a single 'discipline' that clearly treated such classifications, of decisive importance to a modern observer, as marginal or non-existent. This leads us to examine several more occurrences in which this nexus manifested itself more or less clearly, in the first place the already mentioned Homer's island of Syriê, 'above Ortygia, where are the turning-places of the sun (*tropai êelioio*)', which was already in the preceding chapter identified as the location where the sun turns at the solstice.[1] Moreover, it was rather casually suggested that the Ortygia/Syriê complex was part of the mythic model accounting for the annual solar voyage involving the island of sun's turning. This claim will be discussed in some detail in the present chapter.

In opposition to the myths addressed in chapter 4, the set of stories and accounts discussed in this section all clearly reference the sun and its movement, which brings us back to a safer level of justification for their inclusion into the class of myths referencing solar movement, comparable to the level of certainty exhibited by the tradition studied in chs. 1–2. At the same time, in contrast to previous chapters, even though my discussion will pivot on a Homeric concept and its interpretation, the subject-matter treated in this chapter will compel me to focus on a section of the continuum of ancient knowledge of the solstices that consciously distanced itself from myth, i.e., on later tradition that was less reliant upon personalistic causality and is considered genuinely scientific by modern scholars. However, this tradition is equally part of the Greek ethnographic context and should be studied in this light, as argued for in the introductory chapter. In any case, when analysing ancient testimonies on the solstices, in practice no less than in theory it is extremely hard to determine where the category of mythic stops and the category of scientific begins. It would appear that the safest approach is to include all the available testimonies on solstices emanating from the Greek ethnographic context in the inclusive category of cosmology. Within this umbrella-term, various instances of references to solstices irrespective of their level of speculativeness would be encompassed. In the case of the material assessed in this chapter, it is specifically the non-personalistic accounts that will come to the fore, but their connections with myths will also be studied at some length.

5.1. The concept of solstices in early Greek tradition

In order to assess the development of the understanding of the concept of solstices in Greek tradition we need to turn once more to the *Odyssey*. The expression *tropai êelioio* in book 15 of the *Odyssey* undoubtedly in some way designates the solstices.[2] Thus in Hesiod *êelioio tropês* and *tropas êelioio* bear the meaning of 'summer' or 'winter solstice', respectively,[3] and that meaning remained more or less stable during the entire antiquity.[4] Other earliest occurrences of the word *tropai* in the meaning of 'solstices' date from the 7th and 6th century BC, which is not far removed from Homer's dates (for which see chapter 2). It must be borne in mind that solstice is a phenomenon that manifests itself in various ways, both on a temporal and spatial level, and it is sometimes possible to gauge the exact sense in which the term *tropai* was used by individual authors; indeed, I will attempt to determine the specific sense in which it was used in the *Odyssey* at the end of this chapter. Of earliest authors, both Hesiod and

[1] *Od.* 15.404. The following is an abbreviated and slightly modified version of Bilić 2016a.

[2] Thomson 1948: 37 with n. 1; Kahn 1970:113 n. 50; 1979:109, 140, 313 n. 133; Vlastos 2005[1975]: 34 n. 21; Schibli 1990: 5; O'Grady 2002: 149. *Pace* Dicks 1966: 31; 1970: 32–33.

[3] *Op.* 479, 564, 663. Nilsson 1920: 316; Pannekoek 1961:95–96; Dicks 1966: 31; 1970: 34–35, 37; Kahn 1970: 113; Vlastos 2005[1975]: 34 n. 21; Ballabriga 1998: 107; Kidd 1997: 10, 13; Evans 1998: 4–5, 56; O'Grady 2002: 149; Kremer 2020: 190.

[4] Martin (1879), cited by Heath 2004[1913]: 10 n. 1.

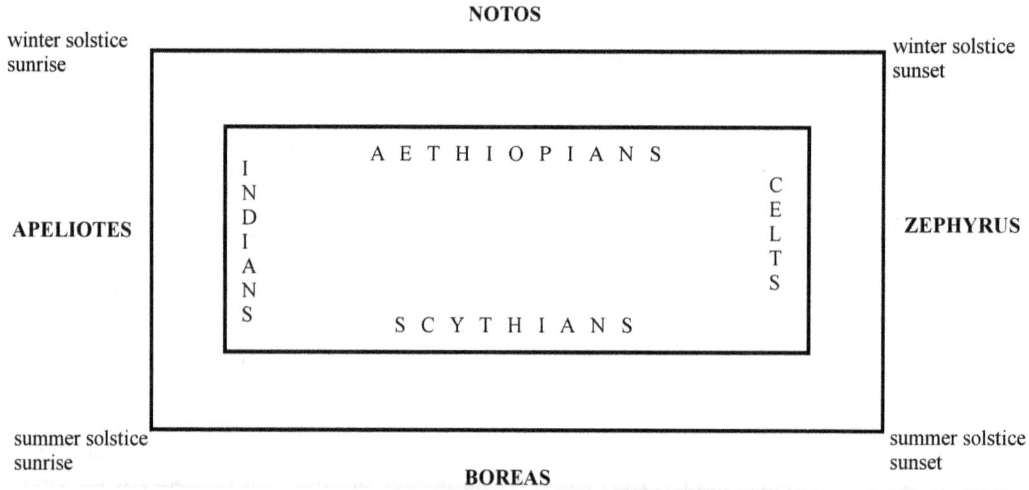

Figure 5.1 Ephorus' schematic diagram of the *oikoumenê* (drawn by the author)

Alcman[5] used the term in a temporal sense.[6] Cleostratus,[7] on the other hand, is said to have observed the solstices from Tenedos with Mount Ida as a natural foresight, thus referring to the extreme horizon positions of the sun. The testimonies on Thales[8] are either non-committal (1, 3, Th 265 Wöhrle) or temporal (17),[9] while in Anaximander the references are certainly spatial (position of the sun on the horizon or at the meridian,[10] the sun's extreme declination in general).[11]

It is entirely possible that Anaximander's map (as well as Hecataeus', derived directly from it) was determined by sunrise and sunset solstice points,[12] Ephorus' parallelogram being only a later rendition of the same concept (Figure 5.1).[13] In this case, Anaximander's concept of *tropai* would definitely encompass the notion of the extreme horizon positions of the sun. Moreover, this notion would be of decisive importance for the making of 'Ionian' maps, which would in this way be closely associated with the coordinate system exemplified in the position of Homer's Syriê, when the latter is understood in terms of the horizon position of the sun at solstices (I will argue against this particular interpretation below).[14] Yet the making of the map probably required an additional step, the recognition of solstitial 'latitudes' derived from solstitial azimuths (on this more below). Another contemporary testimony, that of Anaximenes, discusses the *tropai* of the *astra*, which could refer to the planets,[15] and thus also to the sun, and had the same meaning as in Anaximander.[16] Later, the

[5] Alcm. 17.5 *PMG*.

[6] As did the author of the *Little Iliad*, if the conjecture of Grafton & Swerdlow 1986: 215–16 and West 2013: 211 is right (cf. Fowler 2013: 544; it is, however, highly speculative and based on Dion. Hal. 1.63.1).

[7] Cleostr. 6A1 D-K (also mentioning the astronomers Matricetas, Phaeinus and Meton).

[8] Thales 11A1 (Diog. Laert. 1.23 = Eudemus fr. 144 Wehrli = Th 46, 237 Wöhrle with Th 7, 8, 15 Wöhrle, Diog. Laert. 1.24 = Th 237 Wöhrle), 3 = Th 578 Wöhrle, 17 D-K (Eudem. fr. 145 Wehrli = Th 47, 93, 167 Wöhrle), Th 265 Wöhrle. Cf. Zhmud 2006: 239–41, 244–45 (O'Grady 2002: 147–50 should be used with caution).

[9] Cf. Bowen 2002: 311 n. 10.

[10] This can be inferred from the group of testimonies mentioning Anaximander's association with the gnomon. He apparently used the instrument for marking (12A1 D-K = Favorin. fr. 28 Mensching *ap.* D. L. 2.1, from Eudemus?, Zhmud 2006: 249) or distinguishing (12A4 = Eus. *PE* 10.14.11) the solstices (cf. 12A2). This most probably points to the observance of noon shadow lengths or to the observance of sunrise and sunset points, which was utilized for creating a map of the *oikoumenê* defined by these azimuths (Heidel 1937: 57–58; Hahn 2001: 8, 38, 44, 201, 205–206 with fig. 4.12 on p. 207, 208; Couprie, Hahn & Naddaf 2003: 52, 194–95; Naddaf 2005: 109; Couprie 2011: 80 with Fig. 6.1, 84). In any case, this group of testimonies refers to the spatial aspect of solstices.

[11] Anaximand. 12A27 D-K with Diogenes 64A17 D-K (Arist. *Meteo.* 353b7–9 and Theoph. fr. 221 FHS&G *ap.* Alex. *Meteo.* p. 67.6–7 Hayduck), also Arist. *Meteo.* 355a24–26 = Diogenes 64A9 D-K (Kahn 1960: 67; cf. Zeller 1881: i.277 n.1; Heath 2004[1913]: 33 n. 3; Cherniss 1935: 135 n. 544; Guthrie 1962: i.97 n. 3) with Alex. *Meteo.* p. 73.19–22, not cited by D-K, and perhaps also Arist. *Meteo.* 354b36–355a1.

[12] Heidel 1937: 17–20, 22, 33–34, 42, 47–48, 51, 53–54, 57, 133; Thomson 1948: 97–98; Ballabriga 1986: 147–49; Hahn 2001: 8, 201, 205–206, 285 n. 140; Naddaff 2005: 109–10; Couprie 2011: 80–82, 84; Couprie, Hahn & Naddaf 2003: 52–53, 195–97; Irby 2012: 89.

[13] Heidel 1937: 17–20, 33–34, 42, 47–48; Thomson 1948: 97–98; Ballabriga 1986: 147–49; Naddaff 2005: 109–10; Couprie, Hahn & Naddaf 2003: 52–55. For Ephorus' concept see *BNJ* 70F30abc (for an accompanying diagram illustrating this concept see Ballabriga 1986: 148; Aujac 1987a: 144 Fig. 8.12; Irby 2012: 97 Fig. 6; Kominko 2013a: 35–41, 240–41 (*CT* 2a–c); 2013b: 78–80; Arnaud 2014: 46 Fig. 6; Geus 2016: 83–84, 2018: 406–408; Couprie 2018: 23–24 with Fig. 3.3). Ephorus is our first extant source that systemised the concept of four solstitial horizon points into a coherent structure (Heidel 1937: 16–20, 30, 45–46; Dilke 1985: 27; Ballabriga 1986: 147–49; Irby 2012: 96; Roller 2015: 82).

[14] Heidel 1937: 59.

[15] Heath 2004[1913]: 42–43; Dreyer 1953: 17 n. 1; Guthrie 1962: i.135; Bicknell 1969: 54, 64; Kirk *et al.* 1983: 155; Mansfeld & Runia 2009: 555–56. Cherniss 1935: 135 n. 544 opts for the sun and moon only.

[16] 13A15 D-K *ap.* Aët. 2.23.1 (= Stob. *Ecl.* 1.25.1d). In ch. 23 of the second book of Aëtius' *Placita* as reconstructed by Mansfeld & Runia 2009: 552–62; 2020: 1012–13 the word *tropai* appears only in the chapter heading; this reconstruction respects the text of [Plut.] *Placit.* 2.23, while Stobaeus, besides in his chapter heading (1.25), has the word in the *doxai* of Anaxagoras (1.25.3a = 59A72 D-K), Empedocles (1.25.3e = 31A58 D-K), Democritus (1.25.3h = 68A89 D-K) and Cleanthes (1.25.3i = *SVF* i.501, ii.658), besides in that of Anaximenes. But Stobaeus' addition of the word in several of the *doxai* is most probably the result of his compilatory procedure, although it is possible that in the case of the *doxa* of Anaximenes the phrase was already in Aëtius' text (Mansfeld & Runia 2009: 554 with n. 403). With respect to the Presocratics, the phrase is otherwise attested for Anaxagoras, Democritus and Diogenes (59A42, 68A99 (with Agathar. *FGrHist* 86F19 and Anon. Flor. *De Nilo FGrHist* 647F1.4), B14.2–3, 8, 64A9, 17 D-K), but not for Anaximenes, Empedocles and 'Pythagoras' (it is hard not to presume that most of them, if not all, used some variant of the phrase when referring to the

explicit observance of the points on the horizon where the solstices occur is attested for Eudoxus,[17] while the observation of the position of sunrises on the local horizon was recommended to a physician upon arriving to a city he is unfamiliar with by the author of the Hippocratic treatise *Airs, Waters, Places*.[18]

The most easily recognizable manifestations of the solstices are indeed the extreme points of solar risings and settings on the horizon. Additionally, the spatial aspect of the solstices can be perceived as a projection of the maximum solar declination onto the earth's surface, in terms of spherical earth corresponding to the latitude of the fixed arctic circle, but also applicable, if less precisely and more arbitrarily, on the flat earth, in which case it corresponds to a latitude determined by the solstice azimuths as observed from Greece.[19] As mentioned in chapter 2, this cosmological location was known to the Greeks from Pytheas' time onwards as Thule, where the summer tropic and the always-visible circle become one. Indeed, it could be claimed that the sun turns at (the latitude of) that island.[20] But already Hesiod claimed that the sun during winter goes to the territory of the black men (using the phrase ἐπί... στρωφᾶται, 'turns on'),[21] while Herodotus believed that it is driven to farther Libya,[22] and in the Hippocratic *Airs, Waters, Places* the sun is described as coming nearest to Scythia when it reaches its summer turning-point.[23] This is how the sun and its turnings can 'be' at some geographical position or latitude. The importance of this notion, already current in the Archaic period, will become clear later in the discussion of the location of Syriê, which is decidedly associated precisely with the spatial aspect of solstices, although most probably not with—or not only with—solstitial azimuths.

solstices). All *doxai* in this chapter refer to the spatial aspect of the sun's annual motion.

[17] Frs. 63b (cf. 63a), 124 *ap.* Simpl. *Cael.* 493.15–17, 125 Lasserre *ap.* Alex. *Meteo.* p. 703.32–34 (cf. Dekker 2013: 9). Cf. Porph. *De antr. Nymph.* 23.14–16, cf. 28.1–3.

[18] Hippocr. *Aër.* 1.

[19] Although it is impossible to derive a precise latitude from the azimuths, it seems safe to presume that the later 'standard' latitude for the limit of the *oikoumenê* (54°) is closely connected to this concept (cf. Bilić 2012a: 304–306, 312–13). Comparably, for Strabo (1.1.6) the arctic circle (presumably the always visible circle for the standard latitude of Greece, i.e., Rhodes, at 36°) touches the earth at the northernmost limit of the *oikoumenê*. Naturally, this only applies—as a rule of thumb—to the latitude of Greece. Incidentally, Ptolemy took 30° as the angular distance of solstitial rising and setting points from due east/west, which is the approximate value of solstitial angular distance for the latitude of 36° (Berggren & Jones 2000: 15; Kominko 2013a: 39); this means that the summer solstice azimuths correspond to the latitude of 66°, i.e., approximately the latitude of the geographical arctic circle. However, the same qualification also applies to this derivation of latitude from azimuths.

[20] See below in this chapter.

[21] *Op.* 527–28; Ballabriga 1986: 20; 1998: 108. Cf. Dionys. Per. 586, *GGM* ii.141.

[22] Hdt. 2.24.1, 25.1, 26.2.

[23] *Aër.* 19. Cf. *De victu* 2.38 (a generation later than *Airs*, Jouanna 1999: 409).

5.2. *Heliotropia* and the localisations of Homeric *tropai êelioio*

A variant of the phrase *tropai êelioio* was also applied to several analog devices that were used to mark or measure the precise time of the solstices. These devices were called *heliotropia*, and we know of their existence at Syracuse outside the islet of Ortygia,[24] on Syros (Pherecydes' instrument, perhaps different from the 'cave of the sun' on the island)[25] and on the Pnyx Hill at Athens (Meton's instrument),[26] with an actual preserved 4th-c. device from Itanos on Crete.[27] The name of these instruments certainly refers specifically to the annual turnings of the sun, more precisely to their manifestations on the spatial level.

Two of these *heliotropia* have certain 'Homeric' resonances. Thus, Homeric Syriê was regularly identified with the Cycladic Syros,[28] but it will be shown later that the Ortygia/Syriê complex can be interpreted as a mythic or cosmological location of 'the island of the solstice', which makes these identifications examples of secondary rationalisations. The identification was surely motivated by the similarity or homonymy of names, supported by the strong tradition that identified Ortygia with Delos (see chapter 4), but perhaps also by local cultic tradition. The cult of Sirius is indeed attested on Ceos, a Cycladic island to the west-northwest from Syros, in the historical period (the 5th c. at the latest). Its aetiology involves ritual offerings to the Dog Star made by Apollo's son Aristaeus.[29] Some sources add that heliacal rising of Sirius was observed on Ceos and that Aristaeus was the first to observe the heliacal rising of Sirius at the solstice,

[24] Mosch. *FGrHist* 575F1; Plut. *Dion* 29.2–3, 5; Kirk *et al.* 1983: 54–55; Schibli 1990: 5; Wenskus 1990: 39–40.

[25] 7A1 D-K = fr. 15 Schibli; Σ D (Ernst 2006: 312), QV *Od.* 15.404; Eust. *Od.* 15.404, ii.105; Bowen & Goldstein 1988: 73 with n. 169; Boutsikas 2020: 3.

[26] Aristoph. *Daitaleis* in Achill. *Isag.* 28; Philoch. *BNJ* 328F122; Ael. *VH* 10.7; Bowen & Goldstein 1988: 73–74, 76–78; Lehoux 2007: 96; Hannah 2009: 5–9, 71, 569; Jones 2017: 77–78, 80; Kremer 2020: 202. Meton's teacher Phaeinus used Mount Lycabettus—aligned with the summer solstice sunrise as observed from the Pnyx—to establish an alignment with the solstice sunrise point (Theoph. *Sign.* 4 = 6A1 D-K; Rehm 1941: 135–39; Pannekoek 1961: 107; Huxley 1963b: 98–99; Bowen & Goldstein 1988: 80; Jones 2017: 79).

[27] *ICret* III.4 11.4–6, 13–14 = *SIG*³ 1264; Halbherr 1890: 585–86; Lorimer 1950: 81; Kirk *et al.* 1983:54–55; Wenskus 1990: 39; Isager & Skydsgaard 1992: 163; Boutsikas 2020: 3.

[28] E.g., Pherec. 7A3 D-K = fr. 3 Schibli; Andr. *FGrHist* 1005F4 (Schibli 1990: 5 n. 11); Wilamowitz-Moellendorff 1926: 125–26, further cited in Lorimer 1950: 80–81 and Wenskus 1990: 39; Ballabriga 1986: 17–19.

[29] Arist. fr. 611.27 Rose = Titel 143,1.9.27 Gigon (from Xenomedes, Jacoby 3B.288; Huxley 1965: 236; K. Meister in *BNP* s.v.; Fowler 2013: 512); Theoph. *De Ventis* 5.14; Callim. *Aet.* fr. 75.32–7 Pfeiffer (surely from Xenomedes, thus *EGM* fr. 1 Fowler, Fowler 2001: 372, although this part of Callimachus' poem is not included in *BNJ* 442F1; see Huxley 1965: 237–38, Hollis 1991 and Fowler 2013: 511–12 on fr. 75.58–59); A. R. 2.516–27 (later than Callimachus' account, but using additional sources, Cameron 1995: 256–57) with Σ A. R. 2.498–527qtvw, pp. 171–72 Wendel; etc. Xenomedes most probably wrote in the second half of the 5th c. (Huxley 1965: 235, 244; Fowler 2001:370; cf. Jacoby 3B.288, K. Meister in *BNP* s.v. and Fowler 2013: 733; Cameron 1995: 258 n. 103 argues for the early 4th c., cf. Jacoby 3B.288). For Apollo Aristaeus on Ceos see *CIG* 2364; Athenag. *Pro Christ.* 14. For Apollo = Aristaeus identification see Hes. fr. 216 M-W; Pi. *P.* 9.64–66; Σ A. R. 2.498–527b, p. 169 Wendel; cf. Σ Pi. *P.* 9.107.

but these reports seem secondary and derivative.[30] In addition, although Eratosthenes attached the Icarius/Erigone story to the Ceoan cult of Sirius and introduced strong Egyptian echoes into it,[31] it is also evident that the role of Sirius is not singularly of Egyptian origin,[32] and the association of Etesian winds with the heliacal rising of Sirius and summer solstice could have been observed by the Aegean population independently of Egypt. Thus, a cult associated with summer solstice is indeed attested in the Cyclades in a period not much removed from Pherecydes', which could have supported the Syriê/Syros identification. Early solar worship is attested on Syros by an EC II (c. 2500 BC) Cycladic diadem depicting two sun discs, bird-headed adorants not unlike Mycenaean ψ-idols, and two—probably—dogs.[33]

Another 'Homeric' resonance is reflected in the attempts to identify Homeric Ortygia with the Sicilian,[34] together with Homeric Syriê with Syracuse.[35] The early testimonies probably belong to a general tendency to identify Odysseus' wanderings with the locations in Magna Graecia, as already noted with respect to the Laestrygonians.[36] But since Syracuse was founded in c. 733 BC,[37] it cannot be much, if at all, later than the *Odyssey*. Thus chronology cannot solve the problem whether Homer derived his description from the existing geographical reality or the locations from the *Odyssey* influenced the naming of these toponyms.[38] In any case, the connection with Dionysius' *heliotropion*, sometimes adduced as an argument for the connection of Syracuse with Syriê, cannot be taken as relevant, since nothing actually suggests a link between the instrument, most probably a sundial or a solstice-marking stele such as the contemporaneous example at Itanos, and the Sicilian localisation of Homeric toponyms.

[30] Heracl. fr. 141 Wehrli = fr. 126 Schütrumpf; Σ A. R. 2.498–527w, p. 172 Wendel; Pomp. Trog. in Just. 13.7.10. Cf. Boutsikas 2020: 4, dating the rite to the 4th c. at the latest.

[31] Maass 1883: 102; Pfeiffer 1922: 112; Solmsen 1947: 256 n. 14, 259 n. 30, 269; Merkelbach & West 1964: 186–88.

[32] Merkelbach & West 1964: 185.

[33] Goodison 1989: 16, 52, 76, fig 27a, b; Kaul 1998: 1.286; Green 1991: 117; West 2007: 204–205 (all offering solar interpretations of the depiction). Any connection of the dogs with Sirius and, thus, with summer solstice is untenable.

[34] For the Sicilian Ortygia see Hes. fr. 150.26 M-W; Pi. *N.* 1.1–4, *O.* 6.92, *P.* 2.6; Tim. *FGrHist* 566F164; Eratosth. IB3, fr. 6 Roller; Hermesianax fr. 7.72 Powell; Str. 6.2.4; Verg. *Aen.* 3.692–96; Ov. *Met.* 5.499, *Fas.* 4.471; Sil. Ital. 14.2, 515, etc.

[35] Perhaps already by Hes. fr. 150.25–27 M-W (Ballabriga 1986: 23; 1998: 99); Delphic oracle Q27 Fontenrose in Paus. 5.7.3 (Ballabriga 1986: 23–24; 1998: 103–106); Eratosth. IB3, fr. 6 Roller (Lorimer 1950: 81); Merry *et al.* 1886: 407; Wackernagel 1916: 248–49; Ballabriga 1998: 99, 100, 102–103, 106, 118; West 2014: 84.

[36] E. g., Romm 1992: 184–85, Podossinov 2013: 222 and Malkin 1998, *passim*.

[37] Dunbabin 1948: 13–17, 48–52, 435–38, 442–50; Graham 1982: 105–106, 162; 2001[1988]: 162; Malkin 1998: 78, 86.

[38] See especially Malkin 1998, *passim* (esp. pp. 160, 190–91) for the primacy of the *Odyssey*. For West, who argues for an extremely late date for the *Odyssey*, chronology is no obstacle, and he freely derives the toponyms in the *Odyssey* from the Sicilian ones (West 2014: 84).

5.3. The localisations of Homeric *tropai êelioio* in the context of solar movement

This western localisation of Homeric Syriê and its *tropai* was sometimes associated with the diurnal solar movement, i.e., with the sun's setting-point.[39] Alternatively, it was associated with the sun's rising-point,[40] or with either of these points in its diurnal movement,[41] or with its meridian passing.[42] However, it must be emphasized that the interpretation of *tropai* as diurnal turnings of the sun is most probably incorrect. It was claimed that 'this sense of *tropai* is absolutely unparalleled and highly improbable'.[43] My discussion at the beginning of this chapter on the earliest attestations of the phrase, including that in the *Odyssey*, agrees with and supports the first part of this assertion. Concerning the second, the incredibility shows itself most clearly in the fact that, according to this interpretation, the sun would have to move eastward following the turn. Here, if anywhere, one should cautiously apply the principle of charity, rather than to ascribe the paradoxical to an inherent incoherence of a system being discussed.[44] The application of this principle does not mean that every instance of incoherency or irrationality should be effaced from the analysed text, but rather that one should approach its interpretation with a presupposition it is coherent and rational.[45] An interpretation that arrives at the conclusion that Homer believed in an eastward movement of the sun across the sky does not seem very convincing in this light. That does not mean it is completely out of the question, but that a plausible interpretation preserving the text's coherence, if one is available (and in this particular case one is indeed available), should be preferred. Additionally, the phrase suggests that the sun would be visible after the turn, which is only applicable if the 'turning' is an annual one.

Indeed, other scholars understood in the 'turnings of the sun' precisely a reference to the annual solar motion, associating Syriê/Ortygia with the annual solar turnings, i.e., with the sunrises or sunsets on either of solstices.[46]

[39] Hesiod, according to Wyatt 1966: 628 n. 18; Σ BHQ *Od.* 15.404; cf. Eustath. *Od.* 15.404, ii.105; Hesych. o 1339 (Martin (1879) cited in Heath 2004[1913]: 10 n. 1); Heath 2004[1913]: 9–10; Dörpfeld 1925: 241; Dicks 1970: 31; Wenskus 1990: 39; *LSJ* 1996: 1826 s.v. τροπή Ia; Kominko 2013a: 41.

[40] Lorimer 1950: 80, 82–84 (further cited by Huxley 1960: 17–18, Kirk *et al.* 1983: 55 n. 2, Ballabriga 1986: 19 n. 39, Brown 1985: 293–94).

[41] Dicks 1970: 32–33.

[42] Ballabriga 1986: 20–21 (for criticism see Wenskus 1990: 39 n. 107; Ballabriga himself later rejected his earlier interpretation (1998: 108 n. 2)); Brown 1985: 294.

[43] Kirk *et al.* 1983: 55.

[44] Lloyd 1990: 18.

[45] For a cautiously charitable approach to the interpretation of Greek myth see Buxton 2009: 58–61, 250–51.

[46] Völcker 1830: 24; Martin 1892: 477a; Berger 1904: 8–9; Wackernagel 1916: 247–48; Nilsson 1920: 316; Heidel 1937: 59; Hoekstra 1957: 218; 1989: 257; Huxley 1960: 18–23 (esp. 20); West 1978: 291–92; Kirk *et al.* 1983: 55–56 n. 3; Ballabriga 1998: 108–109; Hunger & Pingree 1999: 75–77, who further associate the Homeric *tropai* with the account of the solstices in ^{mul}Apin II.i.9–24, which are here defined by the position of the sun at the eastern horizon and the change in the direction of the position of successive sunrises (II.i.11–12, 17–18, ii.3; Hunger & Pingree 1989: 72–73, 75, 92); Anghelina 2010: 69 n. 10.

It was also suggested that Delos (here apparently identified with Ortygia) lies in the direction of the winter solstice sunrise when observed from Syros, which explains the Homeric description.[47] When observed from the northernmost part of Syros the sun on the winter solstice indeed rose almost aligned with the southernmost tip of the neighboring island of Rhenea (azimuth 120°13' in 700 BC). Furthermore, Rhenea was occasionally identified with Ortygia,[48] which supports this interpretation, but it demands the turning-places of the sun to be located at Ortygia, defining the direction from a point of observation in the west (i.e., Syros), although Syriê is located 'beyond' Ortygia only if looking from the east.[49] In this case Homer's description would be formulated somewhat awkwardly: 'Syriê is to the west of Ortygia; the latter lies in the direction of winter sunrise when observed from the former'.

Typically, all these interpretations take into consideration the spatial level of manifestation of the solstices defined by the extreme azimuths of solar risings and settings. However, there is, as already indicated, another way of looking at this problem, and I will argue that this line of interpretation originated with Pytheas.[50]

5.4. Pytheas' Thule and the turnings of the sun

As we have seen, Pytheas located Thule a six-days' sail to the north of Britain, which is precisely the distance from the island of Aeolus to the land of Laestrygonians. At the same time, Pytheas introduced another notion concerning Thule into the corpus of Greek scientific knowledge: there the summer tropic and arctic circle become one. The description of a 24-hour day on the summer solstice in Geminus[51] immediately follows Pytheas's account of a 21- or 22-hour day,[52] and is probably derived from his writings, although Geminus does not explicitly mention Thule.

We have already seen in chapter 2 how Crates associated the location of the land of Laestrygonians with a latitude with the longest day of 23 hours.[53] We have also seen that he perhaps compared the 23-hour day in the north with the situation on Thule,[54] while elsewhere he associated the 20-hour longest day with the island.[55] Finally, he seems to have also associated the region under the head of Draco, with the shortest night of less than a third of an hour, with the Laestrygonians.[56] These observational or computational facts introduced by Pytheas and ultimately used by Crates in his Homeric criticism became something of a commonplace in geographical descriptions of the north,[57] occasionally blurred by erroneous interpretations, such as that which finds a six-month day on Thule,[58] or even a continuous night.[59]

It can thus be concluded that Thule was located—from the time of Pytheas onwards—with reference to the concept of *tropai êelioio*, more precisely, on the very sun's turning-place, thus earning the name of the 'island of the solstice'.[60] Pytheas' report suggests that in his account he combined an actual voyage to whatever place in the 'far north' he had visited with a specific Homeric reference. An island of the solstice situated a six days' sail to the north of Britain conforms well with Homer's account in the *Odyssey*: there we find Aeolus' floating island, situated a six days' sail from Telepylus,[61] where the paths of night and day are close together, thus of a similar character to the island of Syriê, where the *tropai êelioio* are located.[62] Of course, Homer does not in any direct way associate Syriê with the land of Laestrygonians, but the two pieces of mythic geographical lore associated with these two locations seem to belong to a similar tradition concerned with the meteorological conditions obtaining in the far north, more precisely, at what would later be understood as the fixed arctic circle defined by the annual solar movement, but what Homer would only recognize as the sun's northernmost reach.[63] This tradition could belong to a pre-Homeric Argonautic context, given that Laestrygonia and Syriê (actually, the nearby Ortygia) were juxtaposed in Hes. fr. 150.26 M-W (admittedly not in an Argonautic context), suggesting that they indeed belonged to a similar setting. In addition, one could recognise in Homer's account of Syriê certain paradisiacal and eschatological references (as noted in the preceding chapter), which

[47] D. Pingree in Wenskus 1990: 39 n. 107.

[48] Str. 10.5.5, probably also *hAp.* 15–16 (cf. *Hymn. Orph.* 35.4–5); Ballabriga 1986: 16–17.

[49] Lorimer 1950: 81 n. 1; Ballabriga 1986: 17–18.

[50] The following is an abbreviated and slightly modified version of Bilić 2012a: 319–21.

[51] *Elem. Astron.* 6.13, Manitius p. 72.21–24; cf. 5.21, p. 50.3–7, 32, p. 54.7–11, 38, p. 56.7–13.

[52] Fr. 13a Bianchetti = fr. 8 Roseman = fr. 9a Mette *ap.* Gemin. *Elem. Astron.* 6.9, pp. 70.21–72.2, cf. 10–12, p. 72.2–20.

[53] Crates fr. 37d Mette.

[54] Crates fr. 37b Mette = Prisc. Lyd. *Sol. Chosr.* 4, 67.9–15 Bywater. However, this is highly unlikely, see Mette 1936: 85 with n. 1.

[55] Crates fr. 37h Mette (without Mette's emendation, Mette 1936: 87 with n. 1, 271); cf. Σ Dionys. Per. 582, *GGM* ii.451; Ptol. *Alm.* 2.6, i.1.114.9–11 Heiberg, *Geog.* 7.5.16, 8.3.3 Nobbe; Steph. Byz. *s.v.* Θούλη; Eust. Dionys. Per. 581, *GGM* ii.329.30–33.

[56] Crates fr. 37i Mette, with the emendation *unaquaque—brevissima* (Mette 1936: 87, 272).

[57] Pyth. fr. 8f Bianchetti = T 23 Roseman = fr. 11b Mette (with Solin. 22.9; Serv. Dan. *G.* 1.30; Σ Juv. *Sat.* 15.112; Mart. Cap. 6.666), fr. 12a Bianchetti = T 27 Roseman – fr. 14 Mette, Pomp. Mel. 3.6.57; Stat. *Silv.* 5.2.54–55; Dionys. Per. 580–86, Lightfoot 2014: 228, cf. 394–95; Ach. Tat. *Isag.* 35, p. 71 Maass; Avien. *Descr. orb. terr.* 760–63, *GGM* ii.184; Prisc. *Perieg.* 588–91, *GGM* ii.195; Procop. *Bell.* 6.15.6–7 (with Jord. *Get.* 3.19–20); Eustath. Dionys. Per. 581, *GGM* ii.329.18–22.

[58] Pyth. fr. 8f Bianchetti = T 23 Roseman = fr. 11b Mette, fr. 9a Bianchetti = T 18a Roseman = fr. 13a Mette, fr. 9b Bianchetti = T 18b Roseman, fr. 12b Bianchetti = T 29 Roseman; Crates fr. 37c Mette; *HN* 6.219.

[59] Isid. *Etym.* 14.6.4. For some later resonances of Isidore's account and full references to the solstice on Thule in medieval authors see Bilić 2012a: 315–17.

[60] Cf. Isid. *Etym.* 14.6.4. Thule was rather explicitly associated with the concept of the sun's turning in the poem of Dionysius Periegetes (Dionys. Perieg. 580–86, Lightfoot 2014: 228, cf. 394–95, cf. Avien. *Descr. orb. terr.* 764–67, *GGM* ii.184; Eust. Dionys. Perieg. 581, *GGM* ii.329.33–37).

[61] *Od.* 10.80–81.

[62] *Od.* 15.404.

[63] I believe it is clear I am not attempting to fabricate here some Homeric 'doctrine' from these two more or less incidental remarks (cf. Skinner 1969: 7, 12), but simply suggest that they both probably refer to the same phenomenon.

were shown to be characteristic of some accounts of the diurnal and annual solar movement, including a reference present in the description of Laestrygonia, once more connecting these two at first glance unconnected passages. Hence Syriê should be located with respect to the setting it belongs to, i.e., the turnings of the sun in the *Odyssey* are to be interpreted as the place where the sun 'turns' in the north when it reaches the northernmost point in its annual voyage, most probably—expressed in terms belonging to a frame of reference that would only be created much later—at the latitude of 54° or 66°. The sun is thus believed to be 'visiting' a certain location at its turning, which is similar to the notions expressed by Hesiod, Herodotus and the Hippocratic author adduced in the early part of this chapter. This concept is less specific than the observation of extreme solar azimuths, although it incorporates the notion of solstitial horizon-points as one of its defining characteristics (we have seen in chapter 3 that the model involving the Laestrygonians might have actually incorporated the solstitial azimuths in its elaboration). The 'location' of Syriê is thus specific geographical latitude defined by the annual solar movement. Hence it seems conceivable that Pytheas was not only a great explorer but perhaps also a fine Homeric scholar.

It is now widely believed that Pytheas did not actually observe the height of the sun or the length of daylight on the winter solstice at various latitudes but rather calculated it with the help of his geometrical knowledge.[64] Indeed, he might have known beforehand, by learning the geometry of the sphere, of a latitude with a 24-hour summer solstice day when the sun did not disappear under the horizon, and he placed Thule precisely at this location.[65] Still, he might have determined the latitudes with observing the elevation of the pole,[66] which was Hipparchus' practice,[67] perhaps already Eratosthenes'.[68] It is even possible that he had actually undertaken his voyage in order to verify theoretical models of celestial phenomena, whether Eudoxus' or someone else's, especially the behaviour of the sun in these high latitudes, or to obtain the latitudes for the far north.[69] His motivation was certainly in part scientific curiosity.[70] Pytheas' scientific achievements are not negligible,[71] and it was actually claimed that he was the first to recognize the difference between the always-visible and the fixed arctic circle.[72]

Thus, it is almost certain that Pytheas' voyage was partly motivated by the wish for gathering specific scientific data. On the other hand, his data on the altitude of the solstice sun came rather through calculation and he knew of a 'Thule' before the voyage itself even commenced. This indeed makes the voyage, in terms of a theory-free data gathering, superfluous, especially if we take into account the Homeric parallels adduced above. Pytheas' voyage might be understood as an empirical confirmation of certain geometrical postulates, a deductive rather than inductive procedure; he empirically tested the theories concerning the phenomena in high northern latitudes that had already been theoretically predicted by the astronomers, combining his actual experience with a specific Homeric reference(s). This is actually very similar to a situation described in Antonius Diogenes *Wonders beyond Thule*. Here one of the characters, Deinias, claimed to have actually witnessed 'beyond Thule' some phenomena the astronomers only speculate about (for instance, the various length of night, from less than a month to even a whole year).[73] Thus Deinias 'empirically' tested the theories concerning the phenomena in high northern latitudes that had already been theoretically predicted by the astronomers, which echoes what Pytheas probably also did in real life.

5.5. Conclusion

Pytheas' solstice island, described in scientific terms and with reference to a spherical earth—where the summer tropic and the always-visible circle become one, i.e., their respective declinations are identical—is thus comparable with Homer's mythic island 'where are the turning-places of the sun'. Both accounts are part of a single ethnographic context, and both refer to the same phenomenon, although in different terms, and with a different underlying cosmological outlook, illustrating the transformation of Greek understanding of the solstices through different periods. At the same time, the fact that Pytheas in part turned to Homer for inspiration when discussing meteorological

[64] Carpenter 1973: 176, 191–92; Whitaker 1981–82: 160; Aujac 1987b: 151; 1993: 245, 296.

[65] Bunbury 1883: i.614; Dilke 1985: 30; Aujac 1987b: 151; 1993: 253–54, 260, 277; Bianchetti 1998: 191.

[66] Nansen 1911: i.46–48; Berggren & Jones 2000: 9 n. 7 (although unsupported by Dicks 1960: 180, 185–87, which they quote); cf. Neugebauer 1975: ii.938, who argues against the use of observations of length of daylight as a method for determining latitude but proposes meridian observations instead.

[67] Hipparch. 1.3.6–7, 12, 11.8 (Shcheglov 2003–2007: 161, 165); Ptol. *Geog.* 1.4 (Berggren & Jones 2000: 62–63), cf. Shcheglov 2003–2007: 165

[68] Galen *Inst. Log.* 12.2 (Geus 2004: 14–15; 2011[2002]: 224). For the elevation of the pole designating latitude see Marinus *FGrHist* 2114F1 *ap.* Ptol. *Geog.* 1.7.4 (Berggren & Jones 2000: 65) and Ptol. *Geog.* 8.2.1 (Berggren & Jones 2000: 120). For the correlation of the height of the north celestial pole with relative geographic position see Gemin. 5.58–61; Str. 1.1.21, 10.2.12; *HN* 2.179; Cleom. 1.3–5 (Bowen & Todd 2004: 44–45, 47, 59–60, 67); Lucan *Phars.* 8.172–84 (Davis 2009: 143, 145–46) (all standard English translations add the 'pole star', the *axis* of verse 175: Riley 1853: 301; Ridley 1896: 233; Duff 1928: 449; Graves 1956: 154; Widdows 1988: 190; Joyce 1993: 203; Braund 1992: 157; cf. Arnaud 2014: 49–50).

[69] Heidel 1937: 108; Thomson 1948: 153; Dilke 1985: 29–30; Aujac 1987b: 151; 1993: 11, 82, 245, 254, 258, 277, 300; Roseman 1994: 145; Bianchetti 1998: 45, 65, 155, 181, 188; Roller 2006: 63 n. 52; Janni 2015: 33; cf. Diller 1975: 225; Roller 2006: 74. For Pytheas as a pupil of Eudoxus see Müllenhoff 1870: i.234–35; Berger 1903: 336; Heidel 1937: 99 n. 212, 108; Bianchetti 1998: 37–38.

[70] Müllenhoff 1870: i.312; Tozer 1897: 154; Holmes 1907: 220; Nansen 1911: i.45; Walbank 1948: 173 (= 2002: 45); Thomson 1948: 143; Van Paassen 1957: 332; Dicks 1960: 187; Hawkes 1977: 7, 26, 29, 31, 37, 39, 45; Whitaker 1981–82: 162, 163 n. 90; Aujac 1993: 11, 244; Nicolet 1991: 60; Bianchetti 1998: 65; Cunliffe 2002: 154; Roller 2006: 63, 74; 2015: 87–89; McPhail 2014: 251; Janni 2015: 33–34; Roller 2018: 316.

[71] See, for example, Van Paassen 1957: 332; Tierney 1959–60: 195–96; 1967: 32, 116; Hawkes 1977: 44–45; Aujac 1993: 81, 245, 257, 259, 296–97, 300–301; Roseman 1994: 148, 155; Bianchetti 1998: 45–46; Käppel 2001: 11; McPhail 2014: 247, 251.

[72] Bianchetti 1998: 43, 153.

[73] Phot. *Bibl.* 166.110b39–111a2; cf. Romm 1992: 209.

conditions obtaining in the far north and that he indulged in an ingenious piece of Homeric scholarship in the form of an interpretation of the Laestrygonian episode as referring to the latitude of the fixed arctic circle shows that the concept of ethnographic context is not only a profitable modern hermeneutic tool created in order to facilitate understanding of ancient societies but also a concept corresponding to the way in which the Greeks themselves understood the relation of what we classify as mythic and scientific material. In this way, a Homeric solar myth (i.e., a story incorporating a reference to the annual solar movement) was incorporated into a study of the annual solar movement and related phenomena by the scientific tradition as a respectable source of information, without any prejudice towards its narrative character and the fact that it used personalistic causality as its preferable method of approach to the phenomenon of solar movement.

Although the focus of this chapter was indeed on later tradition that treated the solstices and related phenomena in a non-personalistic frame of reference, both with respect to a flat and spherical earth, the resonances of mythic tradition were still strongly present in this tradition, while Homeric Syriê certainly belongs to a class of myths referencing the annual solar movement. Since this reference is both clear and uncontroversial, I did not argue at any length for its inclusion into my corpus, but rather concentrated on its interpretation both in Greek tradition and in modern scholarly literature. This analysis has facilitated an accurate understanding of the proper reference of the Homeric description of Laestrygonia and its relation to Pytheas' account of the north, at the same time revealing certain aspects of the Greek ethnographic context from the perspective of the ancient Greeks themselves.

6

Pytheas and Hecataeus: Britain and Hyperborea

διὰ δὲ τὴν ἄγνοιαν τῶν τόπων τούτων οἱ τὰ Ῥιπαῖα ὄρη καὶ τοὺς Ὑπερβορείους μυθοποιοῦντες λόγου ἠξίωνται, καὶ ἃ Πυθέας ὁ Μασσαλιώτης κατεψεύσατο ταῦτα τῆς παρωκεανίτιδος, προσχήματι χρώμενος τῇ περὶ τὰ οὐράνια καὶ τὰ μαθηματικὰ ἱστορίᾳ.

It is because of men's ignorance of these regions that any heed has been given to those who created the mythical Rhipaean Mountains and Hyperboreans, and also to all those false statements made by Pytheas the Massalian (fr. 8g Bianchetti = T 16 Roseman = frt. 6h Mette) regarding the country along the ocean, wherein he uses as a screen his scientific knowledge of astronomy and mathematics.

Strabo, *Geography*, 7.3.1, translation by Jones in *LCL* 50 (1924)

Pytheas appropriation of Homeric poetry in his account of the meteorological conditions obtaining in the far north discussed in chs. 1 and 4 will in this chapter be further contextualised by a comparison with Hecataeus of Abdera's treatment of similar subjects.[1] This comparison will aid us in understanding the various ways in which traditional material was exploited by writers with completely different authorial goals. At the same time, it will bring into sharper focus the different treatment of these two authors by modern scholarship. A seafarer, astronomer and geographer on the one side and a philosopher and historian on the other,[2] Pytheas and Hecataeus of Abdera were two authors only a generation apart. They both exhibited a considerable interest in the far north; but while the older one is firmly assimilated into scientific tradition and hailed as a daring explorer and an accomplished scientist (both empirical and theoretical, see the preceding chapter), the other is often labelled an utopian writer and an untrustworthy author. However, it seems that the reality is not so unambiguous and that both occupy a place in the continuum of Greek geographical/meteorological/astronomical knowledge, even though their precise positions on this continuum are perhaps not adjacent to each other. What brings them closer together is, on the one side, the fact that Pytheas' approach to meteorological conditions obtaining in high northern latitudes was not, as we have seen, as straightforwardly scientific as often maintained; on the other, Hecataeus apparently manifested a lot of attention to the information derived precisely from Pythean tradition on the same subject, even if he was more interested in concocting a story and not in offering an unbiased account of these phenomena. It seems worthwhile to analyse their respective treatments of the far north and in this way to position them on the continuum of geographical/meteorological/astronomical knowledge in their proper places. Since Pytheas' treatment of the far north was already thoroughly discussed in chs. 1 and 4, while Hecataeus' was only touched upon intermittently, much of this chapter will focus on the latter's account of Hyperborean regions.

6.1. Pytheas and the northern barbarians

Besides the already studied examples of the mixture of scientific and more traditional approaches to the meteorological conditions obtaining in the far north, another—from a modern perspective—concoction of this type appears in Pytheas' report, as recorded by at least one trustworthy witness; this part of his account specifically dealt with Britain. This interpretation is in line with the previously recognised presence of Homeric references in Pytheas' account of both the position and meteorological conditions obtaining at Thule. Quite apart from his historically attested visit to or circumnavigation of Britain, Pytheas' voyage necessarily introduces it into my discussion, since in his wake it was regularly associated with meteorological conditions characteristic for Thule itself. But it is probable that already Pytheas' story how 'the barbarians' showed him 'where the sun goes to sleep' refers precisely to Britain.[3] He describes how around those places the night can be as short as two or three hours so that only a short interval elapses from sunset to sunrise,[4] which cannot refer to Britain as such, but the information probably came from the inhabitants of the island or at least Pytheas credited them as his source. Elsewhere Pytheas describes how in the farthest north he was shown 'the sun's bed' by the barbarians, since there 'the sun always spent its nights' with them.[5] Thus Pytheas once again writes of a nightless day, still most probably referring to Britain. Again, we must presume that this description applies primarily to the days about the summer solstice,[6] while the descriptions of Thule and Britain resemble the conditions at Telepylus. The sun's 'sleep' and its 'bed' are certainly not the terms expected from an author who otherwise confidently used

[1] The following chapter is a slightly modified version of Bilić 2020a.
[2] Their respective 'professional' designations in *BNP* s.v. Pytheas 4 and Hecataeus 4.

[3] Fr. 13a Bianchetti = fr. 8 Roseman = fr. 9a Mette.
[4] Heath 1991[1932]: 132; Bianchetti 1998: 103; Evans & Berggren 2006: 162.
[5] Pyth. fr. 13b Bianchetti = fr. 9 Roseman = fr. 9b Mette; Wolska-Conus 1968: 398; Roseman 1994: 143; Bianchetti 1998: 103. The description only makes sense if it is referring to a model of diurnal movement of the sun involving its nightly route behind a large mountain in the north (for which see chapter 9), which makes little sense in terms of spherical astronomy with which Pytheas was more than familiar.
[6] Bianchetti 1998: 190–92

the postulates of the latest developments in geometry and astronomy but is perhaps another concession on his part to his predecessors who used mythic discourse in accounting for the phenomena he was interested in, or at least did so in his interpretation.[7] It appears that this was Pytheas' usual method of approaching and giving account of these phenomena.

6.2. Britain in the wake of Pytheas

At the same time, the tradition proceeding from Pytheas' report of a more or less tangible region, admittedly not particularly clear in the first place, fast became distorted and imprecise. Most importantly, Britain and Thule were merged into a location in the far north with meteorological conditions characteristic of high northern latitudes from an area well south of the geographical arctic circle to the very north pole.

Thus Caesar mentioned some unnamed islands around Britain, of which it is written that at the winter solstice the night there lasts for 30 consecutive days, while Plutarch described an island named Ogygia, situated five days' sail west of Britain 'near the place[8] of the summer sunset', where the sun sets 'for less than one hour for thirty days in succession'.[9] These reports are certainly in some connection with Procopius' and Jordanes' accounts of 'Thule' (actually Scandinavia) and it's 40-days continuous daytime and night about the solstices,[10] and when Caesar and Plutarch associate these specific meteorological conditions with Britain, in light of both Pytheas' voyage and report, as well as these parallel accounts, a link with Pytheas' Thule immediately comes to mind.[11] All these testimonies could thus represent resonances of Pytheas' work, more or less removed from their ultimate source.

Indeed Britain itself was loosely identified with Pytheas' Thule in the Roman period.[12] More to the point, its long daylight hours and bright nights in summer were often emphasised.[13] This tradition was occasionally presented in precise numerical day-length values: the longest day in Britain was reported as consisting of between 17 and 19½ hours, which corresponds to a latitude from 54° to 61°30'.[14] It was thus couched in mathematical terms and offered as empirically verified—and thus scientific—information.

In addition, there are also several mythic accounts that associate Britain with Thule. Of these the most interesting—or at least the best attested one—is Hecataeus' account of Hyperborea. His city of Cimmerians[15] was probably associated with his Hyperborean Helixoia that most likely represents—fictionalised—Britain.[16] It is not impossible that his *On the Hyperboreans*, although mostly fantasy, might have contained certain resonances of Pytheas' report,[17] and the most likely type of data for his appropriation from Pytheas would be the information pertaining to the sun's behaviour at the solstices and the accompanying phenomena of daylight/nighttime. Similarly, the account in *Orphic Argonautica* places the sun-deprived Cimmerians in the far north-west, on the way from the Northern Ocean towards the British Isles.[18] Indeed, the Cimmerians were sometimes identified with the Celts that attacked Delphi, originally inhabiting a region near the western Ocean, or with the Cimbri living at the north pole by the outer sea.[19] It is clear that Homer's Cimmerians were at some point identified with the Celts of north-western Europe, and at the same time with the furthest north. It is the apparent latitude of their abode, more precisely, the meteorological phenomena characteristic for this latitude, rather than anything else that was decisive in their localisation. This Celtic association once more

[7] Cf. Beaulieu 2016: 7–8.

[8] Thus Romm 1992: 204; Cherniss & Helmbold 1957: 181: 'in the general direction'.

[9] Caes. *BGall.* 5.13.3; Plut. *De fac.* 26.941A, D.

[10] Procop. *Bell.* 6.15.6–7; Jord. *Get.* 3.19–20. Cf. for similar accounts of the north in later tradition Paul. Diac. *Hist. Lang.* 1.4–5; Tzetz. *Chil.* 12.844–48, p. 474 Kiessling; al-Gharnātī, *Murib* 14–15 (Bolshakov & Mongait 1971: 32); Adam of Bremen *Descript. ins. aquil.* (4.)37 (Waitz 1876: 185); Anon. *Konungs skuggsjá* 7 (Larson 1917: 98–99); Ranulph Higden *Polychronicon* 1.31 (i.326 Babington); Anon. *Eulogium Historiarum* 4.92 (ii.79 Haydon); Laskaris Kananos (Lundström 1902: 15; Blomqvist 2002: 46); Olaus Magnus *Hist. gent. sept.* 1.2, p.11.

[11] For Pytheas as Plutarch's source see Romm 1992: 204.

[12] Sil. It. 3.597, 17.416–17 (Wijsman 1998: 319); Stat. *Silv.* 3.5.20 (Wijsman 1998: 318), 4.4.62, 5.1.91 (perhaps also 5.2.54–55); Tac. *Agr.* 10.4 (Wijsman 1998: 319: the Shetlands); Claud. *De Cons. Stil.* 3.52–58 (Wijsman 1998: 323 n. 24), 155–56, 4.32 (Wijsman 1998: 323); cf. Thomson 1948: 235; Wijsman 1998: 318–19, 323. Cf. Rutil. Namat. *De reditu suo* 500–501 and Venant. Fort. *Vita S Mart.* 3.493–94, cf. *Carm.* 8.1.16, 18, who associate Britain with Thyle, but clearly distinguish between them.

[13] *HN* 2.186; Tac. *Agr.* 12 (cf. Jord. *Get.* 2.13): one can hardly distinguish between 'the end and the commencement of light' (*ut finem atque initium lucis exiguo discrimine internoscas*), which should be compared to Arat. *Phaen.* 61–62, *Od.* 10.86 and Pyth. fr. 8c Bianchetti = fr. 6 Roseman = fr. 6c Mette, as well as to *Germ.* 45.1; Iuv. *Sat.* 2.161 with Serv *Aen.* 6.265; Dio Cass. 77.13.3; Cleom. *De motu circ.* 1.4.197–207 Todd; *Pan. Const. Aug.* 6(7).7.1, 3, 9.2–5, pp. 205–208 Bährens (Lieu & Vermes 1996: 80–82), with a reference to the passage in *Agricola*; Serv. *Georg.* 1.247; Bede *Hist. Eccl.* 1.1. The phenomenon was also noticed for more southern regions, Hipparch. frs. 57 (= V 15c Berger), 58 Dicks.

[14] Hipparch. fr. 61 Dicks = Str. 2.1.18 = Pyth. fr. 11 Bianchetti = fr. 6b Mette = T 5 Roseman, more than 19 hours (Dicks 1960: 190; Roseman 1994: 44–45); *HN* 2.186 (cf. Mart. Cap. 6.595); Cleom. *De motu circ.* 1.4.197–207, 2.1.443 Todd; Ptol. *Synt.* 2.6, i.112.8–114.8 Heiberg, *Geog.* 8.3.4–11; sundials from Rome and Crêt-Chatelard (Diels 1920: 189–91 with Abb. 63–64; De Solla Price 1969: 254–55 Table 1; Talbert 2010: 266 Table 16.1).

[15] *BNJ* 264F8 = Apollod. *FGrHist* 244F157a, Eratosth. fr. 8 Roller (IA6 Berger) (cf. Geus 2011[2002]: 268 with n. 42).

[16] *BNJ* 264F7, 11a; Hawkes 1977: 38; Bridgman 2005: 133, 136–37, 139–40; Sulimani 2017: 230, 242 (with other references noted in Winiarczyk 2011: 55 n. 63). This was surely Timaeus' interpretation, *FGrHist* 566F164 *ap.* D. S. 5.21.1, cf. 3 (cf. Geus 2000: 71). On this see more below.

[17] Hawkes 1977: 38–39; Geus 2000: 71, 73; Bridgman 2005: 98–99, 101, 103; Roller 2006: 66 with n. 85; *pace* Winiarczyk 2011: 62–63.

[18] *Orph. Argon.* 1120–27. For a land of the dead in the British Isles see Claud. *In. Ruf.* 1.123–28; Procop. *Bell.* 8.20.42–58, cf. Tzetz. (Σ) Lycoph. 1200, ii.346 Scheer, *ad* Hes. *Op.* 169bis (Bianchetti 2014: 124–25); a mediaeval *mappa mundi* reproduced in Edson 1997: 62–63 with Fig. 4.2.

[19] Σ BH *Od.* 11.14 (ii.479 Dindorf); Plut. *Mar.* 1.9–10. For Cimmerians = Cimbri see Posidon. fr. 169 Theiler = *FGrHist* 87F116, fr. 272 E-K = 44a Theiler = *FGrHist* 87F31. For Cimmerians at terrestrial poles see Crates fr. 54 Brogiatto = fr. 37a Mette = Gemin. *Elem. Astron.* 6.16–20, pp. 74.10–76.15 (cf. Bilić 2012a: 298–99).

suggests a link with Britain, which will be discussed more thoroughly below.

Thus Britain, due to its position in the north and its proximity to Thule was both in Hellenistic geography and myth, perhaps already by Pytheas or certainly in the wake of his report, included into a model of the annual solar movement. More specifically, as a solstice island it was incorporated in a model that emphasised the phenomenon of long daylight hours about the summer solstice and, conversely, long nights about the winter solstice, in high latitudes. Its importance and attractiveness thus primarily lay in its cosmological setting, rather than in its geographical position (although the former is a direct consequence of the latter).

6.3. Hecataeus' Hyperborea

Another interrelated mythic model accounting for the annual solar movement involves, together with Britain as the island of solstitial turning, the fabulous northern people, the Hyperboreans, the constellation of Ursa Major, the north wind Boreas and the Rhipaean mountains, all concepts already documented in previous chapters as the markers of the annual solar movement, i.e., as mythic conceptualisations of a phenomenon—or a set of phenomena—in physical reality. Interestingly, this particular model combines the solstice island of turning with the mountain-barrier blocking the sun's advancement, not unlike the pre-Homeric Argonautic tradition of '*Mountain* of Bears' solstice island, although without any apparent connection between the two. The model is instantiated in, and was most likely concocted by, Hecataeus of Abdera, a utopian writer a generation younger than Pytheas.

I have rather dogmatically stated in the previous paragraph that Hecataeus' Hyperborean Helixoia most likely represents Britain, adducing in support only the opinion of Timaeus (and that of some modern scholars). Now it is time to supply some convincing arguments to support this claim and to develop the notion of a model accounting for the annual solar motion that involves Britain, apparently engineered by Hecataeus, at least in the form in which it reached us.[20] Hecataeus, who, as we have noted, might have included certain data from Pytheas' report in his account of the Hyperboreans, described a large island situated 'towards the Bears', beyond the land of the Celts, inhabited by the Hyperboreans.[21] This terminology is echoed by no less a scholar than Hipparchus in his account of the relative positions of Britain and 'Keltikê',[22] while Britain was indeed placed 'under the Bear' by Timaeus.[23] Hecataeus further described how Leto was born on this island and that Apollo was the most venerated deity there, his spherical temple and sacred precinct, as well as a city and its priesthood.[24] He also mentioned the Rhipaeans, whence the Hyperborean priests of Apollo, sons of Boreas, 'summon' the swans in order for them to join in the singing of the hymns of praise to Apollo.[25] The fact that the moon appears to be near the earth when viewed from the island probably belongs to the complex of ideas of the moon being very near some regions of the earth, especially those in the far north.[26] Hecataeus called this island Helixoia, probably in association with the Ursa Major constellation, *Helikê*.[27] Furthermore, he called the inhabitants of Helixoia Hyperborean Carambycians, after the name of a river Carambyca.[28] This name has to be connected to the Carambis promontory in Paphalgonia, situated opposite 'Helikê the Bear' and exposed to Boreas.[29] It was previously argued (chs. 1 and 2) that both Helikê and Helixoia derive their name from the noun *helix* or the verb *helisso* and that Hecataeus used the name 'the island of turning' in the meaning 'the island of solstice'. This 'turning' is most probably derived from the annual solar movement and the semantic range of the verb *helisso*

[20] The following is an abbreviated and modified version of Bilić 2012b: 519–23.

[21] *BNJ* 264F7 = 73B5 D-K *ap.* D. S. 2.47.1. For an account of islands in the land of the Hyperboreans see further Simias of Rhodes fr. 1 Powell (Bolton 1962: 68; Bridgman 2005: 77).

[22] Hipparch. frs. 57 (= V 15c Berger), 58, 61 (= Str. 2.1.18 = Pyth. fr. 11 Bianchetti = fr. 6b Mette = T 5 Roseman) Dicks (Dicks 1960: 190; Roseman 1994: 44–45). This echo most likely reflects Hipparchus' and Hecataeus' common source, i.e., Pytheas.

[23] *FGrHist* 566F164 *ap.* D. S. 5.21.6. If he meant precisely the arctic circle his source was certainly not Pytheas.

[24] D. S. 2.47.2–3. In another fragment Hecataeus emphasised that Apollo is revered by the Hyperboreans in their lands (*BNJ* 264F10 = 73B4 D-K) and he interpreted Apollo's 19-year intervals between his visits of the island in terms of Metonic cycle (D. S. 2.47.6), which means he must have associated Apollo with the sun.

[25] *BNJ* 264F12 *ap.* Ael. *NA* 11.1. Cunliffe (2002: 121) believes that local prehistoric communities on the Hebrides and Orkneys, whose interest in astronomy and, more specifically, the summer solstice sunset is well documented (Maes Howe), are responsible for the discovery of Thule/Iceland. These observers, in wanting to explore solar and other celestial phenomena in the far north, have followed the flocks of whooper swans in their flight to the north and thus discovered Iceland. Similarly, Hawkes (1977: 34, 38–39) recognized the migration-line of whooper swans from the Hebrides to Iceland and hypothesised of Pytheas' interest in prehistoric solar observations performed in the region, but without joining these two observations. Naturally, not a single claim of these hypotheses can be proven. But a connection of swans with the annual solar movement was indeed assumed by other European prehistoric populations (see chapter 8).

[26] D. S. 2.47.5; Phot. *Bibl.* cod. 166.111a4–11 (J. R. Morgan 1985: 477–78): 'beyond Thule'; Plut. *De fac.* 24.937E.

[27] See chapter 2.

[28] *BNJ* 264F11ab (Steph. Byz. s.v. Ἑλίξοια = 73B1 D-K and Καραμβύκαι). Cf. Hdn. iii.1.281.13–14 Lentz for the river Carambyca associated with Helixoia and *HN* 6.34 for the river Carambucis associated with the Hyperboreans and Ripaean (sic!) mountains.

[29] Ephor. *BNJ* 70F41; A. R. 2.360–62 with Σ 2.360–63ab, p. 157 Wendel; cf. Str. 2.5.22, 12.3.10. The UMa is also connected to the north wind in the [mul]Apin texts (II.i.68, Hunger & Pingree 1989: 87).

includes both solar movement[30] and chariot-riding.[31] Euripides used it in a verbal construction associated with the movement of Helios' chariot, thus employing the model based on chariot-racing in order to account for the diurnal solar movement.[32] Having the standard interchangeability of elements used in building models of diurnal and annual solar movement in mind, the use of chariot-race models to account for both is unsurprising.[33]

At the same time, Hecataeus might have understood the name of the island (he himself invented?) with respect to the movement of UMa/ Helikê, presumably its *revolution* about the north celestial pole, in addition to its derivation from the annual *turning* of the sun. Indeed, the analysis of the origin of the word *arktos*, 'the bear' (in the meaning of *Ursus arctos*, *UMa* and *Mi*), showed that PIE **Hr̥tḱo-s* derives from the root *E_2*ret-* (through *E_2*r̥t-ḱo-*), meaning 'roll, turn' (cf. PIE noun *E_2*rot-o-*, 'wheel'—from which Old Indian *ratha* and Avestan *raθō*, 'chariot', derive—and the verb *E_2*ret-o*, 'to turn (about)'), and that the origin of the constellations' names must be associated with the circling path of the celestial Bears (*E_2r̥t-*ḱo-* = 'roller, (re-)turner').[34] Thus, the very name of the constellation of Ursa Major would be connected with the revolving of circumpolar stars around the north celestial pole. These two derivations—from the movement of the sun and that of UMa—need not be mutually exclusive and could both refer to the northernmost reach of the sun at the summer solstice, one referring to the movement of the sun and its change of direction both on the horizon and on the meridian, the other referring to the approach of the daily orbit of the sun to the orbit of Ursa Major in terms of its connection with the fixed arctic circle (what Pytheas described in terms of identity of the arctic circle and the summer tropic) or simply to the sun's approaching the region in which Ursa Major revolves. In Hecataeus' construction the name Helixoia would thus at a single stroke designate both the solar movement in a helix and the name of the constellation Helikê, two concepts already drawn together in Greek tradition.

Ursa Major was regularly associated with the annual solar turning in the north. We have already discussed (chapter 2) Heraclitus' description of the Bear as 'forming the limits (*termata*) of morning and evening'. In addition, the model of chariot-racing can be recognized here in the word *terma*, which has a double meaning: 'turning post' or 'turning point'—actually the only meaning present in the *Iliad*—and 'boundary' or 'endpoint'.[35] Heraclitus' account seems to suggest that he was describing an arctic circle as defined by the Ursa Major constellation, which thus presented a boundary for the sun's annual movement in the north. Heraclitus at the same time described the southern 'limit of morning and evening', i.e., the southern boundary of the annual solar movement opposite the Bear.[36] Both boundaries are subsumed under the concept of the sun's 'proper limits', which is certainly a designation of the solstices/tropics. The *metra* of diurnal (night and daytime) and annual (the seasons) solar motion are also mentioned by Diogenes of Apollonia,[37] while Plutarch's elaborate discussion of the annual movement of the sun perhaps echoes some of Heraclitus' ideas—as it does his vocabulary—on the annual solar motion, its limits and the influence it exhibits on nature.[38] After Heraclitus, as we have also noted in chapter 2, Ursa Major was recognised as the northern limit of the annual solar voyage and/or involved in the interdependent concept of the extent of the *oikoumenê* by several authors including Aristotle.[39] This tradition was followed by Avienius, who associated the concept of the constellation of Ursa Major as the boundary of the annual solar movement[40] with the continuous 24-hour daylight at Thule on the day of the summer solstice.[41] The latter passage is a translation of Dionysius the Periegetes' verses, where 'the pole of the

[30] Herodorus of Susa (SEG VII, 14.8, Merkelbach & Stauber 2005: 74–75; Potts 2016: 360); Heracl. *Quaest. Hom.* 44.4; Mesomed. fr. 2.14, 25 Heitsch; SGO 06/02/27 (Merkelbach & Stauber 2005: 75); Orph. fr. 236 Kern = *PEGr* fr. 539 F; [Callisth.] β 1.30, 33, pp. 45, 50 Bergson; *PGM* iii.222 (according to Merkelbach & Stauber 2005: 75), iv.439, 1960 Preisendanz (in Preisendanz 1928: i.87, 133 and Betz 1996: 46, 72 the texts is rendered 'around the great pole', but *polos* here surely means 'heaven'). Cf. a *helix* resulting from the dual planetary (in general) or solar (in particular) motion (Pl. *Tim.* 39A; [Eudox.] (Leptines?) *Ars Astron.* P. Pap. 1, coll. 9.1–11, 20.17–21.2 Blass (Tannery 1893: 289, cf. 293); Epicur. *Ep.* 2.93 p. 40.14 Usener; Callim. fr. 191.61 Pfeiffer (Burkert 1972: 420 n. 106); Hermesianax fr. 7.86 Powell; Aët. 2.23.8–9 Mansfeld & Runia 2009 = 2.23.6–7 Mansfeld & Runia 2020 (lemma 8/6 attributed to the Stoics by Mansfeld & Runia 2009: 555, 557–58; 2020: 1018–19; lemma 9/7 attributed to Empedocles by Couprie 2020: 9–10, 18–19); [Tim. Locr.] *De Nat. Mund. et An.* 29); see further Beck 2006: 241 with n. 2.

[31] *Il.* 23.309, 466.

[32] Eur. *Phoen.* 3.

[33] For the chariot-race metaphor applied onto both diurnal and annual solar movement see Bilić 2016a: 212–15. It seems that the earliest attestations of the utilisation of the model for the annual solar movement appear in Theodectas fr. 17 Snell and Archestratus fr. 33.1–2 Brandt = *Suppl. Hell.* 164.1–2.

[34] Huld 1999: 125–26.

[35] In another example of applying the model of chariot-racing onto (here) the diurnal solar movement, Johann. Gaz. *Ekphras.* 2.199–200 uses *terma* and *nussa* as synonyms for the western diurnal turning-point of the sun; the *terma* of SEG VII, 14.10 (Herodorus' hymn to Apollo-Helios, Merkelbach & Stauber 2005: 74–76; Potts 2016: 360), on the other hand, can equally apply to the diurnal and annual solar movement.

[36] Kirk 1962: 289; cf. Kahn 1979: 109.

[37] 64B3 D-K.

[38] In *Plat. Quaest.* 8.4.1007DE Plutarch associates the sun with orderly motion having measure (*metron*), limits (*perata*) and periods. The sun is the overseer and sentinel that delimits (*horizô*) changes (*metabolai*)—perhaps in the meaning of 'changes' or 'turnings' in the motions of heavenly bodies causing or marking seasons (the whole discussion is a part of a larger discussion on the heavenly bodies as instruments of time), or the accompanying seasonal changes affecting nature—and the *hôrai* themselves (cf. *De anim. procr.* 33.1030C where he mentions *hôrais kai metabolais metron* affecting the earth). At this point Plutarch introduces a short quotation of Heraclitus on the *hôrai* (22B100 D-K).

[39] Aeschin. *In Ctes.* 165; Arist. *Mete.* 362b3, 9, fr. 248 Rose, p. 195.5–6 = fr. 695, p. 749B.37–38 Gigon = *BNJ* 646F1.6, cf. *Mete.* 350b6–7; [Arist.] *Pr.* 942a1, 4; Ptol. *Tetr.* 2.2.6.

[40] *Or. Mar.* 649–50. Kiessling (1914: 852.38–854.6) believes that Avienius here described the diurnal rather than annual path of the sun. The passage is admittedly ambiguous, and the context suggests that Avienius had the diurnal movement in mind here, but 'limits of the Bear' cannot in any way be associated with the diurnal solar movement (see the next quotation from Avienius).

[41] *Descr. orb. terr.* 761–63 (*GGM* ii.184).

Ursae' also has the meaning 'the fixed arctic circle',[42] and naturally reflects the data originally introduced by Pytheas. Another poet of late antiquity, Nonnus, spoke of the *nussa* ('turning-point', another term borrowed from the chariot-race imagery) of the zodiac, identical with the tropical points[43] and of the *Boreias nussa* beside which the Bears move,[44] which could be associated with the 'higher' *nussa* in Cancer.[45] Thus again the Bear would define the northern limit of the sun's voyage, described in terms of a model accounting for the annual solar motion originally derived from chariot-riding.

We can safely conclude that Helixoia meant to Hecataeus 'the island of turning', that is, 'the island of solstice', combining in its nature the association both with the constellation that marked the location where 'paths of day and night meet',[46] and with the sun's annual turning round a solstitial *terma*, a model clearly originally involving the metaphor of chariot-race.[47] One can thus recognise in Hecataeus' account a reference to a model of annual solar movement involving Britain, the Hyperboreans, the constellation of Ursa Major, Boreas and the Rhipaean mountains. Except for Britain, all other elements could be styled traditional or primarily cosmological, i.e., not attached to a specific tangible location that could be identified on a geographical map. It is highly probable that Britain was introduced into this nexus of mythic references to the northern limit of the annual solar movement by Pytheas himself, with Hecataeus, in addition inspired by Homer's account of the Cimmerians and the description of cape Carambis as it appears in Ephorus, systematising all these elements into a narrative of Hyperborean Helixoia.

6.4. Hyperboreans, Apollo and Celts

However, Hecataeus could not have found the Hyperboreans in Homer, and it does not seem particularly likely that he borrowed them from Pytheas.[48] It seems that they were an addition of his own, perhaps motivated by their close association with (a solar) Apollo—thus in the same time with solar movement—or their identification with Celts, which placed them in the region at close proximity to Britain. Otherwise, their close connection with the northern limit of the annual solar movement, exemplified in the reports of Pherenicus and Pliny (chapter 4), which are probably later than Hecataeus, perhaps featured in a pre-Hecataean tradition and was the reason for their placement on Helixoia.[49]

Apollo's connections with the Hyperboreans are well documented and date from the early period of Greek history; Hecataeus was following here an established tradition. The Hyperboreans were considered especially dedicated to Apollo worship and were his sacred people from at least the early fifth century,[50] perhaps even the seventh, if Aristeas indeed belonged to this period, as now seems likely.[51] His connections with the land and the people of Hyperboreans were numerous and diverse,[52] but they are especially emphasized in the story of Hyperborean maidens and offerings.[53]

At the same time, the identification of the Hyperboreans with Celts was a natural development of the tradition that placed both of these peoples in the far north or west of Europe. The Hyperboreans were already known to Hesiod,[54] who associated them with the Eridanus and amber, thus precisely with the extreme west or north.[55] Pindar had located them somewhere around the source of the Ister,[56] while other early authors, as we have seen (chapter 4), added a further association with the Rhipaean Mountains. The source of the Ister was in its turn located with reference to several related toponyms, all associated with the Celts.[57]

---

[42] Dion. Perieg. 582 (Lightfoot 2014: 228–29); Lightfoot 2014: 394; *pace* Eust. Dion. Perieg. 581 (*GGM* ii.330.8–10).

[43] 1.454, 3.35, 38.277–79, 284–85, 327; cf. 38.259, where the meaning is doubtful.

[44] 25.398, 38.406–407.

[45] 38.284–85.

[46] *Od.* 10.86.

[47] Cf. Macurdy 1920:140–41.

[48] It is possible that he, like Herodotus (4.32), believed in the possibility of the Homeric authorship of the *Epigoni*, where the Hyperboreans were indeed mentioned (fr. 5 West), but this does not seem particularly likely.

[49] Pherenicus is probably a Hellenistic poet, Fornaro in *BNP* s.v. *Pherenicus* 2; Cameron 1995: 268–69 with n. 35 does not exclude a Roman date for Pherenicus.

[50] Pi. *O.* 3.13–16, *P.* 10.29–48; Bacchyl. 3.58–62 Snell; Callim. *Aetia* frs. 186.8–10, 492 Pfeiffer; Simias of Rhodes fr. 2 Powell; Apollod. *FGrHist* 244F126a; etc.

[51] Hdt. 4.13, 15–16; Theopom. *FGrHist* 115F248; Max. Tyr. *Diss.* 38.3d4–5 Hobein; Orig. *CC* 3.26; Bolton 1962: 4–7; Bridgman 2005: 31; Dowden *BNJ* 35 *Biographical Essay* (620–580 B.C.).

[52] Apollo's temple in Hyperborea (prior to Hecataeus: Pi. *Pa.* 8.63–64 = fr. 52i P. Oxy. 1791; Heracl. Pont. fr. 51 Wehrli = fr. 24 Schütrumpf (Gottschalk 1980: 94 n. 21, 123 is noncommittal whether the entire story comes from Heraclides); uncertain date: Pheren. fr. 671 *SH*); Leto arrived to Delos from Hyperborea (Arist. *HA* 580a17–18; Antig. Caryst. *Mir.* 61; Philosteph. fr. 32 *FHG* etc.); Hyperborean Apollo and Pythagoras (Arist. *De Pythag.* frs. 1B, 1D, 2 Ross = frs. 156, 173 Gigon); Apollo's wanderings as far as the land of the Hyperboreans (D. S. 3.59.6); his bride a daughter of a Hyperborean king (Steph. Byz. s.v. Γαλεῶται); the arrow Apollo used to kill the Cyclopes hidden among the Hyperboreans (Heraclid. Pont. fr. 51 Wehrli = fr. 24 Schütrumpf); the Hyperborean (Hdt. 4.36.1; Pl. *Charm.* 158B; Heraclid. Pont. fr. 90 Wehrli = fr. 55 Schütrumpf; Hecat. *BNJ* 264F7; Lycurg. *Or.* 14 fr. 5ab Conomis, cf. Suda α 18, Phot. *Lex.* s.v. Ἄβαρις; Apoll. Para. *Mir.* 4; Nicom. *FGrHist* 1063F1 with 31A13 D-K, etc.) Abaris, a priest of Apollo worshipped in Hyperborea (Nicom. *FGrHist* 1063F1, etc.), arrived from Hyperborea on account of a plague and obtained mantic power from Apollo after serving him (the story is found in Lycurgus, partly in Suda, Photius and Harpocr. s.v. Ἄβαρις, the latter two implying it was also known to Hippostr. *FGrHist* 568F4 and Pi. fr. 270 Snell).

[53] Hdt. 4.32–36, esp. 33.1–3 (in 35 he probably followed Olen's report) (Sale 1961: 75–76, 78, 82–83; Robertson 1983: 147; Graf 1993: 107; Asheri, Lloyd & Corcella 2007: 604–607); [Pl.] *Axioch.* 371A; Callim. *hDel* 278–99, *Aetia* fr. 186.8–15 Pfeiffer; Phanod. *FGrHist* 397F5; Paus. 1.31.2, 43.4, 5.7.8 (citing Olen and Melanopus of Cyme) etc.; some 4th c. inscriptions from Delos (*ID* 98.A68–69; *ID* 100.49 = Durrbach 1911: 5, 11, l. 49; *ID* 104.3.A8 = *IG* ii².1636 A8) explicitly mention the offerings, while others only specify expenditures for them or the sacred objects brought by the Carystians (Bridgman 2005: 58).

[54] Fr. 150 M-W P.Oxy. 1358 fr. 2 col. i.21–24; Gisinger 1929: 325–26; cf. Hdt. 4.32. They were also known to the author of the cyclic poem *Epigoni* (fr. 5 West).

[55] Herodotus' Eridanus empties into the northern sea and he mentions it while describing the *eschatiai* of the west (3.115.1).

[56] Pi. *O.* 3.13–16 with Σ BCDEQ Pi. *O.* 3.25ab (i.112 Drachmann), cf. *O.* 8.46–47.

[57] Hdt. 2.33.3, 4.49.3 (the Celts outside the Pillars of Heracles and the city (*sic*) of Pyrênê); Arist. *Meteo.* 350b1–2 (the mountain of Pyrênê in Keltikê); Timag. fr. 1 *FHG*; Posidon. *FGrHist* 87F116; Procop. *Bell.* 8.5.30, *Aed.* 4.5.9 (the Celtic Mountains in Keltikê); [Scymn.]

Thus, the earliest tradition associates the Hyperboreans with certain well-defined group of geographical toponyms (Eridanus, Ister, Rhipaeans, 'Keltikê'), as well as with the god Apollo. Apparently, Apollo was further associated with the Hyperboreans and amber from the Eridanus by the Celts themselves, if we were to believe Apollonius Rhodius.[58] Certainly it seems more plausible that it was Apollonius himself who identified the Celts with the Hyperboreans, a usual practice by his time,[59] rather than to believe that the Celts had their own myth of 'Apollo' and 'Hyperboreans', together with amber and the 'Eridanus', which was somehow transmitted to Apollonius or his source(s); his immediate source could perhaps be identified as Heraclides Ponticus, who could in fact be the inventor of the entire story or its core.

In any case, the sources associating some elements from the Hyperborean nexus of myths, such as amber or the river from which it originates, whether called Ister or Eridanus, with the Celts, who, according to Ephorus' systematisation of the inhabitants of the *eschata* of the *oikoumenê*, should properly live between the summer and winter sunsets, the north being reserved for the Scythians, either used a different classification of peoples, in which the Celts (also) occupied the north, or transferred the entire mythic nexus from the north to the west.[60] In general, Strabo claimed that 'the ancient' Greek prose writers (οἱ παλαιοὶ τῶν Ἑλλήνων συγγραφεῖς) called all the peoples towards the north 'Scythians' or 'Celtoscythians', but even earlier authors (οἱ δ' ἔτι πρότερον) used the names 'Hyperboreans', 'Sauromatians', and 'Arimaspi' for those peoples living 'above' the Euxine, the Ister, and the Adriatic.[61] Even though we do not know who these authors were, Hellanicus, a contemporary of Herodotus, might have been among them,[62] which would push this particular evidence (although it remains unclear which specific part of the evidence) back to the 5th c. BC. In any case, in some Greek prose authors (geographers, historians, ethnographers etc.) the Celtoscythians and Hyperboreans could refer to the same people, while the Celts could be associated, in addition to west, with the far north.

191–95 (*GGM* i.203), cf. 773–77 (*GGM* i.227); Arrian *Anab.* 1.3.2 (the (farthest) Celts; unfortunately, the meaning of Pseudo Scymnus' text is rather unclear); Anon. *Geograph. Compend.* 30, *GGM* ii.502 (the west); at Dion. Hal. 14.1.1 the Ister rises from the Alps and possibly bounds Keltikê from the north and flows through the entire continent that lies 'below the Bears'.

[58] A. R. 4.611–17. Cf. Σ A.R. 4.611–17 (p. 289 Wendel).

[59] Protarch. *ap.* Steph. Byz. s.v. Ὑπερβόρεοι (*FHG* iv.485) and Herod. iii.1.115 Lentz; Heraclid. Pont. fr. 102 Wehrli = fr. 49 Schütrumpf = *FGrHist* 840F23; Philistus (?) *ap.* Steph. Byz. s.v. Γαλεῶται (Malkin 1998: 248); Asclep. Trag. *FGrHist* 12F19 (Bridgman 2005: 68).

[60] For other examples of such practice see Stat. *Theb.* 2.280–81 (Poynton 1963: 259; Diggle 1970: 9 n. 1); for examples of a reverse practice see Pi. *P.* 10.29–48 (cf. Simias fr. 1.1–2 Powell) with a scornful comment in Σ Pi. *P.* 10.46(72b) (Bolton 1962: 62; Bridgman 2005: 16); Aesch. (?) *PV* 795–97 (for the incoherence of the geography in the *PV* see Finkelberg 1998: 120, 140).

[61] Str. 11.6.2 (cf. 1.2.27) with Clarke 1999: 279 and Bridgman 2005: 79 (for Strabo's acknowledged sources for his account of the region see Clarke 1999: 375–76). For the Celtoscythians see also Plut. *Mar.* 11.7 (Posidonius fr. 191 Theiler with commentary on pp. 112–13). Müllenhoff 1887: 169–71, 186–87 attributed both passages to Posidonius (for the *Marius* passage attributed to Posidonius cf. Jacoby 1926: 182); indeed, Posidonius seems to have written of 'a Scythian and Celtic zone' (fr. 49.94–96 E-K = 13 Theiler with Theiler 1982: 2.24 and Kidd 1988: 234–35), which goes some way to support Müllenhoff's attribution.

[62] Indeed, Jacoby cited Str. 11.6.2 under Hellan. *FGrHist* 4F185 (*Scythica*) and Anon. *De Scythia FGrHist* 845F2 (cf. Van Paassen 1957: 228). For Hellanicus' date see Montanari in *BNP* and Fowler 1996: 66.

6.5. Conclusion

The title of this chapter has paired Pytheas with Hecataeus and Britain with Hyperborea. It seems that both these authors played a role in the creation of a model of the annual solar movement that assimilated the island at the edge or outside the *oikoumenê* to the mythic land of Hesiod and Aristeas in the far north. Greek tradition unsurprisingly suggests a northern or north-western provenance of Hyperborea, a paradigmatic region of the far north. In this context it was occasionally more precisely identified with either 'Keltikê' or Britain, with the latter identification certainly postdating Pytheas and probably originating with Hecataeus. Besides, several mythic concepts and relations were also added to this nexus (Hyperboreans, swans, Apollo, Ister, Rhipaeans, etc.). In addition, a reference to solstices and the annual solar movement in general was also recognised in the description of Helixoia, another element in this mythic nexus, and one more specifically created by Hecataeus. This reference is coupled with one to the movement of certain circumpolar stars, most notably the Ursa Major, in terms of its role in the delineation of the fixed arctic circle (described by Pytheas in terms of identity of the arctic circle and the summer tropic) or simply as signifying the summer sun approaching the region in which the constellation revolves. Finally, the presence of the model based on chariot-racing accounting for the annual movement of the sun—which was more often applied onto the diurnal solar movement, due to the interchangeability of the motifs used in building models accounting for both types of solar motion—was also recognised in this mythic nexus. All these elements combine to form a model of the annual solar motion with an emphasis on the solstices expressed in mythic terms, with mythic discourse characterised by anthropomorphic causality and accounting for phenomena in terms of stories. A reference to the annual solar movement was certainly not Hecataeus' sole motivation for his writing of *On the Hyperboreans*, or even the most important one, let alone had he structured his account upon it;[63] yet the presence of such reference is rather apparent in various motifs he employed in his account of Helixoia, which allows the inclusion of this myth in the corpus of stories referencing the solar movement, in this particular case, the annual motion of the sun. Some of my interpretations of particular motifs used by Hecataeus in concocting his Hyperborea as solar references could be questioned, but they are always supported by a context that is rather unambiguous (such as Hecataeus explicit reference to the Metonic cycle in his account of Hyperborea).

[63] For a full account of Hecataeus' *On the Hyperboreans* see Winiarczyk 2011: 45–71.

With respect to Pytheas, it was unambiguously demonstrated that he should not be treated as 'hard' scientist dismissive of non-scientific sources of information, which is apparent in his appropriation of Homer and his use of traditional terminology (either actually related to him by the natives or attributed to them by Pytheas himself) in accounting for the sun's movement at the time of the summer solstice in high northern latitudes. Even though he was well versed in latest developments in astronomy and geometry—his work on the geographical arctic circle being a prominent example of his proficiency—he was certainly receptive to more traditional sources on the meteorological conditions and the behaviour of the sun in the far north. Hecataeus, in his turn, was equally receptive to the latest geographical and meteorological information available, which he freely utilised in his utopian writings, offering a description of a fabulous country anchored in hard scientific facts (by contemporary standards), both empirical/observational and theoretical.

Thus, there appears to have existed a certain not insignificant common ground between the scientist-traveller and the utopian writer. Their interest in the meteorological conditions obtaining in the far north and the sun's behaviour in high latitudes compelled them to turn to all the sources they had at their disposal. In Pytheas' case, he had the powerful tool in the geometry of the sphere, which he combined with early empirical reports, some of which were echoed in Homer, and with his own observations, further supported by the testimonies of the natives of the regions he visited. While we may be certain that he relied mostly on his own observations and the postulates of the geometry of the sphere with which he was well acquainted, he was apparently open to what we would label less reliable sources, such as Homeric poems and the barbarians' accounts of the solar movement. Hecataeus, in his turn, most probably had at his disposal—in addition to the tradition he shared with Pytheas, if he did not share his scientific proficiency—precisely the report of the great explorer of the north. He was apparently receptive to the new theories and reports on the conditions in the far north, which he adapted to his own authorial goals in concocting a story of a utopian land which he located at the edge of *oikoumenê*, a natural place for all utopias. While his main interest might have been to tell a story, this was not solely meant to entertain, but also to educate. By supporting his utopian landscape with information gathered from reliable and credible sources, he might have attempted to uphold his own credibility; but this information served him only as a canvas upon which he painted what was of real importance to him.

I believe I have demonstrated in this chapter that both Pytheas and Hecataeus partake in the continuum of geographical/meteorological/astronomical knowledge engaged in by the ancient Greeks. While there are manifest differences in their expertise, methods and goals, they do share some common ground. Most pertinently to my study, this common ground is exemplified in their interest in solar phenomena, which allowed me to study them in the context of references to solar movement in myth and in the exegesis of myth.

7

Apollo's Hyperborean voyage: a narrative model of solar movement

Hitherto I have had the chance to draw attention to Apollo's solar connections on several different occasions. In this chapter, in accord with what I have outlined in the introductory essay, I will argue that a mythic model accounting for the annual solar movement can be reconstructed from Greek narratives involving specifically the 'Hyperborean Apollo'.[1] These narratives are as a rule set in a calendrical context, resulting in a set of stories strongly concerned with the seasonal behaviour of the sun. This narrative model, I submit, is a rare example of a story structured on an analogy with the annual solar movement, and accounts for the phenomenon in terms of personalistic causality. At the same time, my interpretative strategy does not exhaust other meanings inherent in this mythic nexus, nor does it intend to favour this particular aspect on account of others. Yet the recognisable references to seasonality derived from the solar movement in the accounts involving the 'Hyperborean Apollo' require an interpretation, and the one I am advancing here seems to explain the available evidence in a satisfactory way and, at the same time, does not appear to be contradicted by any of the existing testimonies.

The Hyperborean connections of Apollo have often been used as arguments for the deity's solar function, but as a rule without a detailed analysis.[2] Nevertheless, even though Farnell decidedly claimed that 'no sane criticism can find any solar meaning in the legend of... his [i.e., Apollo's] visit to the Hyperboreans or his periodical absences and returns',[3] the available evidence does not seem to support such a radical dismissal. Even his own formulation 'periodical' suggests a seasonal character of 'absences and returns', which is supported by much of the available evidence. True, 'seasonal' does not necessarily imply 'solar', although one is at a loss in searching for a non-solar seasonal explanation of the mythic and ritual activity associated with Apollo's Hyperborean connections.[4] However, a fresh look at the evidence suggests, if not that Farnell was completely wrong, at least that he was too hasty in discarding any possible solar contents in the Hyperborean complex of myths and cultic activities.[5] His hastiness was framed by his overall programme of discarding any notion of the association of Apollo with the sun, which was in its turn at least partially conditioned by his ideological outlook.[6] But for once it seems that the myth of Apollo's voyage to Hyperborea was actually structured on an analogy with the annual solar movement, rather than merely referencing it. Here an anthropomorphic agent rather clearly exhibits a behaviour corresponding to solar phenomena, with his actions regularly being contextualised in an emphasised calendrical setting. This naturally does not exclude or diminish other contributions to the creation of this mythic nexus, such as the postulated connections with initiation rituals and agricultural festivals. These contributions notwithstanding, this mythic complex certainly belongs to a class of myths involving solar phenomena and thus deserves to be included in this study. Since solar references in myth are the main concern of my study, I will naturally focus on this aspect of the 'Hyperborean Apollo' myths, attempting to bring them into sharper focus; I leave for the others to evaluate the other facets involved in the creation of this mythic nexus.

The presence of solar references in one group of mythic narratives focusing on Apollo should not be surprising to those familiar with the modern tendencies in the understanding of the structure and functioning of the ancient Greek pantheon. Indeed, an evolutionary development of Greek gods through time by the process of amalgamation of various functions and powers is now a generally accepted paradigm,[7] and Apollo in particular is understood by modern scholars as a complex and dynamic entity.[8] In this light, the numerous and unequivocal—in general terms, if not in particular details—testimonies on Apollo's calendrical function (based on a luni-solar or luni-solar-astral calendar) in the Hyperborean context reflect one particular acquisition of a function by this

[1] Earlier versions of this chapter already partially appeared in Bilić 2012b: 517–18, 523–26 and Bilić 2016b: 447–48. A variant version of this chapter has also appeared as Bilić 2021.

[2] For the association of Apollo's solar aspect with his Hyperborean connection see Moreau 1996: 26–28 and Molina Moreno 2001: 62–63.

[3] Farnell 1907: 141.

[4] Burkert 1975: 19 and Versnel 1985–86: 144 = 1994: 316 explain Apollo's annual (Burkert) or annual/seasonal (Versnel) absence from Delphi (and elsewhere) in terms of initiation rituals, but this interpretation still remains within a solar framework, since the god's absences and returns are conditioned by the annual solar movement (incidentally, Burkert's interpretation does not explain Apollo's seasonal absence from Delphi).

[5] Konaris 2016: 231 similarly recognized Farnell's hasty dismissal of any connection between the solar aspect of Apollo's character and his periodical departures and returns and argued that it 'merits more attention than Farnell gave it'. I cannot enter here into the vexed question of Apollo's solar connections; for an overview of the earliest testimonies see Bilić, forthcoming. It appears that the identification of Apollo with the sun was current at the latest in the 6th c. BC but was always of limited popularity.

[6] Cf. Konaris 2010: 496, 498; 2016: 49–50, 121, 202, 212–14, 228, 230, 233.

[7] Parker 2011: 86 refers to it as a 'snowball' theory; similarly, Iles Johnston 2018: 156–59 treats Greek gods as accretive characters, whose various traits were drawn from their different instantiations in myths and other media.

[8] Burkert 1985: 144; Versnel 1985–86: 136 = 1994²: 293; West 2007a, 148; Graf 2009: 146 (cf. Konaris 2016: 282 n. 85).

Olympian deity attained through his several instantiations in a particular set of myths.

7.1. Delphian traditions

Apollo's multi-faceted connections with the Hyperboreans, attested already in the Archaic period, were outlined in the preceding chapter. In this context, his visits to the Hyperborean island of Helixoia at 19-year intervals were recognised as part of a specific model accounting for the annual solar movement. Yet this particular interval is not a reference to a seasonal or even an annual event, thus making little sense in terms of the course of the year, but rather supports the identification of Apollo with the sun in this particular tradition through its unambiguous reference to the Metonic Cycle. However, no explanation of the god's visit to Hyperborea in terms of the Metonic Cycle is offered. His visit to Hyperborea apparently encompasses the period from the vernal equinox to the rising of the Pleiades, thus to the beginning of summer.[9] In terms of Euctemon's calendar the respective dates of these astronomical events are 26 March and 8 May,[10] which would date his departure from Hyperborea and arrival to—presumably—Delphi, as well as his departure from—once again, presumably—Delphi and arrival to Hyperborea at dates not attested elsewhere that make little sense in terms of seasonal absences or presences (late March seems a bit too late for 7 Bysios, which is in any case the date of Apollo's *departure* from Hyperborea in the Delphian tradition discussed below).Thus Hecataeus' story in its calendrical aspect does not seem to reflect, but rather refracts, the cultic reality at Delphi as we can reconstruct it from reliable sources.

This narrative, however, can be compared with a number of testimonies, which date the origin of the tradition of 'Hyperborean Apollo' and its calendrical resonances firmly to the Archaic period. In fact, Hecataeus was in some measure influenced by Alcaeus' report on Apollo's visit of Hyperborea in the context of Delphian oracle,[11] the earliest appearance of this tradition in Greek myth (Alcaeus was born ca. 620 BC, while the poem belongs to the early 6th c. BC); unfortunately, his poem is preserved only in a summary by the 4th-century rhetorician Himerius.[12] Alcaeus described in his poem Apollo's swan-chariot in which the god flew after his birth on Delos to the land of Hyperboreans, whence he returned to Delphi in the middle of summer (θέρους τὸ μέσον). As it stands, this description, even though it provides a rather precise calendrical information, makes little sense in terms of seasonal (as opposed to annual) absences and returns of the deity. In addition, the formulation θέρους τὸ μέσον is frustratingly unclear. If it was indeed used by Alcaeus, it would be the earliest appearance of this particular calendrical or seasonal reference in Greek. It is usually supposed that it refers in Alcaeus to the summer solstice,[13] but an analysis of both its earliest and later appearances does not support such a straightforward interpretation. On balance, it could be claimed that most of the times when the phrase was used it simply refers to the middle of the summer season ('high summer'), with the season starting with the solstice or—more likely—even earlier in the year, with the heliacal rising of the Pleiades, thus with no direct relation to the solstice itself.[14] This usage of the phrase 'middle of summer' can be styled calendrical, since it is derived from the context of a luni-solar-astral calendrical tradition.

In addition, from Himerius' prose summary of Alcaeus' paean and its emphasis on summer heat one could argue for July or August rather than late June (the date of the summer solstice) as the season described by the poet. However, Himerius (also) used his description of the season to discuss Alcaeus' style, thus adopting an involved stance towards the original text; even so, his description would certainly be appropriate for early or mid-July, the 'middle of summer' in calendrical tradition.[15] At the same time, the singing of nightingales and swallows is generally associated with spring,[16] while that of cicadas is regularly associated with the high summer and Dog Days in late July,[17] which makes Alcaeus' description of the season of Apollo's return—as reported by Himerius—rather inconsistent and thus untrustworthy in terms of calendrical precision.

In general, it seems necessary to suppose that either Alcaeus was reckless in his use and report of the existing cultic reality or that Himerius made a mistake in his retelling of the poet's account. As it stands, it would place Apollo's Delian birth and his Delphian epiphany at dates nowhere else attested, whichever understanding of the 'middle of summer' we take into account (i.e., the summer solstice in Attic Skirophoriôn (or Hekatombaiôn)/Delphic Ilaios (or Apellaios) or the calendrical 'middle of summer' in Attic Hekatombaiôn/ Delphic Apellaios). The already noted fact that one should most probably associate Alcaeus' poem with Delphi will be of some help in the continuation of the analysis of this conundrum.

[9] *BNJ* 264F7 = D. S. 2.47.6.

[10] Pritchett & van der Waerden 1961: 36.

[11] Winiarczyk 2011: 60–61.

[12] Alc. fr. 307c Lobel-Page. On the date of the poem see Furley & Bremer 2001: i.99. For Alcaeus' poem associated with Delphic cultic practice see Furley & Bremer 2001: i.44 (curiously contradicted at pp. 100–101); Rutherford 2001: 27, 28 n. 21; Fearn 2007: 171 n. 28, 172.

[13] Farnell 1907: 104; Fontenrose 1974: 460; Bilić 2012b: 517. Perhaps the English term 'midsummer' (Ger. *Mittsommer*), designating the summer solstice, is at least partly accountable for this particular interpretation.

[14] The most important testimonies which give some indication of the date to which the phrase refers are Hdt. 8.12.1; Euctemon's *parapegma* (Pritchett & van der Waerden 1961: 35); Thuc. 5.57.1, 6.30.1; Democr. 68B17 D-K; Eudox. fr. 221c Lasserre; Callisth. *BNJ* 627F2 *ap.* Ath. 5.196D; Dion. Hal. 1.63.1; Plut. *Dion* 23.3, 24.1, 25.6, 38.1; *Timol.* 27.1; Arr. *Ind.* 6.5. I offer a full analysis of all the sources elsewhere.

[15] Him. *Or.* 48.10–11.

[16] E.g., *Od.* 19.518–19; Hes. *Op.* 568–69; Sapph. fr. 136 Lobel-Page, Voigt; Stes. fr. 174 Finglass; Simon. frs. 586, 597 *PMG* = frs. 294, 307 Poltera; Euctemon in Pritchett & van der Waerden 1961: 35; Ar. *Av.* 713–14, 1417, *Eq.* 419, *Pax* 800, cf. *Thesm.* 1 (with Austin & Olson 2004: 51); Eud. fr. 229ab Lasserre; see further West 1978: 300–301; Austin & Olson 2004: 52; Davies & Finglass 2014: 497; Lightfoot 2014: 382–83.

[17] Hes. *Op.* 582–88; [Hes.] *Scut.* 393–97; Alc. fr. 347ab L-P.

If we read Alcaeus' description as it is transmitted by Himerius and accept the summer solstice as the intended date of Apollo's return, as well as the date of his departure towards the Hyperboreans, or, more plausibly, if we accept the 'calendrical' midsummer as reconstructed above as the intended date, both events would occur in the Delphian month Apellaios, the first month of the Delphian year, beginning with the summer solstice.[18] This Delphian month is surely associated at least in name with Apollo,[19] and it corresponds to the Attic Hekatombaiôn,[20] the month of the summer solstice according to one tradition and the month closely associated with Apollo at Athens, as well as the month associated with an Apollo-sun identification securely attested in later sources, but perhaps already current in Plato.[21] However, apart from Alcaeus' account, there is little confirmation for a festival in Apellaios in the recorded ritual activities at Delphi: the sole evidence comes from the famous Labyadai inscription, but with no connection with Apollo's departure or arrival.[22]

The apparent connection of Alcaeus' paean with Delphi necessitates its comparison with the Delphic cultic reality as reported in reliable sources. According to Alcaeus, Apollo was apparently born a full year before his arrival to Delphi, thus at the summer solstice or the 'calendrical' middle of summer of the preceding year (thus in Delphian Ilaios or Apellaios/Attic Skirophoriôn or Hekatombaiôn). But the birthday of Apollo was celebrated in Delphi on 7 Bysios, which was formerly the single day of the year when oracles were pronounced.[23] The month of Bysios fell between late January and mid-March;[24] it was most probably the month of Apollo's return to Delphi after his winter absence.[25] This absence was described by Plutarch, who claimed that the god left his sanctuary at 'the beginning of winter' (ἀρχομένου δὲ χειμῶνος), which, in the light of the above cited evidence, must mean for Plutarch the winter solstice.[26] Since he further claims that Apollo is absent from Delphi only for a shorter part of the year encompassing three months,[27] it seems reasonable to suppose that he returns to Delphi on his birthday, after somewhat more than two months have passed.[28] It is probable that this day corresponds to the Delphic Theophania mentioned by Herodotus;[29] Pfister and Petridou believe that the Theophania on 7 Bysios was also the occasion of Alcaeus' paean,[30] but this cannot be reconciled in any way with the 'middle of summer', even though some information gathered from Alcaeus' poem corresponds to this hypothesis, such as the fact that Apollo's birthday is also the date of his return. On the other hand, Plutarch is explicit that his absence is not a year-long event but rather a seasonal occurrence.

It thus seems that several different traditions of Apollo's birthday, absence and return existed at Delphi: according to Alcaeus, he was both born and returns to Delphi in Apellaios (ca. July), while according to Plutarch he was both born and returns to Delphi in Bysios (February/March). But at the same time, there seems to be no evidence for a festival of Apollo's return in Apellaios, while Apollo's rule over the part of the year allotted to him by Plutarch is

[18] Samuel 1972: 74; Trümpy 1997: 211–12. This applies to the tradition involving the summer solstice in Hekatombaiôn, rather than Skirophoriôn (Delphic Ilaios). Otherwise, Apellaios/Hekatombaiôn could have fallen anywhere between late June and mid-August, thus encompassing both the solstice and 'calendrical' midsummer.

[19] Burkert 1975: 11; 1985:144. A μενός Ἀπόλλωνος in *GDI* 1931.1 and *FD* 3.3.38 does not correspond to this month (Trümpy 1997: 204–205, *pace* Kubitschek 1894: 2685.30–31).

[20] *IG* vii.4135.10–12 (Delphian Apellaios = Boeotian Hippodromios); Plu. *Cam.* 19 (Boeotian Hippodromios = Athenian Hekatombaiôn; Kubitschek 1894: 2685.33–39; Samuel 1972: 68; Trümpy 1997: 131, 212).

[21] The summer solstice in Hekatombaiôn: Arist. *HA* 5.11.543b.11–12 (cf. Samuel 1972: 64 n. 1, 72; Pritchett 1999: 79), also *Anecd. Grae.* p. i.247.2–4 Bekker (Byzantine-period *Lexicon of rhetorical terms*) and *EM* 321.10–11, cf. Theoph. *Hist. Pl.* 4.11.3, where Skirophoriôn and Hekatombaiôn are both associated with the summer solstice (Iversen 2017: 168 claims that in practice, the solstice might have been associated with the month following the one in which it actually occurred, i.e., here with Hekatombaiôn); Hekatombaiôn and Apollo: *Anecd. Grae.* p. i.247.1–2 Bekker, *EM* 321.7–8; Hekatombaiôn and the Apollo-sun identification: *Anecd. Grae.* p. i.247.2–3 Bekker; Plato and the Apollo-sun identification: Pl. *Leg.* 12.945e, 946cd, taken as evidence for Plato's equation of Apollo with the sun in Procl. *Theol. Plat.* 6.12, p. 376 Portus and Burkert 1985: 336 (with the festival in the 'common precincts of Helios and Apollo' taking place on the last day of Skirophoriôn).

[22] The festival of *Apellai* in Apellaios is mentioned in this late 5th/early 4th-c. inscription from Delphi, 1A31–32, 36, B7, D3 Rhodes & Osborne, but apparently the sacrifices were offered to Dionysus, 1D 44–45 Rhodes & Osborne (Rhodes & Osborne 2003: 2–7; Farnell 1907: iv.99 n. b notes that neither the festival nor the sacrifices are explicitly associated with Apollo). On the festival of *Apellai* at Delphi see especially Burkert 1975: 10, who interprets it as an initiation ceremony; it was not celebrated generally at Delphi, however (Rhodes & Osborne 2003: 11).

[23] 1D5–6 Rhodes & Osborne, pp. 6–7 (late 5th/early 4th c.); Plu. *Quaest. Gr.* 9.292EF with Anaxandrides and Callisthenes *FGrHist* 124F49; Fontenrose 1974: 383; Trümpy 1997:213; Salt & Boutsikas 2005: 565; Boutsikas 2007:99; 2020: 72–73. Presumably, this would also be the date on which the oracle was founded: according to Boio, the oracle was established by the Hyperboreans Olen, Pagasus and Aguieus (other Hyperboreans were also named in her hymn, Boio frs. 1–2 Powell; Paus. 10.5.7–8; Fontenrose 1978: 215). The names Pagasus and Aguieus clearly evoke Apollo (Detienne 2009: 59–62 = 1997: 12–14), and the whole account might reflect his activities upon returning from Hyperborea. *Et. Mag.* 297.58–298.3 mentions Apollo's feast and birthday on the 20th, but unfortunately neither the month nor the location are specified (Rutherford 2001: 206 n. 2; Fowler 2013: 60).

[24] Salt & Boutsikas 2005: 565–66 (on p. 569 they argue that the 7th of Bysios could fall between early February and early March); similarly, Boutsikas 2020: 73.

[25] Mommsen 1878: 280–82; Pfister 1934: 2133.41–45; Trümpy 1997: 213; Marconi 2007: 199; Boutsikas 2020: 73–74.

[26] Cf. Plut. *Timol.* 27.1.

[27] Plu. *De E* 9.389BC. However, it seems that Plutarch's neat scheme does not fully correspond to Delphic cultic reality: Philodamos' paean to Dionysus (1.3–4, 9.110–11) thus mentions how this god arrives in spring, at the Theoxenia festival (Furley & Bremer 2001: i.121–22, 127, ii.53, 55, 60), which is firmly in Apollo's part of the year (the month Theoxeions following Bysios, Trümpy 1997: 212). In addition, a dithyrambic contest took place at the Pythian Games (11.131–34; Furley & Bremer 2001: i.122, ii.56, 82), held in the summer month of Boukatios (Trümpy 1997: 59; Rutherford 2001: 28; Hannah 2005: 35–36; Suárez del la Torre 2013: 72, 78; Iversen 2017: 177; Boutsikas 2020: 77–78), thus during the period of the year dominated by paeans. Philodamos' paean was performed at the 340/339 *theoxenia* (Käppel in *BNP* s.v. *Philodamus*).

[28] Alternatively, if we take the Euctemonian date for the beginning of winter, i.e., November 8, the three months adduced by Plutarch would date the return of Apollo and his birthday to early February, still well within the range of dates on which 7 Bysios could have fallen.

[29] Hdt. 1.51.2; Mommsen 1878: 282; Robert 1886: 167; Nilsson 1906: 158–59; Farnell 1907: 258 n. c, 292; Pfister 1934: 2133.41–46; Fontenrose 1974: 380–81; Petridou 2015: 276. On what little is known on Delphian Theophania see Mommsen 1878: 280–97 and Derow & Forrest 1982: 83–84.

[30] Pfister 1934: 2133.46–49; Petridou 2015: 276.

compromised by Dionysus' presence at Delphi in spring and summer. There is no plausible way to either explain away one of these two main contradictory traditions or to somehow merge them into a single phenomenon, especially in the light of further qualifications derived from the attested cultic activities at Delphi related to this nexus of ideas. But the variety of traditions emphasizes the capacity and appeal of the concept of Hyperborean Apollo for the inclusion of calendric information and their pivotal role in the structuring of mythic narratives and cultic activities referencing or associated with this complex of phenomena.

7.2. Athenian and Delian traditions

In addition to Delphian, there are other traditions of Apollo's visit to Hyperborea involving calendric information. Thus, the epiphany of Apollo was probably celebrated in Athens at the Thargêlia, since Istrus described the *pharmakoi* ritual accompanying the Thargêlia in the first book of his *Epiphaneiai* of Apollo.[31] At the same time, Apollo's birthday was celebrated on 7 Thargêlion in Athens,[32] which is a significant correspondence with the situation at Delphi, where Apollo's birthday on the seventh day of a month coincides with his epiphany. However, since 22 or 23 Thargêliôn fell 17 days before the summer solstice,[33] Apollo's birthday and his epiphany fell on a date 31 or 32 days before the solstice. Since the latter fell on 27 June in Euctemon's calendar,[34] Apollo's birthday and epiphany would be celebrated on 27 or 28 May, nowhere near the 'middle of summer' on any reckoning. Still, the Athenian tradition indeed associated the epiphany of Apollo with early summer (on Euctemon's reckoning), while the Delphian tradition, as opposed to that recorded in Alcaeus' verses, associated it with the beginning of spring.

An Apollo-sun identification must naturally underlie any structuring of a narrative in which he is involved on an analogy with solar movement. In this context, it is noteworthy that some sources explicitly state that the Thargêlia were dedicated to Helios and Horae,[35] or Helios, Horae and Apollo.[36] Indeed, Apollo and Helios were certainly identified in the Boeotian Daphnephoria, with the bronze sphere used in the procession symbolizing the sun/Apollo and the garlands symbolizing the 365 days of the year, but the Neoplatonic source for this identification, Proclus, is late and belongs to a tradition that tended to interpret myths and cults in terms of celestial phenomena.[37]

The Daphnephoria was celebrated in (early) spring,[38] although late spring—or early summer on Euctemon's reckoning—was also suggested,[39] adducing the parallel of the Delphian cultic reality.[40] It was an enneateric festival resembling the Delphian S(t)epterion,[41] which was indeed celebrated in the summer.[42] Certainly, the Apollo of the Thargêlia had some calendric significance, perhaps one conditioned by the annual solar movement. At the same time, any connection with Daphnephoria, with its (late) interpretation of Apollo as the sun, and with S(t)epterion, with its celebration in summer, seems rather circumstantial.

Similarly to Athens, Apollo's birthday was celebrated on 7 Thargêliôn at Delos.[43] This is perhaps the date of the Dêlia(-Apollônia),[44] although alternatively the festival could have taken place in the Delian month Hieros.[45] The former suggestion (taking into account the correspondence of the Delian Thargêliôn with the Athenian month of the same name) would date the epiphany of the god in late May, in concordance with the Athenian tradition; the latter (taking into account the correspondence of the Delian Hieros with the Athenian Anthestêriôn, in its turn corresponding to Delphian Bysios) would date the epiphany of the god in early February to early March, in concordance with the Delphian tradition, but would also involve the separation of the god's birthday from his epiphany. Certainly, the attested Delian birthday of the god speaks in favour of the former, but the *communis opinio* of the scholars seems to favour the latter. The god first arrived at the island of his birth in his mother's womb, with the gravid Leto coming to Delos from Hyperborea;[46] the Hyperborean maidens Arge and Opis came simultaneously to deliver Apollo and Artemis.[47]

---

31 *FGrHist* 334F50 = Harp. s.v. Φαρμακός, 298.11–15 Dindorf; Farnell 1907: 258 n. c; Parker 2005: 482. On the festival of Thargêlia see Parker 2005:203–204, 481–83, 487; Bremmer 2008: 194–96.

32 Plut. *Quaest. Conviv.* 8.1.717B, D, fr. dub. 103 Sandbach.

33 Dion. Hal. 1.63.1.

34 Pritchett & van der Waerden 1961: 32.

35 Theoph. fr. 584A.46–51 FHS&G (cf. Parker 2005: 204, 417); Σ Ar. *Pl.* 1054 = Σ Ar. *Eq.* 729a = Suda ει 184 s.v. Εἰρεσιώνη = [Eudocia] *Violarium* 333, p. 246 Flach.

36 *SEG* 33.115.12–13, 240s BC, cf. Parker 2005: 203–204, 434 n. 64.

37 Procl. in Phot. *Bibl.* cod. 239, p. 321b18–23 Bekker; cf. Σ Clem. Alex. *Protrep.* 10.10, 299.9–13 Stählin & Treu.

38 Calame 1997: 102; Schachter 2016[2000]: 255, 258, 274; Boutsikas 2020: 78.

39 Müller 1839: 336–37: 7 Thargêliôn; Boutsikas 2020: 78.

40 On Theban Daphnephoria see Schachter 2016[2000]: 255–78.

41 Calame 1997: 101–102; Boutsikas 2020: 78; cf. Schachter 2016[2000]: 269–70, who dismisses the connection between the two as secondary and superficial, cf. pp. 274–76.

42 Fontenrose 1974: 460: the summer solstice; Farnell 1907: 293–94: early summer; Rutherford 2001: 28: high summer, Boukatios; Boutsikas 2020: 78, 113 (Ilaios or Boukatios—June/July or August/September—or, alternatively, late spring-early summer, i.e., the Attic Thargêliôn/ Delphian Herakleios).

43 Apollod. *FGrHist* 244F37; D. L. 2.44; Anon. *Vita Plat.* 389.51–52 Westermann = Anon. *De phil. Plat.* 1, ll.43–44 Westerink.

44 Robert 1886: 161–62, 165–67 (the Apollônia, with the Delia in Hieros); Homolle 1892: 56 (the Dêlia, with the Apollônia in Hieros); Farnell 1907: 278, 289–91; Robertson 1983: 152 with n. 22 (the Dêlia); Bachvarova 2016: 236 (the Dêlia); Boutsikas 2020: 83–84. Farnell (1907: 290) adds that Theophrastus' account of the Apollo in whose honour the Athenians celebrate the Thargêlia with dances round the temple of Apollo Delios refers to this Delian festival (Theoph. fr. 576 FHS&G), but it is more probable it should apply to the Athenian Thargêlia (cf. Bremmer 1983: 319).

45 Nilsson 1906: 145–48; Arnold 1933: 453, 458; Sale 1961: 88–89; Calame 1997: 107–108; Trümpy 1997: 64 (the Apollônia); Lambert 2002: 382–83; Lightfoot 2014: 382 (the Apollonia); Petridou 2015: 275–76 (7 Hieros, the Delia every fifth year, the Apollonia the intervening three); Boutsikas 2020: 83–84.

46 Arist. *HA* 580a17–18; Antig. Caryst. *Mir.* 61; Philosteph. fr. 32 *FHG*; Plut. *Quaest. Nat.* 38; Ael. *NA* 4.4.

47 Robertson 1983: 147; Graf 1993: 107; Furley & Bremer 2001: i.146.

There is another piece of information with respect to Apollo's arrival to and departure from Delos that could be styled calendric, and thus dependent upon the annual solar movement, although in a most general sense. Thus Vergil described how Apollo either winters in Lycia and leaves for Delos in spring or summer, or it is Lycia he leaves in the winter.[48] Servius—apparently opting for the former—claimed he gave responses at Patara for six winter months, and on Delos during the six summer months.[49] In the light of everything adduced above, Vergil's account would make some sense only if Apollo winters elsewhere and appears at Delos—let us presume—in Hieros.[50] Thus, Vergil's/Servius' account raises several new issues: Lycia as the winter abode of Apollo—as opposed to Hyperborea[51]—and a six-month winter interval of Apollo's absence from Delos, as opposed to the three-month winter interval of Apollo's absence from Delphi (Alcaeus' year-long absence makes little seasonal sense, as already noted). But first of all, the tradition exemplified in Vergil's account is apparently much earlier. Thus, Simonides mentioned together Lycian Apollo (1), daughters of the Delians (3), and the arrival of spring (7).[52] It is probable that this is an invocation to Apollo to come from Lycia to Delos at the beginning of spring. This report can be with some confidence associated with Vergil's, and further supports the conclusion that Apollo winters in Lycia. It also suggests the early spring (i.e., Delian Hieros/Delphian Bysios) as the date of Apollo's return to Delos.

With respect to Lycia instead of Hyperborea as the winter abode of Apollo, one may note the existence of an apparent connection of these two regions in the context of Apollo's 'Hyperborean' cultic activities at Delos and Delphi. This connection is exemplified in the person of the poet Olen. Olen 'of Lycia' was an early poet who came to Delos and wrote a hymn for the Hyperborean maidens Arge and Opis.[53] Coming from Lycia, he also wrote a hymn for Eileithyia, who similarly came from Hyperborea, sang by the Delians,[54] and a hymn for Achaeia, for whom he also claimed to have arrived on Delos from Hyperborea.[55] Alexander Polyhistor likewise claimed that Olen was a Lycian, but the Suda, although citing both Callimachus and Alexander, mentioned he could be a Hyperborean.[56] Moreover, Pausanias knew of a Hyperborean Olen, who actually established the oracle at Delphi, and was its first prophet, as well as the first to give oracles in verse; he cites a local poetess Boio, who wrote a hymn to the Delphians in which she sung of the event.[57] Thus it appears that there were two Olens associated with the two greatest Apolline sanctuaries, the one from Lycia ostensibly a historical figure, the other from Hyperborea a legendary character. Further, both had strong ties with the Hyperboreans, and while both Lycia and Hyperborea were firmly associated with Apollo, when speaking of 'wintering' and 'arrivals' one is led to support Hyperborea as Olen's place of origin, with the Delian hymnic poet being modelled upon the Delphian prophetic poet. Another piece of information supporting the case for Hyperborea is Vergil's reference to the Agathyrsi as involved in Apollo's festivities on his return from Lycia, who surely refer to the Hyperboreans.[58] Vergil's source here must have been some learned Hellenistic poet,[59] but why he and Simonides before him insisted on Lycia as the winter quarters of Apollo remains enigmatic, especially if the case of Olen is left out of account.

7.3. Beyond calendar

There are a number of testimonies on Apollo's departure and arrival devoid of any explicit or precise calendrical information, most of them—or all those that are specific enough—appearing in the context of the two great Apolline cult centres of Delos and Delphi. Some of these testimonies are not much later than Alcaeus, and some are perhaps influenced by his account. All of them, I submit, clearly belong to the mythic nexus studied in this chapter, since they all presuppose Apollo's seasonal absence conditioned by the sun's behaviour. Thus Menander Rhetor writing on the apopemptic hymns, says they are delivered over actual or supposed departures of the gods, like Apollo's departure (ἀποδημία) at Delos and Miletus, further adducing Bacchylides as one of the authors of such hymns.[60] Indeed, a section of Bacchylides' dithyramb treats a similar subject, although not with reference to either Delos or Miletus, but to Delphi, in the form of an address to Apollo, performed at Delphi during the period of Apollo's absence and his sojourn among the Hyperboreans, perhaps influenced by Alcaeus.[61] It supports the latter's report of the cultic activities at Delphi, mentioning the swans (l. 6), but also notes the singing of paeans in Apollo's honour at Delphi (l. 8–12), which might refer to Plutarch's account of the three-month absence of such performances during Apollo's sojourn among the Hyperboreans. In the context of Bacchylides' poem, in which Apollo's sojourn at the Hebrus during his absence is also mentioned (l. 5), one

[48] Verg. *Aen.* 4.143–44. Cf. Weber 2002: 323–24 for the latter and Wilamowitz-Moellendorff 1903: 575 and Bryce 1986: 195 for the former.
[49] Serv. *Aen.* 4.143, cf. Serv. Dan. *Aen.* 4.144.
[50] Cf. Robert 1886: 166–67: from Hieros (= Delphic Bysios) to Hekatombaiôn (= Delphic Apellaios) on Delos, the rest of the year in Lycia. Wilamowitz-Moellendorff 1903: 575 n. 2 believes that Apollo returned precisely for the Delia.
[51] A notion referred to as 'strange' by Wilamowitz-Moellendorff 1903: 575 and 'inconceivable' by Bilić 2012b: 526.
[52] Sim. *PMG* 519.55; Rutherford 1990: 177–79.
[53] Hdt. 4.35.3; cf. Call. *Del.* 305–306.
[54] Paus. 1.18.5, 8.21.3, 9.27.2.
[55] Paus. 5.7.8.
[56] Alex. Polyhist. *FGrHist* 273F64; Suda ω 71.
[57] Boio frs. 1–2 Powell *ap.* Paus. 10.5.7–8.
[58] Verg. *Aen.* 4.146; Serv. *Aen.* 4.146; Wilamowitz-Moellendorff 1903: 575 n. 2
[59] Wilamowitz-Moellendorff 1903: 575 n. 2; cf. Lightfoot 2014: 382. For example, Vergil's teacher (Lightfoot 1999: 14–15) Parthenius mentioned in his *Delos* (fr. 12 Lightfoot) the 'distant Beledonii', who are located at the Ocean by Steph. Byz. s.v. (the source of Parthenius' line). It is possible to associate this people with the equally remote Hyperboreans (Lightfoot 1999: 153), but this is highly conjectural; since we know little of Parthenius' *Delos*, any connection with Vergil's account is tenuous.
[60] Men. Rh. *Peri epideiktikôn* 1.4, *Rhet. Grae.* 35, iii.336.8–11 Spengler with Bacch. fr. 1A Maehler.
[61] Bacch. 16.5–12; Jebb 1994[1905]: 221–22; Furley & Bremer 2001: i.127; Fearn 2007: 171; Lightfoot 2014: 382.

could adduce Aristophanes' description of swans on the Hebrus greeting Apollo.[62] This passage was in its turn associated with Apollo's return from Hyperborea,[63] but, in the light of Bacchylides' testimony, the Hebrus could equally represent Hyperborea itself. Bacchylides' contemporary Pindar mentioned the presence of Apollo at Delphi at the moment Pythia was giving an oracle,[64] which surely suggests that it was possible for the god to be absent from his Oracle; presumably he would then be with his holy people, the Hyperboreans.[65]

A much later testimony, but apparently derived from significantly earlier material, comes from Claudian's *Panegyric on the sixth consulship of Honorius*. He reported how Apollo leaves Delphi for Hyperborea and (his) altars only to return from the Riphaean (sic!) on a chariot drawn by griffins; during his absence the oracle falls silent and only revives with the god's return.[66] It seems that Claudian replaced Apollo's swans in Alcaeus with griffins because the latter became in time associated with Scythia and the farthest north—their fight over gold with the Arimaspians in the vicinity of Hyperborea was repeatedly described—and thus became attached to Apollo. Claudian certainly knew of Alcaeus' poem, as did his slightly older contemporary Himerius, or at least of a tradition that described Apollo's arrival (or return) from Hyperborea on a chariot drawn by some mythic animals.[67] Moreover, he is the only author, apart from Alcaeus, who combined these two concepts, i.e., the chariot and the Hyperboreans. Furthermore, he seems to have been aware of the tradition reported by Plutarch, according to which the birthday of Apollo on 7 Bysios was the day when oracles were pronounced. Unfortunately, he does not give any calendric information, so his antiquarian testimony is only a confirmation of information known from earlier sources. Similarly, Procopius of Gaza casually mentions Apollo's absence from Delphi and his return from the Hyperboreans, an occasion for the feast of Apollo's ἐπιδημία.[68] This feast can surely be identified with the Theophania on 7 Bysios.[69] Cicero's remark of an Apollo who came to Delphi from the land of Hyperboreans could refer precisely to the tradition manifested in Alcaeus, since otherwise the god's arrival from Hyperborea would refer to Apollo's advent in his mother's womb to Delos, but it is also possible that he refers to a single advent, rather than to seasonal absences and presences.[70] Another reference to this tradition is perhaps represented by the poem on the arrival (ἄφιξις) of Apollo to the Hyperboreans ascribed to the mythic Abaris.[71] If this was an actual pseudepigraphic work,[72] it might have had as its subject Apollo's voyage from Delphi to Hyperborea and his sojourn among this people, but it would still be impossible to say whether this was a one-time visit or a regular seasonal sojourn. Alternatively, if the 'poem' was only in some way incorporated into a Heraclides Ponticus' dialogue, then this arrival of Apollo might refer to his one-time travel to Hyperborea in order to conceal his crime against Zeus.[73] A reflection of the latter hypothesis might be recognized in Apollonius Rhodius, who described Apollo's sojourn in Hyperborea as a temporary exile on account of his banishment from heaven due to Zeus' killing of Asclepius, at the same time ascribing the story to the 'Celts'.[74] It seems that Apollonius here identified the Celts with the Hyperboreans, just as Heraclides Ponticus did,[75] and very likely followed him in this particular story, ignoring the concealment of the arrow and perhaps adding the aetiology for amber as tears of Apollo.

Interestingly, Apollonius Rhodius also reported how Apollo travelled to Hyperborea starting off from Lycia. Since he did not offer any additional information (neither is any provided by the scholia *ad loc.*), it cannot be determined whether this was a one-time visit or a periodic voyage. At the same time, his account provides another appearance of Lycia in the context of Apollo's Hyperborean voyage, now without any reference to Delos (as in Simonides and Vergil/Servius).[76] Similarly, Boeus and Simmias of Rhodes implied that Apollo visited the Hyperboreans starting off from Babylon, certainly more than once, but it is unclear whether he did it on a regular basis.[77] Since the visit is accompanied by the observation of an ass sacrifice, it seems that a seasonal or annual visit is more likely. Finally, it seems that Callimachus and Apollodorus (the latter presumably citing the former) also represented Apollo visiting (ἐπιτέλλεται) the Hyperboreans on the occasion of the sacrifice of asses;[78] if this rendering is accepted, the same conclusion applies as in the case of

62 Ar. *Av.* 769–83.
63 Sprockhoff 1954: 70; Kothe 1970: 205.
64 Pi. *P.* 4.5: he is οὐκ ἀπόδημος, cf. Σ Pi. *P.* 4.7a, 8.
65 Pi. *P.* 10.33–36. On presence (ἐπιδημεῖν) and absence (ἀποδημεῖν) of deities from oracles—upon which their veracity depends—cf. Σ Call. *Ap.* 1.
66 Claud. *De VI cons. Hon.* 25–38.
67 Griffons are said to be sacred to the sun in India and were represented as pulling its quadriga (Philostr. *VA* 3.48). Philostratus' observation probably does not report an actual Indian belief, but rather reflects contemporary Greco-Roman ideas.
68 Procop. *Ep.* 16, 65 Garzya & Loenertz.
69 Cf. Mommsen 1878: 281; Nilsson 1906: 158; Parker 2011: 183 n. 44.
70 Cic. *ND* 3.57.

71 *BNJ* 34T1 = Hesych. *Epit. Onomat.* 1 = Suid. α 18. According to Dowden in *BNJ* 34, Hesychius' immediate source could have been a Hadrianic lexicon of Diogenianus of Heracleia.
72 Dowden suggests that *BNJ* 34F2 might come from this particular work, which he tentatively dates to ca. 520 BC.
73 In fr. 51 Wehrli = fr. 24 Schütrumpf his *On justice* (fr. 17.1 Schütrumpf) is cited in the context of the story of Apollo's hiding of the arrow with which he killed the Cyclopes, as a revenge for Zeus' murder of Asclepius, in the land of Hyperboreans (Gottschalk 1980: 94 n. 21, 123 is noncommittal whether the entire story comes from Heraclides). Heraclides is also known to have written *Abaris* (fr. 17.57a Schütrumpf), *Concerning the sayings*—or *writings*—*ascribed to Abaris* (fr. 17.57b Schütrumpf), and *Treatise to Abaris* (fr. 36 Voss = fr. incert. 149A Schütrumpf—on the authenticity of this fragment see Zhmud 2016: 453), sometimes adduced as alternative titles to *On justice* (Bolton 1962: 157, 202–203 n. 20; Burkert 1972: 103 n. 32 and Gottschalk 1980: 121–22 argue against the identification of *Abaris* with *On justice*).
74 A. R. 4.611–19.
75 Her. Pont. fr. 102 Wehrli = fr. 49 Schütrumpf = *FGrHist* 840F23.
76 A. R. 2.674–75.
77 Simm. fr. 2 Powell *ap.* Ant. Lib. *Met.* 20.
78 Call. fr. 492 Pfeiffer; Apollod. *FGrHist* 244F126a. See the translations in Trypanis *Loeb* 1958: 255 and Cuenca & Brioso 1980: 296; cf. Fowler 2013: 606.

Simmias: a seasonal or annual visit is more likely than any other alternative.

It remains for us to analyse another report by Himerius in which we find early testimonies for Apollo's voyage in a swan-chariot similar to that found in Alcaeus.[79] However, the preserved summary of these testimonies is less revealing than in the case of Alcaeus' description. Here Himerius claims that Sappho and Pindar decked Apollo with golden hair and lyre and 'send him drawn by swans (κύκνοις ἔποχον) to Mount Helicon'.[80] This translation suggests that Apollo used a chariot drawn by swans,[81] but ἔποχος has the meaning 'mounted upon something', whether a horse or a chariot.[82] However, it is not probable that the god was carried upon more than a single swan, and his swan-chariot seems to explain the passage in a satisfactory manner, even if the parallel evidence from Alcaeus is not taken into account. Thus it seems that both Sappho and Pindar acknowledged the existence of Apollo's swan chariot. At the same time, Himerius clearly states that the god arrived to Mount Helicon, rather than Delphi. This makes it difficult to associate Sappho's and Pindar's accounts with the seasonal absence and return of Apollo with respect to either Delphi or Delos. However, it is possible to argue that Pindar's third paean, to which Himerius most probably refers in fr. 262b Bowra, was performed at Delphi, and that Helicon is only a 'way-station' for Apollo on his return from Hyperborea.[83] This is only hypothetical, but it conforms to all the other evidence we have. Thus, a minimum of information that we can gather from Himerius' report (and Pindar's third *Paean*) is that Sappho and Pindar described Apollo as voyaging upon a swan(-chariot), while we can further infer that they—or at least the Theban poet—refer to the well-known voyage of Apollo in a swan-drawn chariot from Hyperborea to Delphi, with his arrival on 7 Bysios.

7.4. Apollo and the solstice island

It can be plausibly presumed that the tradition of the god's northern voyage derives from Apollo's solar connections, and accounts for the annual solar motion, particularly the solstices. It represents a narrative model, a metaphorical expression of a phenomenon in terms of gods or a story structured on an analogy with annual solar phenomena. Naturally, the 'Hyperborean Apollo' nexus of myths should not be reduced to this analogy. It represents only a single facet of this tradition, with many other elements derived from different conceptual domains, such as initiation, communal gatherings and agriculture. But here I am focusing on the fact that both precise and less specific calendrical pieces of information were often introduced in myths and cultic realities focused on Apollo and his travels to Hyperborea. As a rule, Apollo's departures, absences, returns and sojourns are of a seasonal character, conditioned by the annual movement of the sun and the accompanying meteorological phenomena. The fact that specific calendrical information was regularly incorporated in myths and cults involved with Apollo's voyage to Hyperborea suggests its importance in the formation and development of this complex of ideas. Indeed, the literary sources reflecting the model seem to echo a conflation of two different yet complementary ideas: a seasonal change associated with the sun's return from the south, where it abided during winter months, and its cohabitation with the Hyperboreans in the farthest north, at the time of the summer solstice.[84]

At the same time, Apollo's northern voyage appears to be related to the god's association with Delos/Ortygia as the solstice island—a cosmological rather than actual geographical feature.[85] His departures from and arrivals to Delos were already referred to, but in addition it was believed that Apollo's birth had some connection with an arrival from Hyperborea, and there is some additional information on Delos as a mythic solstice island in this particular context. Thus Hyperborean maidens Arge and Opis were said to have arrived to Delos 'together with the gods themselves', which could refer to the fact that they arrived simultaneously with Artemis and Apollo; in its turn, this would imply that the gods were not born on Delos, but arrived there from somewhere else, presumably Hyperborea.[86] Apollo's and Artemis' arrival to Delos in the company of two Hyperborean maidens is perhaps depicted on an amphora dated to shortly after 650 BC;[87] Apollo is here depicted in a chariot drawn by winged horses, which he shares with the two maidens, but Artemis is depicted in front of the chariot, as if she is receiving him. The interpretation of this depiction must thus remain inconclusive, although if the standard interpretation is accepted, Apollo and Artemis must have been conceived as born elsewhere. The usual interpretation of the passage in Herodotus is, as we have seen, that Arge and Opis came to deliver Apollo and Artemis, with the gravid Leto arriving to Delos from Hyperborea. Delos was thus connected with Apollo's arrivals from Hyperborea in various ways: as a one-time arrival at his birth (whether to be born there or as already born elsewhere, perhaps precisely in

---

[79] Him. *Or.* 46.6.

[80] Sapph. fr. 208 Lobel-Page; Pi. fr. 262b Bowra. Trans. Campbell 1982: 191; Rutherford 2001: 278 n. 11 translates 'carried by swans'. Pindar's third *Paean* (= fr. 52c Snell) should be associated with the prose summary of Himerius, but the information contained in it does not add anything new with respect to Apollo's voyage (Snell 1964: ii.20–21; Rutherford 2001: 275–80). 'Swans' in l. 10 is Snell's conjecture.

[81] Cf. Page 1955: 249.

[82] *LSJ* s.v.

[83] Rutherford 2001: 279.

[84] Olmsted 1994[2019]: 137; Bilić 2012b: 509–10, 515–19, 527; cf. Gernet 1981[1933]: 116; Parker 2005: 417–18 (hesitatingly). Is the voyage of the swan-chariot to Libya (Pher. fr. 58 Fowler) a reference to the sun's winter sojourn in the south (cf. Hes. *Op.* 527–28 and Hdt. 2.24.1, 25.1, 26.2, discussed in chapter 5)? It appears that the story was already present in the *Catalogue of Women* (West 1985: 87; Giangiulio 2001: 122 with n. 20, 130), but this particular detail is only preserved in Pherecydes.

[85] For other instances of Delos understood as a cosmological feature by the Greeks see Munn 2007: n. 29 and the section of his paper titled 'Delos as the Center of the Earth'.

[86] Hdt. 4.35.2; Sale 1961: 75–76, 78, 82–83.

[87] *LIMC* 2.1.304, 2.2.270 s.v. Apollon 1005; Rutherford 2001: 279 n. 16.

Hyperborea) or as a seasonal advent at the beginning of spring, admittedly most often associated with Lycia.

A testimony reported solely by Hyginus supports the validity of the interpretation of mythic Delos as a solstice island, although its apparently late occurrence does not raise much confidence, at least not on first glance. On the other hand, the author of what came down to us as Hyginus' *Fabulae* was certainly not an inventor of the material he transmitted, but was heavily dependent on earlier Greek sources, more precisely, on Greek mythographical tradition, itself transmitting mythic narratives found in earlier authors.[88] Thus, for instance, there is some evidence that he might have consulted Chares' (Apollonius of Rhodes' disciple) work on the mythic stories in Apollonius, or some such similar text.[89] We can thus be certain that the account in *Fabulae* has a more respectable pedigree, probably originally being culled from some earlier Greek author (a poet, tragedian or prose writer). According to this account, Leto had to bear her children at a place not reached by the sun (*quo sol non accederet*).[90] This could, naturally, only mean that any place reached by the sun's rays—or susceptible to the eyes of Helios—does not qualify, forcing her to give birth either underground or undersea. Yet she gave birth to Apollo and Artemis on Delos. Admittedly, Neptune at first covered Ortygia/Delos with waves in order to protect her from Python, but when the immediate threat ceased, the island was elevated by Neptune in its initial position (*in superiorem partem retulit*) and the actual birth took place under the clear sky and in full view of Helios.[91] In any case, Delos seems to have been the only place that qualified for the location not reached by the sun. The concept of a region not reached by the sun is by now familiar from the descriptions of the northern inhabitable zone beyond the Rhipaean Mountains (chapter 4). Since Ortygia/Delos, the mythic island of the solstice and a paradigmatic alternative to the mountain barrier, was located on the very border of the northern inhabitable zone, it was barely, if at all, touched by its 'rays'.

It is the Hyperboreans who were said to live beyond the *dromos* of the sun;[92] it is in Hyperborea that Apollo and Artemis were born, or else Leto came from Hyperborea to Delos to give birth to these deities. This certainly associates them with the rendering of the story of Apollo's birth as found in Hyginus' *Fabulae* and with the northern limit of the habitable zone. Thus, once again a mythic model accounting for the annual solar movement emerges, constructed out of several familiar elements already encountered in the previous discussion (a solstice island, solar Apollo, the Hyperboreans).

Moreover, the Hyperboreans have a lot in common with the inhabitants of Homer's Syriê, both in terms of the association of both locations with the annual solar movement and in terms of certain 'ethnographical' characteristics they share with each other. I have already discussed in chapter 5 how this location seems to be associated with a tradition concerned with solar phenomena and meteorological conditions obtaining at the geographical arctic circle and how it was situated at this cosmological location. The island of Syriê was further described as fertile and abounding in cattle, sheep, wine and corn. Its inhabitants never experienced famine or sickness, and when their appointed time to perish comes, Apollo, arriving with Artemis, gently killed them with his arrows.[93] These are all characteristics of the Hyperboreans.[94]

At the same time, another association of Apollo with 'Syriê' existed in antiquity, with the (is)land apparently retaining its solar traits. Apollo had a son with Sinope named Syros, the progenitor of Syrians in Asia Minor.[95] The god's 'Scythian' name, (G)oitosuros/n is probably derived from the same root, while a variant Oitoskuros appears in an inscription as the name of a composite solar deity together with Apollo and Mithras.[96] These Anatolian Leucosyrians or Syrians, or even Assyrians, lived either on the southern coast of the Black Sea[97] or in Cappadocia.[98] Some have searched for Homer's Syriê precisely at Sinope, where these 'Syrians' used to dwell, and which is located towards the summer solstice sunrise as observed from Greece.[99] It is thus possible that this particular Syria in a general sense represented a *heliotropion*, i.e., a point on the horizon where the sun rose on the day of the summer solstice when observed from Greece. It was in any case located near, and on the way to, the land of the sun, Aea, at least after the latter was identified with the region in the eastern recesses of the Pontus (on which see chapter 11).

Thus, certain parallels emerge: both the Hyperboreans and Homeric 'Syrians' are associated with solstices and a solar Apollo, and similar paradisiacal circumstances obtain

[88] Cameron 2004: 34, 41–42, 307.
[89] Cameron 2004: 63, 65.
[90] *Fab.* 140.2, l. 6 Marshall; cf. Fontenrose 1974: 18 for the general tradition to which Hyginus' account belongs.
[91] *Fab.* 140.4, l. 13–14 Marshall.
[92] Pheren. fr. 671 *SH*. See above chapters 4 and 6.

[93] *Od.* 15.405–11. I have already drawn attention to these eschatological references in Homer's account of Syriê in chapter 6.
[94] Pi. *P.* 10.41–43; Simon. 570 *PMG*; Aesch. *Choe.* 372–74; Hellan. *BNJ* 4F187b–c; Theopomp. *FGrHist* 115F75c; Megasht. *FGrHist* 715F27b; Callim. *hDel.* 281–82; A. R. 2.675, 4.614; Pheren. fr. 671 *SH*; Str. 15.1.57 (citing Pindar, Simonides and Megasthenes); Pomp. Mel. 3.5.36–37; *HN* 4.89; Solin. 16.3–5.
[95] Σ A. R. 2.946–54c, pp. 196–97 Wendel with Philosteph. fr. 3 *FHG*; D. S. 4.72.2; Plut. *Lucull.* 23.6.
[96] Hdt. 4.59.2, *cf.* Orig. *CC* 6.39; Hesych. γ 791; *CIG* III.6013 = IG xiv.114*.
[97] Hecat. *BNJ* 1F7ab, 199–201; Pi. fr. 173.1 Snell-Maehler; Hdt. 2.104.3; Soph. fr. 638 Radt; Ephor. *BNJ* 70F43; [Scyl.] *Peripl.* 89, Shipley 2011: 40, 73; Andron *FGrHist* 802F2; A. R. 2.946, 964 (Assyria); Artemidor. fr. 134 Stiehle (1856: 236); [Scymn.] *Perieg.* 917, 941–43, *GGM* i.235–36; Str. 12.3. 9, 25, 16.1.2; Nicol. Damasc. *FGrHist* 90F46, etc.
[98] Hdt. 1.72.1–2, 76.1–2, 5.49.6, 7.72; Callisth. *FGrHist* 124F53; Maeand. *FGrHist* 491–92F4; Nepos (*Datames*) 14.1.1; Str. 12.3.5, 9, 12, 24–25, 27, 16.1.2; Arrian *FGrHist* 156F74; *HN* 6.9, etc. On the reasons for the application of the name of 'Syrians' to Anatolian toponyms see Rollinger 2006b: 78–82, cf. 75–77.
[99] Huxley 1960: 18–23 (esp. 20). For the identity of Syria and Assyria see Rollinger 2006a; 2006b: 72–75, 77–78, with earlier literature.

at both locations. It is thus reasonable to argue that both locations are mythic or cosmological, and that both are part of a model accounting for the annual solar motion.

7.5. Conclusion

A model involving Apollo and accounting for the annual solar movement was apparently instantiated in several differing—and sometimes competing—versions at various key locations of Greek Apolline worship. In cosmological terms, Apollo's northern voyage could only have terminated at the latitude of the geographical arctic circle, in this particular model apparently in relation to an island located at this latitude. This island was inhabited by fabulous people occupying the region of the earth to the north of this latitude and was variously referred to as Syriê or Ortygia. At one point in the development of the tradition, one iteration of the myth applied the story to an island that can be found on a geographical map, Delos (an understandable development in terms of its strong Apolline cult), with a later iteration opting for another tangible island, Britain (once again, in the light of its respectable Hyperborean pedigree, an understandable choice). The latter can be attributed to an influence of a concrete person and event, i.e., Pytheas voyage and his report, as interpreted by Hecataeus of Abdera, while the former remains unrecoverable; both belong to a hermeneutic approach that could be styled rationalising. In addition, it appears that Hyperborea—perhaps from the start—was not limited to this island and that the term encompassed the entire region north of the 'latitude' reached by the sun at the summer solstice. Since none of the versions of this mythic model represents a one-on-one untainted narrative account of the annual solar movement, unconcerned with other important considerations that are clearly present in differing measures in all iterations of this mythic model (most importantly the multifaceted nature of Apollo and his worship, including the vested interests of his most important cult centres), it would be erroneous to claim for any of these stories the closest correspondence to a postulated original account (even if the concept of an 'original' myth was not rightfully abandoned by modern scholarship). Even within a single local tradition, the Delphian, two irreconcilable concepts existed, one placing the return of Apollo in the early spring, the other apparently about the summer solstice. The Delian tradition shows another divergence in the series of different arrivals of Apollo on the island, with different purposes and motivations, as well as with different departure points. But all these iterations have a common underlying structure formed on an analogy with the annual solar movement, with an anthropomorphic agent accounting for this recognisable physical phenomenon.

My discussion has thus made explicit the structure upon which the 'Hyperborean Apollo' complex of myths was fleshed out. This structure corresponds in important ways to the annual movement of the sun, but the actual narratives are not constrained to this overall framework. Since no 'original myth' appears to have existed, what we have is a number of different, but often overlapping, stories that were built upon an analogy with the annual solar movement. These stories as a rule include several mythic motifs that signal their secondary referent: the departing-and-returning Apollo, the Hyperboreans, the solar island Syriê or Delos/Ortygia, the interest in calendrical information, the utopian character of the regions associated with this nexus, etc. I have focused in this chapter on the assessment of calendrical information present in the stories of Apollo's voyage to Hyperborea since this type of evidence most clearly shows the relevance and importance of solar phenomena and their observations for the formation of the 'myth of Hyperborean Apollo'. In addition, I have attempted to clearly delineate the role of the cosmological island Syriê or Delos/Ortygia in this complex of stories and its pivotal position in the models accounting for the annual solar movement involving Apollo.

My analysis of the 'Hyperborean Apollo' complex of myths has shown that Apollo could have occasionally been recognized as a sun god in Greece, with a special connection with the solstices, at least in the traditions associated with the myths involving Hyperborea. However, this does not exclude numerous other functions he fulfilled in Greek society, or even emphasises this particular feature at the expense of others or discriminates other influences on the long process of the formation of this Olympian deity. Although focused on a single aspect—in line with the main subject of my study—my approach is not irreconcilable with, and fully acknowledges the existence of, other points of view.[100] As noted in the introduction to this section, the prevalent opinion among modern scholars is indeed that Apollo is a complex figure that developed diachronically by a process of synthesis and that his 'fully formed' character can be analysed to its component parts. One of these component parts seems to have been his solar association as exemplified in the 'Hyperborean Apollo' set of myths.

The stories pivoting on Apollo in his role as a seasonal traveller to Hyperborea thus naturally belong to a class of myths referencing solar movement. In fact, the mythic nexus as known from Greek tradition seems to be a rare manifestation of a set of stories based on an analogy with annual solar phenomena. Farnell's all-or-nothing approach thus seems to have been misapplied in this particular case, since from their earliest appearance in Archaic poetry the accounts of Apollo's sojourn with the Hyperboreans seems to show a clear interest in calendrical information and Apollo's behaviour appears structured precisely upon the information derived from solar observations.

[100] For these see the latest authoritative study on Apollo (Graf 2009); cf. the essays in Solomon 1994 and Athanassaki, Martin & Miller 2009; Detienne 1998.

8

'Hyperborean Apollo's' swan chariot

After reviewing Greek literary sources for Apollo's seasonal northerly travels, I will now expand my discussion in order to introduce the evidence of a different type: iconographic testimonies of non-literate European societies. My goal is to offer a new interpretation of a vexed question involving the so-called northern Hyperborean Apollo in the light of the conclusions reached in the previous chapter. Much has been written in the past on the question of the connections between Greek Apollo and a Central and/or Northern European prehistoric solar deity recognized in iconographic representations,[1] frequently called, not without some justification—although only if this name is put inside quotation marks and understood as a provisional term referring to a specific concept—'Hyperborean Apollo'.[2] These connections cannot be treated discriminatively towards other influences on the formation of this Olympian deity, but they do occasionally tend to be over-emphasized. As noted in the previous chapter, the prevalent opinion among modern scholars is that Apollo is a complex figure that developed diachronically by a process of synthesis and that his 'fully formed' character can be analysed to its component parts. The results of these modern studies show that any connection with either Central or Northern Europe is at best secondary, but not irrelevant.

Two types of evidence are normally adduced in the discussions on the 'Hyperborean Apollo': literary sources, when analysing Greek tradition (reviewed in the previous chapter), and archaeological record, when analysing the prehistoric. But it has rarely happened that a coherent methodological basis for certain far-reaching conclusions was offered. This chapter will try to review the available evidence, both literary and iconographic (the explananda), and to propose, building from this evidence, what I believe is a sound methodological foundation for the comparative (i.e., paradigmatic) approach in analysing the objects of material culture alongside literary testimonies, based on structural analysis of complex iconographic arrangements, as well as to offer an interpretation of both types of evidence based on the conclusions derived from these comparisons.[3] The main criterion employed in the process of collating iconographic testimonies for this particular concept will be a syntagmatic analysis of their design, but one which is intimately connected with the comparative approach outlined above. This procedure was selected in order to emphasize that it was the data themselves that instigated the discussion on methodology, rather than a preconceived notion subsequently applied onto the two categories of testimonies. However, interplay between theory and data is necessary in every research, and I will often work my way downwards, interpreting data with the help of a theory, in its turn based on ancient literary and iconographic testimonies—thus employing a form of hermeneutic circle. The main hazard of this approach is the possible fall into circular reasoning, which I will try to avoid by different control mechanisms that will be elaborated at appropriate places.

Another difference from earlier studies using similar approaches is that I have chosen to study a more restricted range of iconographic motifs, effecting a significantly more focused analysis of a single, yet complex, iconographic design. I believe that this method offers additional stability on rather slippery ground, and I am confident that the decisive element in this particular design—namely, the presence of anthropomorphic agent—offers still more evidence for the basic similarity of approaches in the treatment of identical phenomena by different traditions.

Finally, it should be emphasized that the concept analysed here was either directly or indirectly connected with cultic activities, which is discernible from both the iconographic and later literary sources and could thus be classified as 'religious'. But, as will become clear in my discussion, the subject of analysis of this chapter, in line with the rest of the book, are literary and iconographic manifestations of a myth understood as a narrative model representing a specific aspect of physical reality, which can only anachronistically be classified as a 'religious' phenomenon.[4] Therefore, this chapter will support the notion that although myth undoubtedly has much in common with the category of 'belief systems', its independence from the concept of religion or cult, with which it is connected and intertwined, but from which it is certainly distinct, should be respected.[5] This also accounts for the relative stability of the narrative model, as opposed to its appropriation by various belief-systems, which undoubtedly in the process re-interpreted it in their idiosyncratic, now largely hard to reconstruct, manners.

[1] For the derivation of a (hypothetical) North European anthropomorphic solar deity from an (equally hypothetical) Central European one see Kaul 1998: 56, 252.

[2] Krappe 1942a; 1943; 1947; Sprockhoff 1954; Gelling & Ellis Davidson 1972; Ahl 1982; Kaul 1998: 253; Kristiansen & Larsson 2005: 44. An earlier version of this chapter appeared as Bilić 2016b.

[3] For the indispensability of the comparative method in religious studies see especially Segal 2001.

[4] For definitions of 'religion' see Insoll 2004: 6–8; for myth and archaeology see Insoll 2004: 127–31.

[5] For a discussion on the relation of Greek myth to Greek religion see Parker 2011: 22–29.

8.1. Hyacinthus—a convergence of literary and iconographic testimonies

As opposed to the numerous literary sources spanning the entire Antiquity, there is only a single—and rather late—iconographic depiction of Apollo in a swan-drawn chariot: the deity (with a quiver) is thus represented together with Cyrene (?) on a Roman-period engraved gem.[6] Moreover, this is the only non-Etruscan depiction of a swan-drawn chariot that can be associated with the god. The connection of this unusual chariot with Apollo in Etruscan tradition is plausibly established through the appearance of *Hyacinthus* in a chariot drawn by swans. Apollo is indeed sometimes explicitly syncretised with Hyacinthus,[7] perhaps also in iconography,[8] his Laconian appearance as Apollo τετράχειρ[9] can be associated with the hero,[10] while Spartan Hyacinthia were devoted to both,[11] also suggesting a sort of syncretism. With this in mind Hyacinthus' appearance in a swan-drawn chariot closely associates him with Apollo. However, as opposed to Apollo in a swan chariot, this concept is almost exclusively known through iconographic representations. Actually, Philostratus (3rd c. AD) is the only literary source mentioning the hero riding in Apollo's swan-chariot (*Im.* 14), albeit while describing a painting, although not the depiction itself, whereas the theme was rather popular in Etruscan art from the 4th c. BC onwards.[12] Thus the Hyacinthus/swan-chariot association is almost exclusively known through iconographic representations of Etruscan origin. This is not so strange if we take into account that Etruscan iconography occasionally illustrates certain poorly documented versions of Greek myths.[13] On the other hand, it will be shown below that it is perhaps possible to associate this particular Etruscan iconographic motif with a strong local prehistoric tradition of producing structurally similar complex symbolic iconographic nexuses.

8.2. Archaeological evidence

I will now offer an analysis of several prehistoric iconographic arrangements revealing a similar syntagmatic structure, allowing for certain paradigmatic shifts: an anthropomorphic deity (or mythic person) riding in a chariot drawn (or accompanied) by water birds, normally swans, decorated with postulated solar symbols.[14] With respect to these, it appears that often the scholars discussing many of the complex symbolic objects analysed here independently arrived at some rudimentary form of the interpretation that is argued for in this chapter, but failed to precisely define the characteristics of the symbolic structure that was the object of their immediate study, to use comparative pieces of evidence and to contextualise particular manifestations of the concept underlying them in their proper intellectual setting. I will henceforth refer to this set of ideas as the 'Dupljaja concept', not because I believe it is the 'origin' of all other attestations of this matrix, but because it is its best known and, as it will be shown below, one of its most revealing, materialisation. This complex arrangement permits, I believe, the supposition of a similar underlying *muthos*, a story or concept shared by these complex schemes.[15] In any case, as will be shown below, it seems more profitable to compare systems composed of several elements that otherwise appear separately, since single motifs or symbols can cover a wide range of meanings, but their arrangement into a presumably coherent structure—i.e., one that carries some meaning[16]—significantly reduces the number of possible meanings. Because of this I have selected a highly complex and specific system, rather than such simpler matrices as *Vogelsonnenbarke* or *Vogelbarke* or water-bird symbolism in general, onto which most of the earlier analyses focused. In this way I hope to bypass some of the pitfalls of comparative method, especially circularity, i.e., arbitrarily adducing the meaning of a motif and applying this interpretation onto other occurrences of this motif without further discussion.

I have applied two main criteria for selecting the complex sets discussed in this chapter. First, following Renfrew, I have employed a cross-cultural identification of various representations of recurring mythic persons, defined by their specific attributes,[17] although I have studied these 'clues' in the context of the structure of a complete iconographic scheme.[18] I have thus been able to recognize a recurring interrelation of a group of motifs and their syntagmatic arrangement (an anthropomorphic deity/mythic person riding in a chariot drawn/accompanied by water birds, normally swans, decorated with postulated solar symbols), and have provided a syntagmatic analysis[19] of thus

[6] *LIMC* Kyrene, no. 18.

[7] Plb. 8.28.2; Nonn. *D.* 11.220.

[8] *LIMC* Hyakinthos, no. 55.

[9] Hsch. κ 3853, cf. κ 4558; Sosib. *FGrHist* 595F25; Lib. *Or.* 11.204; *IG* V.1.259.

[10] Farnell 1907: 127, 371, n. 45.

[11] Pettersson 1992: 9–41; Graf 2009: 34.

[12] *LIMC* Hyakinthos, nos. 35, 37–40. No. 39 might show a trace of a quiver, which is highly conjectural, but it would associate the figure with Apollo, rather than Hyacinthus.

[13] Nielsen 2002: 183.

[14] The postulated solar symbolism must naturally always remain tentative. As a rule, I have noted under this category only those iconographic elements that were accepted as having some solar connotations by previous authors that discussed the respective objects considered here. An element of circular argument is unavoidable at this point, but the structural analysis should eliminate the arbitrariness of the assumptions adduced here. For solar symbols in the European BA iconography see Müller-Karpe 1978–79: 23 with Abb. 6.1–11 on p. 24.

[15] Throughout this chapter I will use the transliterated Greek word when referring to this specific meaning of the concept 'myth' in order to avoid the negative connotations acquired by the latter, especially with respect to its truth-value. It also recollects the Aristotelian meaning of the term (*Poet.* 1450a4–5, cf. 32–33, 36–39), which is very close to this specific meaning.

[16] Cf. Renfrew 1994: 53.

[17] Renfrew 1994: 53. The emphasis on anthropomorphic character of the main agent in the structure of a complex figure will be explained later; it is not conditioned by Renfrew's discussion.

[18] Occasionally, individual religious or iconographic motif could be tentatively treated as an element in a complex religious phenomenon, and, moreover, as representing some feature of particular natural forces (Kaul 1998: 13). I decided not to apply this hypothesis in my discussion, since I wanted to give additional stability to my argument by relying on structural analogies between complex symbolic structures.

[19] Cf. Hodder 1987:3.

Figure 8.1 Dupljaja model (after Bilić 2016b: 449 Fig. 1. Reproduced courtesy of University of Ljubljana, Faculty of Arts).

collated complex prehistoric iconographic arrangements. This criterion could provisionally be called 'structural'.

Additionally, I have placed one of the accents on the precise (as much as possible) chronological relations in order to contextualize the discussed symbolic structures and thus to eliminate the inherent ahistorical nature of structuralist approach.[20] It could be objected that the complex figures analysed in this chapter cover vast chronological (more) and spatial (less) distances; this is indeed true, but I hope that the following discussion will explain why this is not an unsurpassable problem.[21] The key concept with respect to this possible objection will be the definition of a large-scale context or shared cultural space in which these complex figures appear. In a similar fashion Hansen argued for a spatially and temporally broad perspective in studying formal similarities of Neolithic and Chalcolitic anthropomorphic figurines of the Near East, Anatolia and South-Eastern and Central Europe.[22] This criterion could provisionally be called 'contextual'.

These main criteria, together with their refinement that allows a stronger focus and control, create a solid framework that eliminates up to a reasonable point the possibility of arbitrariness in the selection of the complex symbolic objects used in this chapter.

8.2.1. Dupljaja

Apollo's swan chariot and its connection with Hyperborea were long ago, and still regularly are, associated with the famous BA wagon model from Dupljaja (Figure 8.1). It shows a male god,[23] riding in what is best described as a swan-chariot, defined by three water birds emerging from the vehicle. The base of the hemispherical wagon-box is decorated with a four-spoked wheel, most probably a

[20] For the analysis of the ahistorical nature of structuralism see Vidal-Naquet 1981b: 175–76, 185; Hodder & Hutson 2003: 62–63. For structuralism and history see also Harkin 2009: 39–40, 44, 46–48, 50, 52. For the importance of chronological relations in approaching similar issues see Kaul 1998.10, 15. For the structural analysis of material culture see Hodder 1987: 3; Hodder & Hutson 2003: 45–74, 126, 214–15.

[21] Compare Müller-Karpe's study on BA objects with religious symbolism, covering the area from Egypt to Atlantic and the North Sea during the Middle and Late BA (Müller-Karpe 1978–79).

[22] Hansen 2001: 38–39, 42.

[23] An anthropomorphic solar deity, according to Sprockhoff 1954: 67 (cf. Maraszek 1997: 75); Gelling & Ellis Davidson 1972: 119; Green 1991: 114, cf. 13, 84; Kaul 1998: 254; Kristiansen & Larsson 2005: 150, 307. Holenweger 2011: 223–24, 242–44 also argues for the statuette's role in the solar cult manifested in the ceramic material of the Dubovac-Žuto Brdo-Gârla Mare cultural group. For this particular solar cult cf. Palincaş 2013: 317.

solar symbol.[24] Alternatively, the model can be interpreted as a sun-ship with swan-stems.[25] This paradigmatic shift does not change much in the model's syntagmatic arrangement.[26] In any case, the model could be interpreted as a fairly faithful illustration of the myth of Apollo riding in a swan-chariot (although, I must emphasize, it most certainly is not such an illustration).[27]

It is fortunately possible to date the model with a precision that allows its contextualization with some certainty. It is thus regularly associated to the MBA 'Danubian' anthropomorphic figurines,[28] and is accordingly dated to this period.[29] However, the Dupljaja chariot shows some features that are quite unique in the complex of the 'Danubian' MBA anthropomorphic figurines, in the same time clearly foreshadowing the future Urnfield solar/water bird symbolism.[30] This fact is reflected in the attempts at dating it by a number of scholars, opting for a somewhat later period (the LBA) in comparison to the opinions adduced above.[31] From a semantic point of view, that is, with respect to the interpretation of its meaning, however, it is less relevant whether the chariot should be dated to the Middle or Late BA, as long as its large-scale context is clearly defined: the symbolism of either nascent or already established Urnfield cultural complex. Relatively numerous finds of a type of *Kesselwagen* (cauldron-wagons) incorporating in their design the water-birds symbolism[32] and other objects decorated with motifs such as water birds, sun-symbols, and (occasionally) a barque (especially the *Vogelsonnenbarke*), dating from the Bronze-age Urnfield cultural complex and the later Hallstatt period,[33] distributed throughout the Central Europe and beyond (Italy, Northern Europe, the Aegean), show that the Dupljaja wagon, although unique in its complexity, is not an isolated find, but part of an elaborate and widely distributed set of beliefs.

Thus, the existence of an intricate concept can be recognised in the Middle Danube region in the second half of the 2nd millennium BC, manifesting itself in the Dupljaja figure. This complex structure then occasionally reappears in succeeding periods throughout the Central, South-Eastern, South-Central and Northern Europe, as will become apparent in the following discussion.

8.2.2. Northern Europe

The only direct Northern European structural parallel to the 'Dupljaja concept' is a LBA belt buckle from Floth (Ha B2/3–Ha C1, 9th–8th c.).[34] Two symmetrical *Vogelbarken* with stylized human adorants (the head of one of the figures is formed of two concentric circles, the head of the other of a single circle) are depicted on this buckle, with three solar disks around them—positioned at sunrise, zenith and sunset?[35]—the two flanking ones pulled by water birds. B. Hänsel recognizes in the left corner of the plate another stylized human figure with the head in the form of a solar disk and argues that this depiction suggests the existence of the personification of the sun—an anthropomorphic solar deity—in the north of Europe.[36]

8.2.3. Italy

As already noted, the concept of Hyacinthus in a swan-chariot is exclusively known through Etruscan material. Several other examples from earlier periods similar to the 'Dupljaja concept' were found in Etruria. The earliest is a bronze Protovillanovan razor (12th–10th c.), unfortunately of unknown provenance.[37] The razor is in the abstract form of a female idol (a paradigmatic shift), with a double-axe depicted on the body, inside of which is a mirror-image figure with arms formed of *Vogelbarken*, with two flanking figures of a similar type, this time with *Vogelbarken* forming their lower extremities—or, better, showing a human figure inside a *Vogelbarke*[38]—and four water birds in the corners. Not unlike the Knossian pithos discussed

[24] Petrović 1928–30; Kossack 1954: 11–12; Sprockhoff 1954: 67; Bošković 1959; Letica 1973: 63–64; Coles & Harding 1979: 408; Pare 1987: 58–61, fig. 25; 1989: 84–85, fig. 4; 2004: 357–58, fig. 2; Green 1991: 45, 84, 114, figs. 88a–b on p. 113; Kaul 1998: 254 with fig. 169; Harding 2000: 167, 322, 324; R. Vasić & V. Vasić 2003: 158–60; 2003–2004: 182–83; Holenweger 2011: 223–24, 242–44, 334–35, no. 168.

[25] Sandars 1968: 174; Kaul 1998: 254.

[26] Cf. Kaul 1998: 254.

[27] Compare Sprockhoff 1954: 70–71 and, more cautiously, Kaul 1998: 254; this is one of the reasons that led Vasić 1954 to pronounce it a forgery.

[28] For a lengthy review of previous scholarship on the origin of these figurines see Holenweger 2011: 123–48, 150–51; for their Aegean origin, see now Holenweger 2011: 45–46, 148–53, 188–91, 266–67, 270 (to which add Chicideanu-Sandor & Chicideanu 1990: 69–70, 73; Kalogeropoulos 2007: 263–65; Palincaş 2012: 24); for their Central European origin, see Letica 1973: 53 (cf. Kossack 1954: 9; Sandars 1968: 176, 315 n. 51; Kiss 2007: 125–28).

[29] Pare 1989: 84 (cf. Chicideanu-Sandor & Chicideanu 1990: 57): Br C1–Br C2, contemporaneous with LH IIIA (late 15th–14th c.) Mycenaean Phi idols; Letica 1973: 60: Bz B (1600–1500 BC); the whole complex of these figurines is dated to Br C1–Br C2 (1500–1300 BC) by Majnarić-Pandžić 1982: 53, to Br B1–Ha A1 (1600/1500–1200/1100 BC) by Kiss 2007: 127, or to a somewhat broader chronological horizon (the entire MBA) by A. & B. Hänsel 1997b: 59.

[30] Cf. Kristiansen & Larsson 2005: 307–308.

[31] Sprockhoff 1954: 67, 73; Bouzek 1985: 53, 178, 234 n. 15; Kaul 1998: 254: Bz D (early Urnfield, 1300–1200 BC); Coles & Harding 1979: 408; cf. Green 1991: 147 n. 170: vague attribution to 'the Urnfield sphere', at the same time tentatively suggesting a date as late as 1050 BC or even later; A. & B. Hänsel 1997b: 67: late MBA/early Urnfield period.

[32] Orăštie, Br D–Ha A, 13th–12th c. (A. Jockenhövel & F. Verse in Demakopolou *et al.* 2000: 260–61, no. 176 with p. 134, fig. 2); Skallerup, earlier per. III = Br D, 13th–12th c. (Pescheck 1972: 50; Egg & Kaul 2001: 474, Abb. 55); Peckatel, earlier per. III = Br D, 13th–12th c. (J. Jensen in Demakopolou *et al.* 2000: 261, no. 178 with p. 6, fig. 1); Acholshausen, transition from Ha A2 to B1, c. 1050/1020 (Pescheck 1972: 49–52); the 8th–7th c. examples from Bujoru (Moscalu & Beda 1991: 217, Abb. 11), Delphi (Kilian-Dirlmeier 1974), Marsiliana d'Albegna (Egg 1991: 192–207), Veji, Grotta Gramiccia (Müller-Karpe 1974: 96–97, T. 22.1) and Vetulonia, Circolo dei Lebeti (Egg 1991: 202 Abb. 10).

[33] Bouzek 1985: 178 dates the earliest Vogelbarke in early Br D (after 1250 BC), while Matthäus 1980: 319 and Kaul 1998: 278, 280–81, 283 date it to the much later Ha B1 (c. 1000 BC).

[34] A. Hänsel in A. Hänsel & B. Hänsel 1997a: 133–34 s.v. Floth; B. Hänsel 1997: 20–21, fig. 2.

[35] B. Hänsel 1997: 21.

[36] B. Hänsel 1997: 21.

[37] Bouzek 1985: 216, fig. 103.11; Kaul 1998: 284–85, fig. 180. Jockenhövel 1974 with pl. 19.1 opts for a somewhat later date (10th c.; Jockenhövel 1974: 84, 88) and recognizes in it an eastern Mediterranean influence (Jockenhövel 1974: 87–88)

[38] Perhaps representing 'die Personifizierung der Sonne' (Jockenhövel 1974: 87).

below, this symbolism suggests an association with female cult.[39] In chronological terms this object would represent a transition from the Middle or Late BA Dupljaja model to the somewhat later Italian specimens described below.

A similar representation is depicted on a vessel from Veii decorated in the *Buckeltechnik* and dated to the early Iron Age (early 9th c.). The vessel type to which the Veii specimen belongs is attributed to the late phase of the Central European Urnfield culture.[40] The decoration on this vessel depicts a frieze of water birds (swans?), but not the symmetrical *Vogelbarke*, although one could argue that the bodies of the birds indeed form a boat. Between the heads concentric circles are depicted, most probably designating the sun, and, most interestingly, two human figures with outstretched arms and circular heads.[41] Since both the birds and the sun-rings form a continuous frieze, it is unlikely that two human figures represent twins of any kind.[42] Indeed, the decoration on the Veii vessel could represent a sun deity travelling in a solar swan-boat.

A slightly later group of Early IA fibulae (Ha B3, c. 850–750) from Campania, with a person or a deity in a *Vogelbarke*, almost certainly belongs to the same cultural circle and represents the same matrix,[43] as does a bowl handle from Bisenzio in Etruria (late 8th/early 7th c.), depicting a person or a deity whose hands and feet extend into ornithomorphic forms standing in a circular *Vogelbarke*.[44]

Another example of a similar concept is a *bucchero* vase from Tomba Calabresi (?) in Cerveteri, dated to the later part of the 7th c.[45] It is modelled in the form of a double imaginary being with horses' heads and ornithomorphic body, with two rosettes on its sides, suggesting a solar association and in the same time symbolising a wagon. A man standing on the body of the animal and holding a yoke in his hand is at the same time a rider and a wagon-driver, riding his *Vogelpferde* or being drawn by them in a chariot.[46] The *bucchero* vase could represent a solar deity in his chariot drawn by mythic horse-birds.

Finally, Woytowitsch believes that a bronze 6th c. lid in the form of a large, stylized bird's head (with a small bird on the top) on whose back a human figure drives a *biga* (the wheels are not represented, but only the wagon-box) probably originated in Italy, yet it was found in the Saône river.[47] Thus, this depiction can be added to the Etrurian ones enumerated above.

Apparently, the 'Dupljaja concept' was current in Central Italy during both prehistoric and historical (Etruscan) periods, with a clear continuity between the two, which supports (but does not prove) the proposed continuity with later literary testimonies argued for below.

8.2.4. Eastern Alpine region

The so-called *Kriegerwagenfibel*, a group of Iron Age fibulae with the depiction of a male figure on a horse-drawn chariot on the bow and a bird at the end of the foot, found at several sites in the eastern Alpine region,[48] are composed of identical structural elements, and should be compared to the 'Dupljaja concept'. Regardless of the fact that there is no apparent solar symbolism in these fibulae, they were nevertheless associated with the Phaethon legend.[49] Even though this seems somewhat far-fetched, their solar character could be tentatively recognized in other structural elements present in these artefacts. These fibulae are very similar and chronologically close to the bronze lid found in the Saône river described above.

8.2.5. Possible Central European parallels

Several bronze *Vogelsonnenbarke* pendants found over a wide area (Hungary, Bosnia, Italy, France, Austria), and dated to a long time span stretching from the middle Urnfield[50] to the late Hallstatt period (Ha A2 to D), perhaps portray an anthropomorphic figure, with *Vogelbarke* representing its outstretched hands.[51] In the same time, the *Vogelbarke*, together with the figure's circular head, could represent a transport vehicle of the solar disk. Thus, this representation can be in the same time interpreted as an anthropomorphic figure with hands in the form of a *Vogelbarke*, or as a *Vogelbarke* transporting the solar disk. Due to their extreme stylization, and, more specifically, to the uncertainty regarding the anthropomorphic interpretation of the composition, I am not completely convinced that these artefacts belong to the group of complex symbolic sets discussed above. Nevertheless, it is possible to recognize in them all the structural elements that were present in other manifestations of the 'Dupljaja concept': anthropomorphic figure, solar symbolism (the disk-shaped head of the figure), boat (explicitly present only in the specimens reproduced on Kossack's pl. 11.7 and 10, otherwise suggested by the *Vogelbarke* representing the figure's outstretched hands), and water birds. Moreover, both geographically and chronologically they belong to the same large-scale context (the Urnfield

39 Bouzek 1985: 217; Wachsmann 1998: 195–96.
40 Von Merhart 1952: 12–13.
41 Sprockhoff 1954: 81, 82 fig. 24; cf. von Merhart 1952: pls. 3.8, 23.1; Iaia 2004: 397 fig. 2.10.
42 *Pace* Sprockhoff 1954: 81.
43 For the examples see Lo Schiavo 2010: 880–81 (type 450), pls. 694–98, nos. 8047–60B, 883, 885–86 (types 451.1 and 2), pls. 709, 719–25, nos. 8079B, 8085B and C, 8089–90.
44 Kossack 1954: pl. 13.1; Bietti Sestieri & Macnamara 2007: 2004.
45 Guggisberg 1996: 185, fig. 16.
46 Guggisberg 1996: 186.
47 Woytowitsch 1978.64–65, no. 142, pl. 29.
48 Starè 1954: 188 fig. 15; Guštin 1974: 95–98; Lunz 1974: 139–40, pl. 40.9, 91B; Tecco Hvala *et al.* 2004: pl. 10.A2; Tecco Hvala 2012: 228, 259, 260 fig. 99.2, 262–63.
49 Tecco Hvala 2012: 262 no. 1084, 340–41, 385.
50 Perhaps even to the early MBA (Br B), if the specimen from Nagyhangos belongs to this type (Müller-Karpe 1978–79: 23 with Abb. 6.23 on p. 24; 1980: pl. 318.E8).
51 Kossack 1954: pls. 11.2–4, 7, 9–11, 13, 12.4, 7, 9–18; Bouzek 1985: 171, fig. 86.12, 174, fig. 87.1. Kossack 1954: 42, 44, 53, 58–59 only argues for the anthropomorphic nature of the amulets reproduced on his pl. 12.1–10 (cf. for analogous interpretation of similar amulets Müller-Karpe 1978–79: 26).

symbolism) with other examples discussed here. Thus, I would leave the question of their affiliation open, which equally applies to two Ha D pendants from Vače and Vinji Vrh in Slovenia, where a highly stylised combination of an anthropomorphic figure and a *Vogelbarke* can be discerned.[52]

8.2.6. The Aegean

A similar nexus appears in the decoration of a straight-sided pithos from Knossos. This depiction of a winged female divinity riding in a chariot[53] with birds in her hands suggests that the 'Dupljaja concept' was current on Crete in the Protogeometric period (later 9th c.). This is the only Aegean example of a complex symbolic structure related to the 'Dupljaja concept', suggesting (but not in itself proving) continuity between the pre-literate and literary traditions in this particular region. Cretan adaptations, which can be understood as paradigmatic shifts that do not change the structure of meaning of the matrix, are revealed in the driver's gender and in the appearance of birds, which are depicted identically to earlier illustrations of that kind in Cretan tradition. The seasonal character of the depiction is evidenced in the fact that the pithos shows two contrasting scenes, probably representing summer and winter.[54] While it is not clear from the depiction whether the birds actually draw the chariot, the similarity with the 'Dupljaja concept' is rather obvious.

8.3. Methodological procedure for comparison of literary and iconographic record

I will now try to outline a methodological procedure that would allow a profitable comparison of iconographical with literary sources. This procedure will answer in the positive the question whether the *muthos* behind prehistoric iconography can be compared to the one attested in literary sources. But in order to reach this positive answer, several important steps must be taken with respect to obtaining an insight into a past belief system manifested only in material objects. I will try to systematize these steps into five different, yet interconnected, phases of inquiry, with the purpose of reconstructing the *muthoi* of such a belief system.

8.3.1. Material evidence for past beliefs

Several important questions must be raised with respect to the categorical statements outlined in the introductory section under the heading 'Archaeological evidence' (section 8.2.), which I will try to answer here, while simultaneously building an argument for the possibility of comparison of literary with iconographic sources. One of the basic questions is whether it is at all possible to gain any insight into a past belief system manifested only in material objects. Without acknowledging this possibility, it remains impossible either to assess the content of, or meaning inherent in a system, or to compare it with literary sources. The cognitive-processual approach indeed allows that iconographic analysis, that is, the analysis of symbolic systems, represents a coherent method for a reconstruction of particularities and meanings of a belief system attested in non-literary sources.[55] Thus a cautious interpretation of past beliefs based only on iconography is indeed admissible, representing an important step in attempt to compare them profitably with each other or with written testimonies.

8.3.2. Reading the visual language

In general, iconography can be understood as a symbolic system employing a coherent non-verbal language used for codifying a culture's reflections on the objective reality through a system of associations, with the knowledge of the significance of these symbols being shared by everybody that is 'initiated' into the system.[56] A 'translation' of visual into spoken language, which is a basic act of any iconographic interpretation,[57] is necessary in order to compare iconographic with literary testimonies. In this way the former are treated as equals to the latter, differing idiomatically but not essentially.

8.3.3. Structural analysis of visual language

Acknowledging iconography as an interpretable system of communication is a first step in building a defensible method for the comparison of material and literary sources but is also a crucial move towards allowing the introduction of a profitable structural study of complex iconographic systems, which was actually applied in the preceding discussion. Since language, as the form of its communication, is inextricably linked with meaning,[58] a structural analysis of the language's visual form must necessarily be incorporated into the interpretation of meaning,[59] for it is the structure of iconographic elements that reveals the meaning of a complex image, rather than individual elements taken outside their immediate context. This analysis is therefore focused on finding structural principles behind complex images—what Morgan appropriately calls an idiom,[60] but what could more specifically be understood as 'the content of ideas' that assemble signs into a syntagmatic or paradigmatic set,[61] and which I previously termed a *muthos*. The Dupljaja model is precisely this kind of a complex figure which should be studied in terms of structural analysis

[52] Kossack 1954: 44, pl. 17.2, 4; Tecco Hvala 2012: 298.
[53] Coldstream 1984: 97, 99 argues that the wheeled platform is an abbreviation of a chariot (cf. Marinatos 2000: 126).
[54] Coldstream 1984: 99, figs. 1–2 on pp. 96, 98, who recognizes in this representation the spring arrival of the goddess, and her departure in winter (cf. Marinatos 2000: 126).
[55] Renfrew 1994: 49, 53–54.
[56] Morgan 1985: 7; Renfrew 1994: 53; cf. Harding 2000: 345–46. On some idiomatic differences between textual and pictorial mode of representation see Ornan 2005: 11–12.
[57] Morgan 1985: 6.
[58] Cf. Hodder & Hutson 2003: 160.
[59] Morgan 1985: 6, n. 2.
[60] Morgan 1985: 9.
[61] Hodder 198: .3.

and compared with other analogous nexuses in order to interpret 'the semantic implications of syntax'.[62] This is exactly what I am trying to do in this chapter.

8.3.4. The transfer of meaning

With reference to the problem of context, it is now time to analyse another important question raised in my introduction under the heading 'Archaeological evidence' (section 8.2.): whether the transfers of symbols between cultures could also be accompanied with the transfers of meaning,[63] i.e., whether it is possible to argue for any type of connections between the meaning(s) of these complex arrangements. That is, after I have recognised syntagmatic similarities in the structure of several complex prehistoric iconographic matrices, I must now raise the question of their mutual relations. Are meaningful relations between these complex figures even possible? A transfer of a symbol, however complex, bereft of its underlying meaning is not very revealing for attempting to reconstruct a past belief system, although it could be revealing in other respects. But the recognition of the existence of a shared *muthos*, 'story' or concept behind symbols could allow some insight into past beliefs, even without reconstructing the actual contents of the underlying tradition.

Once again, the cognitive-processual approach allows the recognition of references in iconographic representations to recurrent themes across cultures, and the existence of single coherent systems, both symbolic systems and systems of belief underlying them, where there are significant overlaps between specific symbols appearing at various locations.[64] In other words, the existence of shared cross-cultural belief systems is recognized in the appearance of similar iconographic solutions in different cultural systems. Consequently, under certain conditions, similar iconography could indicate similar beliefs. A concrete example of this interpretation is provided by Hansen's study of formal uniformity of Neolithic and Chalcolitic anthropomorphic figurines, for which he argued that it indicates the existence of distinct ideas behind them, which he classified as elements of a religious belief-system.[65]

A similar conclusion can be reached by a study of post-processualist or contextualist account of diffusion, which is implicitly discussed in the foregoing cognitive-processual account of recurrent cross-cultural iconography.[66] Hodder & Hutson thus argue that objects can be transferred from culture to culture or context to context with their meaning unchanged, or if changed, this new meaning still being based on the old.[67] Thus, once again, under certain conditions similar objects could indicate similar beliefs. However, a 'larger-scale context in which similar meanings are assigned to similar objects' must be defined in order to be able to discuss the more localized contexts responsible for both spatial and temporal variability, if any exists, in various manifestations of both widespread and long-term symbolic structure I have recognized and discussed in this chapter.[68] I have indeed tried to summarily contextualize (i.e, to position them chronologically and culturally) individual manifestations of the symbolic structure discussed in this chapter, which I provisionally called the 'Dupljaja concept', at appropriate points in the discussion, and also to provide a particular large-scale context in which this structure appeared and in which similar sets could have had related meanings.[69] I believe I have managed to avoid circularity in my reasoning—defining a large-scale context by the successive appearances of a similar structure and then interpreting the structure by its participation in that very context—by delineating this large-scale context (when discussing the Dupljaja model) by additional elements that it contains: the Urnfield symbolism. In any case, the existence of the particular symbolic complex (or cultural *koine*) to which the 'Dupljaja concept' belongs is non-controversial and does not in any way depend on its delineation as outlined in this chapter.[70] This large-scale context will be further discussed in the concluding section of this chapter.

8.3.5. The transfer of beliefs

This archaeological large-scale context corresponds to what Graf refers to as an osmotic similarity of cultural space, in which a transfer of analogous ritual and mythic concepts between cultures is as likely an explanation for their formation as is independent origin.[71] Here the emphasis is on the transfer of concepts or beliefs (i.e., *muthoi*), rather than objects, but it is equally plausible to assume that *muthoi* accompanying iconography with cultic significance are transmitted in a precisely identical fashion, especially in light of the discussion in the preceding section. Thus, both narratives and symbolic representations—the latter together with accompanying beliefs—are equally transferrable between cultures, which allows the attempt at reconstructing these very beliefs through a comparison with literary sources.

Transferability is indeed one of the most conspicuous characteristic of traditional stories or myths,[72] and both motifs and entire narratives can migrate between cultures, including their denotative applications.[73] In any case,

[62] Morgan 1985: 16.
[63] Milićević Bradač 2005: 187.
[64] Renfrew 1994: 49, 53–54.
[65] Hansen 2001: 42–45.
[66] I find it a felicitous occurrence that two different approaches to archaeological theory agree on this particular point, although this does not automatically prove its validity. For their respective positions on religion, here used as a broad term encompassing symbolism and belief, see Insoll 2004: 79–87, 94–100; for Hodder's position specifically, see Bredholt Christensen & Warburton 2014.
[67] Hodder & Hutson 2003: 140. Here the authors do not refer to complex symbolic structures, but their observations certainly apply to these even more than on simpler objects.
[68] Hodder 1990: 21; cf. Milićević Bradač 2005: 188.
[69] Cf. Hodder 1987: 8.
[70] See, e.g., Bouzek 1985.
[71] Graf 2004a: 5.
[72] Bremmer 1994: 57; cf. 2011: 540; Wyatt 2005[2001]: 170; López-Ruiz 2010: 128.
[73] Burkert 1988: 12; Graf 2004b: 52.

these borrowings between cultures are never random but rather result from a structural similarity between myths in both 'receiving' and 'giving' culture.[74] Thus a transfer of beliefs—whether accompanied by or accompanying iconographic solution with cultic significance or not—is plausibly ascertained; moreover, it is myths or traditional stories that are particularly prone to diffusion. But other types of 'beliefs' are also culturally transcendent. Thus, the transmission of scientific 'traditions' between cultures is an accepted fact in the history of science. Knowledge or methods (the contents of the 'tradition') are received, adapted and occasionally transformed when being translated between cultures, but they still remain within the culturally transcendent tradition.[75] This fact is tangentially relevant to the contents of the narrative model discussed in this chapter, since the particular *muthos* reconstructed here attempts to give an account of a phenomenon that would later be treated by Greek cosmologists and astronomers, although in a radically different fashion.

With respect to the relation between material testimonies and narratives that treat identical concepts, the former are sometimes indeed of prime importance in studying the latter. The significance of iconography in recognizing oral tradition behind it is emphasized by Burkert, who argues that iconographic treatments of myths 'play a fundamental role in the fixation, propagation and transmission of those myths...'[76] He further maintains that iconography can unmistakably indicate connections between different societies, although in the same time he believes it cannot positively indicate a particular myth's diffusion.[77] He allows for some transfer of underlying meaning, but also recognizes possible deviations.[78] These well-balanced remarks, equally applicable to a pre-literate period (but without the control offered by literary sources), emphasize that one should always have in mind possible misunderstandings and reinterpretations in the transmission of meanings of symbolic representations; the method I have adopted in this chapter, in the first place the focus on a quite specific single, yet complex, iconographic design, of a particular character (i.e., anthropomorphism), is meant to reduce these pitfalls as much as possible, although they can never be completely evaded.

It is possible to raise an objection here that religion is a structured system whose elements derive their meaning through their relations with other elements on various levels, from the arrangement of elements in a nexus to the complete world-view of a society. Change on any of these levels affects all the others, and the meanings of apparently identical elements also change in new circumstances.[79] Indeed, even collective representations of physical environment are argued to be culturally determined, and thus both these representations and their meanings are liable to alterations, which are induced by the changes in other elements of the systems forming a society's world-view.[80] But I have demonstrated that meanings can be preserved within a common cultural space, and I will show immediately below that this is especially the case when complex symbolic structures are involved. This claim is less dogmatic and allows for both modification and preservation of meaning, without a priori rejecting but insisting on arguing for or against either possibility. Moreover, a preservation of meaning does not mean that, for example, the 'Dupljaja concept' was not modified in a number of ways—which are particularly visible in paradigmatic shifts noted at appropriate points—but that both elements (less) and nexuses (more) can keep their meaning relatively unchanged; what is more, these meanings can be reconstructed by following a careful methodological procedure. All the manifestations of the 'Dupljaja concept' analysed here are, naturally, culturally determined, but even so they have a stable referent in the physical environment. Precise roles of these manifestations in the world-views or belief-systems of their respective societies, however, are much harder to grasp, and I have not attempted to reconstruct them here.[81] This is thus not a simple case of dynamism vs. apparent permanency, but rather of dogmatism vs. argued tolerance. My case should be judged by assessing the arguments for continuity of ideas and their expression in material form I have offered and not by an a priori rejection due to some presupposed eternal flux in the content of ideas forming belief-systems. Two additional arguments could be adduced in further support of my position: the inherent conservatism in ancient religious works of art[82] and the more conservative nature of cultural forms in earlier periods.[83] Both these factors speak in favour of preservation of meaning in the concrete transfer of complex symbolic structures argued for here.

8.3.6. Transfer of complex symbolic structures

It is important to emphasize that, on one hand, a simpler form of a symbol makes its interpretation to the 'uninitiated' more difficult[84] and thus more complicated to transmit unchanged and unscathed of its intrinsic meaning between cultures,[85] but on the other hand,

'complex symbolic structures are more likely to maintain their internal meaning unchanged than simpler ones, as their transmission demands the adoption of a corresponding complex knowledge... the parallel transmission of two or more symbolic structures makes it increasingly likely that they maintained their internal meaning unchanged, as

[74] Doniger 2009: 208.
[75] Rochberg 1992: 549–50.
[76] Burkert 1988: 25.
[77] Burkert 1988: 26.
[78] Burkert 1988: 27.
[79] Sourvinou-Inwood 1995: 20–24, 29. Paradoxically, in the same work Sourvinou-Inwood 1995: 19–20, 25, 29 does acknowledge continuity in certain elements of various religious nexuses, although not the nexuses themselves.

[80] Sourvinou-Inwood 1995: 22–23.
[81] But see Kaul 1998 and Kristiansen & Larsson 2005.
[82] Ornan 2005: 10.
[83] Wyatt 2005[2003]: 220.
[84] Morgan 1985.7; Milićević Bradač 2005: 188–89.
[85] Kristiansen & Larsson 2005: 22.

it testifies to a more complex and direct transmission of knowledge'.[86]

This statement is an important argument for the comparison between symbolic structures attempted here. It plausibly suggests that a complex structure—in this case a matrix consisting of a travelling solar deity or, if one prefers, an anthropomorphic agent providing causation of solar movement—might have been transmitted from one cultural complex to another without losing much of its intrinsic meaning. In this sense, it is not unreasonable to presume that certain ideas were transmitted both horizontally (in spatial terms) and vertically (in temporal). Since the meaning of complex symbolic structures is more likely to be transmitted unchanged between cultures, it can be argued that the ideas (*muthoi*) behind them were transmitted together with their manifestations in iconographic arrangements.

8.3.7. The Dupljaja model as a complex symbolic structure accompanied by a muthos

It is possible to recognize in the Dupljaja model an iconographic depiction of a mythic concept that could have played—to paraphrase Burkert—an important role in the latter's fixation, propagation and transmission. Generally, this is hardly a novel conclusion, since a number of earlier authors recognized its importance, although not in such specific sense as argued for in this chapter. Thus the origin of the concept of a solar chariot or boat associated with swans was traced to the Central Europe, from where it arrived both to Greece and Northern Europe.[87] The Dupljaja model itself is described as a vital link in this process.[88] But what is novel in my approach is that while the authors named above all discussed a more general concept (solar chariot/boat associated with water birds), I have taken into account and emphasized the anthropomorphic nature of the main agent in the iconography.[89] While theirs is a legitimate hypothesis, it nevertheless seems safer to draw conclusions from an even more complex and less ambiguous example, such as the one chosen for this discussion. What is more, this symbolic structure was selected precisely for its complexity and specificity, which permitted a more secure 'control' over speculating on various possible semantic interpretations of 'mute' artefacts of material culture in order to, finally, allow the comparison of the content of ideas behind these material objects with literary sources.

8.3.8. Comparison of literary sources with iconography

This discussion finally raises the all-important question: can the *muthos*—in the meaning of an applied traditional tale, a narrative model representing a specific aspect of physical reality—behind prehistoric iconography be compared to the one attested in literary sources? The foregoing discussion demonstrated, I believe, that a positive answer can be plausibly defended, or at least that it is possible to (1) gain insight into a past belief system manifested only in material objects, (2) translate the visual language of iconography into a spoken idiom by (3) performing a structural analysis of complex iconographic symbolic sets in order to enable the interpretation of their meaning, and (4) recognise the transmission of both the meaning and beliefs accompanying symbolic structures between cultures sharing a common cultural space, (5) especially when these symbolic structures are complex. With this in mind, it certainly seems possible to compare the written *muthoi* with those reconstructed in steps (1) to (5).

Several parallel case studies might further strengthen the argument. Thus, Thorne postulated similar hypothesis of iconographic and narrative continuity in the case of the concept of Cretan Zeus. He recognizes that a story or myth of Zeus' birth was 'firmly associated' with a constant iconography—a youthful god—first attested in the Minoan period.[90] The similarity between my thesis and Thorne's lies in the fact that the early attestations of the youthful god iconography are not accompanied by literary evidence, but the interpretation of iconography both in this early and later periods is associated with literary evidence accompanying the latter.[91] Nevertheless, both solely iconographic and iconographic and narrative continuities are emphasized in Thorne's paper, persistently re-appearing or being preserved in differing spatial, temporal and cultural frameworks. This is exactly parallel to my interpretation of the 'Dupljaja concept' and its relation to literary testimonies.

In a similar fashion Sørensen interprets iconographical sources from a largely illiterate period of Nordic history by the help of myths recorded in a later period,[92] basing his comparative method mainly on the recognition of common cultural affiliation of iconographical and literary material, even though the sources are both spatially and temporally dispersed.[93] Goodison likewise freely discusses iconographic sources from an illiterate period of Aegean (pre)history (which she prioritizes) alongside later literary testimonies (focusing on the earliest written

[86] Kristiansen & Larsson 2005: 22.

[87] Sprockhoff 1954: 60, 71–73, 103; Gelling & Ellis Davidson 1972: 119; Ahl 1982: 39; Kaul 1998: 53, 55–56, 75, 93–94, 130, 134–36, 143, 148, 157, 159, 163, 173, 178, 214–15, 242, 244, 251–52, 254–58, 260–61, 276–78, 282–84.

[88] Sprockhoff 1954: 73; Gelling & Ellis Davidson 1972: 119; A. & B. Hänsel 1997b: 67; Kaul 1998: 253, 256.

[89] Kaul 1998: 251256 does discuss the anthropomorphic solar deity in this context, but not in the specified way I am studying it in this chapter.

[90] Thorne 2000: 149–50, 152; the continuity from the BA to the historical period in the case of Cretan Zeus is supported by Furley & Bremer 2001: i.70–73.

[91] Thorne 2000: 150 regards the myth/story accompanying iconography as a constant and does not associate it with 'religious' or 'theological' interpretation, which he finds variable.

[92] Sørensen 2002[1986]: 121.

[93] Sørensen 2002[1986]: 122, 130.

works), arguing that their correspondence actually strengthens her particular claims.[94]

Another comparable case is Kristiansen's study of Divine Twins in Bronze Age belief-systems, whose additional value lies in that it offers a methodological framework,[95] roughly analogous to the one employed in this chapter. He argues for complementarity of literary and archaeological sources from the Bronze Age onwards, and uses the former to formulate a hypothesis, by identifying gods and their characteristic features in textual evidence (both Bronze-Age and later), that he subsequently 'tests' in the archaeological record, by identifying corresponding features in material testimonies. Kristiansen believes to have found a full correspondence between the two types of evidence in the case of Divine Twins and freely uses archaeological record alongside literary evidence in his study of the phenomenon.[96] His method is complementary to my own and, if applied to the concept discussed here, would similarly allow the comparison between the two types of evidence.

These parallel cases show that a profitable parallel study of iconography with literary sources is indeed possible, offering theories that explain a number of otherwise poorly understood phenomena.

8.4. Concluding remarks: large-scale context, anthropomorphism and the contents of the *muthos*

Before proceeding to emphasize and clarify several key suggestions that were proposed in various parts of this chapter, I must point out that it was not my intention to extensively discuss the precise genetic relations between different manifestations of the 'Dupljaja concept', since the method I have employed in this chapter cannot be used for this purpose. Similarly, I cannot discuss here the complicated problem of cultural interrelations during the LBA and EIA between Central and South-Eastern Europe, although this question is certainly raised by my discussion, since its solution is similarly outside the scope of the analytic procedure I have employed here. These connections are discussed throughout this chapter only in broadest terms, but I must emphasize it is generally acknowledged that they indeed existed.

The analysis performed here can, on the other hand, help to recognise the proper large-scale context in which this complex symbolic figure appears. Furthermore, it can suggest its possible meaning, that is, the content of the *muthos* underlying its manifestations, especially when compared with relevant literary sources, which is precisely what is attempted here. Finally, the conclusions reached in this way might reveal something new—or support something only scarcely known from literary sources—with respect to the Greek notion of 'Hyperborean Apollo' and *its* underlying concept.

Thus, in the analysis of material evidence, a large-scale context of various appearances of the 'Dupljaja concept', in which the manifestations of this concept occur and in which structurally similar arrangements have related meanings, was defined: it could provisionally be called the Urnfield symbolic complex, after its most distinctive manifestation. It is not easy to precisely delineate either its spatial or temporal span: it certainly encompassed the entire Balkan (with an outpost in the Aegean) and Carpathian region, together with much of Central Europe, with an offshoot in the Nordic region, and also Italy; furthermore, it could be argued that it spans at least the full millennium from the mid-2nd to mid-1st millennium BC. Within this context the osmotic nature of cultural space provides an alternative model for previously dominant discourse of invasions and migrations,[97] allowing for an inter-cultural transfer of symbols, especially complex symbolic systems, together with their accompanying meanings. In this particular case, perhaps the culturally transcendent term 'tradition' should specifically be applied to what I regularly called the 'Dupljaja concept', since this term emphasizes its realized transferability, longevity and propensity to expand spatially.

The Dupljaja model could be understood as a particularly important object for the understanding of one important part of the symbolism of this large-scale context. I have already noted that it probably had an important role in the spread of the concept of a solar chariot or boat associated with swans from Central to both Northern and Southern Europe. But, more generally, it could be claimed that it also played a decisive role in the formation of the Urnfield symbolic complex,[98] and thus its importance—on a much lesser scale—in the formation of what I have called the 'Dupljaja tradition' is immediately obvious.

Next, an attempt to reconstruct the contents of the *muthos* behind both material evidence and literary sources seems possible at this point. In this context, the anthropomorphic nature of the agent in the complex symbolic structure of the 'Dupljaja tradition' should not be treated simply as a development of earlier, pre-anthropomorphic ideas accounting for similar phenomena. In this respect the present analysis bears an additional weight, since it recognizes a supplementary feature that reveals more information on the concept behind a group of structurally similar material testimonies. I have emphasized in the introduction, following Guthrie, that anthropomorphism is utilized as an almost universal explanatory method, one that was used as a strategy of hypothesizing about, and which attempted to offer a plausible interpretation of, the surrounding world. It can be understood as a model-building on the basis of metaphor and analogy; it has a large

[94] Goodison 1989: 119–23, 131 and 123–68 *passim*.

[95] Kristiansen 2014, foreshadowed in Kristiansen & Larsson 2005: 20–24, 256–57, 263, 265, 316, 329, 368.

[96] Kristiansen 2014.81–82.

[97] For the literature on the introduction of Apollo in Greece by various migration-theories see Bilić 2016b: 459 n. 34.

[98] Pare 1987: 61; 1989: 84; cf. Sandars 1968: 175.

explanatory potential and accounts for a large number of phenomena. It thus cannot be treated as an 'improvement' on some earlier non-anthropomorphic thesis accounting for the same phenomenon and should be studied in its own right. Since it is undoubtedly present in the 'Dupljaja tradition', it must be studied with respect to this tradition in the context of its explanatory qualities, and not merely as one iconographic solution among many employed by those working in the tradition of the Urnfield symbolic complex. My proposed reconstruction of the contents of the *muthos* behind this tradition takes into account this decisive circumstance, which is further strengthened by the use of Greek literary testimonies, allowed by the procedure outlined in steps (1)–(5) above.

All the elements that appear in the structure of manifestations of the 'Dupljaja tradition' point to a close parallelism with the 'solar' myth of the Delphian Apollo: an anthropomorphic figure transported in a vehicle and decorated with solar symbols, drawn by animals with an undeniable seasonal character.[99] As noted in the introduction, the annual solar movement is a time-factored phenomenon, and, apparently, the only means to either describe or account for it in prehistory was through a story or *muthos*.[100] This resulted in the creation of a narrative model representing a specific aspect of physical reality in terms of anthropomorphic causality: the annual solar movement, with an emphasis on the solstices, as observed, recognized, described and interpreted by prehistoric European populations. Anthropomorphic explanatory model, accompanied by an array of relatively stable non-anthropomorphic symbols, suggests a shared tradition accounting for a specific phenomenon, while every piece of evidence gathered either from iconographic or literary sources points to the tradition's seasonal or, more precisely, solar character. While the validity of this particular character cannot be conclusively proven due to the inherent nature of the material being studied, it best explains all the available evidence, while none contradicts it.[101]

Finally, returning to the previous chapter, Alcaeus' 'Hyperborean Apollo'—and the same could be claimed for the Greek notion of this concept in general, as attested in other authors is thus one of the manifestations of this narrative model expressed in a hymn to a deity that was precisely in the period when Alcaeus was composing it in the process of attaining his illustrious Pan-Hellenic status. There is no basis, however, for calling the deity or mythic person attested in archaeological record either 'Apollo' or 'Hyperborean Apollo' without the use of quotation marks. The 'Dupljaja tradition' (in the sense I outlined above), on the other hand, does indirectly suggest, but does not in any way prove, that Apollo could have occasionally been recognized as a sun god in Greece, with a special connection to the solstices, at least in the concepts associated with the Delphian myth discussed in the previous chapter, although it should again be pointed out that this does not exclude numerous other functions he fulfilled in Greek society, or that it emphasizes this particular feature at the expense of others.

In this chapter I have provided an analysis of a complex prehistoric iconographic structure which I christened the 'Dupljaja concept', consisting of an anthropomorphic solar deity riding in a chariot drawn (or accompanied) by water birds. This complex arrangement permitted the recognition of a similar underlying *muthos*; after reconstructing the large-scale context in which this structure appears, which is also reflected in literary testimonies, I have concluded that its underlying *muthos* represents an account of the annual solar movement in terms of anthropomorphic causation.

99 For the birds' seasonal character specifically in the Bronze Age see Teržan 1999: 123. Cunliffe 2002: 121 argued that the communities living in northern Scotland actually associated the seasonal migration of the whooper swan with the annual solar movement (see chapter 6).

100 Marshack 1972: 133, 197, 279, 283, 316, 330; cf. Burkert 1979a: 23.

101 Cf. Sourvinou-Inwood 1995: 133–34.

Part Two

Diurnal solar movement

9

Diurnal path of the Sun in Greek tradition

In the first part of the book the emphasis was on the annual aspect of solar movement and its rendering in myths, but occasionally the diurnal aspect of the sun's orbit was also touched upon.[1] The references to the latter are more easily recognisable and are in general treated as less controversial in modern scholarly literature. I believe that at least part of the reason for this state of affairs lies in the apparent simplicity of the observation of the diurnal solar motion in comparison with the annual from the point of view of a modern classical scholar, even though the latter is equally intelligible to both a dedicated observer and student. Naturally, these two aspects are only two sides of the same coin and one cannot properly understand the one without the other. In addition, both aspects of solar movement were accounted for in Greek tradition by models built out of identical motifs, for example by using the chariot-ride model to account for either of them,[2] or the model involving the high northern mountain-range as a barrier to the sun's movement, whether its annual limit or—the aspect that will be studied in this chapter—a screen for its nocturnal rays (for the former see chs. 3 and 5). It is true that the diurnal solar movement was—as one would expect—treated more often and more straightforwardly in various types of discourses. It was described and accounted for in myths, with different versions representing alternative models of the daily course of the sun. In this context it was treated almost exclusively by employing the usual mythic practice of anthropomorphic causality, but occasionally and perhaps somewhat surprisingly with less emphasis on this crucial aspect of myth (e.g., when ignoring personalistic causality in the model accounting for the nocturnal part of the daily solar movement involving the high northern mountain screening the sun at night). In any case, these mythic models undoubtedly reference and account for solar movement—sometimes being structured on an analogy with the diurnal solar course—and their inclusion in the class of myths studied in this book is surely uncontroversial.

9.1. The high northern mountain

Several groups of sources thus account for the nightly voyage of the sun by recourse to the model involving a large mountain (or 'elevation') in the far north that screens the sun during night.[3] This notion is closely associated with and often indistinguishable from the concept of the inclination of the earth and/or the corresponding inclination of the heavens, which, as a rule, left the northern part of the earth 'elevated' and thus accounted for the invisibility of the sun during night.[4] Either of these two concepts accounts satisfactorily for the diurnal solar movement irrespective of the underlying cosmological framework, i.e., regardless of the accepted form of the earth (flat or spherical), even though both seem much better suited and were certainly initially developed with respect to a flat earth. At the same time, the model naturally presupposes a northerly path of the sun's nocturnal journey, which indeed seems to have been a dominant opinion in Greek tradition. A number of Presocratic thinkers adhered to some form of this notion, most notably Anaximenes, who mentioned higher portions of earth (*hupsêlotera*) in the north.[5] This could be understood as a description of the earthly plane inclined towards the north celestial pole[6] or the acknowledging of the existence of a high northern mountain range.[7] Anaximenes' model accounts for the diurnal movement of the sun, more precisely, for its nocturnal course from west to east by way of north,[8] but both interpretations of his model are plausible with respect to this phenomenon. This model, or some form of it, was widely accepted by other Presocratics, who sometimes argued for the inclination of the 'world', i.e., the heavens, or the earth itself.[9] The *locus classicus* of this theory is Aristotle's account in *Meteorology*, where he discusses the higher parts of the earth in the north,[10] but it

[1] An earlier version of this chapter was published as Bilić 2020b.

[2] The version of the model accounting for the diurnal solar motion is so well-known that it was and will be discussed only when it illuminates some other less familiar aspects of the problem (see the introductory chapter).

[3] Aesch. fr. 68 *TrGF*, Radt = fr. 102 Mette, on the interpretation of Diggle 1970: 27–28 and Bridgman 2005: 44, would be an early testimony for this model, but—as already noted in chapter 4—both the restoration and interpretation of the fragment are extremely tentative, and one cannot build a solid argument upon it.

[4] On this subject see Couprie 2011: 72–77 = 2009: 263–72. He does not associate these notions with each other. Couprie 2015: 16–18; 2018: 29–31 eliminates any notion of the inclination *of the earth* in ancient authors, but his objections are irrelevant for my present discussion.

[5] 13A7 D-K = An 56 Wöhrle; cf. 13A14 D-K = An 4 Wöhrle = Arist. *Meteo.* 354a25, 28, 31 (high(er) portions of earth in the north: *hupsêlotera, hupsêlê*); Alexander of Aphrodisias iii.2.70.23–25, 27, 29, 32–33 Hayduck (higher portions of earth in the north: *hupsêlotera*), 34 (elevation of the earth: *exarma*). Couprie 2015: 23, cf. 2018: 122, 126 argues that Aristotle does not actually refer to Anaximenes but was misunderstood by both ancient doxographers and modern scholars.

[6] Bicknell 1969: 78–79; cf. Couprie 2018: 105.

[7] Cherniss 1935: 343; Bolton 1962: 42, 188 n. 6; West 1971: 103 n. 1; Kirk *et al.* 1983: 157; Graham 2013: 65 Fig. 2.2; Couprie 2015: 20; Geus 2018: 404.

[8] 13A7 D-K= An 56 Wöhrle; cf. 13A14 D-K = An 4 Wöhrle = Arist. *Meteo.* 354a25–32 with Alexander of Aphrodisias iii.2.70.22–34 Hayduck.

[9] Emp. 31A58 D-K (Mansfeld & Runia 2009: 414; 2020: 858; Couprie 2011: 73–74 = 2009: 267; 2015: 11–12; 2018: 28; 2020: 13–14); Anaxag. 59A1 D-K, 67, cf. [Pl.] *Amat.* 132AB = 41A2 D-K (Guthrie 1965: ii.305 n. 3); Archel. 60A4 D-K; Diog. 64A11 D-K; Leucip. 67A1, 27 D-K; Democr. 68A96 D-K.

[10] 13A14 D-K = An 4 Wöhrle = Arist. *Meteo.* 354a25–32 with Alexander of Aphrodisias iii.2.70.22–34 Hayduck. However, see the present discussion on Anaximenes' precise referent, which also applies to Aristotle's account.

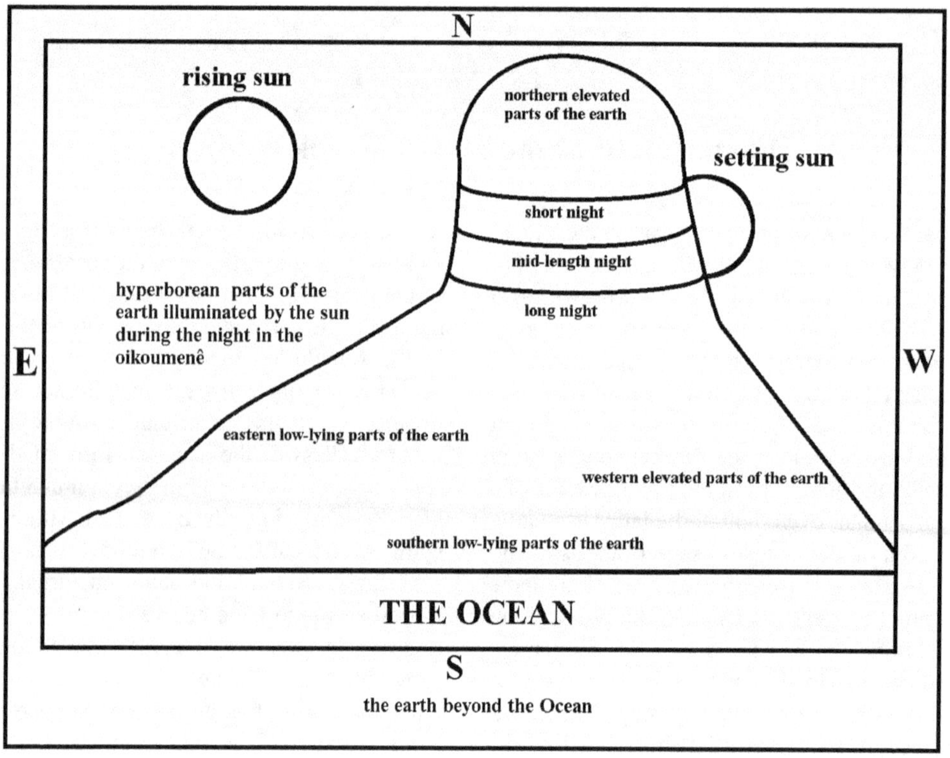

Figure 9.1 Cosmas' conical elevation of the earth and the sun's diurnal movement (drawn by the author)

was repeated by authors both earlier and later than him.[11] In a modified version of this model, Archelaus argued that the earth is upraised (*hupsêlê*) at the circumference, which he further explicitly associated with the diurnal solar movement.[12] Leucippus (?) and Democritus believed that the earth was 'hollow' in the centre of its disk, which presupposes its elevation at the perimeter, as well as some consequences on the daily movement of the sun, even though these concerns remain unelaborated in our sources.[13] It is possible that Epicurus' collocation of the doctrine of certain walls (*toichoi*) in a circle (around the earth) with that of 'those who drive the *astra* around in a circle overhead' represents a reference to Anaximenes,[14] which would then explain later references to a wall in the north. Indeed, according to several late sources, the sun after sunset travels through the northern regions to the place of sunrise.[15] During this part of its course it is concealed 'as if by a wall' (*toichos*)[16] or 'a cone' (*kônos*),[17] or by the part of the earth 'standing upright like a wall' (ὀρθὸν ὡς ἐπὶ τοίχου ὑπάρχον),[18] when it is not simply concealed by the elevation of the earth or mountain(s) in the north (and west) (Figure 9.1).[19] Avienius' description of the Alps that almost completely block the sun and the ensuing description of the sun's nocturnal path around the northern edge of the earth, more precisely around a *meta* in the far north, suggest that he believed that the sun after setting follows a northerly course to its place of rising, while a high mountain blocks it from our view, at the same time employing some elements of the model of chariot-ride in accounting for the diurnal solar movement.[20]

[11] Earlier: Pl. *Resp.* 4.435e; Hippocr. *Aer.* 19 (Kiessling 1914: 853.48–55).

[12] Arch. 60A4 D-K (Couprie 2015: 22 understands Anaximenes 13A7 = 56 Wöhrle in the same way).

[13] Leucippus *ap.* [Gal.] *Hist. phil.* 82 (Couprie 2018: 20, but see Mansfeld & Runia 2020: 1265: a conflation of the Leucippus with the Democritus lemma in Ps. Galen, which eliminates the hollowness of earth for Leucippus); Democr. 68A94 D-K with 59A87 D-K (perhaps also attested in Pl. *Phd.* 99b, Panchenko 1999: 32–33). Cf. 'the Chaldeans' in D. S. 2.31.7 and anonymi at Basil. *Hex.* 9.1 (Mansfeld & Runia 2020: 1269), perhaps referring to either Archelaus or Democritus or both (a disk equally rounded on all sides or a winnowing-basket hollow in the middle). [Bede] *De Mundi Coel. Terres.* 1.23 Burnett 1985 reasonably merged the two models.

[14] Epicur. fr. 26.33.1–2, 6–8 Arrighetti² = Anaxim. fr. 20 Graham; Graham 2013: 127; Couprie 2015: 6, 22; 2018: 121.

[15] It seems that this was the teaching of the school of Antioch, to which Cosmas Indicopleustes also adhered (Kuelzer 2018: 932–33). According to Išo'dad of Merv (who is himself critical), Theodore of Mopsuestia (c. 350–428), a prominent representative of the school of Antioch, believed that the sun is concealed by northern mountains at night (Išo'dad *Gen.* 1.5, van den Eynde 1955: 26; cf. Kominko 2013a: 47; 2013b: 71; for 'L'Interprète' as Theodore see van den Eynde 1955: xvii).

[16] Severian of Gabala, *Hexaemeron*, *ap.* Cosm. Ind. *Top. Chr.* 10.33 Wolska-Conus = *Third Homily on Genesis* 5 (Glerup 2010: 44); Procopius of Gaza, *Comm. Gen.*, *PG* 87.1.90; mid-6th-c. [Caesarius Nazianzus] *Dial.* 1.99 (*PG* 38.964 = 98 Riedinger 1989: 75). All base their interpretation on a reading of Eccl 1:5–6.

[17] Cosm. Ind. *Top. Chr.* 4.11 Wolska-Conus.

[18] Cosm. Ind. *Top. Chr.* 4.16 Wolska-Conus.

[19] Cosm. Ind. *Top. Chr.* 2.34, 4.12, 16, 6.1 Wolska-Conus (*hupsos* of the earth, *hupsêla, hupsêlotata* parts of the earth); [Caesarius Nazianzus] *Dial.* 1.99 (*PG* 38.964 = 98 Riedinger 1989: 75) (*huperochê* of the earth); Philop. *De opif.* 3.10, pp. 135.14–17, 138.3–10 Reichardt, who criticizes the notion of huge mountains in the north (West 1971: 106; Elweskiöld 2005: 96, 99) (also discussing Eccl 1:5–6); Anon. Rav. 1.9–10, pp. 24–25 Pinder and Parthey, who paradoxically places the mountains *beyond* the Ocean, while also bringing Eccl 1:5–6 (p. 22) into discussion (Beazley 1897: i.310–11; Miller 1898: vi.48; Schnetz 1925: 109, 112; Thomson 1948: 389; Dreyer 1953: 222; Podosinov 2018: 204–205, whose translation does not emphasize the transoceanic location of the mountains). See also Kominko 2013a: 46–47; 2013b: 71.

[20] *Or. Marit.* 648–49, 652–73. Cf. Kiessling 1914: 852.43–854.45.

9.2. The southerly path of the sun

As already noted, this model involves the sun after sunset following a northerly path towards sunrise. However, there existed an alternative model, involving the southerly path of the sun after setting, although this model is less easy to recognise in surviving sources. This account presupposes a counterintuitive notion of the cessation of the solar movement determined by the inclination of the ecliptic (i.e., in the northern hemisphere north of the summer tropic the sun appears to follow the southerly path after sunrise which seems to continue by the northerly path following the sunset) and its continuation in a reverse direction. It often involves the concept of cosmic 'nadir' or *axis mundi* located deep in Tartarus, used by the early Greeks as one of the models accounting for the solar phenomena of sunrise and sunset (cf. chapter 4).[21] Identically to the 'northerly path' model, its counterpart the 'southerly path' model equally accounts for the diurnal solar movement irrespective of the underlying cosmological framework, i.e., with respect both to a spherical or flat earth, but seems decidedly better suited to the latter.

The southerly path of the sun's nocturnal journey was explicitly discussed by only a couple of authors that carry much less authority than those arguing for the northerly, especially since the testimony of a more reliable of the two can be said to argue for the southerly path only in a very tentative modern interpretation. Thus, Posidonius' testimony, where he speaks of the *katastrephein* of the sun in the west, was interpreted as a reference to a theory that sends the sun after sunset on a southerly path upon the Ocean to the east.[22] However, this 'theory' does not in any way support Posidonius' immediate argument and it does not seem plausible that the philosopher upheld such an unscientific notion. Thus, Kidd must have made a mistake here in proposing this interpretation.The southerly path of the sun between sunset in the extreme west and sunrise in the extreme east is further mentioned by the much later Pseudo Jerome,[23] perhaps as a reaction against Cosmas Indicopleustes' 'choice' of the northerly return path.[24] Furthermore, Pseudo Jerome actually claims that the sun 'retraces its course along its southerly path' over the sky,[25] and that it does not follow a 'land route', since it 'barely touches the ocean'. Therefore, this account, although undoubtedly referring to the southerly route, is not a particularly clear expression of this model. It could be speculated that the sun remains invisible during the night because of some unspecified obscuration, but the phenomenon remains unexplained in the *Cosmography*. These two testimonies are the only accounts of the sun's southerly path—one most probably only a mirage, the other a muddled, late and probably polemical testimony. They can scarcely be measured against Anaximenes and 'many ancient meteorologists' mentioned by Aristotle. But the adherence to this model can be extrapolated from the writings of a number of respectable sources. In order to evaluate the plausibility of this claim, a detailed analysis of relevant texts is necessary.

9.3. Diurnal solar movement in Homer (Figure 9.2)

Let us begin with the earliest possible source, Homer. His familiarity with the concept of the gates of the sun undoubtedly proves he had some definite notion of the sun's movement subsequent to its setting, but we cannot say for sure what he imagined to happen to the sun once it passes through these western gates.[26] Obviously, it arrives at its rising point in the east, which for Homer must be the location of Aeaea.[27] Elsewhere, however, referring to a different variant of the model accounting for, most likely, the diurnal solar movement, he described the Aethiopians as located both at the place of the sun's rising and setting.[28] Both Ephorus, who claimed that the nation of the Aethiopians stretches from the winter sunrise to sunset,[29] and Crates, who argued for their placing under both tropics,[30] associated the position of the Aethiopians with reference to the annual solar movement.[31] At the same time, the diurnal aspect of solar movement was often associated with the Aethiopians, especially Western, who are regularly associated with sunset,[32] or with the western

[21] Ballabriga 1986: 77, 290–91; Nakassis 2004: 216, 218, 230–31. Both 'nadir' and *axis mundi* are used here somewhat metaphorically, in order to describe a fixed (and thus not observer-dependent) location directly opposite to the uppermost sky above the earth's surface.

[22] Posid. fr. 223 E-K = 66 Theiler = *FGrHist* 87F80 *ap*. Str. 17.3.10; Kidd 1988: ii.804, who refers to this interpretation as 'sheer speculation'.

[23] *Cosmog*. 14, Herren 2011: 14–15.

[24] Herren 2011: li, 67.

[25] Herren 2011: li, 67.

[26] *Od*. 24.12. In the Homeric poems the sun at dawn rises from the earth-encircling Ocean (*Il*. 7.421–23; *Od*. 19.433–34; as does Dawn itself, *Il*. 19.1–2; *Od*. 22.197–98, 23.243–46, 347–48)—alternatively, it rises from the sea (*Od*. 3.1–3)—while at dusk it sets in this body of water (*Il*. 8.485–86, 18.239–40). We are also informed that it sets under the earth (ὑπὸ γαῖαν) (*Od*. 10.190–92) and that it turns back towards the earth (*Od*. 11.15–19; 12.380–81), presumably at the moment it crosses the meridian. For a dismissal of the potential contradiction at *Od*. 10.190–92 with the horizontal nocturnal path of the sun over the Ocean and the apparent identity of the sea at *Od*. 3.1–3 with the Ocean see Dicks 1970: 31–32. At the same time, Homer was indeed aware of the existence of seasonal 'gates', i.e., the gates of the heaven guarded by Horae (Seasons) (*Il*. 5.749–51, 8.393–95, cf. 8.411 (West 1997: 141)). The notion of seasonal gates can be further recognized in the name and nature of Telepylus, which must be associated with solstice and thus the annual movement of the sun (see above chapter 2).

[27] *Od*. 12.3–4. Cf. Dion. Perieg. 587–90 (Lightfoot 2014: 228, cf. 12, 395); Rohde 1925: 82 n. 14; Endsjø 1997: 376; Lightfoot 1999: 444; Käppel 2001: 16, 19, 21 Abb. 1; Marinatos 2001: 396; Nakassis 2004: 223; 2020: 275; Louden 2011: 160; Fowler 2013: 201; 2017: 247, 249; Podossinov 2013: 221; Manoledakis 2015: 302.

[28] *Od*. 1.22–24; cf. *Il*. 1.423–24, 23.205–206, where the Aethiopians are located at the Ocean. For the 'double' Aethiopians see further Timosth. fr. 6 Wagner; Eudox. *ap*. Posid. fr. 49 E-K = 13 Theiler *ap*. Str. 2.3.4–5, etc. The most comprehensive discussion on the topic is found in Str. 1.2.24–28.

[29] *BNJ* 70F30ab; cf. Solin. 30.14; Eustath. *Od*. 1.23, i.11.

[30] Fr. 37 Broggiato *ap*. Str. 1.2.24–25, cf. Gemin. *Elem. Astron*. 16.26–27 (p. 174.11–23 Manitius; Crates' words from 16.26 are cited by Albert the Great, *Liber de natura locorum* 1.7, ix.543 Borgnet) and Eustath. *Od*. 1.23, i.12; fr. 34b Mette. Cf. Gemin. *Elem. Astron*. 16.30 (p. 176.13–16 Manitius) and also 16.28 (p. 176.4–6 Manitius).

[31] Cf. Marinus of Tyre *FGrHist* 2114F1, 4, 6 *ap*. Ptol. *Geogr*. 1.7.2, 8.3, 9.5; Ptol. *Geogr*. 1.9.6–7, 10.1.

[32] Aesch. fr. 192 *TrGF*, Radt (= 323 Mette); A. R. 3.1191–94.

island of Erytheia,[33] in its turn also certainly associated with the setting sun (see below). Homer's association of the Aethiopians with the solar risings and settings, if taken in terms of the diurnal solar movement, which seems a more likely option, certainly favours the southerly route for the sun after its setting due to their general (conceptualised or actual) geographical position.[34] In the remainder of this chapter I will offer my interpretation of the account in the 10th and 11th book of the *Odyssey* that supports this conclusion. At the same time, the association with the Aethiopians does not exclude the possibility that he envisaged a northerly route, although this seems less likely. In any case, it can be extrapolated from later Greek evidence—although still very early—that the concept of 'Helios' cup' might have been involved in a Homeric model accounting for the diurnal solar movement, more precisely, its nocturnal section, even though Homer nowhere refers to it explicitly. This can be inferred once again from the account in the 10th and 11th book of the *Odyssey*, which will be studied below. In its turn, 'Helios' cup' almost certainly presupposes an underlying cosmological concept of a flat earth and a horizontal return voyage, although a voyage underneath the earth's disk cannot be discarded as a possibility.

9.4. The sun's cup and its southerly course (Figure 9.2)

The nocturnal path of the sun as envisaged in early Greek tradition cannot be discussed separately from the concept of 'Helios' cup', which appears both in early Greek poetry and in Presocratic writings. Unsurprisingly, the mythic account is chronologically earlier and it first appears in Mimnermus (second half or last two thirds of the 7th c.),[35] who describes how Helios sleeps in a winged golden couch while being conveyed over the sea to the place of his rising, from the Hesperides to Aethiopia, where his chariot and steeds await him.[36] Here Aethiopia is apparently associated with the diurnal solar movement, perhaps by a reference to a model similar to that employed by Homer, but no indication of course is given. A probably contemporaneous account of the sun god's voyage in a vessel over the Ocean is that given by the author of the *Titanomachy*,[37] followed by a number of early authors.[38] Perhaps Sappho should be added to this group of authors, since a line in a new-found segment of fr. 58 Voigt, the Tithonous poem, which is unfortunately damaged, was read by some as 'Eos mounted the cup (δ[έπα]ς) [of Helios]'.[39] This seems to have been almost a technical term often used to designate the sun's bowl in Greek poetry and mythography,[40] and the origin of the precise usage of this term in a comparable context might lie in Anatolia.[41] The phonetic reading of Hieroglyphic Luvian sign for 'sky' depicted as a bowl (not upturned, as one would expect: ⌣),[42] is *tipas*- [di:bas?], while the Cuneiform Luvian form is *tappas*- [dapas?].[43] We also know of two silver bowls inscribed in Hieroglyphic Luvian, referring to them as *zi/a* CAELUM-*pi* and *zi/a-wa/i-ti* CAELUM-*pi*, respectively ('this **tapi* [**dapi* ?]'), as well as of the name of a vessel, probably a bowl, attested in Hittite ritual context: (DUG)*tapi-sana*. This is probably the origin of both Mycenaean *di-pa* and Greek *depas*, but since from the 7th c. onwards we constantly find an unusual association of a non-inverted 'bowl' ('cup') with the sun's voyage over the heavens/celestial ocean, it does seem probable that both the word and some form of the concept arrived to Greece from abroad. The evidence from Anatolia, however, only offers a connection between the term used for the sky and a bowl, with no reference to the sun. Until further evidence becomes available, it has to be presumed that the Anatolian concept was subsequently significantly redefined by the Greeks into a form that appears in early poetry. On the other hand, it is possible to recognize the influence of the Egyptian concept of solar journey in a boat on the corresponding Greek idea.[44] However, in this case it is perhaps better to speak of parallel developments rather than to posit a direct influence from Egypt.

9.5. Stesichorus' account of the sun's voyage in a cup

None of the sources adduced above suggest either the southerly or northerly route of the sun after setting. However, another early source for the sun god's voyage in a vessel over the Ocean, Stesichorus' *Geryoneis*, adds significant information unmentioned by the authors discussed so far. Stesichorus described how Helios

---

[33] Ephor. *BNJ* 70F129b = [Scym.] 152–53, 157–58 (*GGM* i.200); Dion. Perieg. 558–61 (Lightfoot 2014: 228); Prisc. *Perieg.* 570 (*GGM* ii.195) = Dicuil *Lib.* 7.3; Σ Dion. Perieg. 560 (*GGM* ii.451); Eustath. Dion. Perieg. 558 (*GGM* ii.325–26); cf. Avien. *Descr. orb. terr.* 732–36 (*GGM* ii.183, 738–42).

[34] Dörpfeld 1925: 234–37, 241, 247, 252–56, 263–64, 266 interpreted the nightly path of the sun as envisaged in the *Odyssey* precisely in this way, arguing for its voyage in the solar cup following the southern, 'oceanic', coast of Libya.

[35] Bowie in *BNP* s.v. *Mimnermus*; Nünlist *BNJ* 578 *Biographical Essay*.

[36] Fr. 12.5–10 West, cf. fr. 23 West.

[37] Fr. 10 West; according to Theolytus (see immediately below), the earliest occurrence of the concept.

[38] Pisan. fr. 6 Davies = fr. 5 West (perhaps of a similar date, i.e., the 7th/6th c. (West 2003: 22–23; *BNP* s.v. Peisander 6; Davies & Finglass 2014: 231 n. 6); Pher. *BNJ* 3F18a, cf. [Apollod.] *Bibl.* 2.5.10, where the same cup is mentioned in the same context (also from Pherecydes?; see Jacoby *FGrHist Comm.* i.395, 397 (who is critical); van der Valk 1958: 127, 129; Gantz 1993: 404–405; Fowler 2013: 291, 298–99); Panyas. fr. 7A Davies = fr. 12 West; Aesch. fr. 69 *TrGF*, Radt = 103 Mette, fr. 74 *TrGF*, Radt = 109 Mette; Antim. fr. 66 Wyss, West; Theol. *FGrHist* 478F1; see also much later Macr. *Sat.* 5.21.16, 19 and Mart. Cap. 2.113, 182–84. Perhaps also Herodorus of Susa (SEG VII, 14.22, Merkelbach & Stauber 2005: 74–76; Potts 2016: 360), as claimed by Cumont (cited in Merkelbach & Stauber 2005: 76). On Stesichorus see immediately below.

[39] P. Köln 21351, l. 18; Gronewald & Daniel 2004a; 2004b; Watkins 2007: 306, 317–19; cf. Hammerstaedt 2009; *pace* West 2005b; Obbink 2009; Davies & Finglass 2014: 236.

[40] Besides Sappho (?), this precise word was used to designate Helios' cup by Stes. S 17 = 185 *PMG* = fr. 8a Finglass; Pher. *BNJ* 3F17, 18a with [Apollod.] *Bibl.* 2.5.10, 11; Aesch. fr. 69 *TrGF*, Radt = 103 Mette, fr. 74 *TrGF*, Radt = 109 Mette; perhaps Pisan. fr. 6 Davies = fr. 5 West.

[41] Watkins 2007: 319–22.

[42] CAELUM, Laroche 182 with a variant in Yazılıkaya where it is upheld by two bull-men standing on a sign for 'earth' (⏜) (Relief 28/29 in Chamber A).

[43] See e.g., KARKEMIŠ A6.3: (CAELUM)ti-pa-si, dative *tipasi*.

[44] Kirk *et al.* 1983: 13 (noncommittal); Burgess 1999: 189 (an influence of both Re and Šamaš matrix of ideas); Stephens 2003: 131–32, 221; the sources quoted in Stephens' n. 32 on p. 131, however, do not associate Greek and Egyptian concepts of solar voyage in a boat.

descends from the sky and enters a golden cup, and then traverses (περάσας, from *peraô*) the Ocean to get to his family at 'the depths of holy, dark night'.[45] Thus the sun enters the cup at the point of its setting, but only arrives to a location defined by the presence of night. Several interpretations of this account are possible: one is that Helios turns to the north when he enters the cup, 'the sources of the night' and the realm of night in general being in the Hyperborean/Rhipaean north (chapter 4). But alternative interpretations are possible, one which involves the southerly route and another which involves the concept of 'cosmic nadir' or *axis mundi* located deep in Tartarus, a location that was used in an alternative ('uni-polar') model accounting for the phenomena of sunrise and sunset.[46] The horizontal routes, both northerly and southerly, are supported by the use of the cup and the term 'traverse', but 'the depths of night' suggest a vertical movement and the continuation of the descent after the sunset.[47] In fact, a combination of these two movements (the southerly route and a descent) would not be unprecedented in the context of an otherworldly journey, most notably appearing in the *Odyssey*,[48] and clear resonances of precisely such a journey can indeed be recognised in Stesichorus' account.

These alternative interpretations allow the understanding of Stesichorus' description of the nocturnal solar voyage as (1) following a southerly horizontal route, or (2) alternatively a route below the underside of the earth, in a similar sense as Anaximenes' model relates to the northerly route of the sun during night (if we understand his model as involving the inclination of the earth's plane rather than the high mountain-range in the north), even though the depressed southern part of the earth's plane would do a much poorer job in concealing the sun in comparison to the elevated northern one, or perhaps (3) a combination of the southerly horizontal and vertical routes.

The precise context notwithstanding, it appears that Stesichorus' 'depths of night' correspond to a well-known location from the *Odyssey*, where 'east' and 'west' are so close to each other that they cannot be distinguished from one another.[49] This situation obtains on Aeaea, a solar island *par excellence*, which is otherwise, in terms of the bi-polar model accounting for the phenomena of daylight/nigh exchange, firmly located in the east. In fact, the inability to distinguish between east and west only applies—in terms of the nocturnal path of the sun—to the lower culmination of the sun, i.e., its meridian passage at midnight, or rather its imagined meridian passage upon the surface of the earth in the farthest south. This appears to be the cosmological location of Stesichorus' 'house of Helios' where his family abides at the 'depths of night'.

9.6. Hesiod's house of Night in the light of the 'uni-polar' model

The decisive evidence for accepting any of these possible variants could be provided by the correct localisation of Hesiod's 'house of gloomy Night', which is certainly relevant for the discussion of the corresponding location in Stesichorus. We have already had the chance in chapter 2 to discuss the personalistic model involving the house of Night employed by Hesiod in order to account for the diurnal solar movement (applied by the author of the *Odyssey* to the annual motion of the sun), but at that juncture no discussion of the precise cosmological location of this mythic feature was necessary. The author of the *Theogony* actually offers several signals for the localisation of his house of Night, but these signals will prove to be ambivalent. The abode of Night is thus placed by Hesiod next to Atlas, who upholds the heavens,[50] and whom he elsewhere places at the *peirata* of the earth, near the Hesperides.[51] These are appropriately the daughters of Nux living beyond (*perên*)—in the meaning on the other side of—the Ocean, at the *eschatiê* of night.[52] All these pieces of information apparently point to the far west.[53]

At the same time, the localisation beyond the Ocean suggests an otherworldly nature of this region, since elsewhere it is the location of Hades and the post-mortem abode of souls.[54] Moreover, for Homer it is the southern farther shore of the Ocean to which one should go in order to visit the realm of the dead.[55] Here we return to the dilemma whether the sun follows a northerly or southerly Oceanic route after its setting as envisaged by Homer. It appears that this predicament can be solved by a close reading of Homer's account of Odysseus' visit to Hades. The fact that the hero was sailing 'upstream' (παρὰ ῥόον Ὠκεανοῖο) when going to Hades, and 'downstream' (τὴν δὲ κατ' Ὠκεανὸν ποταμὸν φέρε κῦμα ῥόοιο) on his way back does not mean he was sailing due north and south, respectively,[56] since the Ocean's current is notoriously circular, and the upstream movement of the ship on its way to Hades may suggest that the stream runs south-north at Aeaea. If this island is imagined at the eastern extremity of

[45] S 17 = 185 *PMG* = fr. 8a Finglass. ἱαρᾶς ποτὶ βένθεα νυκτὸς ἐρεμνᾶς cannot be rendered as 'towards the east' and this location cannot represent 'Helios' eastern home' (Diggle 1970: 81; cf. Cerri 2014: 171–72). With δι' ὠκεανοῖο περάσας compare δι' Ὠκεανοῖο περήσῃς in *Od.* 10.508, where Odysseus crosses the Ocean in order to arrive to its farther shore.

[46] As discussed in chapter 5, the early Greeks described the solar phenomena of sunrise and sunset in two alternative ways: according to the 'uni-polar' model the alternation between night and day occurs at the *axis mundi*, represented by and located in Hades/Tartarus, while according to the 'bi-polar' model it rises and sets on the eastern and western horizon, respectively. Both models can equally be applied to a flat—the underlying concept of all early cosmological speculation—or spherical earth.

[47] For '*benthea* of night' suggesting physical depth see *LSJ* s.v. βένθος.

[48] Dörpfeld 1925: 234–37, 241, 247, 252–56, 263–64, 266; Carpenter 1946: 147, 186. For a parallel Mesopotamian tradition see Penglase 1994: 77–78 and George 2003a: i.501, 510 and the discussion in chapters 9 and 12.

[49] *Od.* 10.190–92; Marinatos & Wyatt 2011: 401.

[50] *Th.* 744–48.

[51] *Th.* 517–19 with Verg. *Aen.* 4.480–84 (cf. Romm 1992: 161 n. 93).

[52] *Th.* 215–16, 274–75.

[53] See e.g., A. R. 4.629–30.

[54] *Od.* 10.508, 11.158–59 (the verb used to designate crossing is *peraô*) (cf. 639), 24.11–13 (the movement is indicated by the preposition *para*).

[55] *Od.* 10.507–508, 11.11 (Odysseus' ship propelled by the North Wind). Cf. *Od.* 14.252–58, where the North Wind propels the ship from Crete to Egypt, 'as if downstream' (ὡς εἴ τε κατὰ ῥόον).

[56] *Od.* 11.21, 639.

the earth, at the point of 'sunrise',[57] then the Ocean must have been imagined flowing counterclockwise around the *oikoumenê*. Odysseus thus sailed with the current on his way back and against it on its way towards Hades.[58] This then agrees with the postulated southerly course of the sun towards the east after sunset, if the sun's goblet is imagined travelling with the current, rather than against it.

The concept of 'cosmic nadir' or *axis mundi* accounting for the exchange of night and day can provide further information on the localisation of the Otherworld in early Greek tradition. Tartarus as the cosmic nadir can be recognised in the cosmologies of both Hesiod, where it is the lowest of the three layers of the *kosmos*, a full nine days' fall of a bronze *akmon* distant from the central layer, represented by the earth,[59] and Homer, who envisaged a similar structure of the *kosmos*.[60] Thus neither Hades nor Tartarus apparently had any connection with the west for either Homer or Hesiod, at least not in the more or less complete cosmological models they attempted to present in these passages. Both were located under the earth, but not within it; rather, they were imagined to lie on the other, opposite, side of the earth, whether on the earth's surface as such or above it (in Aristotelian centrifocal sense), representing the lower, i.e., southern, sky.[61] A similar symmetrical picture of the cosmos can be reconstructed in Mesopotamian cosmology of the late (?) 2nd millennium BC, with Esharra, the lowest heaven, and the Apsu, the 'uppermost' underworld, corresponding in position to each other.[62]

In terms of a sun that travels under the earth (i.e., the 'uni-polar' model accounting for the phenomena of sunset and sunrise), Hesiod's house of Night could indeed be localised below the underside of the earth in Tartarus, at a point equidistant from east and west, where the sunrise and sunset coincide.[63] Respectively, this could also be applied to Stesichorus' concept of 'depths (*benthea*) of night', which suggest a physical depth, and thus could easily refer to Tartarus/Hades in its cosmological sense. The conflation of a horizontal voyage and a vertical descent in reaching the otherworld, which, as we have seen, occasionally occurs in Greek tradition, in combination with the existence of the 'uni-polar' model accounting for the diurnal solar movement involving the concept of 'cosmic nadir' under the earth (but not inside it) and a model involving the sun god's nocturnal voyage in a cup traversing the Ocean by way of south, might explain the mixture of motifs that occurs both in Stesichorus' localisation of Helios' palace in the 'depths of night' and Hesiod's localisation of the house of Night.

Two remaining pieces of information introduced by Hesiod in his description of the house of Night remain to be explained in terms of its localisation at this particular cosmological location: Atlas and the Hesperides. Regarding the former, although regularly placed in the far west, it seems that there existed a strong and early tradition that associated him—or it—with the south.[64] At the same time, in his capacity as the *axis mundi*[65] he was indeed sometimes located at the south pole.[66] The association of Atlas with the cosmic region located under the earth, although apparently ancient, was almost completely overshadowed by the idea of Atlas in the extreme west. There are thus only several unambiguous testimonies supporting this notion, where Atlas holds either the earth

[57] *Od.* 12.3–4.

[58] Similarly, Heubeck (1989: 78) believes that Odysseus sailed 'through the Ocean' (cf. Romm 1992: 15; for a different opinion see Hoekstra 1989: 203 in the same *Commentary*) from the farthest east, where Aeaea is located, to the farthest west and the Cimmerians 'following the rim of the earth… *via* the southern perimeter' reaching the limits of the Ocean (*Od.* 11.13), returning by the northern edge of the world. This circumnavigation, however, leaves the upstream/downstream distinction unexplained. Cerri (2014: 168) also argues for a clockwise tour of the Ocean *via* west and north to Aeaea in the extreme east, with Hades on the farther shore of the Ocean (to the NE of Aeaea). Podossinov (2013: 221–22 with Abb. 8 on p. 232) believes that Odysseus followed the Ocean counterclockwise from Aeaea in the extreme NE to the western Mediterranean (he places Hades near Aeaea in the extreme NE, so he does not discuss this part of the itinerary).

[59] Hes. *Th.* 720–25, cf. [Apollod.] *Bibl.* 1.1.2. His symmetrical arrangement of great cosmic regions is perhaps also reflected in his distribution of the posthumous abodes of different generations of mankind (*epichthonioi* and *hupochthonioi*, *Op.* 122, 141) (Rohde 1925: 85 n. 37; West 1978: 182, 186–87; Vernant 1983: 104–106; Calame 2009b: ch. II.2.3.2, V.2.1.2). For *akmon* (together with Hephaestus) as a measuring-device in estimation of cosmic dimensions see further *Il.* 15.18–24 with 1.590–93 (Crates fr. 3 Broggiato = *BNJ* 2113F2 with Porter 1992: 95–96 and Brisson 2004: 47; Σ A *Il.* 1.591ab, 15.18a; Σ D (*Mythographus Homericus*) *Il.* 1.590, 15.18b van Thiel 2014: 78, 460; Eustath. *Il.* 1.591 (i.244), 15.23 (iii.695); Whitman 1970: 37).

[60] *Il.* 8.15–16 with Kirk *et al.* 1983: 9, West 1995: 208 and Gee 2020: 34–35, cf. Dörpfeld 1925: 233. In actual fact Homer describes Tartarus located 'as far beneath *Hades* as heaven is above earth' (cf. *Aen.* 6.577–79 with Gee 2020: 40; the diagram in Σ AT *Il.* 8.13 (i.268 Dindorf; Kirk 1990: ii.297–98)).

[61] Additional passages that corroborate this interpretation are *Il.* 8.478–79 with Σ T *Il.* 8.478–79 (ii.379 Erbse) and Kirk 1990: ii.334 (Tartarus at the lowest *peirath'* of earth and sea); *Th.* 727–29 (cf. *Op.* 17–19), 731, 736–38 = 807–809 with Kirk 1990: ii.334 (Tartarus at the *eschata* of earth below the roots of earth and sea, under the murky darkness, containing the sources and *peirat'* of earth, sea, sky and murky Tartarus itself).

[62] *Enuma Eliš* 4.143–44; Huxley 1997: 190. Cf. the symmetry between the distances between the eastern and western horizon and the zenith and nadir in the ^{mul}Apin (Huxley 1997: 195–96).

[63] Stokes 1962: 17; Ballabriga 1986: 125; cf. Eur. *Or.* 174–76 for Night in Erebus. Similarly, in Indian cosmology the horses of the sun are said to be unharnessed in the netherworld (*R̥V* 5.62.1; Kuiper 1983: 67, 95–96, 160), more precisely in the abode of Varuṇa, who supports both the earth and heaven from his abode at the cosmic nadir (Kuiper 1983: 68, 69 n. 68, 76, 78, 144–45, 149–50).

[64] Hdt. 4.184.3–4 (Thomson 1948: 70; Cary & Warmington 1963: 288 n. 49; Carpenter 1973: 124–25; Arnaud 2014: 39; Lightfoot 2014: 279); Verg. *Aen.* 6.794–97 with Serv. *Aen* 6.795 (cf. Romm 1994: 92) and Ti. Donatus *Interpr. Verg. Aen.* 6.795, i.609 Georgius (Getty 1948: 44; Nadeau 1970: 344), etc. Arnaud (2014: 39) sees a reflection of specifically Hesiod's concept of 'southern' Atlas in the mountain range depicted along the entire southern portion of Africa on the Tabula Peutingeriana.

[65] Arist. *De motu an.* 699a27–29; Σ Hes. *Th.* R2WLZT 507a, R2WLZ 507b, R2WLZT 509, X 518a (Di Gregorio 1975: 78–80); Σ Aesch. *PV* 349ab; Σ NAB Eur. *Hipp.* 747 (ii.93 Schwartz); Hesych. s.v. ἄτλας, etc.

[66] Verg. *Aen.* 4.481–82 = 6.796–98. For *axis* (= πόλος) as 'the extremity of the axis', 'pole' see Verg. *Aen.* 2.512, 6.790, 8.28; Ov. *Her.* 6.106, *Ex Pon.* 2.10.48, 4.14.62, 15.36, *Trist.* 2.190, 5.2b.20 (Hardie 1983: 224–27); for *axis* as the *south* pole see Vitruv. 6.1.3, 4, 5, 7; Manil. *Astron.* 1.375, 577, 589, 613, 624; Luc. *BC* 7.422, 9.542, perhaps 8.175; perhaps also Ov. *Ex. Pont.* 4.10.43 (Hardie 1983: 226).

or both the heavens and earth upon his shoulders.[67] But Atlas from the *Theogony* can be plausibly localised beneath the earth, diametrically opposite to the topmost sky, at the cosmic nadir located deep in Tartarus.[68] Both the concept of the *axis mundi* and its extremities (poles) functions equally well within the framework of a flat earth, which is certainly the context in which it had to appear in Hesiod; naturally, later authors presupposed another framework, that of a spherical earth, retrospectively and anachronistically projecting it onto earlier accounts. Different localisations of Atlas (south and west in the horizontal and cosmic nadir in the vertical model) reflect the parallel existence of two different models (and their variations or combinations) accounting for the diurnal solar movement focusing on the daylight/night exchange.

With respect to the Hesperides, closely associated with the west in the 'bi-polar' model accounting for the daylight/night exchange, they were located by the first half of the 5th-c. author Panyassis in the far south.[69] Panyassis' localisation should probably be understood in relation to the intriguing statement of Pherecydes, who claimed that the Hesperides are to be found 'on Atlas among the Hyperboreans' and 'not, as some have said, in [southernmost?] Libya'.[70] Indeed, Panyassis is the only author, as far as we know, who places the Hesperides 'in Libya', in the meaning of the far south of that continent. Pherecydes' apparently chaotic description of Heracles' search for the golden apples of the Hesperides involves the hero's travel from Egypt 'through Libya to the outer sea' (διὰ τῆς Λιβύης ἐπὶ τὴν ἔξω θάλασσαν; in *BNJ* 3F17 εἰς τὴν ἔξω Λιβύην... ἐπὶ τὴν θάλασσαν τὴν ἔξω κειμένην), where he took Helios' golden cup, crossed 'to the opposite mainland' (περαιωθεὶς ἐπὶ τὴν ἤπειρον τὴν ἀντικρὺ; in *BNJ* 3F17 εἰς πέρην... διὰ τοῦ ὠκεανοῦ) and arrived at Caucasus. If as West claims, Pherecydes here follows Panyassis in having Heracles boarding the cup of Helios in the far south,[71] then both authors certainly believed that the sun travels in his goblet from west to east *via* south. Although this may well be true, it is unlikely that Heracles managed to obtain Helios' vehicle and left him stranded halfway on his journey; it is probably closer to truth to think that Heracles used the cup *after* Helios arrived at the point of his rising.[72] In any case, both Panyassis and Pherecydes most probably conceived the sun's nocturnal voyage as taking place by way of south, especially since the latter's account continues with a landing leading to Caucasus, which is only conceivable in the event that the sun's cup was on a counterclockwise course.

This concept can be further associated with an anonymous fragment, most probably dating from the 4th c., dealing with the nocturnal voyage of the sun.[73] The fragment speaks of 'choir-leaders of the Hesperides' driving their chariot 'along the path of the night'.[74] This could suggest that Hesperides actually transport Helios (albeit in a chariot rather than in a cup, which is a paradigmatic shift) from west to east, to a νε]ότροφον [τρο]πάν,[75] 'new turning-point', where 'Night passes through [or 'changes into' or 'gets in turn'] the light-bearing radiance in the eastern [or 'morgenrötlichen'] air'.[76] The turning point might be located in the farthest east, near the point of sunrise, which is not very likely, or in the 'cosmic nadir', where the Day/Night exchange is emphasized by Hesiod, whether confounded with the concept of horizontal (southerly?) nocturnal movement of the sun or not, and either with reference to a flat or spherical earth. At the same time, the fragment utilises the model of chariot-ride to account for the diurnal solar movement, more precisely, for its nocturnal section, introducing a new and rarely attested (to the best of my knowledge, this is the only testimony of this type in Greek tradition) element in the form of the Hesperides' chariot. In addition, similarly to Panyassis, the fragment seems to associate the Hesperides with the farthest south and/or the underside of the earth, i.e., the underworld location of the cosmic nadir.

The localisations of these two features—Atlas and the Hesperides—in the far south reflect the conflation of

67 Paus. 5.11.5, 18.4 (mid-6th c. chest of Cypselus and painted screens which surrounded Phidias' Zeus in Olympia) (Johnson 1999: 20–21 with n. 39 relativizes Pausanias' testimonies); Eur. *Her*. 403–405 (Ballabriga 1986: 83–84; *contra* Johnson 1999: 20 n. 35); *LIMC* iii.2.8 *Atlas* 11 (cf. *LIMC* iii.1.5–6) (*pace* Johnson 1999: 20), 15 (very dubious). Butterworth (1970: 48) believes that the Atlas on a Laconian cup from Caere (c. 565–550 BC, *LIMC* 3.2.6 s.v. *Atlas* 1, cf. 3.1.4, 13, 7.2.424 s.v. *Prometheus* 54) may be supporting the earth, rather than the vault of heaven; the depiction is certainly ambiguous, since the whole scene, with Atlas supporting whatever he is supporting and Prometheus tied to a pillar actually stand atop a truncated pillar that could represent either the pillar that separates the earth from the heavens (i.e., Atlas' pillar) or even the earth itself, if we accept that the author of the cup was influenced by Anaximander's teachings (see chapter 11).

68 Ballabriga 1986: 77, 290–91. His argument (Ballabriga 1986: 82–83) for *Th*. 334 as a reference to the underworld is rightly rejected by Johnson 1999: 19–20 (cf. West 1966: 258). Similarly, we have just seen that in Indian cosmology Varuṇa supports both the earth and heaven from his abode at the cosmic nadir.

69 Pany. fr. 15 West; West 1979: 145 *pace* Matthews 1974: 24, 70–71. For his date see Matthews 1974: 12–19.

70 [Apollod.] *Bibl*. 2.5.11 (cf. Tzetz. *Chil*. 2.36.357, 372, 377–78; Pediasimus *De Herc. laboribus* 11). For the attribution of most of this account to Pherecydes see Jacoby *Comm*. i.394–95; van der Valk 1958: 123–29; Matthews 1974: 70–71; West 1979: 145; Gantz 1993: 410; Fowler 2013: 291, 298–99, 606. Cf. Pherec. *BNJ* 3F17.

71 West 1979: 145.

72 Cf. Gantz 1993: 404–405, 862 n. 50 and Fowler 2013: 298. He could have continued his voyage counterclockwise in Helios' cup towards the north and then west. But this would hardly work for Panyassis' account, for whom the Hesperides were located in the south; the scarcity of testimonies for Panyassis' poem prevents us from making any firm conclusions on his view of the sun's and Heracles' precise itinerary. Podossinov (2013: 229, 231 with Abb. 8 on p. 232) seems to believe that Heracles, searching for the Hesperides, entered the sun's cup in the extreme west and travelled by way of north to the extreme east (Caucasus) clockwise upon the Ocean. But this interpretation presupposes that Heracles would leave Helios stranded at setting with no nightly transport to his place of rising; it also contradicts the counterclockwise movement of the sun's goblet.

73 *PMG* 1023 = *TrGF* adesp. F692. Alternatively, the fragment was attributed to Euripides or a Hellenistic poet (Alfonsi 1953: 297–99; Merkelbach 1953: 126; 1958: 103–104; Fantuzzi 2006; Snell & Kannicht 2007: 284; Barbantani 2018: 108).

74 Lines 11–13, trans. Campbell 1993: 421.

75 Lines 13–14.

76 Lines 14–15; Merkelbach 1953: 126; Campbell 1993: 421, 423; Fantuzzi 2006. The verb *ameibesthai* 'expresses something of the reciprocal nature of the phenomenon of the succession day/night' (Diggle 1984: 63 = 1994: 293–94; Fantuzzi 2006).

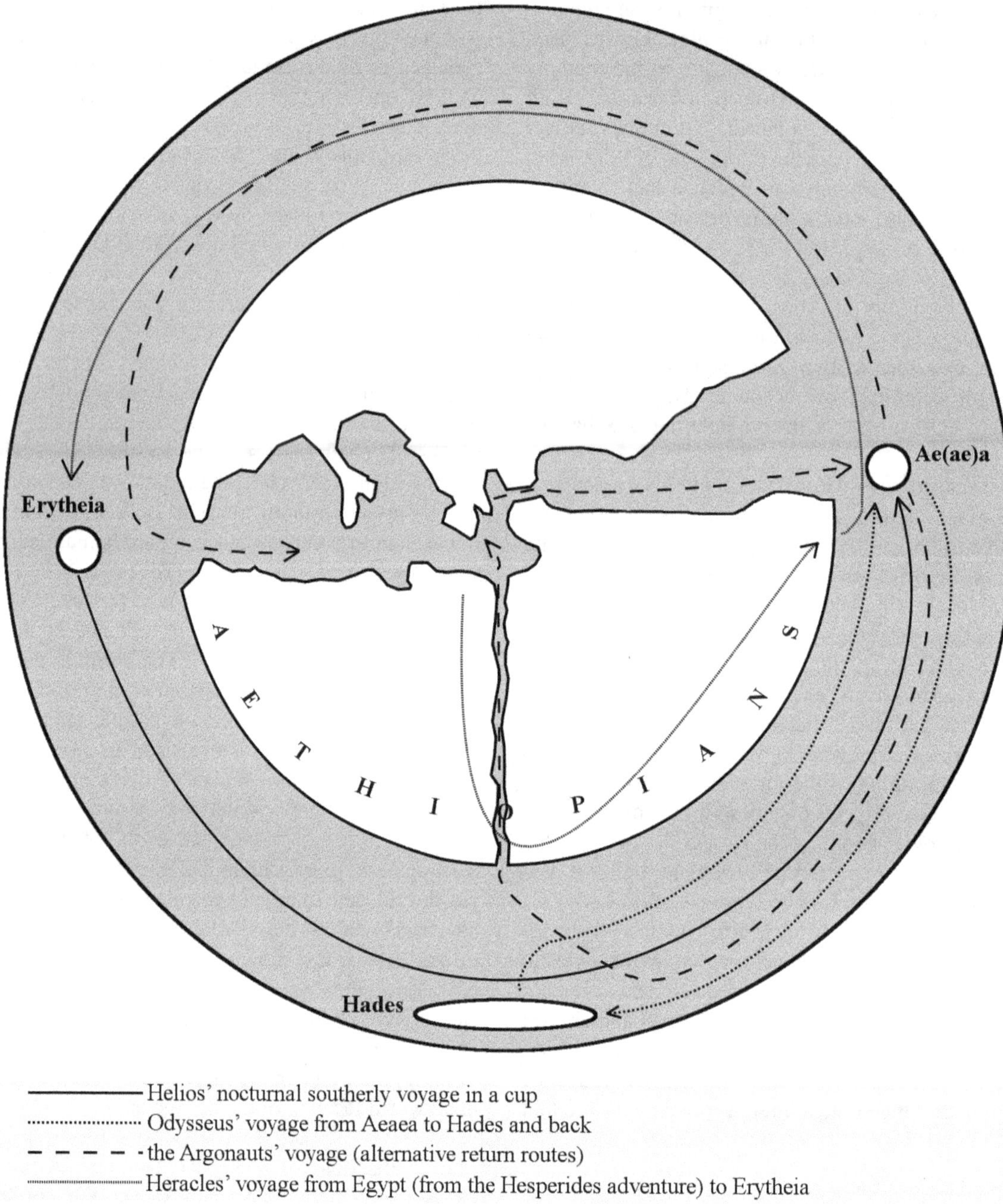

Figure 9.2 Cosmological voyages following the diurnal solar movement discussed in the text (drawn by the author)

two different models accounting for the diurnal solar movement: the vertical *axis mundi* as the location of the daylight/night exchange and the horizontal southerly path of the sun during night, which is properly part of the 'bi-polar' model involving sunrise and sunset on the eastern and western horizon, respectively. This conflation resulted in the localisation of the point equidistant from east and west upon the surface of the earth, rather than in Hades/Tartarus underneath the earth. It is certainly the case that the signals provided by Hesiod for his localisation of the house of Night are somewhat ambiguous. However, it seems that the arguments for a southern/below the underside of the earth setting for this cosmological location are fairly strong and, in any case, the existence of two models accounting for the daylight/night exchange allowed the perpetuation of the ambiguity in the tradition.

Similarly, it could be argued that in Stesichorus the sun took the southerly path after setting, perhaps echoing an understanding of the localisation of Hesiod's house of Night in these terms. The interpretative tradition followed by Stesichorus at this juncture apparently understood the location of the house of Night in the south or at the underside of the earth, rather than in the west. Otherwise, what would be the point of Stesichorus' account if Helios entered his cup at the moment of setting in the farthest west, and then sailed even further west to his family? On the other hand, if this toponym is located halfway to the sun's rising-point, the journey makes perfect cosmological sense. Finally, the house of Helios could be plausibly placed in the farthest east, but the house of Night surely cannot.

9.7. The sun's cup and Heracles (Figure 9.2)

Together with the episode involving the Hesperides discussed above, the cup of Helios was strongly associated with another of Heracles' tasks, his voyage to Erytheia, presumably the sunset island.[77] Heracles' use of the sun's cup in the context of this labour was an important juncture that offered the opportunity to many authors to speculate on the sun's nocturnal voyage. Thus in the fragment discussed here Stesichorus reported how the hero had crossed the Ocean in Helios' cup in order to obtain Geryon's cattle.[78] The story of the sun bestowing on Heracles his golden cup in order to cross the Ocean and arrive at Erytheia is further reported by a number of early authors,[79] and it continued to be an attractive subject throughout antiquity.[80] The island of Erytheia was originally simply an island 'beyond the Ocean',[81] often associated with the Hesperides,[82] but it was rather early identified with the area of Gades/Tartessus or, more generally, with southwestern Iberia.[83] Later sources followed these localisations in associating Erytheia with the region at the mouth of the Guadalquivir.[84]

These interpretations—or rationalisations—notwithstanding, it does not seem particularly difficult to determine the precise cosmological location of the solar Red Island 'beyond the Ocean' in the context of a model accounting for the diurnal solar motion.

From the remains of Stesichorus' *Geryoneis* it is unclear where exactly Heracles entered the sun's cup when going after Geryon's cattle.[85] In the light of S 17 = 185 *PMG* = fr. 8a Finglass, with the sun's cup traversing the Ocean counterclockwise by way of south from west to east, one could infer that the hero entered the unoccupied cup in the far east at the point of sunrise (both geographically and temporally)—i.e., at 'Aea(ea)'—and travelled counterclockwise by way of north during the day to Erytheia in the far west, where he arrived at sunset. I have already discussed the northern and southern alternatives of the sun's path; I have also suggested that the more probable path of the occupied cup is the southerly one. Thus, it seems plausible to suggest that Heracles entered the path of the sun at the point representing the sun's sunrise portal and exited from it at its sunset portal at Erytheia, the 'gate of the sun' already known to Homer (if not under this precise name).

The nature of Erytheia as the western diurnal portal of the sun is further confirmed by other suggestive elements that are present in its mythic mapping. At the same time, another connection of the diurnal solar movement with certain eschatological tenets can be recognised in the development of this mythic nexus, this time within what appears to be a more appropriate setting, the diurnal solar movement. I have already had the chance to mention Erytheia's most telling characteristic in this respect, that is, Helios' herds of cattle living on that island (see chapter 2). But together with Geryon's and Helios' cattle, Hades' herdsman Menoetius was also present on Erytheia,[86] thus offering a strong connection with the underworld. Menoetius was a Titan thrown into Tartarus,[87] and he usually tended the death lord's cattle in Hades.[88] Additionally, Heracles was said to have killed Geryon's herdsman Eurytion 'in the misty (*êeroeis*) farmstead beyond the famed Ocean', a description suggestive of Tartarus/Hades,[89] while Tartessus—regularly identified with Erytheia—was sometimes even identified with Tartarus.[90] Furthermore, Geryon's watchdog Orthrus was a brother of Cerberus.[91] Geryon's connections with Hades

[77] Cf. Davies 1988: 280; 1992: 221; Davies & Finglass 2014: 261; Beaulieu 2016: 48.

[78] Stes. S 17 = 185 *PMG* = fr. 8b Finglass.

[79] Pis. fr. 6 Davies = fr. 5 West (in Athenaeus' report Erytheia is not actually mentioned); Pany. fr. 7A Davies = fr. 12 West (cf. Macr. *Sat.* 5.21.19); Aesch. fr. 74 *TrGF*, Radt = 109 Mette (Erytheia is not explicitly named); Pher. *BNJ* 3F18a, cf. [Apollod.] *Bibl.* 2.5.10, Macr. *Sat.* 5.21.19, Joan. Pediasimus *De Herc. labor.* 10, *Myth. Gr.* i.257 Wagner (for Heracles using Helios' cup in another context in Pherecydes see *BNJ* 3F17 with [Apollod.] 2.5.11, probably Agath. *De mari Eryth.* 7 *ap.* Phot. Bibl. 250.443a, mentioning a *lebês* rather than *depas*, with the above discussion); Antim. fr. 66 Wyss, West (according to Allen 1993: 95).

[80] In later authors certain paradigmatic shifts occur: Euphor. fr. 52 Powell = fr. 72 Lightfoot, cf. Brown 1968: 62 (bronze *akatos*); Alex. Eph. fr. 38 *SH ap.* Eustath. Dion. Perieg. 558 (*GGM* ii.325–26), cf. Brown 1968: 62 (bronze *lebês*); Julian *Or.* 7.219d (golden kylix); *aerea olla* in Serv. *Aen.* 7.662; Myth. Vat. 1.67 Kulcsar, 2.175 Kulcsar, 3.13.6, p. 249 Bode; Albericus *De deor. imag. lib.* 22, ii.328 *Myth. Lat.*; *navis aenea* in Serv. Dan. *Aen.* 7.299. A rationalising interpretation of this story can perhaps be found in Pausanias' account of the Tyrian Heracles' voyage on a raft of planks from Tyre to Erythrae in Ionia (Paus. 7.5.5–8).

[81] Hes. *Th.* 293–94 with Σ ad 290, cf. 981; Pany. fr. 13 West; Hecat. *BNJ* 1F26 (negative); Eustath. *Il.* 7.157–60, ii.426.18–19.

[82] Stes. S 8 = fr. 10 Finglass; Σ A. R. 4.1396–99d (p. 317 Wendel); Avien. *Descr. orb. terr.* 738–42 (*GGM* ii.183). For a Hesperid Erytheia see Hes. fr. dub. 360 M-W; A. R. 4.1427 (Erythêis); probably [Arist.] *De mirab. ausc.* 133.843b28 (Erythê at b30–31); [Apollod.] *Bibl.* 2.5.11. She is said to have been Geryon's daughter (Paus. 10.17.5 with Steph. Byz. s.v. Ἐρύθεια; cf. Hell. *FGrHist* 4F110 and Myth. Vat. 1.67 Kulscar with Höfer 1897–1902: 1215.28–42; Müller-Graupa 1942: 1496.15–23).

[83] Stes. S 7 = 184 *PMG* = fr. 9 Finglass; Hecat. *BNJ* 1F26 (negative); Pherec. *BNJ* 3F18b, cf. [Apollod.] *Bibl.* 2.5.10 and Joan. Pediasimus *De Herc. labor.* 10, *Myth. Gr.* i.257 Wagner; Hdt. 4.8.2; Philis. *FGrHist* 11F3 (= Pherec. fr. 33h *FHG*?).

[84] Ephor. *BNJ* 70F129ab; Pyth. fr. 4 Bianchetti = T 11 Roseman = fr. 8 Mette *ap.* Eratosth. IIIB122, fr. 153 Roller, etc.

[85] Cerri (2014: 171) believed that Heracles entered the unoccupied sun-god's boat in the far east but travelled *via* the south (clockwise) to Erytheia in the far west; he further argued that both for Mimnermus and Stesichorus the sun travels in his boat *via* the north during night, but that the boat travels unoccupied during the day *via* the south, thus traversing the rim of the *oikoumenê* in a clockwise direction (Cerri 2014: 170–71). We have seen that the sources rather point to a different itinerary.

[86] [Apollod.] *Bibl.* 2.5.10.

[87] Hes. *Th.* 510, 514–15; [Apollod.] *Bibl.* 1.2.3.

[88] [Apollod.] *Bibl.* 2.5.12; cf. Tzetz. *Chil.* 2.396–97 (p. 55 Kiessling); Croon 1952: 31; Gantz 1993: 415.

[89] Hes. *Th.* 294 with Davies 1988: 279 n. 16 and Gantz 1993: 402. A *stathmos* of Ocean's four-legged bird is mentioned by Aesch. *PV* 394–96. For the expression *hupo zophon êeroenta* designating Tartarus, Hades or the land of the dead see *Il.* 21.56, 23.51; *Od.* 11.57, 155); Hes. *Th.* 653, 658, 729, fr. 280 M-W = Minyas fr. 7* West; *hHerm.* 256–57; *hDem.* 80, 337, 402, 446, 464, 482; Anon. 925e.9–10 *PMG*; but *poti zophon êeroenta* could also refer to the region opposite the dawn (*Il.* 12.240; *Od.* 13.241). See the discussion in chapter 4.

[90] Ar. *Ran.* 475; Str. 3.2.12; Avien. *Or. marit.* 225–44.

[91] Hes. *Th.* 309; [Apollod.] *Bibl.* 2.5.10; Σ A. R. 4.1396–99d, p. 317 Wendel; Poll. 5.46 = Eudokia 239, p. 165.11; Σ Pl. *Tim.* 427, p. 948 Baiter *et al.*, etc; cf. Palaeph. 39 for Cerberus and Orthrus both guardians of Geryon's cattle (Höfer 1897–1902: 1215.15–25; Müller-Graupa 1942: 1495.43–49; Croon 1952: 13 n. 2; Davies 1988: 281). Its name was associated with the rising of the sun, Σ R2WLZ Hes. *Th.* 293a (Di

are rarely emphasized, but do appear occasionally: as CERUN he appears together with Hades and Persephone on the 2nd half of the 4th c. depiction in Tomba dell'Orco II in Tarquinia,[92] early Imperial authors explicitly place him in Hades (after being dispatched there by Heracles?),[93] he was worshipped as a hero on Sicily,[94] and he had an oracle at Patavium.[95] Thus it seems that Erytheia was at the same time a sun's portal, that is, a place where he or it descends from the sky, and a region under some kind of sovereignty of Hades,[96] which makes Geryon and his cattle a form of Helios or Hades—or both.

Indeed, it was not unusual to collocate the cattle of Helios with (at least) the gate of the Underworld.[97] Certainly the most famous example of such practice was Cape Taenarus, where Helios' cattle had pastures,[98] at one of the most famous entrances to Hades in antiquity.[99]

In this way Erytheia belongs to a group of mythic features that seem to exhibit a dual nature, at the same time being associated with the far west and with the underworld. Moreover, its association with the diurnal solar movement seems irrefutable. As already noted, this duplication of localisation and/or characteristics of a feature reflects the existence of two models accounting for the diurnal solar movement in early Greek tradition. In particular, it reflects the application of the concept of cosmic 'nadir' or *axis mundi* located deep in Tartarus/Hades in order to account for the solar phenomena of sunrise and sunset alongside the more obvious model involving the risings and settings at the eastern and western horizon. Both models were indeed present in Greek tradition alongside each other, even if from a modern perspective the 'bi-polar' model seems much more appropriate, since it appears to correspond to the observed phenomena more adequately.

9.8. Iconographical testimonies for the sun in a cup

Besides the literary testimonies, there is some iconographic evidence for the concept of sun in a vessel—whether a real ship or a bowl—in Greek tradition.[100] The iconography, however, unfortunately cannot be used to distinguish between the sun's southerly and northerly path after setting. Sometimes Helios is represented together with his chariot in a boat,[101] occasionally with a reference to his nightly voyage.[102] In a unique Etruscan representation, the sun god is actually represented during his nightly voyage in a boat sailing upon the Ocean.[103] These examples show that the concept of sun in a boat was known to Greek artists from at least the late 6th c. onwards, although it is fair to admit that this motif did not figure very prominently in their repertoire. After discussing—with qualifications—the solar function of Apollo in chapter 7, two representations of this deity in a tripod traveling over the sea on two Attic vases (the earlier is dated to the third quarter of the sixth century) can safely be included in the group of iconographic testimonies for the nocturnal solar voyage.[104] This depiction seems to be an interpretation of the sun's voyage in his cup, adapted to the iconography of Apollo. This odd way of transport is nowhere mentioned in Apollo-related myths, so it can be safely attributed to Helios.[105] Finally, several 6th and 5th-c. depictions of Heracles in Helios' cup also belong to this group of representations, since they imply the sun's ownership of the cup.[106] Thus iconographic sources support the accounts in literary texts.

9.9. The Presocratic tradition of the sun's bowl

We can safely say that the concept of the nightly voyage of the sun in a cup over Ocean from west to east is fairly well attested in Greek mythic tradition from at least the 7th c. onwards. But an 'unmythical' concept of a sun-in-a-bowl is attested in Greek Presocratic tradition from the time of Heraclitus. He believed that the stars are formed by earthly exhalations gathered in heavenly bowls or basins (*skaphai*),[107] which also applied to the sun and the moon.[108] In addition, Alcmaeon and Antiphon also believed that the moon is in the shape of a bowl.[109]

Gregorio 1975: 58), which seems odd in the context of a sunset island but emphasises its solar association.

[92] *LIMC* 4.1.189, 4.2.107 s.v. *Geryones* 25, 4.1.395, 4.2.226 s.v. *Hades/Aita, Calu* 6; cf. Croon 1952: 27; Davies 1988: 279 n. 16.

[93] Verg. *Aen.* 6.289; Hor. *Carm.* 2.14.7–8 (cf. Σ ad 2.14.7, p. 79.11–14 Botschuyver).

[94] D. S. 4.24.3.

[95] Suet. *Tib.* 14.3. See Burkert 1979a: 179 n. 2; cf. Croon 1952: 23 with nn. 42–43.

[96] Cf. Croon 1952: 32; Davies 1988: 278–82; 1992: 221; 2006: 196–98, 200, 204; Fowler 2013: 304.

[97] Cf. Nakassis 2004: 227. See especially the discussion of Pylos in chapter 10.

[98] *hApoll.* 411–13. Taenarus was originally Apollo's domain (Ephor. *BNJ* 70F150, cf. Paus. 2.33.2), possibly in his solar function explored in chapter 7.

[99] Hecat. *BNJ* 1F27; Pher. *BNJ* 3F39 (Fowler 2013: 306 n. 159); Pi. *P.* 4.43–44; Soph. *Epi Tainaröi Satyroi* (esp. frs. 230–34 Pearson); Eur. *Her.* 23–25; Ar. *Ran.* 187; Men. fr. 785 Körte-Thierfelder, cf. Σ Pi. *P.* 4.76d; Palaeph. 39; A. R. 1.101–102 (cf. Σ A. R. 1.101–104a, p. 15 Wendel); Lycoph. 90 with paraph., i.10 Scheer, etc.

[100] Cf. Allen 1993: 97 and Davies & Finglass 2014: 237 for some of the evidence.

[101] *LIMC* 5.1.1008, 1015, 5.2.631 s.v. *Helios* 2, 99 (= 2.2.678 s.v. *Astra* 61), 100 (cf. Tiverios 1982: Pl. 29.5).

[102] *LIMC* 5.1.1016 s.v. *Helios* 102 = L. Beschi, *LIMC* 4.2.597 s.v. *Demeter* 459

[103] *LIMC* 5.1.1043, 5.2.657 s.v. *Helios/Usil* 30; Krauskopf 2013: 517.

[104] *LIMC* 2.1.233 s.v. *Apollon* 381, 382.

[105] Beaulieu 2016: 144 interprets this depiction as an illustration of Apollo's annual voyage from Delphi to Hyperborea; but the god actually used his swan-chariot for this voyage (Alc. fr. 307c Lobel-Page).

[106] *LIMC* 5.1.1015 s.v. *Helios* 101; *LIMC* 5.1.81, 5.2.92 s.v. *Herakles* 2550 (cf. Boardman 1974: 232), 2551, 2552 (cf. Pugliese Carratelli 1996: 214).

[107] 22A1 D-K = D. L. 9.9–10.

[108] 22A1 D-K = D. L. 9.9–10, A12 D-K; Σ Pl. *Resp.* 498a2 (Greene 1938: 240–41); Ach. *Isag.* 19, p. 46.28 Maass. Cf. Kouremenos *et al.* (2006: 156), who suggest the reading κάδδος or κύκλος (= σκάφη) in PDerv. iv.7, where Heraclitus is cited, with reference to the sun's nature and (Leptines?) *Ars Astron.* [= P. Pap. 1] col. 12.9 Blass (Mansfeld & Runia 2009: 598).

[109] 24A4 and 87B28 D-K. For a tentative identification of the moon, especially when depicted as a recumbent crescent, with a bowl in various traditions (Greek, Mesopotamian, Indian, Christian) see Butterworth 1970: 103–104, 106–107, 111–12, 114–16, 118, 120–21, 132, 136, 198; for the recumbent crescent as a (bowl-)boat see Butterworth 1970: 123, 134–35 (here it is claimed that it was actually serving as a nocturnal solar barque), 197, 220. Butterworth 1970: 223 associated this (tentative) reconstruction with Heraclitus' concept.

It was argued that Heraclitus' theory is actually derived from the mythic account,[110] but it would perhaps be better to say that the developments were parallel, since the mythic account existed side-by-side with the scientific one.[111] It is by now little surprising that the scientific tradition used similar models to account for solar movement as did the mythic, since both belonged to the same ethnographic context and often approached the phenomena in the physical world in a comparable fashion. However, while myth explained the diurnal solar movement in almost exclusively anthropomorphic terms (an agent travelling in a chariot and a boat), the Presocratics employed a non-personalistic description of this phenomenon nevertheless using a comparable model of a solar vessel.

9.10. Non-Greek traditions of the sun travelling in a boat

The particular mythic model accounting for the diurnal solar movement by involving the sun(-god)-in-a-boat motif was not only characteristic for Greek tradition but appears in other ethnographic contexts as well. When I discussed in chapter 4 the presence of the notion of a northern region not visited by the sun in several non-Greek traditions, I emphasised that I do not wish to argue for genetic connections or mutual influences between any of these traditions, but only to document the presence of parallel evidence suggestive of the existence of a shared cross-cultural model of accounting for solar movement, and to provide an outside perspective on Greek understanding of this phenomenon. My intentions with respect to the present model are identical, and the model of the sun(-god)-in-a-boat appears to have been a common explanatory strategy in order to account for the diurnal solar movement in various traditions. It thus appears that both the annual and diurnal solar movement were tackled by shared cross-cultural traditions, which is what one would expect in the light of close connections that exist between these two phenomena (or rather two aspects of a single phenomenon). In addition, many of these non-Greek models show clear signs of interactions with eschatological domain, which is especially pronounced in the Egyptian tradition, but is also present elsewhere, thus offering another connection with the Greek ethnographic context, where this practice is also present.

The best documented case is of course the Egyptian; it hardly needs referencing at all.[112] It was also very prominent in Mesopotamian tradition. In the *Epic of Gilgameš* it is stated that only Šamaš the sun-god can cross the Ocean,[113] and thus he can be found 'at the other side of the sea', according to an Early Dynastic text from Ebla,[114] in which 'Šamaš' boat/raft' also appears.[115] Presumably he crosses the sea/Ocean in some vehicle, most probably a ship, such as represented on numerous Early-Dynastic and Akkadian seals that depict Šamaš in his boat.[116] An Early Dynastic seal with a representation of the sun god in his boat in front of which a Scorpion-man is depicted can be associated with the *Epic of Gilgameš*.[117] Moreover, it is believed that this type of scene represents the sun's nocturnal voyage over the waters of the Otherworld.[118] Thus Utu/Šamaš certainly crossed (part of) the heavens/celestial Ocean in a boat, most probably during the nocturnal section of his daily course.

Elsewhere, the journey of the Ugaritic sun-goddess Shapsh might be associated with the sea,[119] which could be further tentatively associated with Re's subterranean nocturnal journey and interpreted analogously.[120] The Manichaeans also believed that the sun and moon were ships or ferry-boats,[121] and there are some faint traces of the concept in Vedic mythology, where the sun becomes Varuṇa when it enters the waters,[122] and Varuṇa crosses the nocturnal sky in a boat.[123] Finally, Latvian tradition knows of the sun traveling both in a chariot and in a barque,[124] as well as sleeping through the night in a golden boat.[125]

Furthermore, a concept similar to the Helios' chariot-in-a-boat depictions seems to appear in the Nordic Bronze Age, both in rock pictures and in decorations on bronze razors. These depictions show various combinations of sun wheels or disks, sometimes drawn or accompanied by horses, occasionally incorporating in their design anthropomorphic characters and chariots, upon a boat.[126]

[110] Kirk 1962: 269–70, cf. 278 and Guthrie 1962: i.485 and Kahn 1979: 339 n. 440. Lloyd 1966: 321–22 and Marcovich 2001: 333 are sceptical.
[111] Cf. Allen 1993: 99.
[112] Egyptian tradition will be discussed *in extenso* below in Appendix 2.
[113] *EG* 10.81–82 (George 2003b: 78; 2003a: i.683); cf. Heidel 1949:74; Speiser in *ANET*³ p. 91; Kovacs 1989: 86; Foster 2001: 76.
[114] *ARET* 5, 6: vii.1b–4 = *OIP* 99 326 (+) 342 iv.13–16 (Lambert 1989: 19–20, 33; Krebernik 1992: 83).
[115] *ARET* 5, 6: iv.4b–5 = *OIP* 99 326: iii.4–5; see also Lambert 1989: 11, 33 (Steinkeller 1992: 258 n. 38).
[116] Frankfort 1934: 18–19 with Pl. III.e–g; 1939: 67–70 with Pl. XVj, n for Early Dynastic examples, and 108–110 with Pl. XIXe, f for Akkadian examples; Van Buren 1955: 1; Böhmer 1965: 79–82 with Taf. XXXIX.466–XL.480; Steinkeller 1992: 256, also Pls. 4–5 with ten examples; Black & Green 2000: 45; Woods 2004: 69–71 with Figs. 33–37.
[117] Frankfort 1939: 68 (Pl. XVj); cf. Woods 2009: 192 with fig. 22 on p. 236.
[118] Frankfort 1934: 18; 1939: 68, 109–10; Böhmer 1965: 79; Steinkeller 1992: 257.
[119] KTU 1.6 vi.51–53; Smith in Parker 1997: 175–76 n. 206.
[120] Wyatt 2005[1987]: 20–21, 51 n. 1.
[121] *Acta Archel. cum Man. disp.* 8; Epiph. *Adv. Haeres.* (66) 9.8–9, 26.6; Parthian *Hymn to the Third Messenger*; *Manichaean Psalms* 267.7–9; *Kephalaia* 158.31–32.
[122] *Kauṣītaki Brāhmana* 18.9 (Keith 1920: 448). Cf. *ṚV* 2.38.8, 7.87.6, as well as *AV* 13.3.13 and *ṚV* 10.8.5 (Kuiper 1983: 67 with n. 56, 71).
[123] *ṚV* 7.88.3–4 (Parpola 2004–2005: 25); yet the text only mentions the fact that Varuṇa travels in a boat over the sea. For Varuṇa's realm identified with the celestial ocean, in its turn identified with the night sky, see Kuiper 1983: 33, 74–76, 78–79, 144, 148, 150, 159–60. Varuṇa is the lord of the (primeval/subterranean) waters/(western) ocean/netherworld, as well as death (Kuiper 1983: 16–17, 66–70, 96–97, 183).
[124] Barons & Wissendorf no. 33811.
[125] Barons & Wissendorf nos. 33860, 33878, 33908, 33910; West 2007a: 209.
[126] Sprockhoff 1954: 61, Abb. 16.1 on p. 62 = Gelling & Davidson 1972: 92, Fig. 44c on p. 93 = Green 1991: 80, Fig. 61 on p. 79 = West 2007a: 203, 207–208; Kaul 1998: i.200–206, 208–209, 215, ii.43, 74–75, 82, 86–87, 91–92, 99–100, 103–104, 127, 144–45, 156 cat. 95, 176, 199, 210, 220, 221, 223, 243, 256, 317, 353, 381; Kristiansen 2010: 106 Fig. 6.8 (bottom left) = Bradley 2006: 380 Fig. 5 (centre), 108 Fig. 6.10 (bottom right), 109 Fig. 6.11 (bottom right).

If these illustrations indeed represent depictions of a solar chariot on a boat, then this concept is comparable to the Greek one. However, due to uncertainties in the interpretation of these depictions, this assumption must remain tentative.

Likewise, it is possible to adduce in this context a bronze 6th-c. BC lid found in the Saône river. The lid is in the form of a large, stylized water-bird on whose back a human figure drives a *biga* (the wheels are not represented, but only the wagon-box).[127] The cultural context in which the lid belongs—the Urnfield symbolic complex—suggests that this is a depiction of a solar boat carrying a *biga*, which might represent the sun's movement over (celestial) waters.

Another complex of iconographic depictions of sun-in-a-boat was recognised in Minoan Crete. Here a solar disc is sometimes represented above a boat on wall paintings and prism seals, perhaps representing the sun's nocturnal journey.[128] One depiction on a larnax represents a boat above which descends a horse-drawn chariot, with addition of solar symbols, suggesting a concept not unlike the Helios' chariot-in-a-boat motif discussed above.[129]

With this example I will conclude this short overview of mythic models accounting for the diurnal solar movement in the form of anthropomorphic divinities riding a boat. Even the summary treatment offered here shows the wide distribution (both spatial and temporal) of this type of model across a number of different traditions, which thus truly represents a shared cross-cultural model of accounting for solar movement. Its contextualisation within these traditions is in general very secure, and any speculation on the interactions between various iterations of the model seems futile. But these variants show that Greek tradition, here as elsewhere, should not be studied in isolation from its wider Mediterranean context, and that sometimes the evidence from other ethnographic contexts can shed some additional light on some less clearly attested Greek concept, as the example of Anatolian resonances of Helios' cup indeed suggests.

9.11. Conclusion

The question of choosing between a northerly and a southerly course of the sun moving horizontally upon the earth's plane—or, in personalistic terms, navigating upon the surface of the Ocean—following its 'setting' can be solved by an interpretation of a number of surviving early sources. However, some additional caution is in order here, due to the fact that the latter alternative is often conflated with the *axis mundi* or 'cosmic nadir' model for the daylight/night exchange, involving the notion of the nocturnal voyage of the sun below the underside of the earth. Homer's signal in the form of the Aethiopians in association with sunrise and sunset convinced Ephorus and Crates to understand his verses as referring to the southern portion of the earth, although both interpreted these lines in the context of the annual solar motion, and their testimonies cannot be taken into consideration when discussing its diurnal movement; but if the reference is understood in terms of the diurnal solar movement, the Aethiopians strongly suggest the sun's southerly path during its nightly course. The decisive element in this discussion is the concept of Helios' cup, used in the god's nightly voyage from west to east upon the Ocean. The concept, attested in both literary and iconographic sources, shows some early connections with Anatolia, but was apparently adopted and adapted in an early period by the Greeks. It is closely associated with the sunset island Erytheia, which is, in its turn, closely associated with Hades. These interactions of mythic locations signal the southerly nocturnal course of the sun in combination with another model accounting for the diurnal solar movement involving the vertical *axis mundi* as the location of the daylight/night exchange. At the same time, the involvement of Erytheia/Hades in this model is another example of the well-attested eschatological resonances of the diurnal solar movement. Parallels for this concept, including the eschatological element, exist in several traditions in communication with the Aegean, as well as in others for which this contact cannot be ascertained. From Mimnermus onwards (perhaps even earlier) the poets' testimonies tentatively suggest the southerly course of the sun's nightly movement (with or without a contamination with the vertical model), without any explicit or implicit indication of the northerly route. On the other hand, somewhat surprisingly, the explicit testimonies for the southerly course are almost non-existent in that part of the Greek tradition that specifically engaged in the problem of the sun's nocturnal movement. This tradition clearly preferred the northerly route involving the elevation of the earth's plane in the north or vast northern mountains screening the sunlight during nighttime in the *oikoumenê* in order to account for the diurnal solar movement in its entirety.

References to and discussions of the diurnal solar movement in Greek tradition show strong resemblances and continuity between mythic and philosophical-scientific approaches to this phenomenon. One could even go so far to claim that no decisive breech can be discerned between these two approaches other than their respective recourse to personalistic and non-personalistic interpretative frameworks (the model involving the elevation of the earth or high mountains in the north screening the sun being a good example of the latter and Helios' cup a nice example of the former). Both approaches equally accepted the respective cosmological frameworks involving the flat and spherical earth (although for chronological, rather than ideological reasons, the flat earth cosmological framework was dominant in mythic discourse) and both engaged with

[127] Woytowitsch 1978: 64–65 no. 142, Pl. 29.

[128] Goodison 1989: 37–38, 94 with figs. 60ab, 148c, cf. 147, 154 with figs. 293g (Attic cup), 314 (Geometric fibula). The same interpretation of the depictions on Goodison's figs. 60c, 293c, 298b, cf. 293b, 295d, 296, 298a, 299c is highly questionable (Goodison 1989: 37–38, 146–48).

[129] Goodison 1989: 94–95.

similar difficulties in attempting to account for the diurnal solar movement utilising comparable models.

The diurnal solar movement was thus often referenced and accounted for in myths and these particular myths certainly belong to a class of narratives studied in this book. Moreover, the myth of Helios' cup is transparently structured on an analogy with solar movement precisely in order to account for it in anthropomorphic terms, here for once in an almost one-to-one correspondence between a physical phenomenon and a mythic story. In this way, it corresponds to the 'Hyperborean Apollo' set of myths discussed in the preceding chapter, with a significant difference in that the latter accounted for the annual solar movement, while the Helios' cup was involved in a model accounting for the diurnal course of the sun. Related phenomena were thus tackled by similar hermeneutic strategies. Other myths discussed in this chapter, such as Odysseus' and Heracles' particular adventures that clearly involve cosmological features either simply contain references to the daily movement of the sun or are only partially structured on an analogy with it. Even so, and notwithstanding the fact that these narratives contain numerous other elements for which the hermeneutic framework used to recognise and interpret myths referencing and accounting for solar movement is inadequate, they too clearly belong to a class of myths studied in this book. In addition, the myths partially structured on an analogy with solar movement also demonstrate an interest in eschatological domain characteristic for the accounts of the daily course of the sun. This is of some importance since I have noted in chapter 5 how the narratives structured on an analogy with solar movement often show a strong concern for eschatology. The significance of eschatological considerations for drawing the results of solar observations into mythic narratives and structuring these narratives upon these data thus seems strengthened by the discussion offered in the present chapter.

10

Liminal imagery in the accounts of solar movement assimilated to the world of the dead

The importance of eschatological concerns for both incorporating the results of solar observations into mythic narratives and for structuring these narratives upon them will be further analysed in a discussion of liminal imagery, i.e., of the features in mythic narratives that designate boundaries and/or transitions. This imagery is amply represented in a specific group of myths that assimilates the diurnal solar movement to the world of the dead. The discussion will require revisiting some already studied mythic narratives, but with a greater emphasis placed on the hitherto neglected or insufficiently studied features present in these narratives.

10.1. Hesiods' concept of the daylight/night exchange

In the preceding chapter a localisation of Hesiod's house of Night below the underside of the earth in Tartarus was discussed in terms of a model accounting for the diurnal solar movement involving the sun that travels under the (flat) earth. In the context of this model, the house was localised at a point where the sunrise and sunset coincide, equidistant from east and west. But the setting of Hesiod's Hesperides and Atlas in the far south was explained as reflecting a conflation with a different model accounting for the diurnal solar movement, i.e., a model involving a horizontal southerly path of the sun during night. Moreover, as I have already pointed out in chapter 2, Hesiod's description of the house of Night—a clear example of a personalistic account structured on an analogy with a natural phenomenon—cannot be treated separately from the account of the land of Laestrygonians in the *Odyssey*, even though the two narratives refer to different aspects of solar movement.[1] A parallel analysis of these two accounts, with an emphasis on the liminal imagery amply present in both, will provide further information on the early Greek conceptualisations of solar movement.

Prosaic herdsmen greeting each other while one enters and other exits the city gates of Telepylus—it can be safely presumed that they are both passing through the 'Distant Gates' themselves—thus find an interesting parallel in the conduct of cosmological entities Night and Day: at the house of the former Night and Day 'approach and greet each other' as they cross (*ameibomenai*, alternate at?) the great bronze threshold (*oudos*).[2] However, while Hesiod here undoubtedly gives an account of the diurnal solar motion, I have argued that the narrative of the *Odyssey* refers to the annual. This shows that identical mythic motifs could have been interchangeably used in building models accounting for different solar phenomena (we will see that the same conclusion perhaps applies to Mesopotamian tradition), a practice that was explained as a consequence of the similarity and interdependence of two characteristic movements of the sun (diurnal and annual) and a limited number of motifs out of which mythic models accounting for each of these could be built. In line with the nature of mythic discourse, these motifs had to be formed on an analogy with appropriate typical action patterns from human experience; apparently, the number of action patterns utilised in building models accounting for solar movement was extremely modest (anthropomorphic deity travelling in a boat or chariot being the most common choice) and we testify to a repetition of such patterns both across various ethnographic contexts and within separate traditions. I have already argued in chapter 2 for the primacy of Hesiodean concept, with Homer's variant interpreted as an example of a rationalisation of a (more) mythic account, although it would be more precise to claim that the author of the *Odyssey* here rationalised a model already applied onto the annual solar movement—most probably—in the pre-Homeric Argonautic tradition, with the actual original application of the model being equally plausible for either type of solar movement, or to both simultaneously.

While Homer, in a humorous manner, saw in the unusual meteorological conditions he describes a chance to double one's wages, Hesiod approached the consequences of the shortest of meetings of his cosmological entities more seriously. According to the poet of the *Theogony*, when one of the pair descends (*katabainô*) into the house, the other comes out through the door (*thuraze*), and thus it never holds them both simultaneously (*amphoteros*) inside. One of them is always outside the house turning about (*epistrephô*) the earth, while the other waits inside the house until her time to depart comes.[3] Thus we have a picture of a house at whose threshold Night and Day pass by each other; when Night enters it, Day exits the house. This can be interpreted in terms of the diurnal solar movement in the way that the house corresponds to the

[1] Cf. Germain 1954: 521–24; West 1966: 366; 2014: 33; Nakassis 2004: 224; Burgess 1999: 198; 2016: §§11–12.

[2] Hes. *Th.* 744–45, 748–50. West (1966: 365, 367), athetizing the lines 744–45 in which the house of Night is mentioned, opts for the *domos* of Night and Day in 751–52 as the location of the threshold. For my present purpose, this amounts to the same. For night and gates see further A. R. 4.630 (farthest west) and Mart. Cap. 1.45–47, 49 (northern portion of the sky, Weinstock 1946: 104, 109); also, a spurious fragment of Hesiod (fr. 394 M-W), formerly attributed to Anaxagoras (59B20 D-K), where a *porta vesperis* is apparently mentioned, curiously—as the text stands—referring to Sirius (Schultz 1911: 323, 328; Kraus, Schmidt, Kranz 1952: 219, 221).

[3] Hes. *Th.* 750–54, cf. 755–59.

lower hemisphere, the threshold to the horizon, and when Night rises over the threshold in the east Day descends over it in the west.[4] Yet this does not explain how they approach each other, since this interpretation requires that they must always be equally apart. This seems like a concession on the part of the poet to the account's explanatory potential or intelligibility at the expense of its correspondence to physical reality, or an emphasis on one aspect at the expense of another (i.e., a moment of daylight/night exchange is privileged in comparison with the gradual nature of the process, which is thus condensed to a single event).

Within Greek ethnographic context, Hesiod's description of Night and Day approaching and greeting each other at the brazen threshold can be further compared to the scene on a ca. 500 BC black-figure lekythos by the Sappho Painter (see the cover image).[5] The scene illustrates two chariots facing, rather than diverging from each other.[6] The drivers are labelled Nux and Eos, respectively; the location of the scene is emphasized by the presence of a 'rocky elevation' which seems to represent an opening of a cave with a dog inside, surely Cerberus.[7] Night and Dawn appear to ride towards the cave on two fluid and curved paths.[8] Thus it seems that the artist imagined the meeting of Night and Day/Dawn at the gates of Hades. This is complementary with Hesiod's placement of the brazen threshold at the house of Night in Tartarus, but it cannot be claimed that the artist tried to illustrate Hesiod's description from the *Theogony* (as exemplified in the name change); rather, he depicted a concept or a mythic model which was obviously well-known to the Greeks of Archaic age in several similar variants.

10.2. Corresponding models in Mesopotamian tradition

This Greek tradition (exemplified in Hesiod's account, the postulated early Argonautic, Homeric and Sappho Painter's) is not an isolated model accounting for the daytime/night exchange and, in a wider sense, diurnal solar movement. Corresponding accounts exist in ethnographic contexts invoked on several previous occasions, which apparently treated the diurnal solar movement in terms of mythic models similar to the Greek ones. While the majority of scholars now accept the influence of Near Eastern—particularly Anatolian—traditions upon Hesiod (with the particular details of this influence, including its extent, still a debated issue), I would not engage here in this matter, but would prefer to continue speaking in terms of parallel developments and shared cross-cultural traditions. Still, I must observe that the parallels between Greek and Near Eastern treatments of this particular topic are especially pronounced, which is why they will be subjected to a more in-depth analysis. A parallel to Hesiod's description can thus be recognized in an Ugaritic myth describing El going to the shore of the sea/ocean/deep (thm), where he meets two women alternately raising themselves up.[9] El takes them both as wives, and they give birth to šḥr ('Dawn') and šlm ('The Peaceful One', Dusk).[10] 'The alternate rising and sinking of the two mothers obviously prefigures the habits of the sons'.[11] This account thus seems very similar to the Hesiodic one.

Parallels to Hesiod's account also exist in Mesopotamian tradition. Thus a Babylonian incantation mentioning two daughters of Anu between whom a clay wall was built, so that 'sister goes not / never goes to sister',[12] might be a sort of a riddle, with an answer 'Night and Day'.[13] Furthermore, in the *Enuma Eliš* it is described how Marduk allotted the days to Šamaš and fixed the *maṣrāt* of night and day.[14] The translators had to choose here between *maṣṣartu* ('watch, astronomical observation, watch of the night')[15] or *miṣru*, pl. *miṣrātu* ('border, border line, boundary, extremity').[16] In the fifth tablet of *Enuma Eliš* it is also described how Marduk drew the *mi-iṣ-ra-ta*,[17] which is regularly translated as 'boundaries' or the like.[18] Thus it is more appropriate to translate *Enuma Eliš* 5.46 as 'the boundaries of night and day', which is a concept familiar from Greek tradition, and it is possible that here we have a Mesopotamian counterpart of this concept. In this case, 'boundaries' would refer to the annual solar movement, i.e., the northernmost latitude at which the sun rises on the day of the winter solstice or sets on the day of the summer solstice, or the northernmost rising and setting point of the sun at the summer solstice at the latitude near the geographical arctic circle (see chapter 2). Otherwise, this could be another example of the practice of interchangeable use of identical motifs in models accounting for either the diurnal or annual solar movement, with the Homeric account referring to the latter and the Mesopotamian to the former.

[4] Cf. West 1966: 366–67.

[5] Ferrari Pinney & Sismondo Ridgway 1981: 141 n. 6; Nakassis 2004: 218 n. 11.

[6] Ferrari Pinney & Sismondo Ridgway 1981: 143.

[7] Ferrari Pinney & Sismondo Ridgway 1981: 141, 143.

[8] Burgess 1999: 197; 2016: §11.

[9] Over an ảgn, which could be rendered as 'basin', perhaps to be associated with Pherecydes' Ogênos/Ocean (7B4 D-K, cf. 7A11 D-K) (West 1997: 146–47, cf. 1971: 50). Lewis in Parker (1997: 210) renders ảgn as 'firestand', cf. l. 36; Gaster 1977: 427 has 'basin'.

[10] *KTU* 1.23.30–32, 51–53 (Lewis in Parker 1997: 210, 212).

[11] West 1997: 146–47.

[12] Landsberger & Jacobsen 1955: 16.

[13] West 1997: 300.

[14] *EE* 5.45–46; Landsberger & Kinnier Wilson 1961: 158; Horowitz 1998a: 117; *CAD* i, j 7.135 s.v. *immu* c, ml 10.1.338 s.v. *maṣṣartu* 3d; Lambert 2013: 100–101.

[15] *CAD* ml 10.1.337–38 s.v. *maṣṣartu* 3; cf. Deimel 1912: 60. Thus *CAD* i, j 7.135 s.v. *immu* c; ml 10.1.338 s.v. *maṣṣartu* 3d; Horowitz 1998a: 117; Dalley 2000: 256; Foster 2005: 464; Lambert 2013: 101.

[16] *CAD* m2 10.2.113–14 s.v. *miṣru* A1; cf. Deimel 1912: 60. Thus Landsberger & Kinnier Wilson 1961: 159; Sandars 1971. Grayson in *ANET*[3] p. 501 has 'precincts'.

[17] *EE* 5.3; Landsberger & Kinnier Wilson 1961: 156; Horowitz 1998a: 114; 2014: 1; Lambert 2013: 98–99.

[18] Heidel 1951: 44; Koch-Westenholz 1995: 116; Dalley 2000: 255; Lambert 2013: 99: 'divisions'; Speiser in *ANET*[3] p. 67: 'zones'; Landsberger & Kinnier Wilson 1961: 157: 'borderline'; Sandars 1971: 'a beginning and an end'; Van der Waerden 1974: 64; Foster 2005: 463: 'boundaries'; Gaster 1977: 450: 'limitations [sc. of seasons]' (he further adduces Ps 74:17 where borders of the earth and seasons are mentioned together, Gaster 1977: 449–50); Horowitz 1998a: 115, 165, 256; 2014: 1: 'boundary-lines' in the sky.

A further interesting parallel to Hesiod's concept is found in a late-Babylonian text of the Seleucid-Parthian period.[19] In this texts we read how on the 11th of Du'ūzu (IV 11), 'when the nights have become short' the 'daughters of Esagila' go to the temple of Nabû Ezida, described as *bīt mu-ši*, the 'House of Night', while on the 3rd of Kislīmu (IX 3),[20] 'when the days have become short' they go to Marduk's temple Esagila, *bīt* u_4*-mu*, the 'House of Day'. In the ideal Mesopotamian calendar used in the Neo-Assyrian period the summer solstice fell on IV 15 (15th of Du'ūzu), while the winter solstice fell on X 15 (15th of Ṭebētu).[21] The text explicitly refers to the shortening and lengthening of days during the year, and the festival described in it surely had some solstitial associations.[22] Another text mentioning this ritual, referring to the month Ṭebētu,[23] has the goddesses 'pass by each other (as) they walk along the path' in a Hesiodean fashion.[24] Thus in this ritual the 'House of Day' and 'House of Night' were associated with the annual movement of the sun, that is the summer and winter solstice, respectively, rather than with its diurnal movement. As we have just seen, perhaps this also applies to the passage in the *Enuma Eliš* mentioning 'the boundaries of night and day'. In this way it seems that both Greek and Mesopotamian traditions simultaneously applied the imagery of daylight/night exchange with reference to (unsurprisingly) the diurnal and (less obviously) the annual solar movement, which is another instance of the widespread practice of interchangeability of motifs used in building models accounting for both these phenomena.

These Mesopotamian parallels confirm—if any confirmation was needed—the cosmological nature of Hesiod's account, and further support the similar character of the Homeric description.

[19] Livingstone 1986: 255 (BM 34035 1–8 = *Sp.* I 131).
[20] In Cohen 1993: 7, 319, 336 it is on the 3rd of Ṭebētu (X 3) that the second exchange of goddesses takes place (after Unger 1931: 271). According to Astrolabe B I iii.1–10 Kislīm is the month of Nergal and in this month Nergal rises from the underworld (Horowitz 2014: 36, 85, with parallels on p. 86); according to BM 34035 51–53 Nergal comes up from the underworld on the 28th of Kislīm (IX 28; he goes down on the 18th of Du'ūzu (IV 18)), and he is one with Šamaš (Livingstone 1986: 256–57; cf. Horowitz 2014: 87). Thus Šamaš/Nergal descends to the underworld three days after the summer solstice in the ideal Mesopotamian calendar according to the Neo-Assyrian tradition and ascends from it more than two weeks before the winter solstice (according to the Old Babylonian tradition, he descends about a month after the summer solstice, and ascends about two weeks after the winter solstice).
[21] mulApin II.1.9–13, 16–18 (Hunger & Pingree 1989: 72–75, 151; for other attestations see Horowitz 2014: 130, cf. 132); this tradition, according to the 14th tablet of the *Enuma Anu Enlil*, the Old Babylonian tablet BM 17175+ and the Astrolabes, dates from the Old Babylonian period (Van der Waerden 1974: 84, 86; Hunger & Pingree 1989: 163; Rochberg 2004: 282; Horowitz 2014: 16–18, 130), with the solstices on III 15 and IX 15.
[22] Livingstone 1986: 255; Cohen 1993: 7, 318–19, 336; Kingsley 1995: 393; Rochberg 2004: 232; Horowitz 2014: 87 n. 511, 133 n. 918.
[23] *SBH* 8 v.44–47.
[24] Cohen 1993: 319; cf. Livingstone 1986: 255; Horowitz 2014: 87 n. 511.

10.3. Homer, Hesiod and the liminal features in Hades

Returning to the Greek ethnographic context, further parallels could be recognised between Hesiod's account of the house of Night and Homer's description of Laestrygonia that are specifically derived from the association of both with the solar movement. In addition, further resonances of these narratives could be recognised in later Greek tradition. In the light of Homeric Telepylus ('Distant Gates') and its association with solar movement (on which see chapter 2), as well as in its capacity as the gates of Beyond,[25] the threshold of the house of Night—and with it the necessarily accompanying doors (*Th.* 750)—are in a true spirit of mythic redundancy paradigmatically reiterated throughout Hesiod's description of Tartarus and Hades. Thus, the poet of *Theogony* mentions the bronze *thurai* in the wall enclosing the Titans, the bronze enclosure (*herkos*) surrounding Tartarus, the *pulai* of the 'vast chasm', i.e., Tartarus, the *pulai* of Hades' houses guarded by Cerberus, and the bronze threshold (*oudos*) and gleaming (metal) *pulai* in or of Tartarus.[26] It appears that some of these liminal markers can be identified with each other. Thus Johnson associates those from 726 with 732–33 and 740–41,[27] West associates 726 and 732–33,[28] but distinguishes from them those of 740–41 and 811, which he identifies with each other, and also from 773,[29] while Stokes also distinguishes between those of 726 and 732–33 and those of 740–41.[30] The extent to which any of these liminal features can be associated with the bronze threshold of the house of Night remains unclear, but, in any case, the emphasis seems to be on such features, of which the threshold of the house of Night appears to be only a single instalment. Additionally, Homer's description of iron *pulai* and bronze *oudos* in Tartarus,[31] suggests that Hesiod's frequent practice of placing gates, thresholds, etc. in Tartarus was a common feature of early epic poetry. Numerous authors, both earlier and later, used similar imagery, and the *pulai* of Hades were, starting with Homer, the most frequently used example of such imagery.[32]

Thus, although Hesiod does not mention them explicitly, 'the gates of the house of Night' can be safely included in his cosmology, and they can be further associated with Homer's 'Distant Gates' in that that both belong to a stock of motifs common for the construction of models

[25] Fowler 2017: 248.
[26] *Th.* 732–33 (var. 732 *pulai*, West 1966: 362; Johnson 1999: 15), 726, 740–41, 767, 773, 811.
[27] Johnson 1999: 15–16.
[28] West 1966: 362.
[29] West 1966: 362, 364–65.
[30] Stokes 1962: 15–16.
[31] *Il.* 8.15 (adduced as a parallel for *Th.* 811 by West 1966: 378).
[32] *Il.* 5.646, 9.312, 23.71; *Od.* 14.156; Theogn. 427; Aesch. *Ag.* 1291; *Cert.* 79 (v.228 Allen); Arist. fr. 195 Rose = fr. 157 Gigon = 58C3 D-K. For associated phrases see *Il.* 8.367, 13.415, 23.74; *Od.* 11.277, 571 (cf. Nagy 1973: 140; 1990: 226); Theog. 709; Pi. *P.* 4.44; Soph. *OC* 1569; Eur. *Hec.* 1, *Hipp.* 1447; Ar. *Ran.* 163, 460, 462, *Gerytades* fr. 149–50.1 *CAF* = fr. 149.1 Edmonds; vase dated to 370–350 BC in *LIMC* vii.1.89 *Orpheus* 83, cf. vi.1.992 *Hekate* 27, iii.1.390 *Dike* 15, iii.1.828 *Erinys* 12 (cf. Dietrich 1908: 159–60; Harrison 1912: 520–22; Bernabé & Jiménez San Cristóbal 2011: 98); Pl. *Resp.* 10.615de; [Pl.] *Axioch.* 371b; Men. fr. 785 Körte-Thierfelder, etc.

accounting for both the diurnal and annual solar movement. As often noted, these gates were further described in the proem of Parmenides' *Peri Physeos*, where we read of the 'gates of the paths of Night and Day'.[33] Parmenides' gates are associated with the 'house of Night' from which the 'unveiled' daughters of the Sun (Hêliades) went forth to lead Parmenides' horse-drawn chariot.[34] There is some supporting information for the existence of Parmenides' gates aside from these verses but they do not throw much light on his understanding of this concept.[35] The imagery employed in Parmenides' description of this episode, however, strongly suggests a solar context as one aspect of the journey described in the proem.[36] This is quite in line with Hesiod's account of the house of Night with its emphasis on the daylight/night exchange.

Furthermore, Parmenides deliberately used motifs drawn from Hesiod, Homer, or both.[37] In addition to his appropriation of Hesiod's house of Night, his line 11 differs from *Od.* 10.86 only in 'case variation in the final word and ἔνθα πύλαι for ἐγγυς γαρ at the beginning' and represents a conscious reference to the Homeric text.[38] Therefore, line 86 of the 10th book of the *Odyssey* can be associated both with Parmenides' proem which describes the sun's path through the 'gates of the paths of Night and Day' and with Hesiod's description of the house of Night with its brazen threshold in Tartarus—both clearly supporting the argument that Homer's reference is, and was understood as, solar.[39] Furthermore, Hesiod's Night and Day that alternate (*ameibomenai*) at the great bronze threshold can be compared with Parmenides' alternating (*amoibous*) keys or bars of the Gates which Dike possesses, and the doors that 'revolve' or 'swing open in turn' (*amoibadon*).[40] The fact that Parmenides' gates of the paths of Night and Day open in turn probably means that Day and Night, in a Hesiodic fashion, alternatively inhabit the house of Night.[41] Thus Ferrari Pinney and Sismondo Ridgway translate 1.5.14 Coxon as 'the keys which let the one [Night] in and the other [Day, and *vice versa*] out', in this way making Parmenides' construction almost identical to Hesiod's, while West understands the revolving doors in 1.5.19 Coxon as associated with the fact that day and night alternately go forth through them.[42] While Parmenides conception is surely derivative, it does show how the philosophers (even one as rigorous as Parmenides) understood earlier mythic models as having some cosmological potential, even if they were wide of the mark with respect to their correspondence to truth, at least the truth as Parmenides conceived it.

10.4. Gates of the otherworld assimilated to the gates of the sun

Therefore, Homer's (Telepylus) and Parmenides' gates certainly have strong solar resonances and were both appropriated from the models accounting for solar movement, as exemplified in the *Theogony*. It is reasonable to suppose that Hesiod's Tartarean gates—a concept split into various liminal features, including the house of Night's threshold—share this characteristic, since the alternation of night and day is obviously dependent on the diurnal solar movement and the very natures of Day and Night are determined by the sun's absence or presence. In general, it appears that numerous liminal markers used in the descriptions of the Otherworld occasionally have at least implicit solar connotations. This is especially evident when they are located with respect to sunset, and the presence of the 'house of Night' in this context is another argument supporting this conclusion. Although they almost exclusively refer to the diurnal solar movement—i.e., they were predominantly used in models accounting for the daily course of the sun—liminal markers are occasionally used in models accounting for the annual (as in the Homeric account), in line with the repeatedly noted compatibility of motifs used in building models accounting for both aspects of solar movement, conditioned by the similarity and interdependence of these two movements and the limited number of appropriate typical action patterns extracted from human experience out of which mythic models accounting for solar movement could be built.

[33] Parm. 28B1.11 D-K (cf. the gates' *megala thuretra* in l.13). For this translation (with slight variations) see for example Kirk *et al.* 1983: 243; Steele 2002: 584; Burkert 2004: 97; Morgan 2000: 37, 68, 73, 76–78, 80, 87; Graham 2006: 202; 2010: i.211; Mourelatos 2008: 15, 146; Palmer 2009: 363; McKirahan 2010: 145; Trépanier 2010: 289; Tor 2017: 253, 255, 355; for similar translations see Kingsley 1995: 392; Marinatos 2001: 396; Steele 2002: 583; Henn 2003: 23–24, 42; Coxon 2009: 50, 275–76, 328.

[34] 28B1.9–10 D-K.

[35] Two gates: Parm. T133 Coxon = Numen. fr. 31 Des Places (Diels & Kranz 1952: i.229 *ad* 28B1.6ff; Taran 1965: 14); the name of Parmenides' goddess *Hypsipyle*: Procl. *In Parm.* p. 640.39 (Cornford 1952: 120; Morrison 1955: 59 n. 39; Taran 1965: 16); a figure standing by a door in Hades depicted on a crater reproduced in *LIMC* vii.1.89 *Orpheus* 83, cf. vi.1.992 *Hekate* 27, iii.1.390 *Dike* 15, iii.1.828 *Erinys* 12, might tentatively represent Dike as Parmenides' gate-warden (Dietrich 1908: 159–60; Harrison 1912: 520–22).

[36] Cornford 1912: 214–15; 1952: 118 with n. 1; Harrison 1912: 518 n. 1; Guthrie 1965: ii.7–9; West 1971: 225; Steele 2002: 584 with n. 4; Palmer 2009: 56–57; Tor 2017: 256–58, 269, 347, 356–58; 2020: 78–79; cf. Taran 1965: 23–24 (with criticism).

[37] Bowra 1937: 103; Highbarger 1940: 56; Morrison 1955: 59; Dolin 1962: 96; Guthrie 1965: ii.10, 12; Taran 1965: 31; Havelock 1966a: 62; 1983: 39; Kirk *et al.* 1983: 243–44; Ferrari Pinney & Sismondo Ridgway 1981: 142; Kingsley 1995: 392; Most 1999: 354; Morgan 2000: 76–77; Graham 2006: 94, 152, 155–56, 202; Coxon 2009: 9, 275; Naddaf 2009: 115; Palmer 2009: 54–55, 58; Tor 2017: 254–55, 257, 264–65, 347, 351–52, 354, 357, 359; 2020: 77–78; cf. West 1971: 220 n. 1; Burnet 1920: 172 n. 2.

[38] Frame 1978: 60; cf. Havelock 1958: 139; 1978: 270; Vos 1963: 28; Guthrie 1965: ii.12 n. 1; West 1966: 367; Ballabriga 1998: 138; Mourelatos 2008: 9; Coxon 2009: 9, 275; Tor 2017: 354 n. 22; Mansfeld 2021: 25.

[39] Cf. Highbarger 1940: 56; Woodbury 1966: 611. Vos (1963: 34) and Ballabriga (1986: 125–26) reject the association of the passage in *Theogony* with the one in the *Odyssey*, in my opinion, not very convincingly, while Coxon (2009: 275) rejects the association of Parmenides' verses with Hesiod's.

[40] *Th.* 748–50; Parm. fr. 1.5.14 Coxon (for this translation, with slight variations, see Burkert 2004: 60; Palmer 2009: 363; McKirahan 2010: 145 n. 3; Trépanier 2010: 297; Tor 2017: 253, 255; 2020: 77; cf. variant translations in Morgan 2000: 73, 76, 78; Henn 2003: 10, 24; Coxon 2009: 50, 278–79; Graham 2010: i.211); Parm. fr. 1.5.19 Coxon (for translations see Morgan 2000: 73, 76, 78; Henn 2003: 9, 24, 116; Mourelatos 2008: 13; Coxon 2009: 50, 279; Palmer 2009: 363; Graham 2010: i.211; McKirahan 2010: 125; Tor 2017: 253, 255; 2020: 77).

[41] Cf. Parm. fr. 1.5.9 Coxon.

[42] West 1971: 225; Ferrari Pinney & Sismondo Ridgway 1981: 142.

One important feature of the myths referencing solar movement that surfaced in this chapter is the apparent interchangeability of the diurnal solar movement and the world of the dead with respect to liminal imagery. Alternatively, the relation between these two features could be conceived in terms of assimilation of one to the other. Within the framework characterised by the cosmological localisation of the world of the dead defined by the nocturnal section of the daily course of the sun (for which see more in chapter 12), it appears that the realm of the departed was accessible through the solar gates in the furthest west. This effectively identifies the sunset gates with the entrance to the world of the dead within the framework of this mythic model.

This identity can be further corroborated by several case-studies of specific manifestations of such practice. One such example is the application of the specific adjective *kuaneos* ('dark-blue') on features associated both with the world of the dead and solar movement, specifically those of liminal character.

Sometimes the gates to the Otherworld are indeed *kuaneas* (dark-blue),[43] as are the famous Rocks from the Argonautic tradition possibly placed at the entrance to the Otherworld.[44] The latter in any case represented an (almost) impassable limit between two realms. The adjective was further applied in a number of contexts pertaining to the world of the dead, sometimes explicitly referring to its liminal features,[45] but also to Night and her house,[46] and in an early reference to the annual solar movement—more precisely, to its annual limit—when it was claimed that the sun during winter goes to the territory of the *kuaneoi* men.[47] This is perhaps 'simply an elevated synonym of μέλας', designating the black-skinned inhabitants of the far south of Libya (in the Greek sense of the term).[48] However, in addition to the testimonies adduced here for the otherworldly character of *kuaneas*, at least one ancient author clearly distinguished *kuaneos* from *melas*. Pausanias, in describing the colour of a demon from Hades, claims it was something between dark-blue (*kuaneos*) and black (*melas*), thus explicitly distinguishing the two in an appropriate context.[49] Therefore, the term perhaps signifies—in addition to the clear meaning of 'far south'—an otherworldly association of this location, as discussed in chapter 9.[50] This can be supported by the fact that Hesiod, in giving an account of how the sun goes to (using the phrase *epi... strôphatai*, 'turns on') the territory of the black men, renders the word here translated as 'territory' by the phrase 'the people and the city of' (*demon te polin te*), which is a phrase used in the *Odyssey* to describe the 'territory' of the Cimmerians and Phaeacians.[51] This circumstance gives Hesiod's phrase a certain otherworldly resonance. Concerning the expression 'turns on', it can also further be associated with the annual solar movement even more precisely, designating the turning of the sun at the winter solstice among the fabulous Aethiopians, in this way in close connection with the winter solstice sunrise and, consequently, the south-east,[52] or it could equally designate the winter solstice sunset, since the Aethiopians were found at both locations.[53]

It seems that the term *kuaneas* could equally be applied to the entrance to the world of the dead and to several related features, and at the same time to the concepts involved in the accounts of solar movement, such as the house of Night (part of a model accounting for the daylight/night exchange) and the annual solar turning in the far south (whether exemplified in the winter solstice horizon azimuths of the sun or its southernmost declination in general).

Another example of the identity of solar gates with the entrance to the world of the dead can be recognised in an Iliadic myth with a possible solar and, at the same time, otherworldly setting, which can in its turn be connected with a cluster of myths with clear solar resonances.

10.5. The Pylian gates

When discussing the liminal features associated with the world of the dead, I purposely left out one significant passage from the *Iliad* that reflects the liminal imagery in Hades, where it is described how Heracles wounded Hades 'by the gates among the dead' (*en pulôi en nekuessi*).[54] Later authors tried to rationalise this story into Heracles' war with the Pylians, who had Hades fighting on their side, while there existed as many as three cities named Pylos in the western Peloponnese. [55] However, it is probable that Hades was wounded at the gates of the world of the dead

[43] Theog. 709.

[44] Hdt. 4.85, 89; Soph. *Ant.* 968; Eur. *Med.* 2, 1263, *IT* 241, 746, 889, *Andr.* 864; Davies 1992: 219–20; Podossinov 2013: 214–15, 220; Beaulieu 2016: 46, 85.

[45] [Hes.] *Scut.* 249; *hDem.* 347; Eur. *Alc.* 261–62; Pl. *Phaed.* 113BC; Tim. *FGrHist* 566F164 = D. S. 4.23.4, 5.4.1–2; [Simonid.] 9.2 Page (*FGE*) = Anth. Pal. 7.251.2; [Saph.] fr. 158D Lobel-Page = *Anth. Pal.* 7.489.2; Hegesianax fr. 466 *SH* = p. 8 Powell (Cherniss & Helmbold 1957: 38–39; Lloyd-Jones & Parsons 1983: 238; Maxwell-Stuart 1981: 89; Irby-Massie & Keyser 2002: 57) etc.

[46] *Th.* 745 (an interpolation according to West 1966: 365, 367); Simonid. 543.11–12 PMG; Anon. *PMG* 931J.

[47] Hes. *Op.* 527–28; for Dion. Perieg. 586 (Lightfoot 2014: 228) the sun also goes south to the *kuaneoi* after turning at the summer solstice.

[48] West 1978: 292; 1985: 131 n. 13.

[49] Paus. 10.28.7.

[50] In the same vein Kingsley (1995: 96–97 n. 3) distinguishes the meanings of these two words, although he concedes that occasionally they could have been used interchangeably.

[51] *Od.* 6.3, 11.14; cf. 14.43 (an unnamed foreign people).

[52] Ballabriga 1986: 20.

[53] Thus Eustath. Dion. Perieg. 581 (*GGM* ii.330); cf. Σ Dion. Perieg. 586 (*GGM* ii.452). For the sun turning back (*hupostrepsas*) from the (summer) solstice and going to the Aethiopians see Arist. *Aegypt.* 63 (ii.283.12–13 Keil). For 'double' Aethiopians see chapter 9.

[54] *Il.* 5.397.

[55] Pany. fr. 26 West (see Matthews 1974: 52–57, esp. 55); [Hes.] *Scut.* 359–67; Pi. *O.* 9.33–35; [Apollod.] *Bibl.* 2.7.3; Sen. *HF* 560565; Paus. 6.25.2–3; Σ T *Il.* 5.397 (Erbse ii.62) (H. wounded at P., despite citing Aristarchus to the contrary, see below); Σ D (*Mythographus Homericus*) *Il.* 11.690 van Thiel 2014: 403, cf. Σ A *Il.* 11.690b; Σ D (*Mythographus Homericus*) *Il.* 5.397 van Thiel 2014: 244; Σ ACDEHQ Pi. *O.* 9.44a (citing Didymus), CDEQ 9.46 (critical to the interpretation that Nestor's P. is to be associated with this fight, see below); Tzetz (Σ) Lycoph. 39,

he was the warden of.[56] Perhaps this incident can further be associated with the dragging out of Cerberus out of Hades. A Corinthian cup (590–580 BC) depicts Heracles preparing to throw a stone at Hades in the context of this labour,[57] while several sources indeed associated the fight with Hades with the Cerberus episode.[58] Yet how is this concept—the 'Pylian' gates of the Otherworld—associated with the sun? Harrison related the Elean Pylos, where she located—adding the testimony of Pausanias—the fight with Hades and the gates of the underworld with what she styled the 'Gate-place of the setting sun'.[59] Given the geographical position of all three Pyloi in the westernmost Peloponnese, this seems a plausible explanation—the gates of sunset could have indeed been at some early point associated with this region (cf. the case of Leukas below). In addition, this interpretation can be supported by a mythic nexus centred on Pylos that has clear solar resonances, thus allowing the inference of its precise positioning on the sun's daily course. Nevertheless, these is a rather speculative conclusion and the solar resonances of the 'Pylian' gates of the Otherworld should not be understood as conclusively proven; but the value of the hypothesis lies in the fact that all available evidence seems to support it, or at least none of this evidence contradicts it.[60]

The complex of myths centred on Pylos all involve a cattle-raid and a cave at (a) Pylos. In this cave the cattle of both Neleus and Nestor lived.[61] The former, according to the majority of sources, were brought to Pylos by the seer Melampus from Phylace, and they were owned by king Iphicles.[62] According to another version of the myth, Neleus and his sons, all except Nestor, had actually stolen Geryon's cattle from Heracles.[63] This could have been the cause for the war against Neleus, in which one of his sons, Periclymenus, was especially prominent.[64] Since his name seems to designate Hades himself,[65] it appears that the god of the world of the dead eventually (re-)captured the undoubtedly solar cattle (cattle located at the sunset point in the farthest west) that Heracles dragged away from Erytheia.

The latter cattle were brought to Pylos by the hero himself. Nestor led a marauding expedition on Elis,[66] a region in the westernmost Peloponnese ruled at that time by Helios' son Augeas.[67] The latter's name is probably related to the word *augê*, 'radiance (of the sun)',[68] and it is possible that Augeas was originally *Halíeios*, Helios' son, which was converted by the epic tradition into *Waleîos*, 'Eleian'.[69] His solar associations go further than his name and ancestry: he was the owner of the cattle among which were twelve swan-white bulls sacred to Helios, whose leader was named Phaethon;[70] further, he had a daughter named Agamede, well-versed in pharmacopoeia,[71] who, as Perimede,[72] was linked with both Circe and Medea.[73] Thus there is an obvious similarity between Medea (Circe)/Aeëtes on one hand and Agamede/Augeas on the other.[74] These facts support the solar interpretation of Augeas' cattle, i.e., as the cattle belonging to a sun-god in some mythic model accounting for the solar movement.

Furthermore, when discussing the solar cattle kept in a cave at Pylos, one cannot escape the connection with Apollo's cattle, here certainly *solar* Apollo's, hidden in this (same?) cave.[75] A very similar description of a cattle-raid is thus found in the *Homeric Hymn to Hermes*, where Apollo's cattle were driven by Hermes to a cave at Pylos.[76] A Pylian cave was certainly the place where stolen cattle might be laid down—even the cattle of Helios' son or Geryon's cattle from the sunset island—and Apollo's cattle in this

ii.32 Scheer. Three cities: Aristoph. *Eq.* 1059 with Σ *ad loc.*; Str. 8.3.7, etc. (cf. Meyer 1959: 2136.20–26; Nakassis 2020: 272).

[56] Aristarch. *ap.* Σ T *Il.* 5.397 (ii.62 Erbse; cf. Σ D (*Mythographus Homericus*) *Il.* 5.397 van Thiel 2014: 244) (Matthews 1974: 55; Kirk 1990: ii.102); Σ D (*Mythographus Homericus*) *Il.* 5.397 van Thiel 2014: 244; Σ bT *Il.* 5.392–400; Σ ACDEHQ Pi. *O.* 9.44a (citing Didymus), Σ CDEQ Pi. *O.* 9.46; Σ A. R. 1.1350 (p. 121 Wendel); Agatharch. *De mar. Eryth.* 7 (*GGM* i.116), etc.; Nilsson 1932: 87–89, 203–204; 1949: 138–39, 238; Highbarger 1940: 48–49, 115–17; Croon 1952: 57 with n. 19; Meyer 1959: 2135.1–9, 2136.7–13; Nagy 1973: 139–40; 1990: 225; Kirk 1974: 191; Fontenrose 1974: 329–30; Matthews 1974: 55–56; Frame 1978: 92–93; 2009: 537 n. 55; Burkert 1979a: 86, 88–89; Dowden 1992: 98; Gantz 1993: 70; Davies 2006: 195; Fowler 2013: 304 (perhaps Neleus' Pylos itself represented the gates to Hades, cf. Matthews 1974: 55); Ogden 2013: 110–11; Nakassis 2020: 273. Ballabriga (1986: 32, 34) emphasizes Pylos' key geographical position with regards to the navigation towards the west, calling it 'the gate of the West'. For an analogous concept in India (the 'Town of Doors', Dvārakā, located in the westernmost part of India on the shore of the 'western ocean') see Kuiper 1983: 73.

[57] *LIMC* 2.1.958 s.v. *Athena* 11 = 5.1.87 s.v. *Herakles* 2553; Boardman 1975: 7; Gantz 1993: 413–14; Fowler 2013: 304; Ogden 2013: 111.

[58] Pany. fr. 26 West, according to Matthews 1974: 56; Σ BCDEQ Pi. *O.* 9.43, ACDEHQ 9.44a, A 9.48; Σ D (*Mythographus Homericus*) *Il.* 5.397 van Thiel 2014: 244; Σ bT *Il.* 5.395–97 (cf. Fontenrose 1974: 330; Matthews 1974: 55–56; Gantz 1993: 70, 414, 416, 455; Fowler 2013: 304; Ogden 2013: 110–11).

[59] Harrison 1908b: 13, cf. 1912: 370.

[60] Sourvinou-Inwood 1995: 133–34.

[61] Paus. 4.36.2.

[62] *Od.* 11.287–97, 15.225–38; Hes. fr. 37 M-W, fr. 261 M-W *ap.* Σ P A. R. 1.118, p. 16 Brunck & Schaefer; Pherec. *BNJ* 3F33, etc.

[63] Isocr. *Or.* (*Archidam.*) 6.19 (cf. Philostr. *Vit. Soph.* 1.17); Agias *FGrHist* 305F1; Philostr. *Her.* 26.3; Croon 1952: 19; Walcot 1979: 335. This theft is perhaps represented on the late 6th-c. Euphronius' cup (*LIMC* 5.1.77, 5.2.89 *s.v.* Herakles 2501, Gantz 1993: 403, 409).

[64] Hes. fr. 33ab, 35 M-W = 31–33 Most; Burkert 1979a: 86.

[65] Hesych. π 1736; Cook 1925: ii.1.233, ii.2.1113; Fontenrose 1974: 328–29, 479–80; Burkert 1979a: 86; Dowden 1992: 98. For Neleus himself as 'another Geryoneus' see Fowler 2013: 304–305.

[66] *Il.* 11.671–84.

[67] A. R. 1.172 with Σ 1.172–73 (p. 22 Wendel), 3.362; Theocr. *Id.* 25.54, 118; [Apollod.] *Bibl.* 1.9.16, 2.5.5; Hyg. *Fab.* 14; Paus. 5.1.9; *Orph. Argon.* 214; Nonn. *D.* 14.44–45; Tzetz. (Σ) Lycoph. 41 (ii.32 Scheer); Joan. Pediasimus *De Herc. labor.* 5 (*Myth. Gr.* i.253 Wagner).

[68] Frame 1978: 88; 2009: 49; cf. Burkert 1979a: 95; Sick 2004: 442.

[69] Burkert 1979a: 95; cf. Paus. 5.1.9 (opposite: Ἠλεῖος turned into Ἥλιος).

[70] Theocr. *Id.* 25.129–31, 139.

[71] *Il.* 11.741 (cf. Str. 8.3.5); cf. Eustath. Dion. Perieg. 322, *GGM* ii.274.

[72] Σ Theocr. *Id.* 2.15–16b.

[73] Theocr. *Id.* 2.14–15; cf. Prop. *El.* 2.4.7–8 (Hainsworth 1993: iii.303).

[74] Wernicke 1896: 2309.41–42; Frame 1978: 88–89; 2009: 49.

[75] For Apollo's solar function see chapter 7.

[76] *hHerm.* 18, 22, 68–86, 101–102, 191–92, 216, 221, 262, 276–77, 308–11, 340–42, 344, 355, 371, 398, 401 (cf. Ant. Lib. *Met.* 23.5), 503. The author of the *Hymn* clearly distinguished Apollo from Helios (Vergados, 2013: 67, 288, 481). For the date of the *Hymn* (second half of the 6th c.) see Vergados 2013: 130, 145–47 (with an extensive review of earlier opinions on pp. 130–45).

context certainly has at least a distant solar undertone.[77] According to a later source, Hermes did actually steal the cattle of *Helios*,[78] although this is probably only due to the later identification of Apollo with the sun.

Thus, the interpretation of a line from the 5th book of *Iliad*, where the town of Pylos on the western coast of the Peloponnese was assimilated to the gates of Otherworld, led to another association of Pylos, that with solar cattle. It seems probable that these two conceptual domains are actually interrelated, and that the geographical position of the city—in the extreme west of the Peloponnese—determined its association both with sunset and the world of the dead defined precisely by the solar movement. This association was more specifically focused on Pylos' role as a liminal marker, both an entrance to the land of the dead and the sun's western portal through which it enters the realm of the departed at night.

10.6. The White Rock and the Odyssey

The analysis of 'the gates of the sun' from the *Odyssey* will further clarify the argument identifying the two portals, the solar and posthumous one.[79] Instead of using the term 'identifying', it would probably be more correct to describe the relation between the two portals in terms of interacting conceptual domains—the solar movement and eschatology—shared by these features.[80] In the so-called second *nekuia* the souls of the suitors, led by Hermes, pass 'the streams of Ocean', then the 'rock Leukas', and finally 'the gates of Helios' and the realm of dreams.[81] Hermes, who has an epithet *pulêdokos*, 'he who receives [presumably departed souls] at the gates', surely the gates of Hades, is especially appropriate in this context.[82] Furthermore, the rock Leukas is here located beyond the Ocean, i.e., on its farther shore, which corresponds to the description of Odysseus' visit to Hades in the *nekuia* (see chs. 7 and 10), while the gates of the sun are still farther away. The 'White Rock' deserves a fuller explanation, although in a summary form, because G. Nagy collected and interpreted with great erudition the ancient testimonies.[83] A non-cosmological occurrence of this toponym appears on the island of Leukas, where a temple of Apollo Leukatas stood on a prominent '*Leukatas petra*'. There the local inhabitants held an annual service in the honour of Apollo, casting down from the Rock a condemned criminal.[84] This same rock was famously used by those who believed they could heal their wounds inflicted by love by jumping from it.[85] Another 'White Rock', not far from Phocaea, was used in a similar way, and it also had a shrine of Apollo.[86] Furthermore, two incidents in Theseus' life seem to be associated with this complex of ideas. He was pushed to his death from the high rocks at the island of *Skyros*, whose name is associated with 'chalk, gypsum'. Moreover, he himself cast *Skirôn* off the *Skirônides petrai*,[87] and one of these rocks was the famous rock of Molouris,[88] from which Ino/*Leuko*thea jumped into the sea.[89] Thus it seems that both rocks encountered by Theseus' in his career correspond in a significant way to the white rocks on Leukas and at Phocaea. There is not much point in elaborating the subject so thoroughly covered by Nagy, and I can only quote his conclusion: 'the White Rock is the boundary delimiting the conscious and the unconscious—be it a trance, stupor, sleep, or even death'.[90] In the context of present study the most interesting association of the White Rock is that with death, since this is clearly the meaning intended by the author of the *Odyssey*. The association of the White Rock with Apollo, here once more certainly a 'solar' Apollo (focusing on the aspect of this complex deity that is characterised by its solar resonances), together with its Homeric localisation in the context of solar gates, as well as its later rationalised localisation in the far west (at that early time) of the Greek world, is probably indicative in determining its mythic background as the cosmological mountain in which the sun descends in the far west. Another similar feature is a 'stone of withering' beyond the Acherusian lake, the boundary between the world of the living and that of the dead.[91]

This rock shares its cosmological function with the rock at the meeting place of the 'two roaring rivers', Acheron and Pyriphlegethon-Cocytus, the latter itself a 'branch of the water of the Styx', beyond the shore of Ocean and next to the house of Hades from the *nekuia*.[92] It seems more than probable that these two rocks both share in interacting conceptual domains: both are located beyond the Ocean in a solar setting and both are associated with the souls of the deceased.[93] While the gates of the sun are not explicitly mentioned in the *nekuia*, the sun is not visible

[77] Frame 1978: 90; cf. Croon 1952: 80; Solomon 1994: 40–42, 44–45.

[78] Σ Dionys. Thrac. 2 (p. 174.2–3 Hilgard; Cook 1914: i.410; Crane 1988: 162 n. 49).

[79] Cf. Highbarger 1940: 3, 6, 9, 32, 36, 49, 52, 89.

[80] Buxton 2017: 55, 59.

[81] *Od.* 24.1–12.

[82] *hHerm.* 15; Nagy 1973: 140; 1990: 226. Hesych. ω 108, where the Ocean is identical to *aêr* and the souls of the deceased depart thereto (Beaulieu 2016: 52, 64), is perhaps inspired by this passage.

[83] Nagy 1973 = 1990 ch. 9. See also Beaulieu 2016: 147–49.

[84] Str. 10.2.8–9; cf. Ael. *HA* 11.8; Ampel. *Liber mem.* 8.4; Serv. Dan. *Aen.* 3.279; perhaps also Ov. *Fas.* 5.630 and Phot. *Lex.* s.v. Λευκάτης.

[85] Stes. 277 *PMG* = fr. 326 Finglass (both under *spuria*) *ap.* Aristox. fr. 89 Wehrli with Rutherford 2015 on Stesichorus' authorship; Anacr. 376 *PMG*; Eur. *Cycl.* 166; Menan. fr. 258 Körte-Thierfelder (for a reconstruction of the play *Leukadia* see Arnott 1997: 220–43), cf. Serv. Dan. *Aen.* 3.279 and Turpilius *Leucadia* frs. XI–XII Ribbeck³, Rychlewska, etc.

[86] Char. *FGrHist* 262F7; D. S. 15.18.1.

[87] Str. 9.1.4.

[88] Paus. 1.44.8.

[89] Nagy 1973: 144–45; 1990: 230–31; cf. Burkert 1983: 178–79, 211; for Skiras/Skiros/Skiron as liminal terms cf. Vidal-Naquet 1981a: 156.

[90] Nagy 1973: 147 = 1990: 234; cf. Beaulieu 2016: 147–49.

[91] Aristoph. *Ran.* 194 with Σ *ad loc.*

[92] *Od.* 10.511–15. One should compare Q 18:63, where a rock is found at 'the meeting place of the two waters' (Q 18:60–61) (Wheeler 1998: 196, 203), with this cosmological complex.

[93] Eustath. *Od.* 10.515, i.393 cited in J. D. Morgan 1985: 229; Merry *et al.* 1886: 438; Rohde 1925: 565 n. 102; Edmonds 2004: 208 n. 140; 2011: 909. A lofty rock from which the Styx issues can be in part compared with this concept (Hes. *Th.* 786–87).

'on the shores of Ocean',[94] suggesting that the poet here probably imagined the land of the Cimmerians *beyond* the gates of the sun. If this is a correct interpretation, than the 'topography' of the *nekuia* is somewhat different from the one in the second *nekuia*: in the former the order is the Ocean—the (hypothetical) gates of the sun beyond which there is no sunlight—the farther shore of the Ocean—the rock, while in the latter the order is the shore of the Ocean—the rock—the gates of the sun.[95]

10.7. Pherecydes' gates

Now the concept of a rock—or, more often, a double mountain—associated with the setting (or rising) place of the sun and the encircling sea is one familiar from various non-Greek traditions, most notably the Egyptian and Mesopotamian (as will be shown in chapter 13 and Appendices 1 and 2). In these two traditions it is further intimately associated with the concept of solar gates (Appendices 1 and 2), but these particular liminal features do not seem to have been of comparable importance in the Greco-Roman mythic tradition, even though they were mentioned by Homer in a rather straightforward context.[96]

However, it is possible that a section of Pherecydes' rather early cosmology can be interpreted with reference to a similar nexus of ideas, although with no recognised accent on the solar nature of the liminal features present in his account. This 6th c. author mentioned in one of his works certain 'recesses' (*muchous*), 'pits', 'caves', 'doors' (*thuras*) and 'gates' (*pulas*).[97] Porphyry, who is our source for this quote, interprets it in a 'standard' Neoplatonic fashion as a description of the 'becomings and deceases of souls'.[98] This interpretation was in general accepted by most modern scholars, who believe that Pherecydes' account in some way refers to the passage of souls to and from earthly existence, with the doors/gates representing the points of ascent and descent.[99] The 'recesses' (*muchous*) here most probably refer to Hades, but otherwise might signify planetary orbits in their capacity of stepping stones on the soul's ascent and descent.[100] Outside Pherecydes, 'recesses' primarily have Hadean or Tartarean connotations,[101] but can also—in a late author—designate the regions to the north and south of the tropics/arctic circles, characterised by the absence of the sun.[102] It is perhaps somewhat awkward to refer to such vast regions, both terrestrial and celestial, as 'recesses', but this could be a consequence of the resonances that this word had with respect to darkness and the absence of the sun. It could be claimed that Pherecydes' recesses were located beyond his gates and that both these terms can be interpreted in terms of solar movement, whether diurnal or annual (judging by later tradition, the gates in the capacity of eschatological points of ascent and descent would favour the annual solar movement); however, the lack of any direct evidence would make this suggestion, however general and non-committal, extremely tentative.

10.8. Conclusion

From Homer and Hesiod through Pherecydes to Parmenides, mythic models accounting for the solar movement were constructed out of a rather limited set of motifs, features and resonances: naturally, the sun(-god), its/his voyage and associated phenomena (daylight/night exchange, darkness/light, means of transport, course etc.); gates, thresholds, rocks/mountains and similar liminal features, including the circumambient ocean; and, especially, the omnipresent and complex association with or even assimilation to the world of the dead. This assimilation of solar movement to the world of the dead and the presence of liminal imagery in the descriptions of both is exemplified in a set of myths focused on Pylos and cattle-raids associated with this city or cities. In their turn, these raids involved solar cattle, i.e., the cattle found at locations strongly and explicitly associated with the diurnal solar movement. It appears that Pylos at the same time represented an entrance to the world of the dead and the sunset portal in the extreme west (from the point of view of Peloponnesian and other Greeks further to the east, probably prior to or during the proto-colonisation period). The realm of the departed was in this way, within the framework characterised by the cosmological localisation of the world of the dead defined by the nocturnal section of the daily course of the sun, in a consistent way made accessible through the sunset portal. Another example of this identification of the world of the dead and its entrance with solar movement and its liminal points is the use of the adjective *kuaneos* to designate both. Finally, the association of the world of the dead with the solar movement and the close association of the sunset portal with the entrance to the otherworld can be recognised in the cosmological

[94] *Od.* 11.12–19.

[95] For the complementarity of the first and second *nekuia* see de Jong 2001: 566–67; for the compatibility of the topographies in the two *nekuia* see Albinus 2000: 84 n. 50; for the 24th book as an integral part of the *Odyssey* see de Jong 561–62, 565–66.

[96] *Hekaterai pulai* of sunrise and sunset were known to Lutatius (*FRH* 32F7 *ap.* Lyd. *De mens.* 4.2); eastern *limina* of the sun to Cat. *Carm.* 64.271; *Hesperia porta* to Stat. *Theb.* 10.1; double *ianuae caelestis* or *ianua coeli* of sunrise and sunset to Macr. *Sat.* 1.9.9 and Ps. Jerome *Cosmog.* 22, 36a, 84b (Herren 2011: 22–23, 36–37, 184–85); *ianua solis* of sunrise to Ps. Jerome *Cosmog.* 18 (Herren 2011: 16–19). Homer's Telepylus is another example of solar gates, but these are decidedly annual (see chapter 2).

[97] Fr. 88 Schibli = 7B6 D-K. For *muchous* cf. fr. 2 Schibli = 7A2 D-K and fr. 60 Schibli = 11A8 D-K *ap.* Eud. fr. 150 Wehrli.

[98] Porph. *De antro nymph.* 31.7–9 Arethusa ed.

[99] West 1963: 169; 1971: 25 with n. 3; Kirk *et al.* 1983: 59; Seaford 1986: 13; Schibli 1990: 118 n. 32, 119 n. 34; Purves 2010: 107.

[100] Schibli 1990: 117–21, 125, 129. Schibli (1990: 23, 26, 38–39, 48 n. 110, 49, 78, 129) in general interprets *muchous* as cosmic regions in the process of formation; for Pherecydes' *muchous* as cosmic regions see also West 1971: 15.

[101] E.g., *Od.* 24.6 (Schibli 1990: 118–19 n. 34); Hes. *Th.* 119; Anacr. 395.9–10 *PMG*; Anon. 925c.13 *PMG* (probably), cf. 925e.16; Aesch. *PV* 433; Soph. fr. 10c.5 Pearson, *Phil.* 1013 (probably), fr. 442.8, *Ai.* 571; Eur. *Heracl.* 218, *HF* 607–608; Ar. *Av.* 698 = *OF* 1 Kern = *PEG* 64 V Bernabé (Dunbar 1998: 66, 301) (possibly); Carcin. fr. 5 *TGrF ap.* Tim. *FGrHist* 566F164 *ap.* D. S. 5.5.1; Demetr. Phal. *FGrHist* 228F35b *ap.* Posid. fr. 240a E-K = 402 Theiler = *FGrHist* 87F48; A. R. 2.737 (cf. 4.1698), etc.

[102] Heracl. *Quaest. Hom.* 51.2. Its association with Phaëthon in Hes. *Th.* 991 with Σ *ad loc.* is more probably chthonic than solar (Nagy 1973: 171 = 1990: 254).

feature of 'rock Leukas' (later rationalised and localised in the extreme west from the point of view of Greece proper, once again probably prior to or during the proto-colonisation period),[103] which is a manifestation of a well-attested concept of cosmological mountains at the *eschata* of the earth, featuring in models accounting for both the diurnal and annual solar movement.

Another omnipresent phenomenon that can be extrapolated from the set of myths discussed in this chapter is the reappearance of identical motifs in the models respectively accounting for the diurnal and annual solar movement, which is a disposition not limited to Greek tradition. This practice is exemplified by Homer's and Hesiod's respective use of similar imagery in describing two very different solar phenomena. The correspondences between these two respective accounts, both on the level of entire episodes (the Laestrygonians / house of Night) and more specifically with respect to their emphasis on the liminal features of the otherworld (unequally pronounced in these two accounts), shows the strength and endurance of this conceptual nexus in Greek tradition. On the other hand, the complementary nature of the decisive moments in both aspects of solar movement (sunrise/sunset and solstices, respectively), in the first place their liminal character, together with the common general reference of mythic models accounting for the voyages of the sun(-god)—i.e., solar movement—and the interdependent relation of its two principal manifestations, dictated this use of similar and interchangeable imagery. The existence of similar models in Mesopotamian (and Levantine) tradition—with a similar ambiguity of their referents as either the diurnal or annual solar movement—testifies to their hermeneutic potential and cross-cultural character.

A clear and unambiguous reference to solar movement can be recognised in all sets of myths discussed in this chapter. Indeed, this was the main—and appropriately rigorous, mainly because of the failures of earlier interpretative approaches to similar evidence (or, better, to what the adherents of such approaches considered to be similar evidence)—criterion for their inclusion in the first place. This consideration alone allows me to include them in the class of myths studied in this book. In addition, the myths discussed in this chapter show a strong interest in eschatological domain or a concern for cosmological regions associated with either deceased humans or deposed gods, which offers strong supporting evidence for the salient association of the solar movement with the world of the dead appearing in various traditions. In fact, I have demonstrated a strong interaction between these two conceptual domains—the solar movement and eschatology—that resulted in certain features sharing in both or being shared by both. This conclusion is especially relevant to certain liminal features, in the first place the solar gate in the west, at the same time functioning as the gate of the Otherworld.

[103] Cf. Nesselrath 2005: 154; Podossinov 2013: 217–18.

11

Aea and the voyage of the Argonauts

I have already had the chance to touch upon the Argonautic tradition in chapter 3.[1] More specifically, building upon the work of earlier scholars, I have proposed the existence of a pre-Homeric account (or accounts) of the Argonautic voyage concerned with the solar and/or meteorological phenomena characteristic for the latitude of the geographical arctic circle. However, a fuller analysis of this tradition in the context of Greek mythic models accounting for specifically the diurnal solar movement is required, since the tradition contains important pieces of information illustrative of early Greek ideas on this subject. In the first place this applies to the goal of the Argonautic voyage in the pre-Homeric tradition, a mythic land of the sun in the extreme east, on or off the hither shore of the Ocean.[2] This early Argonautic tradition was, as we have noted in chapter 3, indeed concerned with the sun and its diurnal movement and Aea thus seems to have been a mythic feature involved in one early model accounting for the daily passage of the sun. In addition, the island of Aeaea was recognised in chapter 9 as a key cosmological location in a Homeric model accounting for the diurnal solar movement in a position corresponding to the Argonautic Aea, i.e., as a location of sunrise.

The land of Aea was commonly identified with Colchis, a country in the extreme east of the Pontus, but this does not seem to be true for the earliest period.[3] The first to make this identification was apparently the author of the *Korinthiaka*,[4] a mid-6th c. epic poem attributed to Eumelus of Corinth,[5] in which he was followed by a number of early authors.[6] This identification of Aea with the region in the recess of the Pontus might reflect a number of Near Eastern features, but this seems very tentative, since the localisation appears to represent a clear example of the rationalisation of myth, or an attempt at placing a mythic location on a geographical map.[7]

According to the pre-Homeric Argonautic localisation, Aea was situated outside (*ektos*) the *oikoumenê* or at its *eschata*.[8] This late Hellenistic or early Imperial phrasing is strongly corroborated by the earliest tradition, from which it is, after all, straightforwardly derived, as well as by the comparable cosmological location of Aeaea. Apollonius Rhodius himself located the Argonautic Colchis at the eastern boundary of the *oikoumenê*—defined by the daily path of the sun—contrasting it with the ends (*peirata*) of Libya.[9] In addition, the extent of the *oikoumenê* was defined by Socrates by the Pillars of Heracles in the west and the Phasis, a river as a rule associated with Aea, in the east;[10] the coupling of these two locations suggests the association of the Phasis with the eastern branch of the Ocean, since its counterpart in the far west is naturally located at the very limit of the western Ocean.[11] Later authors respected the connection of this land with the sun,[12] or both with the sun and Ocean,[13] and Aiêtês, 'man of Aea', son of Helios,[14] is an appropriate ruler of this mythic land. Indeed, it is not impossible that the very name Aea simply means 'Dawnland',[15] which certainly corresponds to its nature as described in Greek myth. All these facts suggest that this Aea corresponded to Homer's Aeaea, where the dawn's abode and the risings of the sun are actually located.[16]

There are further arguments supporting the thesis of a cosmological location of Aea, a 'land' on the Ocean's shore associated with sunrise, thus an important element in an early Greek mythic model accounting for the diurnal solar movement. The place of Prometheus' punishment was thus located in the Argonautic tradition near or in 'Colchis',[17] representing a region located at the eastern

---

1 Some of the subjects discussed in this chapter were also treated in Bilić 2020b.

2 Mimnerm. fr. 11 and 11a West = fr. 11 Alen *ap*. Demetr. fr. 50 Gaede (approvingly) *ap*. Str. 1.2.40 (critically) (cf. 1.2.10); cf. chapter 3. On Aea see especially Lesky 1948 and Ballabriga 1986: 126132, 137–43.

3 Allen 1993: 88–90; Endsjø 1997: 376; Nesselrath 2005: 157; West 2005a: 41; Fowler 2013: 201.

4 Fr. 17 West, cf. Paus. 2.3.10; cf. Σ BDEQ Pi. *O*. 13.74d.

5 West 2003: 109, 130–31; 2005a: 41; 2007b: 193; cf. Allen 1960: 80, 1993: 90.

6 Simon. 519A *PMG* = S 373, 545 *PMG*; Pi. *P*. 4.10–11, 211–13, fr. 172.7; Hdt. 1.2.2, 7.62.1, 193.2, 197.3; Soph. *Kolchides* frs. 336–49 Pearson; Eur. *Med*. 2; Pher. *FGrHist* 3F100 (thus Fowler 2013: 201; perhaps only the scholiast's opinion, Σ A. R. 3.1093, p. 249 Wendel); Xen. *Anab*. 5.6.36–37; Herodor. *BNJ* 31F10; Asclep. *FGrHist* 12F31; [Scyl.] 81, Shipley 2011: 39, 72; Medeius *FGrHist* 129F1 and Cyrsilus *FGrHist* 130F1.

7 I will only mention here a possible connection with the Mesopotamian goddess Aia, the wife of the sun, worshipped by the Hittites (Bremmer 2008[2006]: 311; Fowler 2013: 201; cf. Van Gessel 1998: 1–3 s.v. Ā).

8 Str. 1.2.40, cf. 1.2.10; Σ A. R. 2.413–18c, p. 164 Wendel.

9 A. R. 1.83–85.

10 Pl. *Phd*. 109ab.

11 For the eastern limit of the *oikoumenê* at the Erythraean sea associated with the Ocean see Liv. 36.17.15; 42.52.14–15 45.9.6; Plut. *De def.* 2.410A; Cosm. Indic. *Topogr. Christ.* 2.26 Wolska-Conus (Aethiopian Ocean). The Erythraean Sea in the east is a natural counterpart to the island of Erytheia in the west (see chapter 9). On the other hand, Pindar couples the Nile with Phasis (*I.* 2.41–42), suggesting some kind of a north-south opposition (cf. *I.* 6.23); similarly, Aristid. *Ad Rom*. 82, p. 219 Jebb, cf. *Aegypt*. 3, p. 332 Jebb.

12 E.g., A. R. 1.83–85, 4.131; Eratosth. IIIB76 Berger.

13 [Plut.] *De fluv. et mont.* 5.1 (*GGM* ii.642).

14 *Od*. 10.137–38 with Σ Q *Od*. 10.138, QV 10.139; Hes. *Th*. 956; Eum. fr. 17 West; A. R. 2.1204, 3.309, 598; Dionys. Skytobr. *FGrHist* 32F14 *ap*. D. S. 4.45.1, frs. 21ab Rusten etc.

15 West 2007b: 196–98; 2014: 119 n. 46 (cf. Fowler 2013: 201 n. 19). West (2007b: 195–96) further understands Phasis to mean 'Radiant', and believes it initially represented the mythic river from which the sun rose in the farthest east, the pre-Mimnermian goal of the Argonauts' journey.

16 *Od*. 12.3–4 with Str. 1.2.10, 40. Cf. Endsjø 1997: 376; Manoledakis 2015: 302–303.

17 A. R. 2.1246–59, 3.851–53; Val. Flacc. 5.154–76

eschata of the earth.[18] Prometheus was indeed placed at the earth's farthest limit (*têlouros*), on a rock located at the edge of the world (*termonios*), strongly associated with the Ocean,[19] over which Heracles sailed in the cup of Helios on his way to the chained Titan.[20] These testimonies clearly situate Prometheus' cosmological place of punishment on the shore of the Ocean at the limits of the earth. This conclusion can be supported by Hesiod's account, whose Prometheus is bound in fetters driven through the middle of a pillar (κίων).[21] The term is fairly commonly, from Homer onwards, used to denote cosmological pillar(s) that connect or, rather, divide, the earth and heavens, in general associated with Atlas, both in mythic and in accounts of the philosophers of nature, which here show a clear continuity in subject matter, conceptualisation and terminology.[22] Hesiod further seems to have connected Prometheus' tortures with the fate of Atlas near the Hesperides,[23] pointing to a concept which places Prometheus and Atlas at the opposite edges of the world, the former in the east and the latter in the west.[24] This is nicely illustrated on a Laconian cup from Caere (c. 565–550 BC) that depicts together the tortures of Atlas and Prometheus. The author of this cup was perhaps influenced by the contemporary teachings of Anaximander, once again disregarding the expected boundary between a mythic and an account of a natural philosopher.[25] Thus the association of Prometheus with 'Colchis' (originally probably Aea) further proves its mythic or cosmological nature.

11.1. The return of the Argonauts

The second argument supporting the thesis of a cosmological nature of Aea and its involvement in an early model accounting for the diurnal solar movement is found in some of the varying accounts of the return route of the Argonauts. Several early authors incorporated the Ocean in their return journey, which the Argonauts entered *via* the Phasis, and returned to the Mediterranean by the Erythraean Sea, the Nile or overland through Libya.[26] These sources describe what could be called a 'southerly route', which might indicate the clockwise flow of the Ocean, against my argument outlined in chapter 9. At the same time, there was also a 'northerly' route, which involved the Tanais and the northern and western Ocean,[27] and which could have followed the postulated counterclockwise flow of the Ocean and the path of Helios' cup upon it.

Two out of three features of these cosmological accounts of the return voyage of the Argonauts that involve the navigation upon the circumambient Ocean, a path also followed by the sun at night, remained staple features of classical and medieval geography: the connections of the Pontus/Maeotis through the Tanais in the north and the Nile in the south with the Ocean. As opposed to these two geographical concepts, the third, the connection of the Phasis with the Ocean, does not seem to have had such lasting resonances in later geography and cartography.[28] An important segment of classical and mediaeval geographical knowledge was thus apparently derived from a set of cosmological notions ultimately dependent upon several variants of early Greek mythic accounts of the diurnal solar movement. The appropriation of these mythic models is not particularly surprising (once narrative elements are placed into the background, 'objective' geographical information are readily recognisable), but their longevity and perseverance, with the exception of the Phasis, is remarkable. This makes them worthy of investigation, especially in the light of the interactions and overlappings between mythic and post-mythic approaches to the phenomena pivoting around solar movement outlined in the introductory chapter.

The northern connection was perpetuated outside the Argonautic tradition by Pytheas, who claimed to have explored the whole Oceanic seaboard from Gades to the Tanais, which surely implies the latter's northern outlet,[29] as well as by Polybius and Strabo, who both describe

[18] Str. 11.5.5; cf. 11.2.16, where he cites *TGrF Adesp.* 559 Kannicht, Snell, cf. Eustath. Dion. Perieg. 687 (*GGM* ii.340); compare also Eratosth. IB24 Berger, fr. 23 Roller. Pl. *Phd.* 109ab and A. R. 1.83–85 are also pertinent here.

[19] Aesch. *PV* 1–2, 117, 128ff, 132–33, 286–398; West 1979: 135 n. 24; Finkelberg 1998: 120, 122, 132. The *correctio argumenti* (arg. b) *PV* corrects *hypothesis* Arg. A, where the action is located 'in Scythia by the Caucasus mountain' and locates is 'at the limits of Europe at Ocean' (Finkelberg 1998: 135). For Prometheus' fate compared to the punishment executed at the extreme frontiers (*eschata*) of the country see Pl. *Leg.* 9.855c with Gernet 1981[1936]: 240–42, 244, 246, 248 n. 1, 249 n. 14, 250 n. 33; [1924]: 254, 261; Detienne & Vernant 1991: 86–88.

[20] Pherec. *BNJ* 3F17 with [Apollod.] *Bibl.* 2.5.11.

[21] *Th.* 521–22.

[22] *Od.* 1.53–54; Hes. *Th.* 779; Anaximand. 12B5, A5, 11 D-K; Ibic. 336 *PMG*; Pi. *N.* 3.21, *P.* 1.19; Aesch. *PV* 349; Hdt. 4.184.3; Pl. *Resp.* 10.616b with Procl. *In Remp.* ii.200.5ff Kroll and Eustath. *Od.* 1.52, i.18 (Richardson 1926: 131), cf. Suda τ 365; Ennius *Euh.* 66–67 Warmington; Dion. Perieg. 67, Lightfoot 2014: 202.

[23] *Th.* 517–20.

[24] Cf. Lindsay 1965: 180. This interpretation contradicts the one offered in chapter 9 and presupposes the 'bi-polar' model accounting for the daylight/night exchange. I am thus not completely convinced of the validity of the interpretation of the Hesiodic passage in these terms. Kahn (1960: 139) recognizes in *Od.* 1.53–54 a reference to the pillars standing 'on at least two sides (*amphis*) of the earth', most probably its eastern and western extremity.

[25] *LIMC* 3.2.6 s.v. *Atlas* 1, cf. 3.1.4, 13 = *LIMC* 7.2.424 s.v. *Prometheus* 54; Yalouris 1980: 314; Stückelberger 1990: 71 n. 7; Hahn 2001: 287 n. 166; Couprie, Hahn & Naddaf 2003: 33–34; Naddaf 2005: 93–94.

[26] Hes. fr. 241 M-W; Pi. *P.* 4.26, 251, cf. Σ A. R. 4.257–62b, pp. 273–74 Wendel; Antimach. fr. 76 Matthews (Matthews 1996: 222–23); Hecat. *BNJ* 1F18a = fr. 302c, cf. fr. 18b.

[27] Tim. *FGrHist* 566F85 *ap.* D. S. 4.56.3, with other unnamed sources, both earlier and later (Diodorus considered Timaeus among the latter; Hawkes 1977: 3, 7, 9 n. 24 believes Ephorus is the most prominent of those sources); Scymn. *BNJ* 2047F4; *Orph. Argon.* 1037–245; the 'route' is repeated by Antonius Diogenes' heroes (Phot. *Bibl.* cod. 166, 109a; Bolton 1962: 190–91 n. 22). Perhaps Soph. *Skuthai* frs. 546–53 Pearson (Bacon 1925: 116), *pace* Σ A. R. 4.282–91b, p. 281 Wendel and Valerius Flaccus (Summers 1894: 7; Bacon 1931: 182; Wijsman 1998: 320; *contra* Bacon 1925: 112).

[28] Hes. fr. 241 M-W; Hecat. *BNJ* 1F18a = fr. 302c, fr. 18b. Any connection of the Phasis with the Ocean seems to have died off rather early, or it was incorporated in complicated arrangements of rivers, subterranean connections, lakes and seas that connected the Pontus with the Ocean (Hec. *BNJ* 1F195 or *BNJ* 264F13; Arist. *Met.* 350a24–25, etc.; Bacon 1931: 175, 178–79; Bolton 1962: 59; Green 2007: 297; Dan 2016: 258–59, 265 with n. 46, 268).

[29] Pyth. fr. 21 Bianchetti = fr. 7a Mette = T8 Roseman *ap.* Eratosth. fr. 14 Roller (Bunbury 1883: i.595; Tozer 1897: 158 (critical); Bolton 1962: 59; Cary & Warmington 1963: 53; Hawkes 1977: 2, 12; Dilke 1985: 136; Bianchetti 1998: 33–34, 65–67, 167–69; Roller 2006: 67).

the extent of the Oceanic coast of Europe in the north to the Tanais (both perhaps dependant on Pytheas).[30] The Pontus-northern Ocean connection was also advanced by other classical authors,[31] and the concept entered medieval maps, especially with the Tanais as the connecting link.[32] A further variant of this tradition is reflected in the connection of Pontus or Maeotis with the Caspian Sea, itself regularly conceived as a branch of the Northern Ocean.[33]

The southern connection of the waterways within the *oikoumenê* with the circumambient Ocean was accomplished by way of the Nile, which was supposed to issue forth from the (southern) Ocean.[34] Occasionally it was postulated that it issues from the Eastern/Indian Ocean or the Red Sea,[35] but in a comparable and probably derivative tradition the Nile was commonly believed to issue from the Western/Atlantic Ocean.[36]

Thus, both the southerly and northerly return route of the Argonauts are to be associated with several layers of geographical lore, which are all focused on the connection of the mythic/cosmological land of Aea with the earth-encircling Ocean, thus acknowledging and perpetuating the earliest cosmological tradition derived from and based upon the models accounting for the diurnal solar movement.

Both the location of Prometheus' punishment and the southerly and northerly return routes of the Argonauts point to this mythic or cosmological localisation of Aea, the land of sunrise. The importance of this particular mythic model (or a set of models) accounting for the diurnal solar movement is reflected in the rich geographical lore that developed around it through centuries, even millennia, to the point that both classical and medieval (both western and Oriental) geography and cartography were partially structured (with respect to the specific features and concepts concerned by the model(s) discussed above) by a basically cosmological model formulated in terms of anthropomorphic causality developed in several narratives that display strong correspondences between them.

From the surviving testimonies it is possible to reconstruct two—perhaps successive—earliest layers present in the narratives involving a voyage to Aea and reflecting the mythic account(s) of the diurnal solar movement, differing to some extent with respect to the goal of the Argonautic voyage: either the Argonauts sailed directly to the shore of the Ocean, where Aea was situated (whether on proximate or farther shore of the earth-encircling body of water), or they came to Aea and sailed up the river Phasis, which was somehow connected to the earth-encircling Ocean.

Any connection of these two variants with actual geographical situation must reflect the fact that the Greeks were aware from the earliest period of the 'closed' eastern shore of the Mediterranean, which conditioned that any plausible story of a long voyage to 'where the sun rises' could only have been located into the Pontus, which was probably believed at one point to extend to the Ocean.[37] According to Strabo, in Homeric times the Pontus was considered 'another' Ocean, surely in opposition to the western, and it was believed that those that travelled it left the *oikoumenê*, just as those that sailed outside the Pillars abandoned the inhabited world. The Pontus received its name—i.e., *pontos*, 'sea'—precisely because it was the greatest sea in 'our' part of the world.[38] Even though this is only a conjecture on Strabo's part, it seems concomitant with the Argonautic tradition as reconstructed from early sources, mainly based on the initial cosmological location of Aea. Subsequently, it was believed that the connection of the Pontus with the Ocean is possible through the Phasis—probably reflecting a newly acquired Greek knowledge of the Pontus as a closed sea—whether identified with the Caucasian Rion or the Don/Tanais.[39] The apparent identification (or confusion) of the Phasis with the Tanais,[40] however, necessitated the northern

[30] Plb. 3.38.2, cf. 37.8 (cf. Bunbury 1883: i.596; Roller 2006: 68 n. 104); Str. 17.3.24.

[31] Pomp. Mel. 3.33 with Roller 2006: 89 (the Vistula); Luc. *BC* 3.277–79; *Peripl. mar. Eryth.* 64, GGM i.304 (Casson 1989: 90–91, 241) with Podossinov 2013: 214; *HN* 2.168 with Podossinov 2013: 214; Max. Tyr. *Diss.* 26.3 with Podossinov 2013: 214; Mart Cap. 6.619 with Podossinov 2013: 214; Isid. *Etym.* 14.3.34; [Aeth. Ist.] *Cosmog.* 1.28, p. 84 *GLM*; perhaps also Posid. *FGrHist* 87F116 = D. S. 5.25.4 (the Ister).

[32] Already on the Tabula Peutingeriana, chapter 7 east (http://www.euratlas.net/cartogra/peutinger/medium_zoom/original_map_27_1.html), chapter 8 west (http://www.euratlas.net/cartogra/peutinger/medium_zoom/original_map_29_1.html) (cf. Rathmann 2015: 357 n. 72).

[33] Arist. *Met.* 350a24–25, 351a8–16, cf. Str. 11.7.4 (with the testimony of Polycleitus of Larissa *FGrHist* 128F7); Curt. *Hist. Alex.* 6.4.18; Arr. *Anab.* 7.16.2. Perhaps Aeschylus (thus Dan 2016: 266–69); Xenophon (thus Dan 2016: 257); Q. Metellus Celer *ap.* C. Nepos *ap.* Pomp. Mel. 3.45 and *HN* 2.170 (thus Roller 2003: 243 n. 111); Procopius (thus Dan 2016: 259); [Jerome] *Cosm.* 67c (Herren 2011: 150–51), *pace* Herren 2011: 251.

[34] Hecat. *BNJ* 1F302b; Romm 1989: 100; 1992: 172 (Herodotus does not name Hecataeus, but he is probably referring to his theory, cf. Van Paassen 1957: 133; Fowler 2011: 47, 54); Egyptian priests in Agathar. *FGrHist* 86F19 *ap.* D. S. 1.37.7 (cf. Theoph. Sim. *Hist.* 7.17.6) = Hecat. *BNJ* 1F302a, Anon. *De Aegypto* 665F62; Democr. 68A99 D-K; Luc. *BC* 10.255–57; perhaps Theoph. fr. 214b FHS&G.

[35] Aristot. (?) *De inundatione Nili BNJ* 646F1.4 = fr. 248 Rose, p. 193.13–15 = fr. 695, p. 749A.23–26 Gigon (Gehrke 2015: 86); Marinus *FGrHist* 2114F5 with Diogenes *FGrHist* 2204T1 *ap.* Ptol. *Geog.* 1.9.1, cf. 15.11 = *FGrHist* 2114F29 (Berggren & Jones 2000: 162; Geus 2013b: 229; Roller 2015: 183–84); [Aeth. Ist.] 2.12, 93.12–13 *GLM*, cf. 13, 93.26 *GLM*; Oros. 1.2.28, cf. 32.

[36] Euthymenes of Massalia (Ephor. *BNJ* 70F65f *ap.* Aristid. *Or.* 36.85, ii.290 Keil = ii.353 Jebb (cf. Aristides' comments in *Or.* 36.86–96, ii.290–94 Keil); Aristot. (?) *De inundatione Nili BJN* 646F1.6 = fr. 248 Rose, p. 195.3–4, 8–10 = fr. 695 Gigon, p. 749B.34–36, 40–43 (thus Aubert 2016; otherwise, this discussion could actually refer to Thales' theory); Anon. Flor. *De Nilo FGrHist* 647F1.5 = Ath. *Epit.* 2.87E (i.131.17–25 Meineke), 2 = Aët. 4.1.2 Mansfeld & Runia 2020; Sen. *Quaest. Nat.* 4A.2.22 ~ Lyd. *De mens.* 4.107 (p. 145.5–12 Wünsch), 24–25; Cary & Warmington 1963: 62; Romm 1992: 200; Roller 2003: 190; 2006: 16–17); Dicaearch. fr. 126 Mirhady = 113 Wehrli = Lyd. *De mens.* 4.107, p. 147.1–3 Wünsch = Sen. *QN* 4a fr. 6 Gercke.

[37] West 2005a: 41. An alternative would be a voyage by way of Sinus Arabicus (modern Red Sea), but this would necessitate an—however short—overland passage.

[38] Str. 1.2.10. Cf. Danoff 1962: 951.15–24 with Hecataeus *BNJ* 1F196, 214, 216, cf. 18b, 203, 213 and Hdt. 4.8.1–2, 10.3, 24, 38.2, 46.1, 85.1, 86.2–4, 87.1, 89.1, 95.1, 99.3, 7.36.1–2, 55.1, 95.2, 147.2 for the Black Sea at first called simply 'Pontos'.

[39] Cf. Bolton 1962: 55–59; S. West 2003: 155 n. 16; West 2005a: 41 with n. 11; 2007b: 195.

[40] Aesch. fr. 191 *TrGF*, Radt *ap.* (and with the discussion in) Arr. *Peripl.* 29, *GGM* i.394 (cf. Procop. *Bell.* 8.6.15 and Σ Dion. Perieg.

return route, or else the route motivated the identification. Finally, the Ister/Danube was introduced into the story, which now as a whole has become far (although not completely) removed from any cosmological resonances it might have contained earlier. These resonances can now be recognised only in the incorporation of some elements that on their own undoubtedly belong to the conceptual domain of cosmology. The most important of such elements, and the one most clearly containing a reference to the diurnal solar movement, is the appearance of Circe and Calypso in the Argonauts' return voyage involving the Ister/Danube route.

11.2. Circe, Calypso and the Argonauts' return voyage

According to the report of Apollonius Rhodius, who seems to have followed Timagetus' account (1st half of the 4th c.),[41] the Argonauts have encountered both Calypso and Circe on their return voyage. This seems particularly strange with respect to Circe, since they were returning precisely *from* her mythic abode. Thus, they could not have encountered her on their return voyage in a less derivative account, since they could not have visited the land of dawn on the Ocean' shore (again) on their way back to the Mediterranean. This kind of blatant contradiction and/or incoherence is unacceptable under any form of principle of charity. In line with my (and others') reconstruction of the earliest Argonautic tradition, it is only in later accounts of the Argonautic return voyage that this location appears on their itinerary in this context. Even then, it was apparently felt that an explanation of this secondary localisation must be provided. The account was certainly influenced by and modelled upon the *Odyssey*, since there it is described how the hero meets both these goddesses on *his* return voyage.

According to Apollonius' (Timagetus'?) account the Argonauts encountered Circe in the far west, in Tyrrhenia.[42] Her transfer from the far east was explained by recourse to her father's chariot,[43] presumably following his daytime path across the sky from east to west, which represents a clear reference to the diurnal solar movement and apparently goes back to the Hesiodic author.[44] As a daughter of Helios,[45] she was in an early period firmly associated with the far east, which left some trace in the Colchian toponyms Circaean plain and Circaeum; these toponyms are surely derived from an early tradition, but necessarily postdating the Colchis/Aea identification.[46] On the other hand, her association with the western part of the Mediterranean, the tradition followed by Apollonius (and Timagetus?), is clear from the locations chosen for her abode,[47] as well as her offspring of 'western' provenance.[48] This localisation is certainly part of a general trend of localising Odyssean toponyms in the western Mediterranean, which started already in the Archaic period (see chs. 3–4), rather than having any independent connection with the Argonautic tradition.

At the same time, Apollonius' (Timagetus'?) account locates Calypso in the west,[49] which is concordant with her usual setting.[50] Her 'Oceanic' genealogy supports this presupposition,[51] as does her offspring,[52] perhaps also the fact that she is Atlas' daughter.[53] Atlas was undoubtedly associated with the farthest west from the earliest period (although here one needs to have in mind the alternative localisation of Atlas in the light of different models accounting for the daylight/night exchange specifically discussed in chapter 9).[54] Calypso's western provenance can also be inferred from the sailing directions given by the nymph to Odysseus and his ensuing voyage, since she told him to keep the Bear (whether Ursa Major or Minor is irrelevant) on his left, thus instructing him to sail in an easterly direction.[55] This is one of rare exact pieces of information concerning both the length of the

10, *GGM* ii.431) (cf. Dan 2016: 266–67); Hdt. 4.45.2 (cited in Procop. *Bell.* 8.6.13–15); Agathem. 1.3, *GGM* ii.472; [Plut.] *De Fluv. et Mont.* 5.1–2 (*GGM* ii.642–43 = Ctesippus *FGrHist* 844F1) (cf. Dan 2016: 268); Procop. *Bell.* 8.6.2, 4–8, *Aed.* 6.1.7. In Xen. *Anab.* 5.7.7 one, sailing from the Bosporus to the north, arrives at the Phasis.

41 Timag. *BNJ* 2050F1ab. For his date see Gärtner in *BNP* s.v. *Timagetus* (Meyer *BNJ* 2050 is non-committal).

42 A. R. 4.590, 659–62, 850.

43 A. R. 3.309–13; cf. Dionys. Scyt. *FGrHist* 32F14 *ap.* D. S. 4.45.5 (Rusten 1982: 118, 151 believes that this is Diodorus' insertion in Dionysius' narrative); D. S. 4.45.6; Val. Flacc. 7.120, 218–19, 234, 262.

44 The fragment is printed as spurious by Merkelbach & West ([Hes.] fr. 390 M-W *ap.* Σ A. R. 3.309–13b (p. 229 Wendel)) but could actually be Hesiodic (Malkin 1998: 188–89).

45 *Od.* 10.137–39; Hes. *Th.* 956–57 (also cited in Σ QV *Od.* 10.139), cf. 1011; probably Hecataeus fr. 35A Fowler; [Hom.] *Ep.* 14; A. R. 3.310–11, 4.590–91, 683–84; Plaut. *Epid.* 604; Cic. *ND* 3.48, cf. 54; Verg. *Aen.* 7.10–11, cf. 12.164; Hyg. *Fab.* Prol., 156; [Apollod.] *Bibl.* 1.9.1 etc.

46 Tim. *FGrHist* 566F84; A. R. 2.400, 3.199–200; *HN* 6.13; Val. Flacc. *Argon.* 5.327, 6.426, 7.544; Dion. Perieg. 692, Lightfoot 2014: 234; Avien. *Descr. orb. terr.* 877, *GGM* ii.185; Prisc. *Perieg.* 674, *GGM* ii.196; Eustath. *DP* 689, 692, *GGM* ii.340–41.

47 To the sources cited in previous notes add Aesch. fr. 2 West (cf. *HN* 25.11); Verg. *Aen.* 3.386, 7.10–24; Str. 1.2.10; Ov. *Fas.* 4.69, *Met.* 14.8–10, cf. 348; Hyg. *Fab.* 127; [Apollod.] *Bibl.* 1.9.24 etc. for the Tyrrhenian coast; for the mountain promontory Circeius and the Roman town Circeii see Theophr. *Hist. Pl.* 5.8.3, 9.15.1; [Scyl.] 8, Shipley 2011: 25, 55; [Arist.] *De Mirab.* 78.835b33–836a6; Dionys. Scyt. *FGrHist* 32F14 *ap.* D. S. 4.45.6; Σ HQ *Od.* 11.134 with Lycoph. 1273–76 and Tzetz. (Σ) Lycoph. 1273, 1274, 1276 (West 2013: 302–303); Varro *ap.* Serv. *Aen.* 3.386, cf. 7.10; Cic. *ND* 3.48 etc.; for other Italian locations see Eur. *Tro.* 438; [Scymn.] 222–25 (*GGM* i.205); Parthen. *Narr.* 12; Hyg. *Fab.* 125; outside the Pillars of Heracles: Str. 1.2.10; *Orph. Argon.* 1207–43 (Bacon 1925: 114; 1931: 178).

48 Hes. *Th.* 1011–16 (l. 1014 athetized by West 1966: 434–35; 1988: 33; 2013: 302 n. 15, but see Malkin 1998: 86, 161, 180–91); *Telegonia* Arg. 3–4 West, frs. 5 (=Σ V = D *Od.* 11.134), 6 West; Hypoth. *Od.* (i.6 Dindorf) (from Sophocles?, Pearson ii.109–10); Arist. *Poet.* 1453b33–34; Euphant. *FGrHist* 74F1 *ap.* Eustath. *Od.* 12.70 (ii.12); Xenag. *BNJ* 240F29 = 840F17; Cn. Gellius fr. 9 *HRR*; [Scym.] 227–28 (*GGM* i.205); Cic. *Tusc. Disp.* 2.21 etc.

49 A. R. 4.574–75, cf. Steph. Byz. s.v. Νυμφαία (Wilamowitz-Moellendorf 1884: 114, n. 2).

50 For which see Bilić 2009: 74–77 (the earliest explicit source seems to be Hes. fr. 150.30–31 M-W; Meuli 1921: 61; West 1966: 435).

51 *Od.* 1.52–54; Hes. *Th.* 359 (*pace* West 2005a: 62, n. 81); *hDem.* 5, 422; Eur. *Cyc.* 264; [Apollod.] *Bibl.* 1.2.7

52 [Apollod.] *Epit.* 7.24.

53 *Od.* 1.52; A. R. 4.574–75; Lycoph. 744 with paraph., i.65 Scheer etc.

54 Atlas' western provenance: Hes. *Th.* 517–20, 744–50 (cf. fr. 150.25 M-W); Aesch. *PV* 350–52; probably Pherec. *BNJ* 3F17 (see the discussion in chapter 10); Polyeid. 837 *PMG*; *Orph. fr.* 215 Kern = *PEG* 319F Bernabé *ap.* Procl. *Tim.* i.172.11–12, 31–173.8, 18–22 Diehl. Cf. Pl. *Tim.* 24e–25a, *Crit.* 108e, 114a–b.

55 *Od.* 5.270–81.

voyage (without using a 'typical number')[56] and the starting and ending points in the whole *Odyssey* and it might be as old as the pre-Homeric Argonautic—or some other travel-related—tradition, but its context remains unrecoverable. For seventeen days Odysseus sailed from Ogygia to Scheria, which places Calypso's island that many days' west from the land of the Phaeacians. Since the latter was regularly from the earliest period identified with Corcyra/Corfu,[57] perhaps—but unlikely—already by Homer himself,[58] it can be deduced that Ogygia was imagined to be 17 days to the west of that island.

However, contrary to all what has hitherto been said on the location of Calypso's island and her genealogical connections, her unequivocal western provenance comes into question if we examine all the available information on the nymph, almost unknown apart from her memorable appearance in the *Odyssey*, and compare it with Hesiod's Styx. The resemblances between Homer's Calypso and Hesiod's Styx are striking; it is almost impossible they should represent two completely independent creations and the most plausible explanation of these resemblances is that both figures are derived from an identical or at least similar tradition or source, or that one (presumably Calypso) was modelled upon the other (Styx). The points of resemblance are thus summarized by Frame:[59]

1. both goddesses live apart from the other gods (*Od.* 5.80 : *Th.* 777);
2. both are rarely visited by the messengers of the gods (*Od.* 5.88 Hermes to Calypso : *Th.* 780–81 Iris to Styx);
3. both are reached by crossing the sea (*Od.* 5.50–55 : *Th.* 781);
4. Calypso lives in a cave, Styx in a house 'roofed with long rocks' (*Od.* 1.14, 5.57–58, 77, 226 : *Th.* 778);
5. the house of Styx is supported by columns reaching to the sky; these can be compared with the sky-supporting pillars, and consequently with Atlas, who is the father of Calypso, and who holds the great columns which keep heaven and earth apart (*Od.* 1.52–54 : *Th.* 778–79).

These are the most striking parallels between the two goddesses; but there are also some etymological connections between them:

6. a god who swears false on the water of Styx is for one whole year 'wrapped' in a comatose state (*Th.* 798); the word used is *kaluptô*, 'to cover', from which derives the name Calypso;
7. finally, the epithet Hesiod used to describe the water of the Styx, *ôgugios*, 'primeval' (*Th.* 805–806), corresponds to the name of Calypso's island, Ogygia.

I am inclined to add to Frame's list several supplementary parallels between these two figures:

8. according to Hesiod they are actually sisters, both being the daughters of Ocean (*Th.* 359, 361);
9. both were present when Hades abducted Persephone (*hDem.* 422–23; cf. *PEG* 387.9–10 F Bernabé);
10. (continuing upon 7). according to a reading of Parthen. fr. 11 Lightfoot (ὠγενίης Στυγὸς ὕδωρ) Styx is named as a daughter of Ogenos,[60] who is probably to be equated with Ocean.[61] Therefore Styx and 'primeval' Ocean/Ogenos seem to be related, and both appear to be further associated with Ogygia and Calypso.

Once again, the existence of two models accounting for the diurnal solar movement provides a coherent explanation of the apparent duality in the localisation of the Calypso/Styx figure: according to the 'bi-polar' model she is located at the point of sunset in the extreme west (Calypso), while according to the 'uni-polar' model she is located at the *axis mundi* in Tartarus/Hades (Styx). Thus, a single mythic figure was 'doubled' in order to fit into two differing cosmological models concerned with the daylight/night exchange.

The correspondences between Calypso and Styx can be extended to include Circe. Indeed, the similarities between Circe and Calypso were often emphasized.[62] Already in antiquity it was sometimes hard to distinguish between the two,[63] which is also exemplified in both being named as a mother of identical mythic figures in different traditions.[64]

[56] De Jong 2001: xix, *pace* 231.

[57] Alc. 441 Lobel-Page; Hecat. *BNJ* 1F104; Acus. *BNJ* 2F4; Hellan. *BNJ* 4F77; the Corcyraeans themselves in Thuc. 1.25.4, cf. 3.70.4; [Scyl.] 22, Shipley 2011: 27, 58; Arist. fr. 512 Rose = fr. 517.1, 3 Gigon, probably also fr. 611.56 = Titel 143,1.56 Gigon; Heracl. Pont. fr. 157 Wehrli = fr. 109 Schütrumpf; Demetr. Phal. fr. 192 Wehrli = fr. 146 Stork, Ophuijsen & Dorandi; probably Tim. *BNJ* 566F79 etc.

[58] He thus described Scheria as not far from Thesprotia (*Od.* 14.321–26, 17.525–27, 19.270–72), while one of royal chambermaids was snatched from her country Apeirê (*Od.* 7.8–9), which might designate Epirus (thus Σ EPQT *Od.* 6.204 and Σ PTV, V, P, B 7.8) (Shewan 1918: 328–29). However, modern scholarship does not support this interpretation, see e.g., Hainsworth 1988: 320 and Frame 2009: n. 2.158, although Malkin 1998: 131–32, 152 argues that the identification of Scheria with Corcyra is indeed implied in *Od.* 19.270–87.

[59] Frame 1978: 166–69. I have left out some of Frame's—in my opinion—less convincing arguments.

[60] West 1966: 201; 1971: 18 n. 4; but see Lightfoot 1999: 152 ('ancient'). Perhaps also Antim. fr. 114 Matthews (Matthews 1996: 295, who is sceptical; Lightfoot 1999: 152).

[61] For Ogenos see Pherec. 7B2 D-K = fr. 68 Schibli, B4 D-K = fr. 78 Schibli; Lycoph. 231 with paraph., i.23 Scheer; Hesych. ω 19, 21; cf. Steph. Byz. Ὤγενος ἀρχαῖος θεός…; Suda ω 8 s.v.; Hesych. ω 20 (West 1966: 201; 1971: 18).

[62] Wilamowitz-Moellendorf 1884: 116; Butterworth 1970: 30; Crane 1988: 31, 33–34; Cook 1992: 249; Tracy 1997: 374, 377; West 1997: 404; 2005a: 61; 2014: 127; Abusch 2001: 4; de Jong 2001: 130; Marinatos 2001: 401; Nakassis 2004: 223; 2020: 275; E. West 2014: 144–45, 147, 153, 161, 170; Kozlowski 2018: 24 n. 46.

[63] Proper. *El.* 3.12.31; Hyg. *Fab.* 125; Pom. Mel. 2.7.120; Plot. *Enn.* 1.6.8. Sisters in Dictys *BNJ* 49F8–9; Sisyphus of Cos *BNJ* 50F3; Tzetz. (Σ) Lycoph. 174 (ii.79 Scheer).

[64] Telegonus (or Teledamus, variant, see West 2013: 300): Calypso according to *Telegonia* fr. 4 West, for Circe see above; Auson: Calypso according to [Scymn.] 229–30; Fest. p. 16 Lindsay s.v. *Ausoniam*; Tzetz. (Σ) Lycoph. 44 (ii.34 Scheer); Serv. Dan. *Aen.* 3.171; Steph. Byz.

Another important connection between the two, in the light of Calypso's stygian connections, can be recognized in Aeaea's (and thus Aea's) proximity to Hades,[65] which is a reflection of the 'uni-polar' model accounting for the diurnal solar movement, and at the same time another manifestation of the widely attested connection between eschatological concerns and the daily movement of the sun. This interaction between the two conceptual domains was commented upon repeatedly throughout the book; it ultimately depends on the analogy between an observable physical phenomenon interpreted in personalistic terms and the unobservable human afterlife destiny, explaining the latter in terms of the former.

Indeed, both Circe and Calypso are mythic figures intimately associated with the diurnal solar movement, i.e., they are part of the model (or models) formulated by the early Greeks in attempting to account for the daily movement of the sun. This explains their involvement in the Argonautic tradition, which was itself, as we have seen in chapter 4, from the early period definitely concerned with the sun and its diurnal movement (in addition to its concern with the annual solar movement discussed in chapter 2).

11.3. Conclusion

Aea (with its parallel, the island of Aeaea in the *Odyssey*) is thus a cosmological location, a land in the farthest east and the place of sunrise. It is an important feature of a mythic model accounting for the diurnal solar movement exploited by a number of successive iterations of the Argonautic myth. The importance and perseverance of this model is manifested in the rich geographical lore that accrued onto it from the Archaic to the Hellenistic period and onwards to the end of the European Middle Ages, with important resonances in the closely associated Islamic tradition. Several later geographical models involved with the continental boundaries, the shape and extent of the *oikoumenê*, the courses of several important rivers, most notably the Nile, the possibilities of continental circumnavigations and other similar subjects were directly or indirectly dependent upon this early mythic model, or at least to the lore that was involved in its formation in the earliest layer of Greek tradition known to us. A decisive break between a decidedly mythic and a comparably distinctly scientific approach in the diachronical development of the continuum of knowledge on these subjects does not seem to have occurred, or at least it does not seem apparent to us. The Greek ethnographic context showed equally inclusive towards both the narrative and descriptive/interpretative accounts of the diurnal solar movement. The former were imperceptibly incorporated into—or formed the basis of—the geographical, cartographical or cosmological models that laid a much stronger emphasis on such concerns as the correspondence with reality, were expressed in non-narrative terms and were apparently based on empirical observations. Their development from speculative (mythic) cosmology, which accounted for the daily solar movement in terms of anthropomorphic causality and treated it in a dynamic and evolving set of myths, was no obstacle to their general acceptance as valid geographical, cartographical and cosmological models of reality. Long after the sun-land Aea(ea) and Circe were abandoned as mythic fantasies, the courses of the most important rivers known to the Greco-Roman civilisation, as well as its understanding of the shape end extent of the *oikoumenê* and its limits (*eschata*), were determined by the very mythic models of which these fantasies were integral parts.

Once again, the set of myths involving references to solar movement studied in this chapter shows certain overlapping and interaction between two different conceptual domains, the solar movement and eschatological considerations. However, this phenomenon is not particularly pronounced and was thus studied only incidentally, mainly in terms of the existence of two models accounting for the daylight/night exchange. On the other hand, the subject matter treated in the following chapter will show a much stronger concern precisely with this interaction across different domains.

---

s.v. Αὔσων, for Circe see above, for both *Etym. Magn.* 171.16–17 s.v. Αὔσονες; Eustath. Dion. Perieg. 78, *GGM* ii.232; Latinus: Calypso according to [Apollod.] *Epit.* 7.24, for Circe see above; Nausithous: Calypso according to Hes. *Th.* 1017–18; Eustath. *Od.* 16.118, ii.117, Circe according to Hyg. *Fab.* 125.

[65] *Od.* 10.506–509, 11.1–13, 638–40, 12.1–4.

12

World of the Dead at the Antipodes

The interaction of the conceptual domains of eschatology and specifically the diurnal solar movement is strongly exemplified in a complex of myths and other accounts introducing the region of antipodes in the discussions of the posthumous fate of humans. The antipodes are introduced in these accounts either intentionally or they can be envisaged as an inevitable consequence of their localisation of the world of the dead in terms of solar movement.

When using the word 'antipodes' (I will normally use the capitalised form only when it refers to the inhabitants of the region), one should have in mind its different applications. With the main subject of this book in mind, in distinguishing between these applications I will place a special emphasis on various meteorological (in Greek sense) effects caused by the solar movement on different regions upon the (spherical) earth's surface. Incidentally, in a slightly modified way these distinctions are equally applicable to a flat earth, although a conceptualisation of the antipodean parts of a flat earth seems odd to us, who are accustomed to think in terms of a spherical earth. One's antipodes can thus be latitudinal (when sharing the same longitude, i.e., Tokyo and Melbourne, which have an identical daylight /night exchange but experience opposite seasons), longitudinal (when sharing the same latitude, i.e., Anchorage and Sankt Petersburg, which experience the same seasons but opposite daylight /night exchange) and true antipodes, with both opposite latitude and longitude (i.e., Madrid and Wellington, which have both the seasons and daylight/night exchange reversed). These three positions were later respectively designated as the areas of *antoikoi*, *perioikoi* and antipodes and all could be invoked as a location of the world of the dead, although the *antoikoi* were less frequently imagined as such a location (Figure 12.1).[1]

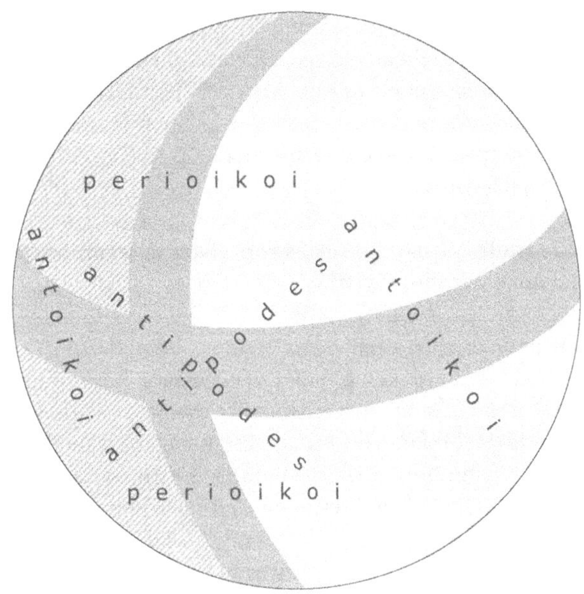

Figure 12.1 The antipodes: mutual relations of the quarters upon the sphere of the earth as understood in antiquity (drawn by the author)

We have seen in chapter 9 that for the author of the section of the *Odyssey* involving a voyage to Hades from Aeaea the realm of the dead was located on the farther shore of the southern section of the circumambient Ocean.[2] According to this cosmological model the Ocean must have been imagined to flow counterclockwise around the *oikoumenê*, and Odysseus sailed with the current on his way back and against it on his way towards Hades, thus following the postulated southerly course of the sun towards east after sunset when returning to Aeaea. This 'horizontal' model accounted for the diurnal solar movement by way of explaining the nightly course of the sun in terms of Oceanic navigation, with Hades in the extreme south of the flat earth in a longitudinal antipodean position in relation to the *oikoumenê*. At the same time, I have also demonstrated that a variant of this model involved the concept of cosmic 'nadir' or *axis mundi* located deep in Tartarus/Hades, used by the early Greeks as another model accounting for the diurnal solar movement in which the sun travels below the underside of the earth during night. This model works equally well for the flat earth as for the spherical, and with the introduction of the new cosmological framework involving the latter it was assimilated with little or no effort.

When involving a reference to the solar movement, the antipodean position of Hades and/or Tartarus was almost

[1] For the idea of antipodes in Greece and Rome see especially Aujac 1993: 235–37, 278–82; Moretti 1994; Hiatt 2008, esp. pp. 14–95; Stallard 2010a; 2010b: 12–49; 2016: 2–11. The author of this quadripartite division of the earth's surface was Crates of Mallos (fr. 37 Broggiato = *BNJ* 2113F13, compare Macr. *Somn.* 2.9.1–6 = fr. 35f Mette; Mette 1936: 75–78; Nicolet 1991: 35, 51 n. 31, 63; Aujac 1993: 14, 28, 32 n. 19, 45, 96, 162–63, 237–38, 279, 306–307, 313, 339, 361; 2015: 319 n. 20, 321; Randles 1994: 6, 10–15, 19–21, 24–26, 31–33, 38, 51, 56–58, 75; Broggiato 2001: liii, 224; Bilić 2012a: 296; 2016c: 129–32; Dueck 2012: 77–78; Geus 2013a: 119–22; Kominko 2013a: 78–80; Roller 2015: 133; Diederich 2018: 92, 106; 2019: 131, 139), who was perhaps in some measure influenced by Eudoxus (Gisinger 1937: 2132.47–2133.38, 2143.50–55, cf. Randles 1994: 10, mainly on the basis of fr. 288 Lasserre, which, however, offers only the slightest support for this thesis). Crates' system was restated in more scientific (i.e., mathematical/astronomical) terms by John of Sacrobosco, the author of the most successful medieval astronomical textbook (Grant 1994: 20), who—instead of an equatorial and meridional ocean—described the quadripartite division in terms of equator and a 'prime' meridian passing through both poles (Thorndike 1949: 110, 138–39).

[2] *Od.* 10.507–508, 11.11.

exclusively defined in terms of the diurnal voyage of the sun. Even though it seems awkward to a modern observer, the 'antipodal' localisation of Hades in these terms can equally well be applied on the flat earth as on the spherical. However, occasionally it was defined with respect to the sun's annual motion as well and sometimes with respect to a combination of both, such as when the Otherworld was placed at the true antipodes, with both the seasons and daylight/night exchange reversed. In this particular case, the localisation of the antipodes fits much better in terms of a spherical earth. Indeed, a location characterised by these meteorological features is hard to conceptualise upon a flat earth.

12.1. Hades at the antipodes conceived in terms of the diurnal solar movement[3]

The applicability of the model upon both the flat and spherical earth makes it rather unnecessary—or does not make it mandatory—to distinguish between these two cosmological frameworks when discussing the testimonies for the localisation of Hades/Tartarus in terms of the solar movement. For several testimonies it seems impossible to determine whether they refer to a spherical or flat earth, although the former almost always seems the more likely (I am aware that this is perhaps only a modern bias). My first example appears to be the only exception: Pindar, a contemporary of the first half of the 5th c. Presocratics, described how the righteous in Hades 'always have the sun in equal nights and equal days'.[4] This can be understood as meaning that the dwellers in Hades have the same sun as we do ('we' always refers to the inhabitants of the Greco-Roman *oikoumenê*), but it shines for them in reverse order.[5] This interpretation becomes more likely if it is associated with an apparently complementary fragment in which Pindar describes the abode of the pious in Hades as having daytime when we experience nighttime and, most probably, *vice versa*.[6] It seems that Pindar envisaged the position of Hades—or at least its section reserved for the righteous—in terms of the diurnal solar movement, with a strong possibility that its occupants experienced the daytime/night exchange contrary to the inhabitants of the *oikoumenê*. This would place them, in terms of a spherical earth, at our *perioikoi* or antipodes proper (if we accept the reading of the second *Olympian* as referring to an equatorial localisation—having daytimes and nights of equal and identical length throughout the year—this would place them precisely between the two antipodal regions). In terms of a flat earth, which is probably how Pindar himself understood his description, they would spend their days in a region upon the underside of the earth through which the sun passes during night.

Furthermore, it could be argued that Philolaus' (c. 470–380s BC) counter-earth (*antichthôn*) might be interpreted in terms of an antipodean hemisphere.[7] This interpretation is feasible on the condition that Philolaus' central fire is understood as the earth's interior.[8] Alternatively, the *antichthôn* could be identified with Hades.[9] However, these identifications must be credited to post-Aristotelian readings of this concept, when a geocentric interpretation of the system was introduced.[10] At the same time, (pseudo-)Pythagorean teachings on the Antipodes—with the earth being 'inhabited all around' (περιοικουμένην)—might be associated with the early account of the *antichthôn*,[11] since the latter was also inhabited according to Philolaus, presumably also by the living.[12] Thus Philolaic-Pythagorean and later Pythagorean traditions seem to have expounded several different conceptions with respect to the counter-earth, the antipodes (i.e., the lower hemisphere defined by the diurnal solar movement) and its inhabitants. Some of these possibly placed Hades at the antipodes, and, since the *antichthon* represents a counter-world where everything is reversed (as it should be with the antipodes with respect to the solar motion),[13] its inhabitants surely experience daytime and nighttime contrary to us, perhaps also the seasons. However, both the obscurity and multiplicity of sources seem to allow a rather simple concoction of a doctrine one is inclined to recognise, including an antipodal Hades defined by the diurnal (and annual) solar movement; for this reason, I cannot claim that this concept ever existed in this form in Pythagorean thought, however widely we define it.

There are several early sources—in addition to Pindar—that do expound such a concept: thus, Aristophanes mentioned that only the initiated enjoy the light of the sun in Hades,[14] and that the same light shines upon the deceased and us.[15]

---

3 An earlier version of this section appeared in Bilić 2019: 23–24, 48–49.

4 *O.* 2.61–62 (= *PEGr* fr. 445 V Bernabé).

5 Rohde 1925: 444 n. 38; cf. Woodbury 1966: 600 n. 5. Otherwise, Woodbury strongly opposes this interpretation, cf. Ballabriga 1986: 120–21; Boutsikas 2020: 113, 164; for an argument for 'perpetual light' in this passage see Edmunds 2009: 670.

6 Pi. fr. 129.1–2 (= *PEGr* fr. 439 V Bernabé); Rohde 1925: 444 n. 38; Scodel 2003: 85; cf. Ballabriga 1986: 121 and Edmunds 2009: 670. Woodbury (1966: 601 n. 5) and Ballabriga (1986: 121–22) unconvincingly recognize here the concept of perpetual light.

7 Cf. Zeller 1881: i.453, 460; Röth's similar conjecture in Zeller 1881: i.453 n. 1; Zeller 1881: i.454 n. 3 and Burch 1954: 275 n. 10 cite Böckh's conjecture which is remotely similar; Burch 1954: 290, 294 actually offers a hypothetical interpretation of Philolaus' system in these terms. For the references for the Pythagorean *antichthôn* see Kingsley 1995: 91–92 n. 12, 172–73 with n. 3.

8 Simpl. *Cael.* vii.512.9–14, cf. Σ Arist. *Cael.* 504b42–505a5 Brandis = 58B37 D-K; Procl. *Tim.* iii.143.26, 144.8; Damasc. *Phd.* 1.534–36 (Zeller 1881: i.453–54; Burkert 1972: 232–33; Huffman 1993: 242; Kingsley 1995: 180–82, 193–94).

9 Kingsley 1995: 185–87 (his Hades is decidedly subterranean, but in this context, it should rather be identified with another hemisphere).

10 Huffman 1993: 242 with n. 5, 245 n. 7. At the same time, Dicks (1970: 229 n. 80) strongly rejects the reading of the Simplicius passage resulting in a geocentric cosmology attributed to the Pythagoreans, which would make the Hades/*antichthôn* identification unfeasible.

11 Alex. Polyhist. *BNJ* 273F93 (*ap.* D. L. 8.25) with *Commentary*; Zeller 1881: i.452–53; Kahn 2001: 80. The text excerpted by Alexander Polyhistor, the *Pythagorean Notebooks*, can only be dated to a period from ca. 350 BC to ca. 100 BC, but it exhibits some 'archaic', i.e., Presocratic and pre-Platonic features (Kahn 2001: 79–80, 83; Zhmud 2012: 10, 90, 194, 423 opts for a late 2nd – early 1st c. BC date).

12 Philol. 44A17a D-K (Huffman 1993: 238; Mansfeld in Mansfeld & Runia 2010: 91 n. 166).

13 Burkert 1972: 347–48.

14 *Ran.* 454–55.

15 *Ran.* 155.

These two descriptions are contradictory, but it seems that Aristophanes placed at least some of the deceased in an 'antipodean' hemisphere defined by the solar movement. Aristophanes' contemporary, the anonymous author of the Hippocratic *On Regimen*, stated the following: 'daylight for Zeus (i.e., daytime sky), darkness for Hades; daylight for Hades, darkness for Zeus'.[16] This could only refer to a reciprocal daylight/night exchange between the upper- and the under-world. Finally, it is possible that Callimachus, when he described the sun after setting as shining upon the sons of Ophion, also meant that it visits Hades,[17] while Herodorus of Susa similarly claimed that Apollo-Helios illuminates both the upper- and under-world, presumably reciprocally.[18] A number of later sources expound the notion of the sun in an otherworldly realm of the dead, thus only implicitly testifying to its antipodal location. At the same time, these testimonies as a rule do not explicitly commit to a specific shape of the earth, which was of no interest for their immediate purpose; for the vast majority, if not all, an underlying notion of a spherical earth can be presupposed.[19] Admittedly, this conclusion is solely based on the late date of these sources.

In the category of testimonies on the antipodal setting of Hades that programmatically refer to a spherical earth, the familiar name of Crates of Mallos reappears. He thus localised the Homeric sunless Cimmerians, whose name he read as Cerberians, presumably together with the nearby Hades, underneath the south celestial pole.[20] In addition, the meteorological characteristics of their land as described in the *Odyssey* were recognised as corresponding to those at the north or south pole.[21] Similarly, meteorological features under the poles were also compared with Homer's description of Tartarus and perhaps also with Hades.[22] Contrary to all the testimonies studied hitherto, these interpretations exceptionally define the location of the Otherworld with respect to the annual solar movement; a juxtaposition of these two concepts was already recognised in the account of the Laestrygonian Telepylus (chapter 2) and Syriê (chs. 3–4), but with no emphasis on any kind of antipodal nature of the world of the dead (both these cosmological locations being firmly associated with the north and thus belonging to the 'upper' hemisphere). In addition, an antipodal localisation of Hades and Tartarus—in terms of a flat earth—can be recognised in Homer's and Hesiod's symmetrical vertical arrangements of great cosmic regions (chapter 9), although with no reference to solar movement, which is why it is only mentioned here in passing.

12.2. Later testimonies for an antipodal Hades conceived in terms of solar movement

Some later sources are more explicit on the antipodean localisation of the world of the dead. Thus a lower hemisphere, opposite to the inhabited world, and more precisely the south pole, was explicitly the location of the otherworld described by Vergil.[23] This 'counter-*oikoumenê*' has its own sun and stars, i.e., their visibility is contrary to that in the *oikoumenê*, which explicitly defines the Otherworld with respect to diurnal movement of the sun.[24] Perhaps following and explicating on Vergil's localisation of Tartarus, Valerius Flaccus placed Hades opposite the north celestial pole, the things pertaining to the upper world (*supernae*), and the 'toppling' sky (*ruens polus*), most probably at the south pole, but with no explicit reference to solar phenomena.[25] On the contrary, Seneca compared the pleasure-loving Romans precisely with the Antipodes and their contrary alternation of day and night, further comparing them with the dead. He does not explicitly place the realm of the dead at the antipodes, here defined exclusively by the diurnal solar movement, but suggests his familiarity with such a tradition.[26] Following Vergil, Servius also placed the realm of the

[16] 1.5 (on the date see Jouanna 1999: 409).

[17] *Aet.* fr. 177.5–7 Pfeiffer; West 1971: 23 n. 2; Stephens 2003: 222 n. 140.

[18] SEG VII, 14.9–10 (Merkelbach & Stauber 2005: 74–75; Potts 2016: 360). Merkelbach & Stauber 2005: 74, 76 date Herodorus' hymn to the 3rd/2nd c. BC.

[19] *IEph* 1625A = Kaibel 228b.7–8 (Apollo; for Apollo's solar nature see chapter 7); [Plut.] *De Hom.* 2, ll. 1007–18 Kindstrand (Keaney & Lamberton 1996: 163); Apul. *Met.* 11.23 (uncertain); *Orph. Hymn.* 34.13–14 (Apollo); Serv. *Aen.* 3.93 (cf. Porph. in Serv. *Ecl.* 5.66; *Myth. Vat.* 2.18, p. 80 Bode, 3.8.16, p. 209) (Apollo); Porph. *Imag.* fr. 7 Bidez *ap.* Eus. *PE* 3.11.109cd, cf. Lyd. *De mens.* 4.137 (cf. Iambl. in Lyd. *De mens.* 4.149) (Pluto); *PMG* i.33–34, iv.444–45, 447, 1600–601, 1695–96, 1959–60, 1962, v.248–49 (cf. Carey 1994: 24 with n. 64), viii.80 (all referring to Helios), i.316 (Apollo-Helios) (Faraone 2004: 214, 227–28, 231–32; i.346 suggests further syncretism with Re, Betz 1992: 12 n. 61; Faraone 2004: 232).

[20] Crates fr. 54 Broggiato = *BNJ* 2113F23a *ap.* Gemin. *Elem. Astron.* 6.16–17 with Mette 1936: 84 n. 1, 265–66 and Abel 1974: 1054.60–66; Mette 1936: 84 n. 1, 88, 92, 265–66; Abel 1974: 1054.51–66; Broggiato 2001: xlix, liv, 222; *Commentary* to *BNJ* 2113F23. For Crates' Cimmerians/Cerberians see fr. 53 Broggiato = *BNJ* 2113F22 = frs. 38ac Mette. Mette adduced several additional sources where Crates' name is not explicitly mentioned (frs. 38bde, Hesych. κ 2298, Phot. *Lex.* κ 593). Cf. Aristarch. in Σ H *Od.* 11.14 (= Crates fr. 38f Mette) and Ephor. *BNJ* 70F134b. See also Lehmann-Haupt 1921: 426.24–36 and Bilić 2012a: 300–301.

[21] North pole: Plut. *Mar.* 11.9–10 (from Posidonius?, Olbrycht 2000: 88–89). South pole: Crates fr. 54 Broggiato = *BNJ* 2113F23a (Gemin. 6.15–20) with fr. 57 Broggiato = *BNJ* 2113F24 (Hades of the *Odyssey* in the southern hemisphere).

[22] Crates fr. 7 Broggiato = *BNJ* 2113F3. Mette adduced two additional sources where Crates' name is not explicitly mentioned (fr. 39b, where Tartarus is associated with the 'dark portion of the *oikoumenê*', which is probably only a negligent use of terminology, and fr. 39c, where Hades is interpreted as the 'dark portion of the *earth*').

[23] Verg. *Georg.* 1.240–51. As a rule, I will henceforth use 'lower hemisphere' for that half of the earth with has its centre opposite to the imagined centre of the Greco-Roman *oikoumenê*, i.e., the half of the earth corresponding to the equivalent hemisphere of the sky effectively below the horizon for an observer standing at this centre. It includes the region of the south pole, since the north celestial pole is significantly elevated to such an observer.

[24] *Aen.* 6.640–41, *Georg.* 1.250–51 (cf. Manil. *Astron.* 1.242–45; Serv. *Georg.* 1.243 glosses this account as a reference to the antipodes). Cf. 6.535–36, where it is described how Aurora (i.e., the sun) crossed the meridian (i.e., it is past noon), and the night is *ruit* (539); this is usually taken to mean that the night is approaching, but Casagrande-Kim (2012: 127) understands the phrase to mean that the night is now nearing its end *in Hades* (where the midnight has now passed), claiming that the world of the dead is antipodal to the world of the living with respect to the diurnal course of the sun, i.e., it is located in the lower hemisphere.

[25] Val. Flacc. 1.827–32.

[26] *Epist. ad Luc.* 20.122.1–3 (citing Verg. *Georg.* 1.250–51); Moretti 1994: 246–47.

dead—or the realm of allocated souls—in the antipodean hemisphere, implying the contrary alternation of daylight and night in their region.[27] A similar opinion was held by Tertullian, who placed Paradise in the southern temperate zone.[28] This, however, does not automatically mean that this location was defined by the diurnal solar movement, since it remains unclear whether he had in mind the entire southern hemisphere or only the section of the southern temperate zone in the lower hemisphere. Finally, Claudian in a highly eclectic description of the location of Hades combined underground,[29] western[30] and 'antipodean' features. The lattermost are manifested in the fact that Hades is placed under another stars (*altera sidera*), which follow other courses or skies (*orbes alii*), and another, Elysean sun.[31] These stars and their courses are presumably those unseen from the poet's latitude, which strongly points towards the southern hemisphere, but the whole description is a real pastiche of concepts and different models accounting for the diurnal solar movement (i.e., the 'uni-' and 'bi-polar' model) that refuse to be neatly disentangled.

In any case, it appears that the idea of an antipodean world of the dead defined by the diurnal solar movement was rather widespread in antiquity, although its explicit elucidation is encountered less often. This is unsurprising due to the nature of our sources, consisting mainly of narrative, mostly poetic, accounts (with a possible exception exemplified in the Pythagorean tradition) that were regularly focused on other considerations. Even so, the interpretation of the cosmological location of the world of the dead in terms of solar movement and, more specifically, in terms of the antipodes, seems to have been of some concern for the Greeks, who were as a rule strongly inclined to allow an interaction between the conceptual domains of eschatology and solar movement.

12.3. Hades at the celestial 'antipodes'

Another closely associated concept is the notion that the world of the dead is located in the lower (unseen) hemisphere of the *sky*. This concept is naturally also concerned and interacts with the solar movement. Its earliest attestation is found in the Platonic *Axiochus*, where the heavenly gods obtained the upper hemisphere of the spherical sky for themselves, while the lower was given to 'those below', where the souls after death depart to Pluto's realm, an underground *adêlos topos*.[32] Similarly, 'the Chaldeans' apparently held that the invisible half of the (thirty or) thirty-six stars (twelve zodiacal, twelve located to the north of zodiac and twelve to the south) were adjacent to the dead, while the visible ones were assigned to the world of the living.[33] This account, although not completely clear, suggests that the invisible half of the heavens represents the world of the dead; furthermore, since the origin of this concept is apparently Mesopotamian, it is best not taken as a testimony on a Greek belief. But for Manilius, as for the Platonic author, the Tartarus either commences at the western intersection of the zodiac with the horizon and extends to the opposite intersection or is reduced to the 'lowest sky' at the 'lower-hemisphere' intersection of the meridian with the zodiac.[34]

For the Neoplatonists a similar concept was symbolised by the goddess Persephone, but here it stands in relation to the annual movement of the sun. It is thus claimed that Pluto, the sun that passes under the earth through the lower hemisphere towards the winter tropic, and whose helmet is the symbol of the unseen pole, abducted Kore (i.e., the seeds), who is thus taken to abide under the earth.[35] Here the celestial lower hemisphere represents the six signs of the zodiac south of the celestial equator through which the sun passes during the winter half of the year and Proserpina is indeed the goddess of the lower (southern) terrestrial hemisphere of the antipodes.[36]

Thus, there appear to have existed two complementary but distinct ideas concerning the localisation of the world of the dead in the lower celestial hemisphere in relation to the solar movement: when it was conceptualised with respect to the diurnal, 'lower' hemisphere represented, roughly, the longitudinal or lateral unseen half of the sky; but when the determinative factor was the annual solar movement, it represented the southern half with the celestial equator as the dividing line. The former is certainly the earlier and corresponds to the dominant localisation of the world of the dead upon the surface of the earth's lower hemisphere defined with respect to the diurnal solar movement.

12.4. The antipodal world of the dead in non-Greek traditions

Once again (see chs. 3 and 7), a concept that appears in a Greek ethnographic context seems also to have been present in the traditions of other civilisations in some contact with the Aegaean, thus representing a shared cross-cultural explanatory strategy, here applied in order to account for the diurnal solar movement and/or the posthumous fate of humans. Since the nocturnal section

[27] Serv. *Aen.* 6.532, *Georg.* 1.243 (cf. *Aen.* 6.127, where the notion is discarded); also *Myth. Vat.* 3.6.25, p. 189 Bode; Moretti 1994: 253–54, 256.

[28] *Apolog.* 47.13.

[29] Claud. *De rap. Pros.* 2.151–203, esp. 156–58, 167–73, 186–87, 193–98.

[30] Claud. *De rap. Pros.* 2.284–93.

[31] Claud. *De rap. Pros.* 2.282–84 (cf. 2.169, 194 and Verg. *Aen.* 6.641). Gee 2020: 89, 94–95, 244 recognizes in the description of Proserpina's tapestry from book one a collapsing of the spherical and stratified cosmos, precisely on account of the position of Hades.

[32] *Axioch.* 371ab.

[33] D. S. 2.30.6–7, 31.4 with van der Waerden 1949: 22–23 and Horowitz 2014: 5, 119–20, 163.

[34] *Astron.* 2.793–95, 798–800 with Housman 1912: ii.xxvi–xxviii. Cf. the imagery in *Astron.* 2.864–70, 948–58 with Housman 1912: ii.xxix–xxxi (the *Ditis ianua* at the western intersection of the zodiac with the horizon controlling the bars of death and *porta laboris* related to both intersections of the zodiac with the horizon). See also Firm. *Math.* 2.17, 19.3, 9, 20.2.

[35] Porph. *Imag.* fr. 7 Bidez *ap.* Eus. *PE* 3.11.109cd; cf. Lyd. *De mens.* 4.137.

[36] Macr. *Sat.* 1.21.1–3; cf. *VM* 3.11.17, pp. 238–39 Bode (Remigius).

of the diurnal solar movement, however elusive, remained much easier to explain in comparison with the posthumous human fate, and since solar motion appears to have been an alluring analogue for this fate, it was apparently used as an explanans for the latter, i.e., the afterlife was explicated in terms of the solar movement. This seems to have been the primary reason that encouraged the widespread interactions between these two conceptual domains.

The notion of the sun illuminating the underworld is indeed present in non-Greek traditions, but the ramifications of such claims—i.e., a precise cosmological localisation of the world of the dead it suggests and allows—are hardly ever brought to the fore. It seems that such cosmological explications were of little concern for those who worked within these traditions and concerned themselves with exploring the analogy between the solar movement and the posthumous fate of humans. The reasons for this phenomenon are most likely the same as those postulated for the similar occurrence observable in Greek tradition, i.e., the nature of our sources.

The best documented case is once again the Egyptian; it hardly needs referencing at all.[37] In Mesopotamia the sun god Utu was believed to bring light into the darkness of the netherworld.[38] Moreover, he is said to have 'turned the dark palace into day', suggesting the reversal of day and night in comparison to the world above.[39] In Ugaritic tradition, the sun goddess Shapsh apparently travels through the underworld at night, where she rules the deified ancestors and gods, while her companions are the divinities and the dead—all these groups apparently represent the deceased.[40] In Vedic tradition the sun is found in Varuṇa's netherworld abode,[41] from where it (and Dawn) actually rises.[42] Finally, it seems that this particular cross-cultural explanatory strategy was also present in Irish tradition, with the sun travelling at night 'on the under-side of the earth', where it shines upon various otherworldly localities;[43] it is likely that this particular tradition was derived from classical sources, perhaps in combination with some local beliefs in an underground world of spirits.

Thus, a conceptualisation of the antipodean nature of the otherworld is also attested for other mythic systems besides the Greek, even though it is as a rule hard to argue for a truly antipodean localisation, i.e., one placed on the underside of either a flat or spherical earth, but rather only for a location defined by the nocturnal part of the sun's voyage. Consequently, the concept as attested in Greek tradition seems to be the most developed one in the sense of the one most concerned with ramifications of the idea of the sun illuminating the world of the dead on the structure of the cosmos—or at least the one most completely preserved in this respect.

12.5. Navigating to the Otherworld in Greek and non-Greek traditions[44]

The above discussion suggests that the idea of the antipodes was apparently introduced as—or was incorporated in—an elaboration of a mythic model accounting for the diurnal solar voyage, especially its nocturnal section. This model—or models—understood and treated the apparently associated question of the location of the world of the dead, a question from a conceptual domain that regularly interacted with the solar movement, as an(other) explanandum. This mythic nexus was in its turn from the earliest period intimately connected with the idea of sailing over the circumambient Ocean (already mentioned at the beginning of this chapter), and its natural extension was the question whether it is possible to actually sail to either the world of the dead or the antipodes—or both, in cases when the two were identified. It appears that this is another cross-cultural tradition, most probably intimately related to the equally shared model of the sun(-god)-in-a-boat, used in order to account for the diurnal solar movement, with many instantiations of this model containing an eschatological element (see chapter 9).

Even though Odysseus complains how no one ever came to Hades by a ship,[45] it is clear that the poet of the *Odyssey* believed this was indeed possible. This is clear since Odysseus himself was soon to perform this feat, which simply had not been achieved until that point. Hades in the *Odyssey* is located beyond the Ocean and the hero has to

[37] Egyptian tradition will be discussed *in extenso* below in Appendix 2.

[38] Jacobsen 1976: 134; Heimpel 1986: 146. On this subject see a more extensive discussion in Appendix 1. Utu/Šamaš's visit to the Underworld in order to bring Enkidu back to the world of the living (*Bilgames and the Netherworld* 238–43, George 2003a: ii.758–59, 773–74, 2003b: 187; *EG* 12.79–87, George 2003a: i.732–33), while certainly showing the sun god's ability to visit the Underworld, does not seem to imply his regular appearance in this realm. The same could be said of Helios' threat to descend to Hades and shine upon the dead (*Od.* 12.383); it shows his ability to descend, but at the same time implies it is unusual for him to do so.

[39] Heimpel 1986: 148.

[40] *CAT* 1.161.18–26 (Levine & de Tarragon 1984: 649, 657; Smith in Parker 1997: 175 n. 202); *KTU* 1.6 vi.45–49 (Gaster 1977: 124, 215, 227, 230; Levine & de Tarragon 1984: 657; Smith in Parker 1997: 163–64, 175 n. 202–204); her journey might additionally be associated with the sea (*KTU* 1.6 vi.51–53; Smith in Parker 1997: 175–76 n. 206), which could be further tentatively associated with Re's subterranean nocturnal journey (Wyatt 2005[1987]: 20–21, 51 n. 1); cf. Parker in Parker 1997: 4; Marinatos & Wyatt 2011: 397. For the rejection of Shapsh's connection with the netherworld in both sources see the convincing arguments in Schmidt 1994: 84–88; Marinatos & Wyatt 2011: 397–99 do not believe that Shapsh visits the underworld in *KTU* 1.6 (*pace* Wyatt 2005[1987]: 19–21, 34 n. 8), but argue that she does so in *CAT* 1.161.

[41] *R̥V* 7.88.2 (Kuiper 1983: 70–71, 74, 78–80, 83, 89; Seaford 2020: 251). Varuṇa is the lord of the (primeval/subterranean) waters/(western) ocean/netherworld, as well as death (Kuiper 1983: 16–17, 66–70, 96–97, 183).

[42] *R̥V* 1.48.7, 164.47, cf. 92.3, 113.12, 4.40.5 (Agni), 51.8, 5.45.1–3, 7.76.2; *Aitareya Brāhmana* 7.28.13 (Keith 1920: 342–43) (Agni) (Kuiper 1983: 71–72, 79–80, 83, 224–25).

[43] *In Tenga Bithnua* 64–66 (Carey 2000[1989]: 137, 139; 1994: 14–15, 28–29; 2009: 174–83, 338–53); cf. both Latin and vernacular question-and-answer texts from the *L'Enfant sage* and similar traditions cited in Cross & Hill 1982: 131–32 and Carey 1994: 20 and the Old English *Adrian and Ritheus* 6 (Cross & Hill 1982: 131, 133–34; Carey 1994: 19; Haycock 1997: 40 n. 112; 2007: 67).

[44] An earlier version of parts of this section appeared in Bilić 2019: 44–45.

[45] *Od.* 10.502.

cross it (*peraô*) in order to arrive there.[46] The naturalness of the idea of sailing in or to Hades is expressed elsewhere in Greek literature,[47] but the notion is on the whole rather poorly attested in Greek tradition.[48] But an explicit parallel to this Homeric episode appears in the *Epic of Gilgameš*, as another occurrence of a shared cross-cultural concept involving the solar movement, although this time perhaps exemplifying a more direct connection between these two traditions.[49] Here the king of Uruk arrives at the shore of the Ocean, and is warned by Šiduri that nobody ever crossed the Ocean.[50] There is, however, an exception to this rule, as we have seen in chapter 9: Šamaš the sun-god can indeed cross it,[51] and thus can be found 'at the other side of the sea'.[52] However, even if Gilgameš manages to cross the Ocean, he would not succeed in crossing the Waters of Death.[53] Both these Waters and Utnapištim's abode beyond them have clear eschatological resonances, although they do not correspond perfectly to the Hades of the *Odyssey*. At the same time, Gilgameš succeeded to descend into Apsu,[54] although—once again—it is only Šamaš who descends into Apsu.[55] In the Early Dynastic account referred to above these two feats of the sun god are combined: Šamaš at the same time goes 'to the other side of the sea'[56] and to Ea's Apsu.[57] In both cases he is '(riding) on the ÉREN + X',[58] suggesting that Ea's Apsu—which is a much better, although far from perfect, parallel for Greek Hades—is located 'at the other side of the sea'.[59]

Thus, both Odysseus and Gilgameš arrived to the Otherworld, which was until their exploits a feat reserved for the sun-god (this is explicit only in the Mesopotamian account); both accounts also combine a horizontal voyage with a vertical descent, as argued in chapter 9. In their exploits they must have followed the sun's diurnal path, as we have argued Odysseus has done in chapter 9 and as we will argue Gilgameš has also done in chapter 13. This mythic nexus seems to combine an interest in the diurnal path of the sun upon the circumambient Ocean with one in the position of the world of the dead in relation to this path and the cosmological stream encircling the *oikoumenê*, both always with respect to a flat earth and thus with little specific interest in antipodean regions.

Within the cosmological framework of a spherical earth, however, the question of accessibility of the antipodes and the otherworld by navigation was bound to be somewhat rephrased, perhaps with a certain shift in emphasis, but still remaining current. Servius claimed that this navigation was indeed possible;[60] he further reported the opinion of 'chorographers and geometers' who had located the infernal regions (*inferos*) to lie 'under [the spherical] earth' (*sub terra*), referring to the 'other' antipodean hemisphere.[61] Incidentally, it seems that this interpretation correctly accounts for the lemma it sought to explain, i.e., the possibility to reach Hades by sea, since, as demonstrated above, Vergil indeed located Hades in the antipodean hemisphere, at least theoretically accessible by sea.[62] This particular mythic tradition was very much alive in the medieval period, surely in part deriving from Vergil's influence, but in line with earlier treatments of the subject; I cannot follow this interesting subject here any further.[63]

Maritime accessibility of the world of the dead, without explicit antipodean characteristics other than the fact that it is in some way associated to or identified with the nocturnal path of the sun, which would place it at the underside of the earth (irrespective of its perceived shape), is attested in other traditions besides the Greek and Mesopotamian. This phenomenon strongly suggests that this particular cross-cultural concept, appearing in various mythic models accounting for the diurnal solar movement and/or the posthumous fate of humans, had some non-negligible hermeneutical potential with respect to both these subjects or conceptual domains.[64]

---

[46] *Od.* 10.508, 11.158–59, cf. 639 and 24.11–13 (*para*). Hesychius (ω 108) explains the term *Ôkeanoio poron* as eschatological '*aêr*', which can certainly be associated with this concept of a 'trans-Oceanian' Hades (*poros* here means a 'passage') (cf. Betegh 2004: 198–99).

[47] In: A. R. 4.1699. To: Eur. *Her.* 426–27, *Hipp.* 139–40 (Vermuele 1979: 179).

[48] The Phaeacians as the ferrymen of the dead: Welcker 1832, followed (with qualifications) by Cook 1992: 239–41, 244–45, 250–52, 266–67; Griffith 2001: 222, 232 (with earlier literature in n. 60 on p. 222); West 2014: 130. The account of conveying the souls of the dead in boats to the island of Brittia: Procop. *Bell.* 8.20.53–57; cf. Tzetz. (Σ) Lycoph. 1200, ii.346 Scheer, *ad* Hes. *Op.* 169bis.

[49] For the influence from Mesopotamian ideas on Greece see the introductory chapter.

[50] *EG* 10.79–80 (George 2003b: 78; 2003a: i.683); cf. Heidel 1949:74; Speiser in *ANET*[3] p. 91; Kovacs 1989: 86; Foster 2001: 76; also, the OB version VA + BM iii.28 (George 2003a: 125 (iii.27); 2003b: i.279).

[51] *EG* 10.81–82 (George 2003b: 78; 2003a: i.683); cf. Heidel 1949:74; Speiser in *ANET*[3] p.91; Kovacs 1989: 86; Foster 2001: 76.

[52] *ARET* 5, 6: vii.1b–4 = *OIP* 99 326 (+) 342 iv.13–16 (Lambert 1989: 19–20, 33; Krebernik 1992: 83).

[53] *EG* 10.83–86 (George 2003b: 78; 2003a: i.683); cf. Heidel 1949:74; Speiser in *ANET*[3] p.91; Kovacs 1989: 86; Foster 2001: 76.

[54] *EG* 10.290 (George 2003b: 98; 2003a: i.723); cf. Heidel 1949: 91; Speiser in *ANET* p. 96; Kovacs 1989: 106; Foster 2001: 94. See also *EG* 1.1 (George 2003b: 1; 2003a: i.539), for which see Silva Castillo 1998: 219–21; George 2003a: i.444, naqbu = Apsu; cf. Jacobsen 1976: 111, Naqbu a name of Ea according to *An* = *Anum*, tab. II, *CT* XXIV.14.

[55] *BWL* 128: 58 (Horowitz 1998a: 343).

[56] *ARET* 5, 6: vii.1b–4 = *OIP* 99 326 (+) 342 iv.13–16.

[57] *ARET* 5, 6: xii.3–4.

[58] Krebernik 1992: 83, 85; cf. Steinkeller 1992: 266. In Lambert's translation (1989: 19–20, 33) the parallelism between these two accounts disappears.

[59] Lambert & Millard 1969: 12 argue that Atra-ḫasīs implied he is building a boat in order to reach Enki's realm (*Atra-ḫasīs* III.i.42–49, Lambert & Millard 1969: 90–91; Foster's (2005: 248) translation of line 49 is more explicit). In addition, Enki seems to have invited him to Apsu (*Atra-ḫasīs* III.i.34–35, Lambert & Millard 1969: 88–89; in Foster's translation (2005: 247–48) this invitation is lacking). These allusions can indeed be understood as referring to the accessibility of Apsu by sea (cf. the Late Babylonian version, V.79, Foster 2005: 264, for an oversea Apsu).

[60] Serv. *Aen.* 6.532 (cf. *VM* 3.6.25, p. 189, where Servius' account is repeated together with an explanation of Caeneus' change of sex as an overseas metempsychosis). For the sea separating us, the 'upper ones' (*superi*) from the antipodes, the 'lower ones' with respect to us (*inferi*), see Serv. *Aen.* 8.671.

[61] Serv. *Aen.* 6.532 (cf. *Myth. Vat.* 3.6.25, p. 189); Moretti 1994: 253–54.

[62] Cf. Serv. *Georg.* 1.243 and *Aen.* 6.532 (cf. *VM* 3.6.25, p. 189); also Serv. *Aen.* 6.127, where Servius explicitly rejects the notion of 'infernal antipodes'.

[63] I have treated the question in full in Bilić 2019.

[64] Unfortunately, in the Sumerian text *Ningišzida's Boat-Ride to Hades*, it is not specified whether the journey to the netherworld by a boat is conducted over the sea or by a river or some other watercourse (Jacobsen

The clearest example of maritime accessibility of the world of the dead is provided by the Egyptian eschatological tradition. Here a boat is described which brings the dead to the otherworldly Duat (*d3.t*, i.e., *'Imntt*-Amenta, 'the west'), Field of Offerings (*sḫt ḥtpw*) or Field of Reeds (*sḫt i͗3rw*).[65] It is reached by the celestial Winding Waterway (*mr nḫ3*),[66] and its pilot is one Mahaf (*M3-ḫ3.f*).[67] This path and its destinations can in this way be understood as celestial,[68] but the sailing can also be interpreted as a voyage over underground waters, with the use of the solar boat, terminating in underground destinations.[69] In any case, the ships of the dead were thought to follow the nocturnal path of the sun, while the solar boats of Re, certainly associated with the boats of the dead, were imagined travelling upon celestial waters.[70] This is the clearest exposition of the doctrine of an otherworld reached by a boat, and its cosmic setting and solar associations both indicate and support the similar nature of other comparable concepts, including the classical one. The apparent localisation of the sun's nocturnal path upon the underside of the earth or upon the lower celestial hemisphere—or perhaps along the rim of the earth upon 'the Ocean', a concept poorly attested in Egyptian tradition—suggests its antipodal setting, both in terms of the sun's presence and in terms of the actual physical position of the region it traverses during the night. This assumption, however, remains implicit in the Egyptian texts.

Another example of a similar concept was recognised in the description of the Indic solar deity Pūṣan, who was imaged as a ferryman, which was associated with the concept of the journey of the deceased from sunset to sunrise following the nocturnal course of the sun. In this way it was indicated that the journey is conducted over a watery surface, which is further supported by the fact that Yama, the first person to experience death, reached his otherworldly abode by way of water-courses. In short, both the sun and the dead follow a watery course and travel in a boat.[71] However, there are some problems with this interpretation. First, Pūṣan is associated with boats on the slightest of evidence.[72] Nevertheless, he is described as travelling in golden boats over a celestial ocean, reaching the solar deity Sūrya. Furthermore, another solar deity, Savitr̥, served as a psychopomp to Yama's realm.[73] Thus Pūṣan in his golden boats could be imagined as carrying the deceased to Yama, although the evidence for this is far from solid. However, Nagy adduces some stronger evidence for the solar voyage in a boat in association with the posthumous fate of humans: the forward course of Savitr̥'s chariot, his daily voyage from east to west, is called 'downstream' (*pravát-*), while his backwards course, his nightly voyage from west to east, is called 'upstream' (*udvát-*).[74] Here the diurnal solar voyage seems to be imagined as taking place on a waterway, even though the solar vehicle is explicitly a chariot. This represents a typical paradigmatic shift often encountered in various traditions specifically in the context of the diurnal solar movement. Furthermore, Yama is said to have travelled 'downstream' (with the sun at sunrise),[75] which led Nagy to propose that before sunrise Yama also traversed the nightly path of the sun, i.e., the 'upstream' way.[76] Whether Nagy's conjecture is right or not, Yama certainly travelled 'downstream' at some point, thus in a way he followed a watery path. Finally, Nagy interprets *R̥V* 1.113.16 as the dead being resurrected at dawn and concludes that by night they travel by an underworld solar path from west to east guided by a solar psychopomp and rise with the sun on the 'downstream' path, following Yama, to the realm of the *pitr̥*.[77]

The following conclusions can be drawn from these considerations of Indian texts: the solar voyage in India can be occasionally associated with traversing a watery surface (Pūṣan's boats, 'downstream' and 'upstream' paths); solar deity was believed to serve as a psychopomp (Savitr̥; Pūṣan); Yama, the ruler of the realm of the ancestors (*pitr̥*), was also said to have travelled over a watery surface, perhaps with the sun, but certainly following the solar path; the deceased, Nagy conjectures, also travelled over the underworld watery solar path. Thus, a conjectural, although certainly not implausible, conclusion would be that Yama's realm, the abode of the deceased ancestors, could be reached (exclusively?) by a (solar) boat; any antipodal characteristics of the world of the dead remain implicit in its association with the nocturnal section of the sun's daily course.

12.6. Conclusion

The testimonies examined in this chapter clearly indicate a widespread cross-cultural tradition—also appearing in the Greek ethnographic context—of the world of the dead accessible by a solar boat or, more generally, regularly visited by the sun(-god). This world is typified by a reverse daylight/night exchange and was also accessible to a few selected mortal heroes, such as Odysseus, Gilgameš and Yama, or, alternatively, to the blessed dead of mystic (?) eschatologies. The tradition represents a series of mythic models accounting for the diurnal solar movement characterised by a strong eschatological component and a salient interaction between the conceptual domains of solar movement and eschatology. The Egyptian

& Alster 2000: 318–25; cf. Jacobsen 1976: 67–68; Alster 1986: 27). For a tentative interpretation of the Minoan iconographical record as evidence for an afterlife journey by a boat following the sun's nocturnal path see Goodison 1989: 38, 48, 92–94, 96, 100, 144–48, 151, 157.

65 Griffith 2001: 214–16, 218, 221, 234.

66 Griffith 2001: 217.

67 Griffith 2001: 222–24.

68 Cf. Griffith 2001: 215, 217–18, 225–26.

69 Burgess 1999: 173.

70 Cook 1992: 255–56.

71 Nagy 1990: 97–101, 117–18, 255; Cook 1992: 257–58.

72 *R̥V* 6.58.3 (acknowledged by Nagy 1990: 255).

73 *R̥V* 10.17.4; Nagy 1990: 97–98, 101. Savitr̥ is Pūṣan's correlate solar psychopomp (*R̥V* 10.17.5; Nagy 1990: 93, 97, 118).

74 *R̥V* 1.35.3; cf. 5.31.1, where Indra drives the solar chariot (5.31.11) also downstream (Nagy 1990: 95–96).

75 *R̥V* 10.14.1.

76 Nagy 1990: 97.

77 Nagy 1990: 117–18; for different interpretation see for example Kuiper 1983: 164.

tradition seems to have placed the strongest emphasis on the eschatological component of the model, but other traditions (Greek, Mesopotamian, Indian) also show a clear interest in this conceptual domain. The presence of eschatological considerations is best explained as an attempt to speculate upon the afterlife fate of humans by appealing to an analogy with the sun's daily course, its daytime but especially nocturnal section. The sun was anthropomorphised, and its daily and annual courses were described in and explained by myths, and between this anthropomorphisation and its use as an analogy for eschatological speculation only a short step remained, which was apparently often made. But it is a second-order analogy: it was first necessary to account for the diurnal solar movement in terms of anthropomorphic deities travelling in a boat or chariot, which subsequently allowed the eschatological component to be developed. It seems that this analogy was exploited by a not insignificant number of traditions, which testifies in practice to its appeal and hermeneutic (or at least heuristic) potential.

The localisation of the world of the dead in relation to solar movement was naturally understood in cosmological terms, i.e., it was imagined as a tangible place and part of the physical world. A significant corollary of the use of argument from analogy with the solar movement was the recognition of the existence of an antipodal region underneath the earth (irrespective of its shape) characterised by the nocturnal presence of the sun(-god). This region became the realm of the dead, accessible through solar gates in the furthest west (in Greek tradition exemplified at *Od.* 24.12, and as Erytheia and Pylos—for which see chapter 10) or at some other point in the diurnal course of the sun. Less often it was rather the annual solar movement that was associated with the localisation of the world of the dead, which is unsurprising in light of the interchangeability of motifs used in the models accounting for both the diurnal and annual solar movement. A similar second-order analogy was at work here: just as the anthropomorphic solar god lives continually and resurrects annually, after his winter departure for the far south, so presumably do the humans.

The particular corollary in the form of the existence of antipodal regions underneath the earth was often passed over, not being of much interest to those who exploited the analogy between the solar movement and human life and death—apparently the only partial exception being the Greek tradition—but its existence should nevertheless be noted. This treatment—or its lack—of an obvious ramification of a model combining an account of the diurnal solar movement with the existence of a world of the dead along the nocturnal section of the sun's daily course shows that the model was created in order to account for the phenomena of major importance for its creators (the afterlife fate of humans, the sun's daily course). These phenomena were emphasised, but other aspects of the model deemed less relevant or peripheral were played down (such as the existence of antipodal regions). Later the emphasis somewhat shifted—especially with the change of the underlying cosmological framework, i.e., the recognition and the acceptance of a spherical earth—towards an often-implicit discussion of the 'meteorological' characteristics of the realm of Antipodes, especially in terms of solar movement, even though the relevance of the subject always remained modest.

It was in terms of solar movement that many traditions conceptualised their ideas of the posthumous fate of humans. The interaction of these two conceptual domains was commented upon repeatedly throughout this book and deserves a strong emphasis. The myths devised as the results of such considerations certainly belong to a class of myths referencing solar movement, and antipodal regions are an interesting, if not always explicated, feature of some of these myths, features that resonated in later geography until the modern period.

13

Beyond Odysseus: Gilgameš

On a number of occasions throughout this study I have referred to non-Greek traditions in adducing parallels complementary to Hellenic accounts. This was especially the case in the preceding chapter, where a number of shared cross-cultural models from various ethnographic contexts were evoked. Most often the non-Greek tradition referred to was the Mesopotamian, which is unsurprising due to the well-attested and enduring contacts and interactions between these two civilisations.[1] The universal nature of the phenomena whose treatment in myths is studied here, the diurnal and annual solar movement, together with a modest number of appropriate motifs and action patterns out of which the mythic models accounting for either of these could be built, once that the widely attested process of anthropomorphisation—representing an effective and coherent cognitive strategy—of the sun was accomplished, had as its predictable consequence the appearance of similar models accounting for the solar movement in different traditions. This claim does not exclude the viable possibility of influences exercised by some traditions upon others or even possible genetic connections between various identical or similar models. I am consciously not committing to either of these two possibilities since I believe that both are equally possible in principle and an informed choice between them should be made on a case-by-case basis.

In this and thc following chapter I will focus on two mythic characters (with no regard for their historical existence and exploits), Gilgameš and Alexander, more precisely, on certain segments of their itineraries that can plausibly be interpreted in terms of either the diurnal or annual solar voyage.[2] 'Mythic characters' refers to their involvement in literary renditions of mythic models referencing the solar movement—that they actually existed as physical persons, with all ramifications of this fact on the development of the tradition that grew around them (including Alexander's actual itinerary), although fascinating subject in itself, is of little or no relevance for my approach. Furthermore, I am not interested in the positively attested non-solar mythic developments of their characters, even though this is in itself another interesting subject of study. Their 'solar' itineraries can be conceptualised as derived from mythic models accounting for the solar movement in anthropomorphic terms—a necessary condition for any further consideration, but one amply testified across various traditions, including the Greek and Mesopotamian—and the recognition and acceptance of the possibility that certain exceptional persons other than the sun-god can follow the course of the deity on its path. Already we have encountered the Argonauts, Odysseus and Heracles, as well as Gilgameš himself, in this capacity (see esp. chapters 8, 11–12), and recognised in this practice a shared cross-cultural phenomenon based upon the existence of equally shared cross-cultural mythic models accounting for the solar movement (without speculating on genetic connections between these phenomena or models). Now it is time to elaborate upon the appropriate section of Gilgameš's voyage and to introduce his 'heir' Alexander (on this claim see the following chapter) in his capacity of a traveller upon the sun's path.

Far from claiming that either Gilgameš or Alexander was a 'solar hero' (I take it to mean that they were personifications of the sun and that their voyages reflected the sun's, with every episode interpreted in terms of solar phenomena), I will continue using the criterion applied throughout this essay by studying the evidence that is unequivocal: the hero explicitly following the route of the sun, visiting solar gates (alternatively, places or mountains of sunset or sunrise), or the land characterised primarily by the diurnal or annual absence of the sun, without venturing into any type of strained interpretations. I cannot think of a more rigorous methodology in the treatment of this type of testimonies. The alternative would be to ignore them, since the fact that some, many or all of them are later accretions onto stories that originally lacked them still begs the question of interpreting their presence, which is a fact of crucial importance. Once again, a paraphrase of Parker's remark noted in the introduction seems pertinent: it is not the apparent marginality, however defined, of the practice that matters, it is its very existence that needs to be accounted for.

At the same time, I do not wish to reduce the character of Gilgameš to that of a 'solar hero'. I readily acknowledge that even in the episodes upon which I will focus, i.e., the hero's voyage upon the sun's path and its continuation to Utnapištim's abode and the Apsu, many elements

1 For the influence of Mesopotamian ideas on Greece see the introductory chapter.

2 It appears that the *Epic of Gilgameš* as we know it—a work essentially probably of the Old Babylonian period (Tigay 1982: 11–12, 20–21, 30, 42–47, 54, 139, 242–43)—incorporated in its narrative various originally disconnected themes, episodes or stories. One of these, Gilgameš's journey to Utnapištim, it was argued by Jastrow & Clay (1920: 49), was a 'nature' or 'solar myth' in which the hero represented the sun, more precisely, its annual movement during the winter season (Tigay 1982: 18, 37, noting that this particular suggestion was subsequently neither substantiated nor disproved). Naturally, Jastrow & Clay's interpretation is tainted by the hermeneutic approach characteristic of the earlier scholarship it reflects. Nevertheless, this does not automatically mean that the explicit solar references in Gilgameš's itinerary should be discarded together with this antiquated interpretative strategy. Similarly, Davies 1992: 223–24: 'The idea that Heracles was ever a "solar hero" has long since gone out of fashion, but his use of the Sun's bowl to get to Geryon's island in the West shows that in one specific and defined instance he is capable of taking the Sun's place'. For Heracles' use of the sun's sup see chapter 9.

from different conceptual domains are present and that other hermeneutical approaches can throw some light on the elucidation of these episodes. But I submit that the approach advocated in this study is the most appropriate method for interpreting the elements that clearly belong to the conceptual domain of solar movement.

13.1. Gilgameš breaking a path for Odysseus

The already discussed section of Odysseus' voyage from the Laestrygonians to Aeaea to Hades and back to Aeaea, which was interpreted in cosmological terms, more precisely, in terms of solar movement, represents a convenient background against which we can study the corresponding Mesopotamian material. I must once again emphasise that this interpretation is firmly based on the text itself, with its repeated emphasis on explicit solar features, such as the island of sunrise, Circe's solar ancestry, the lack or excess of sunlight (the Cimmerians and Laestrygonians, respectively), to which we may add the gates of the sun from a similar context. Moreover, the features that can be inferred as solar in character, such as the rock associated with the world of the dead and—in a similar context—the daily course of the sun, the town of 'distant (probably solar) gates' and the hero following the nocturnal section of the daily course of the sun upon the Ocean (together with other complementary cosmological features, such as the circumambient Ocean and the world of the dead), can also be added to this list. The corresponding section of Gilgameš's voyage is also characterised by explicit solar references, which allows its inclusion in a class of myths referencing solar movement. More specifically, this section of Gilgameš's voyage is evidently structured upon the diurnal course of the sun, thus belonging to a small number of myths that exhibit such characteristics. Although this myth belongs to an ethnographic context different from the Greek, its relevance for the latter is widely acknowledged, which is one of the principal reasons for its inclusion in this study.

The section of Odysseus' voyage from the Laestrygonians to Circe can be compared with Gilgameš's travel from mountain Māšu to 'Šiduri the alewife': after visiting the Laestrygonian 'distant gates' where 'the paths of night and day are close together'—which is in terms of mythic geography a location comparable to Māšu, a mountain intimately connected with the horizon solar phenomena—the hero arrives at the island of sunrise, Aeaea.[3] This location also has a close parallel in Mesopotamian tradition outside the *Epic of Gilgameš*, but still in the context of the Gilgameš tradition: the eighth *nagû* of the Babylonian *mappa mundi* is thus described by a verb meaning 'to dawn, to become morning, to rise early' and a noun *handûru*, which may refer to the gate of sunrise in the extreme east.[4] The last word in the line may be restored as *šam-[šá-tu]*, 'sun-[disk]', so the whole translation would be something like 'where the sun-disk dawns at its entrance', clearly referring to an island (or less likely a mountain) of sunrise.[5] On the other hand, *handûru* can be associated with *abul ḫandûri*, which is the name of the western gates of Nineveh, also named ^{d}*šár-ur*$_4$.[6] In its turn, ^{d}*šár-ur*$_4$ is identified with the stars λ or υ Scorpii.[7] Otherwise, in lexical lists *andurû* is equated with *daltu*, 'door';[8] more explicitly, we are informed of the existence of *ha-an-duh* MÁŠ, '*handūhu* of Capricorn', that is, 'gate of Capricorn'.[9] Undoubtedly, the eighth *nagû* is a cosmological location, associated either with sunrise, thus with the diurnal path of the sun, or perhaps with the winter solstice, thus with the annual solar course, or even with certain stars in the Scorpio constellation, and thus with the gates of the Galaxy (Scorpio being a galactic constellation). The lattermost identification suggests the presence of some eschatological resonances in this concept (see below). Since the motifs used in building mythic models of solar movement were freely exploited by both those created to account for the diurnal and for the annual motion of the sun, this apparent ambiguity in the interpretation of the eighth *nagû* most likely represents a manifestation of this phenomenon.

Šiduri's role in the narrative of the *Epic of Gilgameš* corresponds to the role of Circe/Calypso in Homer's epics to such a measure that a direct influence of the Mesopotamian poem on the Greek was actually proposed.[10] In addition, there appears to be some semantic overlap between Calypso's name and Šiduri's description: 'Calypso' is connected with the verb *kaluptô*, suggesting 'darkness' and 'death', and also *kaluptein*, 'to cover'.[11] Her name—'hidden' or 'veiled'—might thus correspond to Šiduri, 'covered with a veil'.[12] One could further compare the unveiling of the Heliades described by Parmenides, which places the detail in a clear solar context.[13]

Their respective roles in the narrative are indeed very similar. Šiduri gives Gilgameš directions to Utnapištim's

[3] West 2005a: 62; cf. West 1997: 406–407, after Germain 1954: 414–17, 421–22; Burgess 1999: 199.

[4] Rev. ll. 24–25. Cf. West 1997: 145. The obverse (l. 10) of the tablet under consideration refers to Utnapištim.

[5] Horowitz 1988: 163–64; 1998a: 39. For *nagû* as islands or mountains see chapter 4. In any case, 'islands' and 'mountains' represent only paradigmatic shifts in the context of accounting for the diurnal solar movement in terms of mythic discourse.

[6] CT 26 32 viii 3; Horowitz 1998a: 39; *CAD* Ḫ 6.79 s.v. *ḫandūru*; cf. Thompson 1940: 90, 13–14.

[7] Gössmann 1950: 208.375; Weidner 1957–71: 77, 80; Reiner & Pingree 1981: 8, 15; Pingree & Walker 1988: 321; Koch-Westenholz 1995: 208; Hunger & Pingree 1999: 108, 110, 275; also, ^{mul}APIN I.ii.31–32 (Hunger & Pingree 1989: 38–39, 138).

[8] Horowitz 1998a: 39.

[9] Rochberg-Halton 1983: 214 with n. 27. On Capricorn as a solar gate in classical (Neoplatonic) tradition see Bilić 2020c: 84–87; it was conceived as such both as a tropical constellation marking the winter solstice and after it was brought into connection—now as a zodiacal *sign*—with the Galaxy, overlapping with the constellation Sagittarius.

[10] West 2005a: 62; cf. 2014: 126–27 with n. 67; 2018: 275–77. For the references to the Šiduri/Calypso connection see Burkert 1992: 200 n. 1; Marinatos 2001: 401 n. 51; Henkelman 2006: 811 n. 12; Kozlowski 2018: 16. On Calypso/Circe overlap see chapter 11.

[11] Frame 1978: 21, 73; Crane 1988: 17, 29 n. 32; Nagy 2007: 72–73.

[12] *EG* 10.4 (George 2003a: 498, 679); West 1997: 410; 2005a: 62; 2014: 127 n. 67; 2018: 277; Marinatos 2001: 397; Kozlowski 2018: 16.

[13] Parm. fr. 1.5.9–10 Coxton = 28B1.9–10 D-K; West 1997: 410.

dwelling-place, while Circe explains to Odysseus the path to Hades (and Tiresias), and Calypso gives him instructions regarding his voyage to Scheria.[14] Šiduri's name is preceded in the Akkadian text by the divine determinative, indicating that she is—just like Circe and Calypso—an immortal.[15] Calypso's island Ogygia is referred to by Homer as 'the navel of the sea',[16] which corresponds fairly well with the description of Šiduri's dwelling by the Ocean's shore.[17] Thus the similarities between these characters are more than convincing. In terms of cosmology, Homer unquestionably placed Circe's Aeaea in the farthest east, since he located the dawn's abode and the risings of the sun there,[18] just as Šiduri's dwelling is near the place where Šamaš rises.

If Šiduri's location parallels that of Circe's Aeaea, and if the land of Laestrygonians can be compared to Māšu, then Utnapištim's abode also must have a parallel in Greek myth. There can be no other candidates than Hades—visited by Odysseus after Aeaea—or Scheria—which he visited after Ogygia. Its precise location, 'the mouth of the rivers', as we shall see, has an exact parallel in 'the meeting place' of the two rivers in Hades, which points to the former as the more probable exact counterpart, even though this does not mean that these two cosmological locations are identical in every aspect.

13.2. The twin mountain

Prior to arriving to Šiduri's dwelling, the king of Uruk in search for Utnapištim arrived to mountain Māšu, 'Twins',[19] which daily observes[20] the rising and setting sun,[21] supports the heavens and whose breast reaches down to the netherworld.[22] This cosmological mountain, whose name suggests it is two-peaked, in its capacity as 'sun-gates' daily monitors the risings and settings of the sun, representing the westernmost and/or easternmost point of the habitable earth in the Sumerian/Akkadian cosmology.[23] It is also depicted on numerous Akkadian cylinder seals on which Šamaš rises behind or out of a double mountain, a mountain with two prominent summits with a deep breach between them. Occasionally he is represented thrusting upwards by propping himself with his hand(s) placed upon the two mountain tops; otherwise, he is depicted, as if climbing, with a raised leg and his foot resting upon one of the summits. He wears a horned crown, designating his divine nature, and usually holds in his hand a serrated knife or a pruning saw, with rays emanating from his shoulders.[24] In addition, a number of texts refer to—whether explicitly or implicitly—the mountains of sunrise and sunset, without committing to their particular shape.[25] A cosmological solar mountain (i.e., a feature that accounts for some phenomena in the solar movement) is thus a salient feature of Mesopotamian models accounting for the solar movement. It was conceptualised as a two-peaked elevation, and its role in the *Epic of Gilgameš* is unequivocal, even if its involvement in the solar movement is somewhat succinctly presented.

A similar concept of a double mountain, although unrelated to the sun according to the present evidence, existed in Ugaritic tradition. Thus, in *CAT* 1.4.viii.1–9 two or twin mountains represent the boundary of the earth/underworld,[26] while in the *Epic of Kirta* Baal's mountains are Saphon and Nani.[27] The liminal position of this double mountain with respect to the underworld suggests its involvement in a model that localised the latter's position in terms of the diurnal solar movement, even if such a model is not explicitly evoked. The iconographic motif of a Storm- or Weather-God (Ba'al Zaphon/Saphon, Tešub) standing or striding upon two mountains or mountain-tops on MBA (and later) Anatolian and North Syrian seals, as well as at Yazılıkaya and in various Hittite iconographic (and textual) sources is another testimony for this tradition.[28] Still, no explicit connection with the sun can be detected here, with only the association with Ugaritic tradition providing some tentative connections with the solar movement. On the other hand, a double solar mountain was also present in a variety of Egyptian models accounting for the diurnal solar movement, perhaps as early as the first half of the 4th millennium.[29] As opposed to the Levantine examples, the Egyptian variant is surely not derived from the Mesopotamian and suggests the existence of a shared cross-cultural model accounting for

[14] *EG* 10.78–91 (George 2003a: i.683, 685: 2003b: 78–79) ÷ *Od.* 10.487–515, 1.160–70, 262–77. Cf. Cook 1992: 263 with n. 89 for earlier literature; Dalley 1991: 2; S. Morris 1997: 620; West 1997: 405, 409; Burgess 1999: 175.

[15] West 1997: 405; George 2003a. i.148–49; Kozlowski 2018: 16.

[16] *Od.* 1.50–51.

[17] EG 10.1, 76, 79–82, 85, etc. (George 2003a: i.678–79, 682–83).

[18] *Od.* 12.3–4.

[19] Wiggermann 1992: 180; Burgess 1999: 178 n. 26; George 2003a: i.492, 669, ii.863, 865; Woods 2009: 191, 197; Marinatos & Wyatt 2011: 393.

[20] The word used is *naṣāru*, 'to observe astronomical phenomena' (*CAD* 11 n2 38–39 s.v. *naṣāru* 5; Horowitz 1998a: 98 n. 5). Cf. *EG* 9.45 (George 2003a: i.669).

[21] Heidel 1949: 65, Speiser in *ANET*³ p. 88, Kovacs 1989: 76, Horowitz 1998a: 98 and Foster 2001: 67 (cf. Woods 2009: 196–97) render the texts in such a way that the mountain daily guards both sunrise and sunset, George 2003a: i.669; 2003b: 71 opts for sunrise only, while Heimpel 1986: 140 does not try to reconstruct this part of the text. The intervention in the text is justified by the fact that the Māšu-inhabiting Scorpion-men guard its gate and observe the sun at both rising and setting (see immediately below).

[22] *EG* 9.39–41 (George 2003a: i.669; 2003b: 71; cf. Heidel 1949: 65; Speiser in *ANET*³ p. 88; Heimpel 1986: 140; Kovacs 1989: 76; Horowitz 1998a: 98; Foster 2001: 67).

[23] Wensinck 1921: 1–2 (who insists on a westerly localisation).

[24] Frankfort 1934: 20; 1939: 98–100, 105–108 with Pl. XVIIIa, c, g, XIXa; Van Buren 1955: 1–14; Böhmer 1965: 71–76 with Taf. XXXII.377, XXXIII.392–XXXVI.439, XXXVIII.464–65; Black & Green 2000: 184 with Fig. 152 on p. 183; Woods 2004: 55, 57, Figs. 21–24 on p. 58; Felli 2006: 36–39, 48–50, Figs. 75–79 on pp. 187–89. For a Kassite variant of this type of seal found in Thebes see Müller-Karpe 1980: iv.2 p. 782, iv.3 Taf. 254C; Porada 1981–82: 45 Pl. 3, 49–51, no. 26; Ornan 2005: 231 Fig. 18.

[25] See Appendix 1.

[26] Gaster 1977: 199; Wyatt 2005[1987]: 46; Smith in Parker 1997: 138; Marinatos & Wyatt 2011: 393

[27] *KTU* I.16.i.6–9, ii.44–47 (Dijkstra 1991: 137 for the plural form; Greenstein in Parker 1997: 31, 35 has sing. 'Baal's mountain').

[28] Dijkstra 1991: 127–31, 136–37, 138 Pl. 1.1–3, 139 Pl. 2.2–3, 140 Pl. 3.1, 3, cf. 2; Klingbeil 1999: 247 n. 258, 248 (with n. 260), 248–49 with. Fig. 80; Marinatos 2000: 6 Figs. 1.8–9; Barrick 2008: 54 with n. 104, 55–57 with Fig. 3, 57–58 with n. 127.

[29] See Appendix 2.

the diurnal solar movement involving a double-peaked mountain(s) of sunrise and sunset.

Returning to the *Epic*, how can the same mountain 'observe' both the rising and setting of the sun? Naturally, the mountain could be imagined as circumscribing the entire rim of the earth and thus 'observing' both sunrise and sunset at different points on the horizon. Its 'twin' aspect would thus be understood as referring to a single peak at each of these points.[30] Otherwise, we could interpret its description in the *Epic* in terms of the uni-polar model accounting for the phenomena of sunrise and sunset, with Māšu located in the cosmic nadir at the point where the sun simultaneously sets and rises. However, iconography and literary evidence, as well as the text of the *Epic of Gilgameš* itself, suggest the existence of two double-peaked Māšu-mountains in Mesopotamian cosmology, one in the farthest east and another in the farthest west.[31] This is compatible with the bi-polar model of the diurnal solar movement accounting for the phenomena of the daylight/night exchange.

In addition, the mountain Māšu had a gate, apparently that through which the sun passes at sunrise/sunset.[32] Indeed, the Mesopotamian tradition was familiar with the concept of gates of the sun through which it passes both at dawn and dusk. Furthermore, the gates through which the sun passes at sunset were closely associated with the underworld, and one can also find testimonies for the existence of the concept of cosmic nadir as a location closely connected with both the nocturnal section of the daily movement of the sun and the otherworld.[33] These phenomena represent another example of the interaction of the conceptual domains of eschatology and solar movement. Similarly, various solar gates are omnipresent in Egyptian tradition, often in an intimate relationship with eschatological themes,[34] and the concept also appears in a number of other ancient traditions, such as Hittite,[35] Ugaritic,[36] Vedic, [37] classical[38] and others.[39]

[30] For a similar case in certain Egyptian interpretations of the *akhet*-sign see Appendix 2.
[31] Cf. George 2003a: i.492–93.
[32] *EG* 9.42, 135 (George 2003a: i.669, 671; 2003b: 71, 73).
[33] For supporting evidence for these claims see Appendix 1.
[34] See Appendix 2.
[35] Güterbock 1958: 240, i.29–31; West 1997: 143.
[36] *KTU* 1.78:34 and 2.1.5 (Schmidt 1994: 88 n. 203; for the former text cf. Hunger & Pingree 1999: 10–11; Smith 2003: 194).
[37] *RV* 1.48.15 (Griffith's and Jamison & Brereton's translation, '*twin/two* doors of heaven' opened by Dawn not in Geldner's translation), 1.113.4, 4.51.2 (West 2007a: 170, 222; cf. Kuiper 1983: 159–60). For 'gates of darkness' opened at dawn cf. *RV* 3.5.1; for Dawn opening the gates of 'closed rocks' (Geldner), 'firm rock' (Kuiper), 'firm-fixed stone' (Jamison & Brereton) or 'firm-set mountain's portals' (Griffith) see *RV* 7.79.4; for the opening of the rock in the sky, Dawn releasing pent-up cows, and the sun opening the portals of men see *RV* 5.45.1 (cf. 3, where 'mountain' is also opened, and 2, where 'enclosure' is mentioned).
[38] See chapter 10.
[39] Persian (*Greater Bundahišn* Vb.1, 3, 9.5 (Anklesaria 1956: 65, 95; MacKenzie 1964: 517; West 1997: 143 n. 186)); Jewish (4Q208–209, the Aramaic *Astronomical Enoch*, with Ratzon 2015: 97 n. 11, 101–104; *1 Enoch* 72.3ff with Neugebauer 1964: 60, also *2 Enoch* 13–14 with Neugebauer 1964: 61; Jer. Talmud *Rosh Hashanah* 2, 58a with Keim 2014: 47 n. 11; *Pirḳê de Rabbi Eliezer* 6 (Friedlander 1916: 37–39); other Jewish sources cited in Friedlander 1916: 37 n. 5 and Ratzon 2015: 104 with n. 33–34); Ethiopic (BN Ethiopic MS 64 with Neugebauer 1964: 50, 52, 54–55, 57).

13.3. Scorpion-men

This gate of Māšu-mountain is guarded (again *naṣāru*, in this context more appropriately translated as 'guard') by Scorpion-men, who observe (yet again *naṣāru*) the rising and setting of the sun.[40] Depending on the model accounting for the phenomenon of daylight/night exchange one accepts as relevant in this context, we can presume they are either present at both postulated Māšu-mountains of sunrise and sunset or there is only a single Māšu-mountain, which accounts for both risings and settings. In any event, Māšu-mountain is here apparently used in a model accounting for the diurnal solar movement. Before proceeding to discuss and interpret it in these terms—and with it, the section of Gilgameš's voyage involving Māšu-mountain, anchored as it is upon this cosmological location—I will assess some potential alternative identifications of 'the Scorpion-men observers' in the context of the cosmic environment in which this section of Gilgameš's voyage apparently takes place. After discussing these alternatives, I will return to the line of interpretation involving the daily motion of the sun.

An alternative explanation of the presence of Scorpion-men both at rising and setting is that which places this concept in the context of the annual solar movement. Even though the narrative of the *Epic of Gilgameš* strongly suggests the context of the diurnal solar movement, the widespread phenomenon of the interchangeability of motifs used in the mythic models accounting for respectively the diurnal and annual solar movement might also be at work here (its currency in Mesopotamian tradition was already noted in chapter 10). If this is taken into account, and the Scorpion-men are understood as an ecliptical constellation (as suggested by their appearance), it was possible for Scorpion-men to be at dawn in the extreme east and at dusk in the extreme west because for about a month the sun rises and sets in or with a particular ecliptical constellation.[41] They would thus travel with the sun and accompany it when its orbit intersects with the horizon. The nature of the celestial landscape discussed in the *Epic* is in this way inferred from the appearance of these creatures, which should predictably be associated with the Scorpio (or perhaps Sagittarius) constellation. Thus, a temporal (seasonal) as well as spatial reference can be recognised in this specific detail. Moreover, it seems to be a crucial reference since at this point in the narrative

[40] *EG* 9.42, 45 (George 2003a: i.669; 2003b: 71). The cosmic nature of the creatures guarding the mountain is further emphasized by their presence in other cosmological accounts: a Scorpion-man, *girtablullû*, is mentioned on the so-called Babylonian *mappa mundi* (obv. l. 5) among the creatures created by Marduk 'on top of the sea' and probably represents one of the creatures that primarily inhabited the *nagû* beyond the ocean (Horowitz 1988: 160; 1998a: 35–36; Lambert 2013: 231–32), while another *girtablullû* is mentioned among the eleven creatures of Tiamat in the *Enuma Elish* 1.142, 2.28, 3.32, 90 (for its presence in related lists of monsters see Lambert 2013: 227–28, texts v–vii, ix–xii).
[41] Huxley 2000: 125, 134.

the 'observances' of solar movement are introduced and repeatedly emphasised (three times in seven lines).

It is possible to be even more precise in determining the particular celestial identity of the Scorpion-men guarding mount Māšu. As already noted, solar gates and their guardians must necessarily be located in the constellations along the ecliptic.[42] M. Huxley suggests that the gates guarded by Scorpion-men are the gates of the summer solstice,[43] but she surprisingly does not associate them with the constellation Scorpio but rather with the constellation that marked the summer solstice during the first millennium BC, Cancer. This is a rather arbitrary assumption, especially since Cancer was not the constellation that marked the summer solstice during the time when the *Gilgameš Epic* was created (the Old Babylonian period),[44] and it is at the same time incompatible with the iconographical evidence. However, it does seem reasonable to connect the Scorpion-men and their gate both with the annual solar movement upon the ecliptic in general and with the extreme points of this movement, the solstices, in particular. At the same time, it seems more plausible to search for their location in the constellation manifestly related to them, i.e., Scorpio or, alternatively, Sagittarius.[45] The latter might be represented by the Scorpion-man archer on the *kudurruu* of Nebuchadrezzar I and Meli-Shipak's *kudurruu*, as well as by the archer with scorpion's tail and a scorpion below its body and front legs on another *kudurruu* of Meli-Shipak.[46] An archer with a scorpion's tail is also represented on the rectangular zodiac of Dendera.[47] Variants of scorpion-men were relatively frequently represented in Mesopotamian iconography.[48] They were considered to be gatekeepers, and they are occasionally depicted on seals as upholding the solar disk of Šamaš.[49] Indeed, from the earliest period they were associated with the sun, both, as we have seen, in the *EG* and on an Old Akkadian seal, where a scorpion-man has rays extending from his body and supports Utu in a conflict.[50]

According to this line of interpretation, Gilgameš appears to have started his voyage towards Šiduri by entering the portal of the sun in the constellation Scorpio (this could also apply in the more likely event that he followed the diurnal path of the sun). In addition, Šiduri herself, together with her Homeric counterpart Calypso (see above), was compared with the goddess *Išḫara tam-tim*,[51] who was in her turn identified with the constellation Scorpio ([mul]GÍR.TAB) and Tiamat.[52] A scorpion, possibly a symbol of the goddess, often appears on the *kudurrus*.[53] This supports the interpretation of this section of Gilgameš's voyage in 'seasonal' terms, with the constellation Scorpio at the same time being present in its capacity of Scorpion-men observers/guardians at the western portal and as Šiduri-Išḫara at its eastern portal. Furthermore, it removes the necessity of postulating a second pair of Scorpion-men (perhaps also a second mount Māšu) in the extreme east, although it does not necessarily make either of these two cosmological features completely redundant (this especially applies to mount Māšu).

As already indicated, it is possible to add another level to this line of interpretation. The constellation Scorpio is located at the crossroads of the Galaxy and the zodiac (this also applies to the neighbouring constellation Sagittarius). Mesopotamian tradition does not attach any eschatological importance to either the constellation or this specific celestial position. On the other hand, the Greek—more precisely, Neoplatonic—tradition placed a great emphasis precisely on the eschatological potential of this celestial location, even though there is some confusion between the constellation/sign Capricorn understood as a solstitial marker and the Galactic constellation Sagittarius that in due time overlapped with this sign.[54] Even if it seems far-fetched to derive conclusions on the character of this celestial position solely on extraneous evidence, the similarity of the mythic models accounting for solar movement in both traditions, including the omnipresent interaction with the conceptual domain of eschatology, as well as the context in which both Šiduri and the Scorpion-men appear in the *Epic of Gilgameš* (on the way to a manifestly otherworldly location), suggests that the fact that Scorpio/Sagittarius

[42] Huxley 2000: 124.

[43] Huxley 2000: 126–28.

[44] Implicitly acknowledged by Huxley 2000: 115.

[45] Cf. Zimmer & Winckler 1903: 580 n. 2; Jensen 1906: 78–79, 92–93, 97–98, 105, 107. In *Enuma anu Enlil* Tablet 51 Text X.20 (Reiner & Pingree 1981: 60–61) the 'Sting of Scorpion' (λυ Scorpii) rises in Kislīm, the month of the winter solstice according to the Old Babylonian tradition (perhaps also in HS 245 obv. 6, rev. 8–9, Horowitz 2014: 228–30) (Horowitz 2014: 237). Thus, in this period the asterism was indeed associated with the solstices.

[46] Hinke 1907: 98–101, 244; King 1912: xv. Nebuchadrezzar I: BM 90858; Hinke 1907: 131 Fig. 49, no. 15; King 1912: 31; Black & Green 2000: 113 Fig. 90; Meli-Shipak (1): BM 90827; Hinke 1907: 98, 244, King 1912: 9; Meli-Shipak (2): BM 90829; Hinke 1907: 98 Fig. 32; King 1912: 19.

[47] *Description* iv.40, Pl. 20.

[48] Green 1985: 75–76 with Pls. VII–X; Black & Green 2000: 161, Fig. 131 on p. 160; Huxley 2000: Figs. 7 and 8 on p. 123, Fig. 13 on p. 126; Ornan 2005: 123–25, 243 Fig. 73, 251 Fig. 97a–b, 268–69 Fig. 161–63, 165.

[49] Frankfort 1939: 201, 210 with Pl. XXXIIIb, e; Huxley 2000: 120, 122 with n. 65, Fig. 9 and 10 on p. 123, 128, 133; Black & Green 2000: 161; Ornan 2005: 124–25, 269 Fig. 162, 165; Woods 2009: 192, figs. 13, 15–20 on pp. 232–35.

[50] Wiggermann 1992: 153, 180–81 (cf. 149, 152 for the cosmic function of scorpion(-men) watching over sunrise and sunset), who notes other seals where—formally somewhat different—scorpion-men are associated with the sun; Woods 2009: 192 with fig. 21 on p. 236 (cf. 192 with figs. 23–25 on pp. 237–38 with scorpion either supporting the sun-god or present in the sunrise scene).

[51] Albright 1920: 260.

[52] Astrolabe B II obv. ii.7 (Horowitz 2014: 38, cf. 104, 106); 'Great Star List' 34–35 (Koch-Westenholz 1995: 188–89); [mul]APIN I.ii.29 (Hunger & Pingree 1989: 38); *Enuma Anu Enlil* Tablet 51 Text X.19 (Reiner & Pingree 1981: 60–61); the list of Ea-stars in V R. 46 no 1: 31 (Prechel 1996: 148); reports 376.rev.4 and 504.5 (Hunger 1992: 213, 281); an incantation *BMS* 7 rev. 1 (Prechel 1996: 152–53); K. 7931 rev. 13' and BM 47799: 4' (Horowitz 2014: 112, 185); cf. Jensen 1890: 71; Zimmer & Winckler 1903: 432 n. 5; Deimel 1914: 149, 1606.6; Weidner 1915: 77; 1971: 77; Gössmann 1950: 31.94.i.5; Reiner & Pingree 1981: 12; Livingstone 1986: 234; Prechel 1996: 147–48, 150–53; Rochberg 2010: 327; Lambert 2013: 245; Horowitz 2014: 112, 149, 175–76.

[53] Hinke 1907: 96 and n. 2, 243; King 1912: xv; Deimel 1914: 149, 1606.6; Prechel 1996: 65–67, 147, 187; Black & Green 2000: 160; Lambert 2013: 234. For Išḫara as a scorpion, cf. the prayer *BMS* 7, rv. 1 (Prechel 1996: 152–53).

[54] On this subject see Bilić 2020c.

stands at the crossroad of the Galaxy with the zodiac might have had some similar resonances in Mesopotamian tradition. But even if this claim is rejected, unsubstantiated as it is by any explicit evidence in Mesopotamian texts, the plausibility of locating the Scorpion-men in the homonymous constellation (and the closely associated adjoining asterism) and their association with both the diurnal and annual solar movement remains unshaken.

With that said, the narrative of the *Epic of Gilgameš* strongly points to the diurnal solar movement as the cosmological setting of the episode. The fact that some, perhaps almost all, motifs employed in its construction can also be interpreted in terms of the annual solar motion is a vivid reminder of the fact that almost all motifs used in building the mythic models of the diurnal solar movement could have been also used as construction blocks in the models accounting for the annual aspect of its motion. It is now time to follow this line of inquiry.

13.4. Gilgameš on the diurnal course of the sun

In terms of the diurnal solar motion, it seems probable that the hero arrived at the western solar portal, guarded by the western mount Māšu, where the western pair of Scorpion-men observes the sun.[55] On the other hand, it is occasionally claimed that Gilgameš actually arrived at the mountain of sunrise in the extreme east.[56] This is sometimes supported by a passage in the OB version in which Gilgameš tells Ur-šanabi/Sur-sunabu that he travelled by way of mountains and adds an unclear reference to sunrise.[57] It is possible to claim from this passage that Gilgameš travelled from east to west after he entered mount Māšu, but this does not seem likely. On the contrary, it is possible to interpret the same passage differently and understand it as describing the voyage *to* the east, rather than *from* the east. This is supported by the passage from the prologue of the *Epic*, where it is asserted that the hero 'crossed the ocean, the wide sea to the sunrise/as far as the sunrise',[58] implying an easterly journey to the 'mouth of the rivers' from mount Māšu.

Furthermore, the OB version reports a dialogue between Gilgameš and Šamaš that could throw some light on the subject.[59] Unfortunately, the placing of this fragment is in doubt. Earlier scholars placed the fragment after the hero's ascent to the magical garden, implying he met the god there,[60] while more recent editors place the dialogue before Gilgameš even arrived to mount Māšu.[61] The OB tablet consists of four columns: the first contains the dialogue with Šamaš, the second and third a dialogue with Šiduri, and the fourth a dialogue with Sur-sunabu.[62] Thus, if the first solution is the correct one, Šamaš is encountered in the magical garden in the extreme east; if second, he is near mount Māšu in the extreme west. Both possibilities place this section of the voyage in a similar cosmological context, but neither of them solves the issue of the exact location of mount Māšu in terms of the bi-polar model accounting for the diurnal solar movement.

This episode can be further compared with a parallel incident in *Inanna's Descent*. In this narrative, the gatekeeper of the underworld asked Inanna to explain her presence at his gates, to which she answered she is traveling eastward to the 'Place of Sunrise'.[63] If this was only an excuse, it would be a pretty bad one if one could not actually arrive at the gate of the underworld travelling on the path to the Place of Sunrise. Thus, it was obviously possible to arrive to the underworld by traveling eastwards, probably following the section of the path of the sun from the western sunset towards the eastern sunrise gates. The underworld seems to have been here at least partly conceptualised in terms of the diurnal solar movement, i.e., in relation to its nocturnal section (for other occurrences of this practice see especially chapter 12).

The continuation of Gilgameš's journey clarifies to some extent the itinerary of his voyage (supporting my interpretation of the journey in terms of the diurnal solar movement), but in other respects extends and even deepens the uncertainty with respect to his exact course. After neglecting the advice of the Scorpion-men, who had told him that 'never did anyone [travel the path] of the mountain',[64] he nevertheless entered the gate of mount Māšu, and travelled for 'twelve double hours' through the darkness. More precisely, it is explicitly stated that 'he took the path of the Sun God', *harrān* ^{d}*šamši*. The expression designates a fixed band in the sky, reached by the stars at night in astronomical texts,[65] towards which the Moon god is commanded to approach,[66] or, occasionally, a course of the sun in the sky at a given time.[67] Thus it is a technical term denoting the course of the sun across the sky. This distinction is important, and it has to be emphasized in order to appreciate the cosmological nature of Gilgameš's voyage: both the 'observation' of the sun and 'the course of the sun' are terms used in observational astronomy, and their use in the *Epic* suggests a programmatic intention of its author, i.e., his intentional use of the terminology of

[55] Cf. Jacobsen 1976: 204; Alster 1983: 54; Heimpel 1986: 141–42; Burgess 1999: 192–93, 201; Marinatos 2001: 409; see below for supporting arguments.

[56] George 2003a: i.492–93, 496; Woods 2009: 197–98; Henkelman 2010: 335–36, 345; West 2018: 277.

[57] *EG* 10.iv.11. The obscurity of the line is reflected in the translations, see Speiser in *ANET*³ p. 90 (10.iv.10–11) with p. 89, n. 152, cf. Burgess 1999: 192–93 n. 72; Dalley 2000: 151; Foster 2001: 77; George 2003a: i.275, 281; 2003b: 125 (10.Si(VA + BM).iv.10–11).

[58] *EG* 1.40 (George 2003a: 495, 541; 2003b: 2; cf. Tigay 1982: 141; Kovacs 1989: 4; Horowitz 1998a: 97, 104; Foster 2001: 4).

[59] *EG* 10A.i (Heidel 1949: 69); 10.i (Speiser in *ANET*³ p. 89); 9.22–31 (Foster 2001: 66–67); 9.Si.i.5–15 (George 2003a: 276–77; 2003b: 71); Horowitz 1998a: 104.

[60] Heidel 1949: 69; Speiser in *ANET*³ p. 89; Jacobsen 1976: 204.

[61] Foster 2001: 66–67; George 2003b: 70–71.

[62] George 2003a: 276–81.

[63] *Inanna's Descent* 81–82; cf. George 2003a: i.500 n. 192.

[64] *EG* 9.81 (George 2003a: i.671; 2003b: 73).

[65] ^{mul}Apin II.i.50, 52, 53, 57 (Hunger & Pingree 1989: 82–84) and Horowitz 1998a: 256 n. 16 for reports and omens.

[66] *EE* 5.21–22.

[67] See some examples in Horowitz 1998a: 257.

observational astronomy in order to portray this section of Gilgameš's voyage.

At the same time, when the Scorpion-men warned the hero that no-one has travelled upon this path, they must have excluded the sun-god, since he obviously used it regularly. This is another example of the already noted pattern (see chapter 12): no-one ever crosses the Ocean from Šiduri's to Utnapištim's abode, except Šamaš and, in his wake, Gilgameš; also, no-one can descend into Apsu, except Šamaš and—predictably—Gilgameš.

For the first eight double-hours Gilgameš travelled upon the course of the sun through a complete darkness, but after nine double-hours he felt the North Wind on his face, and after eleven he 'came out in advance of the Sun'. After twelve double-hours 'there was brilliance' and the hero ascended in a wonderful garden, filled with marvellous trees, bearing all kinds of precious stones.[68] Presumably near the garden Gilgameš found 'Šiduri the tavern-keeper who lived by the seashore'.[69]

This description of Gilgameš's journey corresponds with the path of the sun below the horizon, and decidedly sets mount Māšu in the 'sunset' west, as announced at the beginning of this section. He enters the western portal, travels through the darkness and finally emerges in the garden in the far east, where the sun rises (alternatively, in a uni-polar model accounting for the diurnal solar movement, he enters an underworld location at the cosmic nadir corresponding to sunset and exits it in an action corresponding to sunrise). Thus, the place where the sun rises is located in the narrative before the crossing of the ocean and the waters of death, as well as the confluence of the rivers/Eridu/Abzu (for which see below). This corresponds to the situation in *Inanna and Enki*, that is, if we accept Alster's restoration of Tf. I.5 (Segment F), where he reads 'the place where the sun rises'.[70] This location is mentioned in the context of Inanna's flight from Eridu-Abzu, that is, on the way *from* the confluence of the rivers/Eridu/Abzu towards Uruk. This appears to be the location of Gilgameš's ascent from the road of the sun.

However, the duration of Gilgameš's voyage as we have it is not in line with the interpretation of his route given above: had he travelled for twelve double-hours, he would find himself in the same place where he had begun his voyage, which is obviously not the case. *Dana* or *bēru* ('double-hour') was a unit of distance, degrees of arc and time, and was equivalent to about 10800 m, 30° or 120 minutes. It was also used to express distances between stars, that is, the intervals of time between successive culminations of *ziqpu*-stars, but also the arc distances between rising stars.[71] At the same time, the unit for the daily solar progress used in the [mul]Apin, expressed as 40 ninda (GAR/NINDA, *nindanu*), is to be understood in astronomical context as a fraction of UŠ (1/60 UŠ, itself 1/30 of *dana/bēru*). Here the annual solar movement is described in the context of a schematic daylight increase/decrease model.[72] A Neo-Babylonian astronomical text claims that the sun both travels from the Path of Enlil to the Path of Ea, and then from the Path of Ea to the Path of Enlil, and also from sunrise to sunset and from sunset to sunrise, '12 *bēr* in area (*quaqqaru*, region in the sky around the sun, stars and constellations) is the measurement of the circuit', suggesting that both the annual and diurnal path of the sun consists of 12-*bēru*—360°—circuits.[73] The [mul]Apin describes the path of the sun over the course of the year in similar terms,[74] as do two passages from *Enuma Anu Enlil*.[75]

Therefore, if Gilgameš had travelled for 12 bēru he would either have covered a little less than 130 km (which is clearly not implied here), or his voyage would have taken 24 hours or 360°. In any case, he would find himself at the exact point at which he had begun his voyage in the first place. Since he had started the voyage at one of the gates of the sun, he would also have ended it at that point.[76] I cannot explain this incongruity and have nowhere found a plausible explanation for it.[77] With taking all other available evidence into account, which coherently points to the same direction, we must presume that the hero emerged from the darkness somewhere in the easternmost part of the earth,[78] and leave this particular inconsistency unexplained.[79]

13.5. Gilgameš arrives at the mouth of the rivers

Following his exit from the *harrān* [d]*šamši*, Gilgameš continued his voyage to Utnapištim's dwelling at the *pī nārāti*, 'the mouth of the rivers'.[80] This mythic location

[68] *EG* 8.138–90 (George 2003a: i.671, 673, 675; 2003b: 73–75).
[69] *EG* 10.1 (George 2003a: i.675, 679; 2003b: 76).
[70] Alster 1974: 21, 24.
[71] Van der Waerden 1949: 6–7; 1974: 48, 62–63; Gössmann 1950: 185.373; Rochberg-Halton 1983: 210–11, 216–17; Huxley 1997: 195–96; Horowitz 1998a: 183; 2014: 16, 225, 236; Hunger & Pingree 1999: xiv, 41, 88; Brown 2000: 105, 107, tab. 2, 109–11; George 2003a: ii.817; Rochberg 2004: 238; 2010: 168–71, 187; 2012: 12, 36–37; 2020: 149; Flygare 2006: 23, n. 41; Ossendrijver 2020: 42–44, 47–49.
[72] [mul]Apin II.i.12, 18 (Hunger & Pingree 1989: 73, 75); Rochberg 2010: 171 with n. 9, 358.
[73] K 2077+, rev. ii.1–4; Horowitz 1998a: 190–91.
[74] II Gap A.1–7 (Hunger & Pingree 1989: 88–89).
[75] Tablet 50, Text III.24b (Reiner & Pingree 1981: 42–43) and a parallel text cited in Reiner & Pingree 1981: 43, n. on III.24b; also cited in Horowitz 1998a: 255 n. 14; 2014: 12 and Koch-Westenholz 1995: 24 n. 1 (only the first text).
[76] A similar reconstruction of Aeneas' journey through Hades (entering and exiting it at the gates of sunrise at a 24-hour interval) is offered by Highbarger 1940: 95.
[77] See a detailed discussion in George 2003a: i.492–97, who shares my frustration, although with respect to a wider question of the interpretation of this section of Gilgameš's itinerary (specifically on the 12 bēru duration/length of this section of the voyage see pp. 494–95).
[78] Wensinck 1921: 2–3.
[79] Cf. George 2003a: i.497 (referring to this passage): '... it may be that the inconsistencies observed were not present at the beginning of the epic's life but accumulated over time, as the text was transmitted down the generations'.
[80] Thus Heidel 1949: 88; Speiser in *ANET*[3] p. 95; Gaster 1977: 183; Tigay 1982: 4; Alster 1983: 54; Kovacs 1989: 103; West 1997: 167; Horowitz 1998a: 96; Dalley 2000: 116; George 2003a: i.519–21, 717; Katz 2007b: 579–81; Woods 2009: 200; Lambert 2013: 232; Albright (1919: 161) gives the same translation, but in the meaning of 'source',

corresponds to the Eridu of cuneiform texts.[81] Indeed, Enki's city Eridu (= Abzu) was located between 'the mouth of two rivers' (Sum. ka.min, Akk. *pī nārāti*) in a bilingual Sumerian-Akkadian text.[82] It is inaccessible to mortals, but explicitly associated with Utu/Šamaš. At the same time, a *kiškanû*-tree is planted in this fertile domain of Enki/Ea, i.e., in Eridu/Abzu.[83] Thus a paradisiacal garden associated with the sun at 'Eridu' parallels the wonderful garden into which Gilgameš ascended after traversing the path of the sun.[84]

All the information hitherto discussed, including the nature of the path leading to Utnapištim's abode and its association with Abzu, strongly suggests its cosmological setting. Elsewhere, Utnapištim is indeed mentioned on the obverse of the Babylonian *mappa mundi*, depicting and describing distant landmasses beyond the ocean, surely in a cosmological context.[85] In precisely such a context Ea's city Eridu was identified in Mesopotamian uranography with ^{mul}NUN^{ki}, the star Canopus from the (Greek and modern) constellation Argo or some star(s) in its immediate vicinity.[86] Alternatively, an association of Eridu with the solstitial gates in Capricorn was proposed in modern scholarship, but this claim is highly conjectural and rests upon insufficient evidence, even though it respects the cosmological character of the venue.[87]

Even though, as already mentioned, the mouth of the rivers at Eridu was explicitly associated with Utu/Šamaš, the cuneiform sources do not specify their precise relation.[88] It is also unclear in what way can the star Canopus be associated with this mythic nexus. But Utnapištim's dwelling at the *pī nārāti* can be securely associated with the solar movement by way of Gilgameš's itinerary; it can further be identified with the Eridu/Abzu (in itself in some way associated with the solar movement) and, additionally, with ^{mul}NUN^{ki}, the star Canopus (or some star(s) in its immediate vicinity). All these indications—except for the lattermost—further converge in a single, in this context mythic, location: Dilmun.[89] The analysis of the testimonies on Dilmun that follows will reveal some important aspects of solar resonances contained in this mythic nexus.

13.6. Dilmun

The mythic Dilmun is associated with the sun god in various ways. Thus in a lament for Dumuzid cedar is associated with both Ḫašur and Dilmun,[90] and a number of texts associate (the mountain of) the Ḫašur-tree and/or cedar with the sun: the sun was said to come out from the heaven's interior and pass over the mountain of the Ḫašur(-tree), or it was described as 'rising from Ḫašur/ḫašur trees';[91] at the same time, it was described as rising from the mountain of cedar.[92] Indeed, both the Cedar and Ḫašur mountains are owned by the sun god.[93] It is thus possible to compare the function of Ḫašur mountain with that of mount Māšu in their capacity as cosmological features involved in a model accounting for the diurnal solar movement, more precisely, the moment(s) of sunrise and sunset (depending on the choice between the uni- or bi-polar model of daylight/night exchange).[94] In addition, these sources provide an indirect connection with the cosmological Dilmun. However, the most substantial connection of Dilmun with solar phenomena can be found in various—in particular—Sumerian, and, more generally, Mesopotamian myths treating the Flood.

The earliest reference to Dilmun is found in the context of Gilgameš tradition in a wider sense (notwithstanding the possibility that the Flood episode existed independently

see pp. 168–69; also, 'the source of the rivers' (Foster 2001: 91; Horowitz 1988: 161, 1998a: 36) or 'where the rivers flow forth' (George 2003b: 95). The most plausible explanation is that of Howard-Carter (1981: 220): 'The mouth of the two rivers should be interpreted as being their confluence' (cf. Kramer 1944: 28 n. 41). For other occurrences of the term *pī nārāti* in cuneiform literature see Jeremias 1887: 91; 1902: 45–46; Albright 1919: 163; Wiggermann 1992: 66; Horowitz 1998a: 104 n. 21 (*CT* XVII 26 65–66, 38 30ff; the town of this name is mentioned in *RGTC* 2 153 s.v. Pī nārātum and 3 185 v. Pī nārātim).

[81] Jeremias 1902: 39, 43; Albright 1919: 165; Wensinck 1921: 18; Alster 1974: 33.

[82] *CT* XVI: 46–47.183–98 = MA *Utukkū Lemnūtu* (=*Udugḫul*), Tablet 12, 1–18. Wünsche 1905: 2; Jeremias 1911: i.214; Albright 1919: 164 ('mouths'); Wensinck 1921: 4; Zuntz 1971: 387 (alternative: 'the river with the two mouths'); Geller 1980: 34, 38; George 2003a: i.520 n. 268 ('between the twin mouths of the rivers'); Woods 2009: 200 n. 67 ('mouths'), 221. A parallel Sumerian text from the OAkk. period has ab.šà.ga, 'the midst of the sea', in a corresponding place (Geller 1980: 24, *MDP* 14.91.9).

[83] Albright 1919: 165–66; Geller 1980: 25, 39. For further association of Eridu-Abzu with solar movement see Appendix 1.

[84] Cf. Zuntz 1971: 387–88. On the other hand, the garden and the mouth of the rivers are separated in the *Epic of Gilgameš* by both the Ocean and the Waters of Death.

[85] Obv. l. 10. Cf. Horowitz 1988: 161, 1998a: 36; Flygare 2006: 11; Rochberg 2012: 32–33.

[86] Van der Waerden 1949: 14, 15, 21, 1974: 74, 76; Schaumberger 1935: 335–36 (Carina with Canopus + the stars towards the west all the way to Eridanus); Gössmann 1950: 117.306 lists several identifications offered by earlier authors (Argo or Vela (Bezold), Vela + Puppis (Kugler)); Alster 1974: 33; Reiner & Pingree 1981: 8 n. 11, but Reiner & Pingree believe that ^{mul}NUN^{ki} actually represent the stars of the Puppis constellation, with ζ Puppis as the 'main' star; Hunger & Pingree 1989: 138, but Hunger & Pingree 1999: 272 identify as Eridu 'parts of [the constellations] Puppis and Vela', cf. Pingree & Walker 1988: 319–20, γλ Velae; Koch-Westenholz 1995: 207; Horowitz 2014: 222–23, 247 (Puppis and Vela).

[87] Mander 1999: 102–103.

[88] For a possible interpretation of Eridu-Abzu as a cosmological location corresponding to cosmic nadir, a point equally distant from sunset and sunrise and dependent on the meridian passage of the sun at midnight (in one variant of the model located in the far south), see Appendix 1.

[89] I am emphasising here the mythic—i.e., cosmological—nature of Dilmun in order to distinguish it from the terrestrial location often mentioned in cuneiform texts.

[90] *CT* XV.26.22 (Albright 1919: 180–81; Kramer 1944: 21; 1963a: 281–82; Jacobsen 1976: 70; Livingstone 1986: 110).

[91] *BA* 10.1, 11–14 (Kramer 1944: 19–20 n. 9; Heimpel 1986: 143; George 2003a: ii.864; Woods 2009: 190 with n. 29); *Enki and the World Order* (373, ETCSL 374–75, cf. Black *et. al.* 2004: 223; Woods 2009: 189 with n. 27, 213 n. 120) and a Sumerian hymn to Ninurta (*TCL* XV.7.13) (George 2003a: ii.864; Woods 2009: 190 n. 29); cf. *Lugalbanda in the mountain cave* 228–29 (Woods 2009: 214 with n. 123).

[92] *Tigi Song to Inanna* (Kramer 1944: 20–21; Tigay 1982: 77 n. 11; Heimpel 1986: 144; Horowitz 1998a: 331; Woods 2009: 191); a prayer K.3333+ and Nabonidus' inscription, *OECT* 1.27.iii.10–11 (Tigay 1982: 77 n. 11; Heimpel 1986: 144).

[93] *Incantation to Utu*, OrAnt 8.8:35 (Kramer 1944: 20 n. 9; Heimpel 1986: 144–45; Woods 2009: 191 n. 35). For the fomer cf. *Gilgameš and Huwawa* A11–12, B27–28 (Tigay 1982: 76 n. 10).

[94] Cf. Albright 1919: 179–80, 190.

previous to the creation of this tradition and was actually involved in its formation). Thus, in the Sumerian *Eridu Genesis* it is described how Ziusudra, the Sumerian equivalent of the *Epic of Gilgameš* Utnapištim, was transferred after the Flood to Dilmun, becoming like the gods, who gave him eternal life.[95] The translation of key phrases in the description of Dilmun is, however, somewhat doubtful. Line 11 of the segment, describing whither the gods transferred Ziusudra, reads: kur-bal kur dilmun-na ki ᵈutu e_3-še_3.[96] It is variously translated, and it is best to analyse it word by word. Kur-bal is rendered 'overseas' or 'foreign mountain' or 'land', or 'land of crossing',[97] while kur dilmun non-controversially refers to 'the mountain' or 'land of Dilmun'. Finally, ki ᵈutu e_3-še_3 is translated with respect to sunrise, either positionally or directionally.[98] Whatever the interpretation of kur-bal, Ziusudra's abode is certainly to be looked for in the farthest east,[99] and it is associated with sunrise and thus the diurnal solar movement. This corresponds to the reconstruction of Gilgameš's itinerary offered above, where the hero exits the path of the sun in the furthest east, before proceeding to Utnapištim's abode. Thus, the Sumerian tradition is perfectly consonant with the Gilgameš myth at this particular juncture.

The rendering of kur-bal as 'the land of crossing'[100] suggests another association with Utnapištim's 'mouth of the rivers', which is referred to in the *Epic* as the 'place of crossing', *nēberu*.[101] Dilmun is further associated with *nēberu* in the *Legend of Naram-Sin*, where it is placed 'in the midst of the sea' and mentioned immediately after the (sea) crossing, *nēberu*.[102] However, since the identification of *nēberu* remains controversial,[103] it does not seem prudent to discuss it any further, even though its proposed association with solstice-point(s) would provide another clear association of 'the mouth of the rivers' and Dilmun with the sun's (here annual) movement.[104] On the other hand, even if the rendering 'foreign' or 'overseas' country is accepted for kur-bal, there still remains more than enough evidence to understand Dilmun as corresponding to Utnapištim's abode. Furthermore, the situation of this cosmological Dilmun is by way of its correspondence to the *pī nārāti* comparable with that of the cosmological Eridu (see above). In its turn, Eridu is regularly brought into connection or considered identical with Abzu;[105] often it is labelled Abzu-Eridu, thus emphasizing its cosmological nature.[106] Hence 'Eridu' cannot simply represent a geographical toponym in this context, but rather a cosmological one, functioning as a feature with a role in the 'return' of certain celestial bodies.[107] Therefore, both Abzu and Eridu can be associated with the movement of celestial bodies which agrees with their cosmological localisation as adduced above.

It is further possible to directly compare 'Dilmun', that is, Utnapištim's abode and the goal of Gilgameš's horizontal voyage, with the Abzu, from where he obtains the plant of life in a vertical descent.[108] Furthermore, if Ur-šanabi, Utnapištim's boatman that captains the ferry that crosses the ocean and waters of death, is indeed to be understood as 'man of Ea',[109] based on the equation 40 (šá-na-bi) = Ea,[110] then the association of Utnapištim's abode with Abzu would be further strengthened.[111] It was proposed that Ea's sacred number (40, i.e., 2/3) derives from the relative length of daylight at the winter solstice, associated with the god, and thus relates to the annual solar movement.[112] Moreover, as we have seen above, Eridu can be tentatively associated with the winter tropic in the constellation Capricorn. This points back to the solstices as the most important referents in all mythic concepts or models treating the annual solar motion. Both Māšu (as argued above) and Eridu-Abzu, in their capacity as cosmological features involved in the mythic models accounting for the solar movement, can thus apparently be associated with the annual as well as the diurnal solar motion, in line with the already emphasized interchangeability of the motifs used in the mythic models respectively accounting for the two complementary aspects of solar movement.

An interesting description of Dilmun that can be correlated with the one in the Flood story and its resonances in *the Epic of Gilgameš* is found in the story of *Enki and Ninhursağa*.[113] It begins with what seems to be a description of some kind of primeval conditions prevailing on Dilmun at the

[95] *CBS* 10673, Nippur Segment E = *PBS* V.1.
[96] Col. vi l. 11 = v. 260 or 261.
[97] Kramer 1944: 18–19; 1956: 181; 1961: 98; *ANET*³, p. 44; 1983: 121; Civil 1969: 145; West 1997: 167; Horowitz 1998a: 329; George 2003a: i.519; ETCSL (cf. Black *et. al.* 2004: 215); Katz 2007b: 579; Woods 2009: 202. Jacobsen (1981: 525; 1987: 250) offers the translation 'over the mountains', while Heidel (1949: 105, col vi.7) does not translate the phrase.
[98] The former: Kramer 1944: 18–19; 1956: 81; 1961: 98; *ANET*³, p. 44; 1983: 121; Heidel 1949: 105; Civil 1969: 145; Caspers & Govindankutty 1978: 139; West 1997: 167; Black *et. al.* 2004: 215; Katz 2007b: 579, 581; Woods 2009: 202; the latter: Jacobsen 1981: 525; 1987: 250; Alster 1983: 45; Horowitz 1998a: 329; George 2003a: i.275 n. 141, 496, 519.
[99] George 2003a: i.97, 275, 496.
[100] Kramer 1944: 19; 1956: 181; 1961: 98; *ANET*³, p. 44; 1983: 121; Horowitz 1998a: 329; cf. Spronk 1999: 876.
[101] *EG* 11.248 (Heidel 1949: 90; Speiser in *ANET*³ p. 96); 'ferry landing', Kovacs 1989: 105; 'crossing point', Foster 2001: 93; 'ferry', George 2003a: i.719; 2003b: 97.
[102] *Legend of Naram-Sin* 58–60 (*CT* XIII.44, S.U. 51, 78 + 166; Gurney 1955: 100–101; Westenholz 1997: 314–15; Foster 2005: 351).
[103] Schott 1936: 141–45; Böhl 1936–37: 210–13 = 1953: 303–307; Gössmann 1950: 118–19.311; Landsberger & Kinnier Wilson 1961: 172–74; Koch 1991, esp. 50; Koch-Westenholz 1995: 117; Horowitz 2014: 20–25, 115; Lambert 2013: 182–83.
[104] Jeremias 1913: 72 with Abb. 48 on p. 73 and Weidner 1915: 33, 41 (both partially cited in Schott 1936: 141); Von Soden 1942: 17; Heidel 1951: 59; cf. Böhl 1936–37: 212 = 1953: 306 and Landsberger & Kinnier Wilson 1961: 173 n. 18.
[105] See for example *Enki and the world order* 67 (Black *et al.* 2004: 217); *Hammurabi's Code* i.64–ii.1 (Meek in *ANET*³ p. 164); *Innana and Enki* Tf. I.ii.5–7 (Alster 1974: 20–21).
[106] Alster 1974: 32.
[107] Alster 1974: 32.
[108] Penglase 1994: 77–78; George 2003a: i.501, 510. For a parallel Greek tradition see chapter 9.
[109] Rawlinson *et al.* V.44.iii.48.
[110] Cf. Parpola 1993: 203–204.
[111] George 2003a: i.150–51, 500.
[112] Parpola 1993: 203. For association of Ea with the winter tropic see Jeremias 1913: 74 with Abb. 48 on p. 73 and Weidner 1915: 33.
[113] *PBS* 10/1.1.

time.[114] After the complaint of Ninsikila, Enki supplies Dilmun, 'when Utu stepped into heaven', with fresh water from 'the mouth of the waters (ka.a)' of the underworld Abzu.[115] Besides the obvious association of Enki (and, in a lesser measure, Utu) with Dilmun, it is hard to draw any definite conclusion based on this description, although the 'paradisiacal' environment suits well the conception of Ziusudra's eternal home, even if we accept the concept of Dilmun in this story as some kind of negative utopia.[116] An interpolation of some twenty to thirty lines[117] that follows the irrigation, describing the now flourishing emporium of Dilmun, was probably meant to be contrasted with the primeval conditions before the irrigation occurred. Yet it may only be describing the actual 'unmythic' Dilmun of later times. The most important question for my inquiry is what led the narrator to choose specifically Dilmun rather than some other location in this context? Perhaps he knew of a 'mythic' Ziusudra's Dilmun, and contrasted it with a flourishing contemporary emporium in the Persian Gulf.[118] In any case, any solar connection of this Dilmun is peripheral, which shows that different authors chose different emphases in their renderings of traditional cosmological or mythic material, with the *Epic of Gilgameš* stressing both the solar character of the approach to 'Dilmun' and its intimate connection with the Abzu, while *Enki and Ninhursağa* emphasised only the latter and disregarded the former. What both narratives have in common is a special position of Dilmun in relation to the normal mortal experience and circumstances that condition and accompany it.

13.7. 'The mouth of the rivers' outside the Mesopotamian tradition

The cosmological feature called 'the mouth of the rivers' in Mesopotamian tradition—corresponding to Dilmun in the Sumerian account of the Flood—apparently corresponds to 'the source (mbk) of two streams/oceans/rivers (nhrm), in the midst (qrb) of the springs/streams (ảpq) of two oceans/deeps/seas (thmtm)', the dwelling of the highest god El in the Ugaritic tradition.[119] According to another text, the location is actually 'the meeting-point' or 'confluence' ('dt) of the deeps.[120] This location was certainly taken over from a Mesopotamian source, but was adopted by the dwellers on the Mediterranean coast and adapted and incorporated into their mythic narratives. In that process almost all references to the solar movement have been lost, apart from the fact that the sun goddess is able to visit 'the confluence'. Another interesting parallel to this Mesopotamian cosmological location appears in the *Qur'ān*, where it is described how Moses and his companion travelled to the place called *majmā' al-baḥrayn*, 'the meeting place of the two waters' (18:60, cf. 61).[121] This expression is almost certainly a resonance of the Akkadian phrase describing the location of Utnapištim's dwelling,[122] but here also any reference to the solar movement was lost in the process of adaptation.

Apart from these Near Eastern parallels, it appears that classical tradition was also familiar with this cosmological location. At the same time, I must submit that another interpretation accounts satisfactorily for the appearance of what could be recognised as a resonance of 'the mouth of the rivers' in Greek tradition. The place 'where the two seas meet' (*duô alos en xunochêsi*) at the Cyanaean rocks from Apollonius' *Argonautica* is apparently an appropriate description of Bosporus, where the Pontus and Propontis

114 Kramer 1944: 25 n. 28; 1956: 172–73; 1961: 55; *ANET*^{3} p. 38, ll. 1–30; Alster 1983: 55, 61–63, ll. 1–28; ETCSL, ll. 1–28, after Attinger 1984: 6–9, ll. 1–30; Dickson 2007: 2; Katz 2007b: 572–84.

115 Ll. 40–43, 50–54 ETCSL, after Attinger 1984: 10–11, l. 42–45, 52–56, cf. Katz 2007b: 585–86; Alster 1983: 62–64, ll. 40–44, 51–55; cf. George 2003a: i.520–21.

116 Dickson 2007: 5; cf. Alster 1983: 56; Katz 2007b: 573, 577–79, 584, 587.

117 *UET* 6.1 (Alster 1983: 64–65; Attinger 1984: 12–13; Katz 2007b: 569–70, 587–88).

118 This 'historical' Dilmun was regularly located 'in the midst of' or 'across the' Lower, that is, Eastern sea, occasionally designating the extreme east (*Sargon Birth Legend* 19, 29 (Speiser in *ANET*^{3} p. 119; Foster 1995: 165–66; Westenholz 1997: 42–45); *Letter from Aradĝu to Šulgi about irrigation work* 4–5 (ETCSL 3.1.03); *Sargon's Geography* 41–43 (Grayson 1974–77: 61; Alster 1983: 45; Horowitz 1988: 161; 1998a: 73); *Legend of Naram-Sin* 60 (*CT* XIII.44, S.U. 51, 78 + 166; Gurney 1955: 100–101; Westenholz 1997: 314–15; Foster 2005: 351); an OB copy of a Šu-Sin inscription (Civil 1967: 37; Horowitz 1998a: 88 n. 25); Assyrian royal inscriptions of Tukulti-Ninurta I (Grayson 1972: i.121.782, 78.17; Howard-Carter 1987: 93), Sargon II (Albright 1919: 183; Cornwall 1946: 4; Caspers & Govindankutty 1978: 131, 134; Alster 1983: 44–46; Kessler 1983: 148; Howard-Carter 1987: 93–94; Potts 1990: i.334; Horowitz 1998a: 105), and Sennacherib, Esarhaddon and Ashurbanipal (Albright 1919: 183; Cornwall 1946: 4–5; Oppenheim in *ANET*^{3} p. 297; Alster 1983: 43, 46; Kessler 1983: 149–50; Potts 1990: i.339; West 1997: 378)).

119 Smith in Parker 1997: 95 (*CAT* 1.2.iii.4) = Ginsberg in *ANET*^{3} p. 129 (b.III AB C.4); Smith in Parker 1997: 116 (*CAT* 1.3.v.6–7) = Ginsberg in *ANET*^{3} p. 137 (f.V E.14–15); Smith in Parker 1997: 127 (*CAT* 1.4.iv.21–22) = Ginsberg in *ANET*^{3} p. 133 (e.II AB iv.21–22) = Gaster 1977: 183–84; Smith in Parker 1997: 153 (*CAT* 1.6.i.33–34) = Ginsberg in *ANET*^{3} p. 140 (h.I AB.33–34) = Gaster 1977: 216; Parker in Parker 1997: 62 (*CAT* 1.17.vi.47–48) = Ginsberg in *ANET*^{3} p. 152 (AQHT A vi) = Gaster 1977: 349; Smith in Parker 1997: 148 (*KTU* 1.5.vi); López-Ruiz 2010: 115; Lambert 2013: 447. *thmtm* is dual, so *nhrm* is understood accordingly (Wyatt 2005[2004]: 196, 227 n. 13).

120 *CAT* 1.100.3 (Parker in Parker 1997: 220). In this text, the location is visited by the sun goddess Shaphs (*CAT* 1.100.2). In addition, kṯr-w-ḫss (Kothar wa-Hasis) is styled bnm.'dt, 'Son of the Confluence' (*CAT* 1.4.vii.16; Smith in Parker 1997: 136; Smith & Pitard 2009: 649, 667). It is possible that Utnapištim's and Adapa's epithet Atra-ḫasīs is reflected in the second part of Kothar wa-Hasis' name (Dalley 1991: 8 n. 25), which points to an intimate connection of 'confluence' and 'Utnapištim/Atra-ḫasīs' in Ugaritic myth, certainly derived from the Mesopotamian tradition.

121 Thus Wheeler 1998: 208 and *passim*. Albright 1919: 192: 'juncture of the two seas'; Gaster 1977: 183: 'junction of the two oceans'.

122 Friedländer 1913: 303; Wensinck 1921: 18; Wheeler 1998: 208. For the connections between the Gilgameš story and Q 18:60–82 cf. van Bladel 2008: 176. Further resonances in the Qur'ānic tradition can be recognised in 'a barrier between the two waters' (Q 27:61, 55:19–20), which most probably separates the sea of salt from the sea of fresh water, as does the barrier described in Q 25:53. The 'Meeting-place of the Two Seas' from the *Tale of Buluqiya*, where it is explained—albeit implicitly—that the expression refers to the seas of salt and fresh water (Burton 1897: iv.272 no. 497; the version in al-Tha'labī (Brinner 2002: 602), although not mentioning the meeting-place, does recognize the existence of the two seas with a partition between them), probably reflects the Qur'ānic tradition, but the influence of the *Epic of Gilgameš* should not be excluded, since some of the motifs in this story certainly point to the *Epic* as their indirect source (Dalley 1991; 1994, with criticism in George 2003a: i.65–68 and Henkelman 2006: 829 n. 49; 2010: 341–42 n. 77).

indeed 'meet'.[123] However, initially the Cyanae were a cosmological location on the way to the land of the sun, Aea, in the farthest east (see chs. 8 and 9). It is possible that this detail in Apollonius' poem reflects the presence of the motif in the early Argonautic tradition, which was rationalised by Apollonius or his source (perhaps already by Eumelus, who rationalised the Argonautic voyage into one with the goal in Colchis), and interpreted as referring to the Bosporus. A meeting place of two bodies of water in an intimate connection with the island of sunrise corresponds to the situation as presented in Mesopotamian tradition; at the same time, the late occurrence of the phrase and its evident appropriateness to its actual geographical referent make the cosmological interpretation superfluous. A more secure comparison with the Mesopotamian tradition is 'the meeting place (*xunesis*)' of two rivers near the very entrance of Hades to which Odysseus should have arrived following Circe's instructions after leaving the island of sunrise, Aeaea.[124] In this case we might also consider a pre-Homeric Argonautic tradition as the source of the motif as it appears in this section of the *Odyssey*. Notwithstanding whether either of these two testimonies are accepted as referring to a cosmological location corresponding to the Mesopotamian 'confluence of the rivers', it appears that in the Greek tradition the solar resonances of 'the meeting place' on the one side, and eschatological resonances on the other, remained equally prominent.

Finally, it appears that Indian tradition was also familiar with a corresponding cosmological location. Thus, according to *R̥V* 8.41.2 Varuṇa dwells at *síndhūnām úpodayé*. This is variously translated as the 'Mündung', 'confluence', 'rising', 'origin' or 'source' 'of the rivers'.[125] We have previously had the chance to note that Varuṇa's abode is located at the cosmic nadir,[126] as he is the lord of the (primeval/subterranean) waters/(western) ocean/netherworld.[127] At the same time, we have noted the correspondence of Varuṇa with the sun (at night), which becomes Varuṇa when it enters the waters.[128] Thus Indian cosmology seems to have preserved, however fleetingly, the association of this particular cosmological location with the solar movement.

13.8. From Gilgameš to Odysseus

Since, as we have seen, the cosmological location of 'the mouth of the rivers' in Mesopotamian tradition was as a rule equated with the star Canopus or some star(s) in its immediate vicinity and thus near the south celestial pole, it seems possible to interpret the section of Odysseus' voyage involving the visit to Hades, which parallels in important respects that of Gilgameš (see above on the Laestrygonians-to-Circe and Māšu-to-Šiduri correspondences), in similar terms on a cosmological level.[129] What this means is that a section of Odysseus' voyage can be understood as structured in terms of and involving a reference to certain cosmological features, phenomena and processes (similar to the episode involving the Laestrygonians analysed in chapter 2 that references the annual solar movement). Indeed, I have already interpreted the section of Odysseus' voyage from Aeaea to Hades and back in similar terms of the diurnal course of the sun(-god) upon a flat earth in combination with a vertical descent to the world of the dead (chapter 9), and will now expand on that discussion in the light of the Mesopotamian evidence. This analysis will once again show that alternative interpretative frameworks can be applied in the exegeses of various strands of evidence, susceptible of interpretation both in terms of the diurnal and annual solar movement. This was as much the case for the Gilgameš material, as it is for Odysseus' voyage.

Indeed, the cosmological nature of Odysseus' voyage as a whole was recognized by the Homeric exegete Crates of Mallos, whom we have already discussed in chapter 2, as attested in both literary and cartographic sources.[130] Crates interpreted the section of Odysseus' voyage from Aeaea to Hades in terms not unlike those employed in some exegeses of the *Epic of Gilgameš* noted above, especially—but not exclusively—those that used the interpretative framework of the annual solar voyage. He argued that the 'river ocean's stream', located between 'the sea' where Aeaea is located and Hades, represents 'a sort of estuary or gulf, which stretches from the winter tropic in the direction of the south pole'.[131] Consequently, this would place the Hades itself at the south pole, which is in concordance with Crates' placement of the Cimmerians at this location (chapter 12). This concept is confirmed by a cosmological map whose origin seems to lay in the Cratetean exegesis of the *Odyssey* (Figure 13.1).[132] The map depicts a zonal division of the earth together with several geographical landforms. In addition to these, several features belonging to the domain of eschatology are also discernible on the map (Lethe, Pyriphlegethon, Acherusian Marsh). The presence of cosmological, geographical and eschatological levels of knowledge employed in the making of the map comes as no surprise if 'the frequency… with which maps have been used as teleological instruments, epitomizing the sacred and mythical space of cosmologies as well as the more tangible landscapes of the real world' is recognised.[133]

The correspondences between the map and Crates' system as reconstructed from literary sources are evident both

[123] A. R. 2.318.
[124] *Od.* 10.515 with Σ V *ad loc.* and Σ D 10.515a (Ernst 2006: 220). This cosmological location can be by inference also recognized in the second *nekuia* (see chapter 10).
[125] The respective translations in Geldner 1951: 2.354; Witzel 2000: 299 n. 33, 313 n. 68; Jamison & Brereton 2014: 2.1111; Kuiper 1983: 145; Griffith 1897²: 2.183; cf. Olmsted 1994[2019]: 128.
[126] Kuiper 1983: 68, 69 n. 68, 76, 78, 144–45, 149–50.
[127] Kuiper 1983: 16–17, 66–70, 96–97, 183.
[128] Cf. Olmsted 1994[2019]: 128.
[129] For some recent interpretations of the nature of Odysseus' voyage as cosmological in character see Marinatos 2001; Käppel 2001: 16, 19, 21 Abb. 1; Nakassis 2004; Beaulieu 2016: 53; Fowler 2017: 249–50.
[130] I have treated this subject in Bilić 2012a: 297–301.
[131] Fr. 57 Broggiato = *BNJ* 2113F24 on *Od.* 11.1–2, 11, 21, 12.1–4.
[132] Edson & Savage-Smith 2000.
[133] Harley 1987: 4 (citation); Jacob 1988: 4; Gee 2020: 92.

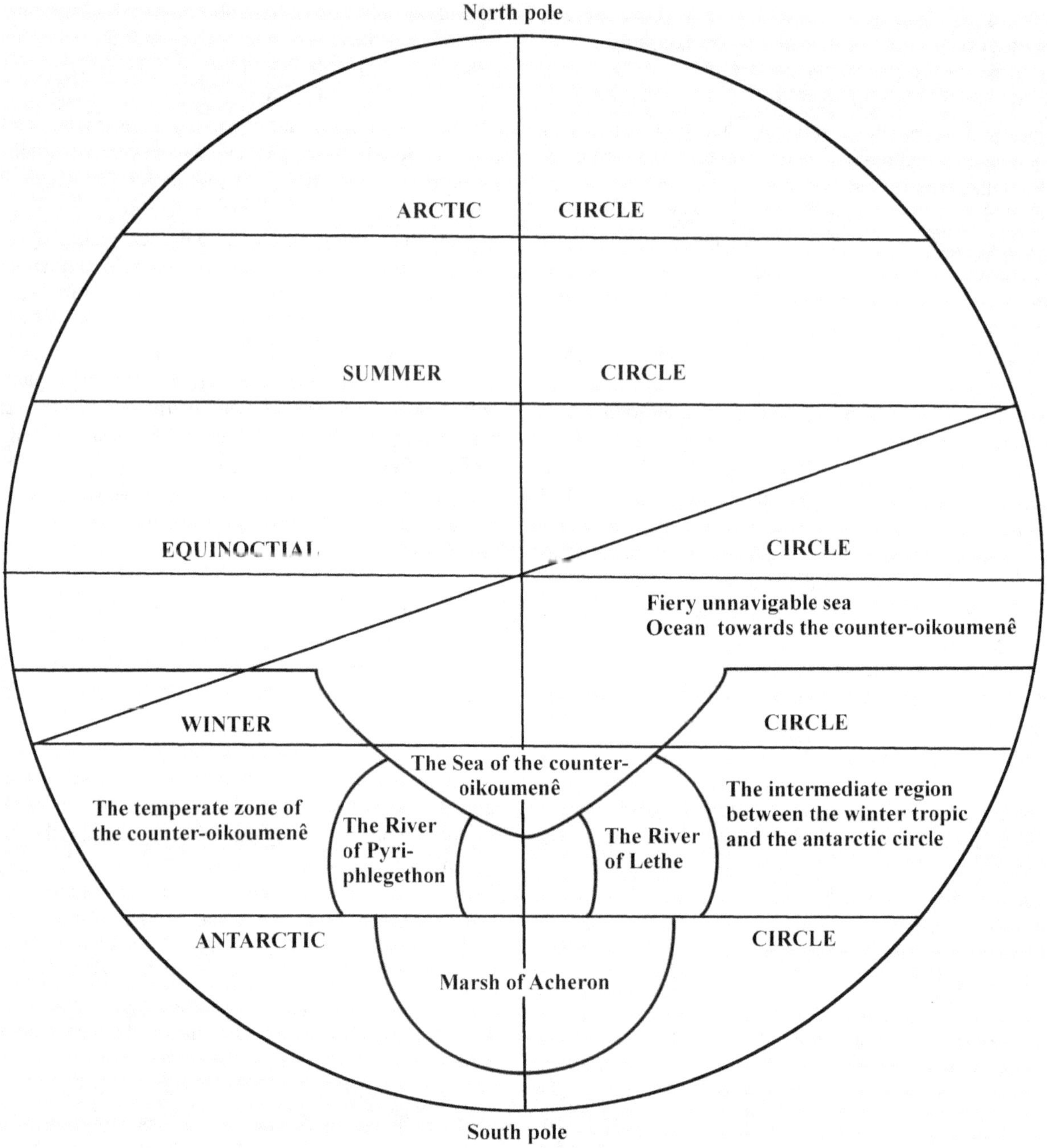

Figure 13.1 Crates' world map (drawn by the author)

on the level of structure and detail. The map's equatorial sea, or, more precisely, the ocean extending between the temperate zones—a distinctive feature of Crates' system[134]—represents the 'sea' where Aeaea is located in fr. 57, while the map's 'sea towards the *antoikoumenê*' represents the estuary/gulf stretching from the winter tropic towards the south pole. This branch of the Ocean, according to both Crates' literary exegesis and the map, divides the landmass of the southern hemisphere into two parts. It seems plausible to presuppose the existence of another branch of the Ocean stretching from the summer tropic to the north pole (which would correspond to the modern northern Atlantic) and a corresponding 'pair' of branches at the opposite (eastern) part of the *oikoumenê*.[135] Both Crates' literary exegesis and its cartographical depiction were made in the context of Homeric criticism, more precisely, in the context of the interpretation of the section of Odysseus' voyage involving Aeaea and Hades. What is not immediately clear from Crates' interpretation of this section of the voyage is whether the Pergamene scholar understood Odysseus as starting off from the western border of the *oikoumenê*—along the southern section of the 'Atlantic' (thus Mette)—or from the eastern, which would be more in line with Homer's placing of Aeaea in the east. The latter was argued for in chapter 9, with Odysseus sailing against the current of the

[134] Bilić 2016c: 129–32.

[135] Mette 1936: 76–77.

counterclockwise-flowing circumambient Ocean, and I submit that this is the correct interpretation.

If Crates' words are taken literally—and presuming that he took Homer's words at what we recognise as their face value, which is a rather unwarranted assumption—according to his interpretation Aeaea would be located somewhat to the north of the tropic of Capricorn (corresponding to c. 24° southern latitude) and Telepylus most probably still further north. This is certainly not in accordance with what Crates had to say on the location of the Laestrygonian land elsewhere. If he truly located Telepylus at 54° or 66° (see chapter 2), he probably imagined Aeaea somewhere between Telepylus and Hades, perhaps at or near the equator, which is compatible with both Crates' literary exegesis and the map.

In both the *Epic of Gilgameš* and the *Odyssey* 'the confluence of the rivers' is intimately connected with some form of cosmological underworld, exemplified in Abzu and Hades (respecting the manifest differences between these two cosmological locations). The latter's strong eschatological character is evident, while the former apparently lacked any such quality. However, Gilgameš quest for immortality and his and Enkidu's later sojourn in the world of the dead as the context in which 'the mouth of the rivers' appears, in addition to Abzu's close relation to the posthumous abode of Utnapištim and his wife at the *pī nārāti*, allow the inference that Abzu could indeed appear as a feature in eschatological considerations. Thus, both the section of Odysseus' voyage from Telepylus to Aeaea to Hades and back, and the section of Gilgameš's voyage from mountain Māšu to Šiduri to *pī nārāti* and Abzu and back share the concurrent involvement in the conceptual domains of solar movement and eschatology. In Crates' cosmological interpretation, the Aeaea-Hades section of Odysseus' voyage partially corresponds to the winter tropic-south pole route (with a probable localisation of the former somewhat to the north towards the equator), while Gilgameš's voyage could similarly be interpreted as encompassing the track between the southern tropic and the furthest south, with a goal in a cosmological location with certain eschatological traits.

Indeed, an eschatological application of the winter solstice-Canopus axis can be recognized in the Egyptian tradition. Thus the temple complex of Amon-Re in Karnak, more precisely, the part of the complex erected in the reign of Hatshepsut and Thutmosis III (18th dynasty), has two axes: the main temple is oriented upon the winter solstice sunrise (with some modifications in the reign of the latter pharaoh), while a secondary axis, defined by the orientation of some features of the Akh-menu (the Festival Hall) and the processional axis of the 7th and 8th pylon, is perpendicular on the primary axis and oriented upon the setting of Canopus during the reigns of Intef II (11th dynasty) and Thutmosis III.[136] This combination suggests a certain connection between Canopus and the winter solstice, and the most plausible explanation of the nature of this connection would be in eschatological terms. It was indeed argued that the Canopus orientation(s) represent a chthonian and Osirian aspect, and were opposed to the 'solar' aspect of the primary orientation.[137] However, it seems more probable that these aspects are complementary, moreover, that they are actually parts of a single complex of ideas. On the other hand, at this point, due to the lack of evidence, it is impossible to discuss any further the Egyptian understanding of this concept.

13.9. Conclusion

The presence of terms such as *naṣāru* and *harrān* ᵈ*šamši* in the *Epic of Gilgameš*—more precisely, in the section of the narrative with clear solar resonances, i.e., the segment explicitly considering the solar motion—suggests the appropriateness of interpreting this series of episodes in the *Epic of Gilgameš* in terms of a cosmological voyage. More precisely, this section can be understood as representing an incorporation of a mythic model accounting for the solar phenomena into a larger narrative of the *Epic*. In addition, it could be argued that this segment of Gilgameš's itinerary is actually a rare example of a story structured on an analogy with the diurnal solar movement, with Gilgameš's itinerary from mount Māšu to Šiduri's dwelling explicitly described as an imitation of the sun god's nocturnal journey from west to east. It is thus clearly a mythic narrative referencing solar movement and belongs to the class of myths discussed in this book.

If the path of the sun is indeed to be connected in some way with Canopus or some star(s) in its vicinity, as it apparently is in the *Epic of Gilgameš* or in its elucidation in the Mesopotamian tradition, then it seems necessary to introduce the annual solar movement—in some form—as the conceptual framework informing this connection, since it is during winter that the sun 'goes' to the south, attaining its southernmost declination on the winter solstice. This evokes the proposed identification of *pī nārāti*/Eridu/Abzu with the solstitial gates in Capricorn, the identification of the corresponding Sumerian cosmological location of Dilmun with *nēberu*, itself interpreted as solstice-point(s), and the postulated association of Ea with the winter tropic. Moreover, it is possible to interpret in similar terms the cosmological location of Scorpion-men and Māšu-mountain, both closely associated with 'the mouth of the rivers'. But Gilgameš's travel upon the sun's path is explicitly structured in terms of the diurnal solar movement, which presents a serious obstacle for accepting this line of interpretation, and an alternative reading precisely in those terms appears necessary. The southernmost visible celestial regions, which include the star Canopus and the south celestial pole—and presumably the terrestrial regions corresponding to them, i.e., their projections upon the surface of the flat earth—

[136] Gabolde 1999; Shaltout & Belmonte 2005: 284–86; Shaltout, Belmonte & Fekri 2008: 201.

[137] Gabolde 1999: 281–82.

should thus rather be associated with the nocturnal section of the sun's daily voyage by way of south upon a flat earth (as discussed in chapter 9), when the sun approaches the southernmost regions of the earth.

In the light of the fact that the diurnal and annual solar movement are closely connected, both recognisably being two aspects of a single phenomenon, perhaps these two notions can be combined: the sun during the night of the winter solstice follows its southernmost course (i.e., one farthest from the hither coast of the circumambient ocean), when it most closely approaches the region 'under Canopus'. Since the fact that the sun takes a longer time to traverse its nocturnal path during winter nights is easily observable, the phenomenon must have been explained either by its slower movement along an identical course every night—which would contradict the observable phenomena during the daylight hours—or by positing its nocturnal route of unequal length, in line with the corresponding daylight one. In addition, the distance between its southernmost rising and setting points on the winter solstice was actually the shortest (measured along the rim of the earth's surface), so some way of prolonging it must have been conceived. A spatial prolongation of the sun's course to the south is a viable explanation of this phenomenon in this particular context.

Thus it appears that the cosmological location 'the confluence of the rivers' (Sumerian ka.min/Akkadian *pī nārāti*) current in several related traditions—Ugaritic (mbk nhrm qrb ảpq/'dt thmtm), Greek (*xunesis duô potamôn* of the *Odyssey* and perhaps the *duô alos en xunochêsi* of Apollonius Rhodius, if the latter belongs to the early Argonautic tradition) and Arabic (*majmā' al-baḥrayn*, together with the Qur'ānic barrier between the two seas), perhaps also Vedic (*síndhūnām úpodayé*)—can be interpreted, on the authority of the Mesopotamian tradition, in celestial terms as the region of the sky around the south celestial pole in general and the star Canopus in particular. At the same time, the corresponding terrestrial locations—the southernmost regions of the earth in general and the south pole in particular—should be understood as the locations intimately associated with the 'confluence of the rivers'. The road that leads to these locations is the path of the sun(-god), notwithstanding which of the two alternatives proposed immediately above one favors: the nocturnal section of its daily path taken as an unchanging course in 'latitude' or the particular course of its path on or about the winter solstice, one that encompasses its southernmost 'latitude'. A direct and unambiguous connection of the solar movement with the 'confluence of the rivers' indeed appears in both the *Odyssey* and the *Epic of Gilgameš*.

Finally, my analysis has again revealed the interaction of conceptual domains of eschatology and solar movement in the stories discussed in this chapter. The eschatological resonances of the *Odyssey* episode are highly pronounced, representing an instantiation of the common overlap between these two domains. At the same time, the *Epic of Gilgameš* approaches this subject somewhat differently, or with a different emphasis, but still within the scope of the conceptual domain of eschatology. In addition, the appearance of eschatological winter solstice–Canopus axis in Egyptian tradition signals a similar complex of ideas, although its poor preservation in otherwise elaborate Egyptian eschatological traditions does not provide much extraneous support to the nexus of ideas exemplified in the *Odyssey* and the *Epic of Gilgameš*.

It remains unclear whether 'the confluence of the rivers' can be associated with the diurnal or annual solar voyage, or perhaps—in the light of the interchangeability of the motifs used in building models accounting for both aspects of solar motion—with both, either within the frame of a single ethnographic context or with different applications in different ethnographic contexts. Even within the two best preserved traditions there is enough ambiguity to allow one to argue for either of these two possibilities. At the same time, the solar references as such in these two traditions are unambiguous and hold an important position within the narrative episodes discussed in this chapter. Moreover, it is apparent that these narratives were at least partially structured on an analogy with solar movement. This makes their inclusion into the class of myths referencing the solar movement difficult to argue against.

14

Beyond Odysseus: Alexander

References to solar movement are also recognisable in several accounts of the exploits of Alexander the Great. Some episodes of his adventures even appear to be structured on an analogy with the diurnal or annual solar movement, i.e., those that explicitly describe Alexander as following the sun's path and thus mimicking the solar course. Naturally, this corresponds to a similar practice already observed with respect to Gilgameš, the Argonauts and Odysseus. In contrast to these, due to its late date, the Alexander romance tradition can be understood in terms of a *nachleben* of certain mythic/cosmological concepts or studied as a reception of earlier texts and traditions. Both approaches, while correct in themselves, belittle its inherent qualities by emphasising its derivativeness. I submit that this is a simplified and one-sided view, and that this tradition should be studied in its own right, without forgetting the contribution of its predecessors and role models. It seems that the character of Alexander attracted like a magnet a vast number of pre-existing mythic episodes, including a fair share of those referencing the solar movement, which were integrated into a cycle of stories centred upon him. His exploits 'formed an ideal track for continuously renewed and updated geographic and ethnographic information about the world' to be presented to the readers.[1] It is relatively clear that the process started already in the Hellenistic period,[2] but it is best known from comparatively later renderings. At some point, certain cosmological notions were joined to the rest of this compendium, it would seem also as early as the Hellenistic period.[3] Among these notions were the concepts of the diurnal and annual solar motion, including the concept of solar turnings, together with a set of ideas frequently accompanying it, such as the limits of the habitable zone, circumpolar constellations related to the limits of the sun's annual movement, hindrance of the sun's annual movement to the north, direction of solar risings and settings, etc. Since these notions were both known to the Greco-Roman scientific tradition and incorporated into its mythic discourse—i.e., they were treated by a number of texts belonging to different genres, but all part of the same ethnographic context—it seems profitable to explore their further development in other cultural settings, to see which elements that entered the Alexander romance tradition certainly originated in classical antiquity and which were added later, and to study the ways in which various later traditions, such as Syriac, Persian, Islamic and Ethiopian, treated these notions. At the same time, the treatment of Alexander's exploits in these traditions, especially the episodes most telling for my study, i.e., those referencing in some way the solar movement, was apparently influenced in a not insignificant way by the traditions focusing upon Gilgameš, initially derived from the *Epic of Gilgameš* and its resonances in the Near East or the resonances of parallel oral traditions that developed around the hero.[4]

Once again, as was the case with Gilgameš, I am not trying to reduce the character of Alexander to that of a 'solar hero'. I am well aware of the presence of numerous features from various conceptual domains even in the particular episodes I am concentrating upon, i.e., Alexander's visits to the Land of Darkness and mount Mûsās, as well as his voyage to the places of sunset and sunrise. These features can doubtless be profitably approached by other interpretative methods, but I submit that the hermeneutic strategy furthered in this study remains the most appropriate method for interpreting the elements that clearly belong to the conceptual domain of solar movement.

14.1. Hellenistic tradition

The cosmological setting of Alexander's exploits current in the Hellenistic period, as well as its connection with the Gilgameš tradition, can be recognised in the Greek version of Pseudo Callisthenes' novel and its Armenian translation. Here the Amazon episode is followed by a voyage 'towards the Red Sea' and the episode with the island City of the Sun and its Ethiopian priest and the 'place of darkness'.[5] Although it is almost impossible to ascertain conclusively where Alexander actually is at this point of his voyage either in terms of geography or cosmology, the far west seems the most plausible suggestion, as suggested by presence of *Atlanta potamon* in Greek and Armenian versions. The Red (Erythraean) Sea, the City of the Sun and its Ethiopian minister could all well be located in the west, even though intuitively they belong to the features characteristic of the furthest 'solar' east, while the darkness unequivocally points to the farthest west. Alternatively, the farthest 'solar' west might be assimilated here with its eastern equivalent in an instantiation of a uni-polar model accounting for the diurnal solar movement and the daylight/night exchange. In any case, a section of Alexander's voyage apparently includes a clear set of references to the sun's movement already in the earliest versions of the narrative of his exploits.

[1] Casari 2011: 95, cf. 2012: 177–78.
[2] Casari 2006: 215–16; 2011: 97, 100; 2012: 179.
[3] Casari 2011: 97, specifically referring to the Land of Darkness in the far north.
[4] Wheeler 1998; Van Bladel 2008; Henkelman 2006; 2010.
[5] Arm. version 258 (Wolohojian 1969: 146), translating the α recension of the Greek novel; [Callisth.] β 3.28 (pp. 174–76 Bergson; Stoneman 1991: 146–47).

Furthermore, the Armenian version of Pseudo Callisthenes' novel, summarising the chs. 2.39–41 of the Greek version, combines the visit to the Land of Darkness ('the place where the sun does not rise') with a construction of the gates that 'sealed up that place'.[6] Since this version here mentions 'Arcturus' as the expedition's guiding asterism, where the Greek version has Ursa major,[7] it is certainly possible that the concept of the Land of Darkness should be associated with the north in both the Greek original and Armenian translation, actually related to an earlier Greek model (the α recension). At the beginning of this section of his wanderings Alexander indeed started off in the direction of Ursa major.[8] This account of a section of Alexander's voyage is also conditioned by the solar movement, but here it seems that it is its annual aspect that shaped the meteorological characteristics of the Land of Darkness, located in the far north beyond the course of the sun. It is possible that this section of the *Romance* originated in the Hellenistic period as an autonomous composition and that it was incorporated into the 'original' *Romance*, only to be removed by a later redactor, before being subsequently reintroduced.[9] In any case, the section shows many signs of an early date.[10] Thus the cosmological, and in particular solar, features found at least in this part of the *Romance* indeed possess an ancient pedigree. Furthermore, the account of Alexander's visit to the Land of Darkness in Pseudo Callisthenes *Romance* β 2.23–41 can be brought into connection with the Gilgameš tradition,[11] which once again points to the former's manifestly cosmological setting and, at the same time, reveals the influence of the Gilgameš tradition upon the creation of the early narratives of Alexander's exploits.

The earliest Alexander romance tradition, dating from the Hellenistic period, thus used the motifs drawn from the solar movement, both daily and annual, in its rendering of Alexander's exploits. And while it would perhaps be too bold to claim that sections of Alexander's itinerary were structured on an analogy with solar movement, it is manifest that they were permeated with motifs extracted from solar phenomena. Whether these were introduced independently of the Gilgameš tradition remains an open question, since the latter seems to have exhibited a strong influence upon precisely these sections of Alexander's itinerary that reference the solar movement. In any case, these Alexander's exploits should be classified in the category of myths engaging with the solar movement, even though they most probably did not originally belong to the Greek ethnographic context.

6 Arm. version 209 (Wolohojian 1969: 115–16). On these northern gates see below.

7 [Callisth.] β 2.40 (p. 134 Bergson; Stoneman 1991: 122).

8 [Callisth.] β 2.32 (pp. 125, 193 Bergson; Stoneman 1991: 115); the Armenian version again has 'Arcturus' (Wolohojian 1969: 112).

9 [Callisth.] 2.23–41of the β recension (pp. 123–34, 193–203 Bergson; Stoneman 1991: 114–23; summary in Henkelman 2010: 329–34); Merkelbach 1977: 63–65, 70–72, cited in Henkelman 2010: 327 with n. 16.

10 Henkelman 2010: 326–28.

11 Henkelman 2010: 341, 350.

14.2. Land of Darkness

The references to both the diurnal and annual solar movement present in the earliest Alexander romance tradition are iterated in later texts, sometimes reflecting much earlier sources, including once again the Gilgameš tradition, as well as the earlier Alexander-tradition itself. Two typical episodes or motifs that contain the references to the solar movement are Alexander's visit to the Land of Darkness in the far north, characterised by a total lack of daylight and most probably reflecting the sun's annual behaviour in relation to latitude (an early example of this episode was mentioned in the preceding section), and his visiting the mountain of sunset followed by a travel upon the way of the sun (both in the wake of Gilgameš). The latter episode or a set of episodes is manifestly structured upon an analogy with the diurnal solar movement. In addition, occasionally the Land of Darkness was located in the far west in the region of sunset, and was thus defined in terms of the sun's diurnal movement (an untypical variant of this motif was also mentioned immediately above). The dual localisation of the Land of Darkness in Alexander-narratives—in both cases in terms of solar movement—appears to be another example of the interchangeability of motifs employed in respective accounts of the diurnal and annual solar movement in various mythic traditions.

This dual localisation is particularly manifest in the Islamic tradition. In one strain of the Islamic Alexander romance tradition the Land of Darkness in which Alexander entered in search for eternal life was located in the far west. Its location was frequently, almost regularly, associated with the solar movement, in this case with its diurnal aspect. Thus in 'Umāra's (BM* Add. MS 5928) Arabic version of Pseudo Callisthenes' novel (late 8th–early 9th c.) Alexander's visit to the Spring of Life follows his voyage to the farthest west and the muddy spring into which the sun sets (cf. Q18:86).[12] Similarly, al-Tha'labī located the Land of Darkness in relation to the point of sunrise or, perhaps, more plausibly, sunset[13]—in any case, in relation to the diurnal solar movement, since it is located 'at the horn of the sun'.[14] Likewise, his near-contemporary Firdawsī described how Alexander, following an encounter with the Amazons, continued his voyage to the west and arrived at a great city.[15] Beyond the city was a

12 From Friedländer's synopsis of this episode it is not clear whether there occurred any change in the direction of Alexander's course at some point, but this remains unlikely (Friedländer 1913: 158; cf. Stoneman 2003: 11; Doufikar-Aerts 2010: 37–38, 40). However, this is not the only place where Umāra describes the visit to the Land of Darkness and the Spring of Life (cf. Friedländer 1913: 142–47); elsewhere he apparently placed both the River of Sand and the Land of Darkness in the east (Friedländer 1913: 140–47; Doufikar-Aerts 2010: 183). Others placed the River of Sand or the Sabbath River in the extreme west (Doufikar-Aerts 2010: 182)

13 See Friedländer 1913: 166 for the former, Brinner 2002: 617 for the latter, and Friedländer 1913: 166 n. 6 for both variants.

14 Friedländer 1913: 165–66 (cf. Brinner 2002: 617: the land on 'the first ray of the sun'), var. on p. 166 n. 1: 'at both horns of the rising sun'.

15 *Shāh Nāma* 1338.49 (Warner & Warner 1912: vi.157), cf. 1342.22 (Warner & Warner 1912: vi.163). For the Amazons on the western Ocean near Atlas see already Dionys. Scytobr. *FGrHist* 32F7 *ap.* D. S. 3.53.1, 4.

'Deep', filled with bright water into which the sun sets, while still beyond rules a complete darkness in which is located the Spring of Life.[16] Alexander proceeds towards the Land of Darkness and the Spring of Life, into which the sun actually set, and which is located beyond another great city.[17] It was claimed that Firdawsī's sunset fountain represents the 'northern' sunset,[18] which would thus relate it to the sun's annual, rather than diurnal 'sunset', but from his description it seems that it more probably refers to the 'standard' western well of sunset. The Land of Darkness was further placed in the extreme west by Ibn Hishām (together with the islands in the Atlantic where the sun sets) and al-Gharnāṭī,[19] in the *Sīrat Al-Iskandar*,[20] and according to Dijārbekrī, who cites earlier work *Janābī'*, both the Land and the Spring are located in the west.[21] This localisation is apparently also followed in the anonymous *Iskandarnāmah*.[22]

Other versions of this episode of Alexander-legend show a similar cosmological setting, i.e., a localisation of the Land of Darkness in terms of the diurnal solar movement. Thus, the Mongolian version (early 14th c.) offers another explicit association of the Land of Darkness with the descent of the sun in the west. According to this version, Sulqarnai sets with Mother Sun and descends to the land of darkness.[23] The fragmentary text of this version consists of only four selected episodes, and much of the translation is doubtful; but the concept of a descent to the Land of Darkness following the sun is reasonably clear. Whether this version derives from an unknown Turkish original,[24] or from a more 'direct' source, it is clear that the concept originated outside Mongolia, probably in some Arabic version of the Alexander-legend, from which the Ethiopian versions (on which see below) also derive. The author of the so-called *Ethiopian Christian Romance* similarly defined the location of the Land of Darkness in relation to the diurnal solar movement. The sun both rises out of and sets into the Darkness, and beyond this Land of Darkness is an antipodal paradisiacal land of the blessed, with reciprocal daylight/night exchange. The Land of Darkness is thus a kind of a belt stretched between the two hemispheres, diurnal and nocturnal, with no special preference shown for the western edge of the earth as the location of this land.[25]

14.3. Mount Mûsās

These sources unequivocally locate the Land of Darkness visited by Alexander in the far west, at a location defined by the diurnal solar movement, more precisely, the region of sunset. At the same time, a conflation of two different but complementary notions—that of the diurnal and annual solar movement—is apparent in a number of sources treating Alexander's exploits (a possible occurrence of such conflation is perhaps also current in Firdawsī's account, on which see above, and in al-Tha'labī's account, on which see below). This conflation is exemplified in the dual appearance of the geographical feature named Mûsās in the Syriac *Alexander Legend* and the Ethiopian version of Pseudo Callisthenes' *Romance*.[26] First, it is described as a mountain (Syriac) or a river (Ethiopian) in the farthest west, near the place where the sun enters the window of heaven (Syriac) or the place of sunset (Ethiopian).[27] Second, it appears as a mountain somewhere in the north.[28] The former is related to the phenomena conditioned by the diurnal, the latter by the annual solar movement, thus epitomising—as a single cosmological feature with a dual role and localisation—the notion of interchangeability of motifs used in building models accounting for either the annual or the diurnal solar movement.

The interpretation of the localisation of the first Mûsās in the west can be further elaborated in terms of the diurnal solar movement. At the same time, this particular account connects fairly straightforwardly Alexander's exploit with Gilgameš's. Thus the scholarly consensus appears to be that according to the Syriac and Ethiopian versions of this episode, Alexander visited this cosmological mountain precisely in Gilgameš's wake.[29] Indeed, both the account in the *Alexander Legend* / Ethiopian version of Pseudo Callisthenes' *Romance* and a similar account in the Qur'ān (18:83–102), describing Alexander's exploits in the far west and east, could derive from the account of Gilgameš's cosmological voyage.[30] The Qur'ān describes how Dhū l-Qarnayn, following what is explicitly described as a 'heavenly course',[31] visited the place of sunset (a foetid spring), the place of sunrise, and the place

[16] 1339.7–20 (Warner & Warner 1912: vi.158). It is unclear whether the Deep and the Spring of Life are the same feature. At first, they are distinguished from each other, but later they seem to have merged.
[17] 1339.35–41 (Warner & Warner 1912: vi.159).
[18] Casari 2005: 236–38; 2006: 219–20; 2011: 98–99; 2012: 181–82.
[19] Doufikar-Aerts 2010: 182.
[20] Doufikar-Aerts 2010: 344. For the date of this work see Zuwiyya 2011: 84.
[21] Friedländer 1913: 165 n. 4.
[22] Southgate 1978: 48, 53, 57, 59–60, 62–65, cf. Wheeler 2003: 181–82. Here a number of elements points to the western provenance of the Land of Darkness, most notably the spring of sunset (Southgate 1978: 59), *pace* Wheeler 2003: 183 with n. 12.
[23] Turfan document *TID* 155, 10r.; Cleaves 1959: 8, 21, 58.
[24] Thus Cleaves 1959: 27.
[25] *Ethiopian Christian Romance* 8–9 (Budge 1896: 472–73, 475–76). The *Romance* is a relatively modern work but draws on much earlier material (Budge 1896: xlix; Kotar 2011: 163, 174–75).

[26] For the editions of the Syriac *Alexander Legend* see Reinink 2003: 150 n. 2. According to Schmidt 2008: 90 with n. 5, the name of the mountain *m-s-s* should be read as 'Masis'.
[27] Syriac, Budge 1889: 148; Ethiopian, Budge 1896: 226.
[28] Syriac, Budge 1889: 149; Ethiopian, Budge 1896: 228. Cf. Van Donzel & Schmidt 2010: 18 = 2009: 19. According to Pseudo Jacob of Serugh's *Song of Alexander* the location of mountain Māsīs is not specified, although Alexander follows a northern course after ascending it (101, 105; Budge 1889: 168–69; cf. Anderson 1932: 26; Henkelman 2010: 331, 336 with n. 56, 345 locates it in the farthest east). Two versions of the *Song* render the name of the mountain somewhat differently: 'Masas' (version II) or 'Masis' (version I) (Henkelman 2010: 331 with n. 32).
[29] Meißner 1894: 13; Ungnad & Gressmann 1911: 161–62, 184; Krappe 1941: 128; 1942b: 343–44; Czeglédy 1957: 242 n. 39, 245, cf. 249; Casari 2006: 225; Van Bladel 2008: 197 n. 6, 198 n. 12. Cf. Schmidt 2008: 90 n. 5; Henkelman 2010: 335; Van Donzel & Schmidt 2010: 18 n. 10 = 2009: 19 n. 11.
[30] Van Bladel 2008: 176.
[31] Q 18:84, 85, 89, 92 (see van Bladel 2008: 177, 182). In this particular case, 'heavenly course' surely refers and corresponds to the 'path of the sun' (*harrān ᵈšamši*), since it is this specific course Alexander is explicitly following.

'between two barriers' beyond which Yājūj and Mājūj dwell, most probably in the far north.[32] The *Alexander Legend* describes a similar itinerary, especially if the description is indeed referring to a 'conduit'—leading from the point of sunset ('the window of heaven')[33] and the mountain Mûsās to the point of sunrise—followed by Alexander.[34] Although the text of the *Legend* is obscure at this point, and van Bladel attempts to synchronise it both with the accounts in the *Gilgameš Epic* and the Qur'ān, representing Alexander's voyage as an exact counterpart of Gilgameš's, it is clear that Alexander's course here indeed follows the diurnal solar movement and that the narrative in which this course is described is thus structured upon this phenomenon. Unfortunately, the Syriac text is frustratingly unclear precisely at the point where it describes the position of Mûsās with respect to the diurnal solar course, leaving open the possibility of its localisation in terms of sunrise, thus in a way replicating to some extent the controversial situation obtaining in the *Epic of Gilgameš*; in any case, it is a toponym closely related to the sun's diurnal course.

The second appearance of mount Mûsās in the north in the Syriac *Alexander Legend* and the Ethiopian version of Pseudo Callisthenes' *Romance* is intimately connected with the Land of Darkness. This northern mountain is further closely associated with Alexander's building of his celebrated Gates, although perhaps not explicitly and/or unequivocally. Possibly the unnamed mountain further to the north[35] where the Gates are actually erected is to be equated with the Caucasus mountain chain as understood in antiquity,[36] which would, if one insisted on a further identification with an actual terrestrial mountain(-chain), amount to a rationalisation of a cosmological itinerary. In any case, according to the Ethiopian version (which now departs from the Syriac *Alexander Legend*), after he erected the Gates protecting the *oikoumenê* from the incursions from the north Alexander next arrived to the Land of Darkness.[37] Similarly, in the Syriac *Song of Alexander* the hero enters the Land of Darkness through a door,[38] after which he erected a mountain gate against the north.[39] Thus the Land of Darkness is here also clearly located in the north.[40] The main meteorological characteristic of this region in the far north, located beyond a mountain named Mûsās, is a complete darkness conditioned by the absence of the sun. It is probably to be understood as a cosmological feature that accounts for the absence of the sun from northern latitudes in winter, especially about the winter solstice. It thus should be distinguished from the homonymous cosmological feature in the far west which accounted for the daily disappearance of the sun at night.

14.4. Mount Mûsās in later tradition

This northern cosmological mountain was part of the account of Alexander's exploits in certain later Greek versions of Pseudo Callisthenes' novel, now with another version—or interpretation—of the name that it was destined to carry during the entire medieval period. Thus the γ and ε recensions of Pseudo Callisthenes' novel mention the *Mazzoi Borra* (Breasts of the North), two mountains in the 'unseen world' (*aphanês kosmos*) at the 'ends of the north' (*perata tou borra*), where Alexander built his famous Gates.[41] Similarly, in β and γ recensions of the novel there is another version of the same episode where the *Mazzoi Borra* are described as two great mountains extending towards the north all the way to the great sea under the ἄνσος[42] and the land of darkness.[43] The same report with identical wording is given in the Byzantine poem *Life of Alexander*, a metrical version of Pseudo Callisthenes' novel.[44]

These treatments of Alexander-legend were derived from Pseudo Methodius' *Apocalypse*.[45] Its author described how Alexander transferred the unclean nations from the extreme east to 'the ends of the north' (*perata tou borra*) where the Breasts of the North (*ubera aquilonis*/Μαζζοὶ Βορρᾶ) are located.[46] The original Syriac version of the *Apocalypse* calls the mountains 'Sons of the North', *bnay garbyā*.[47] Since the usual name in subsequent works is *bezzay garbyā*, 'Breasts of the North', it is plausible that

[32] A large number of sources explain Alexander's epithet Dhū l-Qarnayn precisely with the fact that he has reached both poles or horns of the sun at the eastern and western extremity of the earth (al-Mas'ūdī, i.126, ii.248–49 de Meynard, citing the earliest sources for this interpretation, 'Umar ibn al-Khaṭṭāb and Wahb ibn Munabbih; other sources in Budge 1896: 355, 388; Anderson 1927: 118; 1932: 97; Doufikar-Aerts 2010: 148, 178–79, 301 with n. 45; Brinner 2002: 606, 609; Wheeler 1998: 211; 2003: 181 with n. 4; Van Donzel & Schmidt 2010: 50 n. 1 = 2009: 57 n. 3). All these explanations seem only variants of Alexander's Qur'ānic association with the diurnal extremes of solar voyage ('heavenly course').

[33] Not sunrise, as in Henkelman 2010: 330 n. 31, 335, 336 with n. 56.

[34] Van Bladel 2008: 179, 181, 197 n. 6, 198 n. 12.

[35] Syriac *Legend*, Budge 1889: 149–56; Pseudo Jacob of Serugh's *Song of Alexander*, Budge 1889: 168–69, 171, 176–78, 182–84, 186, 199–200; Ethiopian, Budge 1896: 228–41.

[36] Van Donzel & Schmidt 2010: 18 = 2009: 19; cf. Hunnius 1904: 12; Czeglédy 1957: 242, 243 n. 41.

[37] Budge 1896: 242. Later, the Spring of Life was said to be located in the Land of Darkness near the right side of the rising sun (Budge 1896: 264), which could refer to the south, but in any case, should be associated with the diurnal solar movement.

[38] L. 173 (Budge 1889: 173), in a mountain? (l. 140; Budge 1889: 171). Cf. Friedländer 1913: 52–56. For the editions of the Syriac *Song of Alexander* see Reinink 2003: 150 n. 3.

[39] Ll. 348–50, 352–53, 356, 358–59, 371–72 (Budge 1889: 182–84) (gate); 376–77 (Budge 1889: 184) (north), cf. 432 and 685, referring to Jer 1:14 (Budge 1889: 186, 199).

[40] Cf. Van Donzel & Schmidt 2010: 23.

[41] [Callisth.] γ 3.26a = 3.29 (Parthe pp. 402, 406), ε 39.4, 7, pp. 144, 146 Trumpf; Anderson 1932: 35–36; Stoneman 1991: 186.

[42] Surely ἄρκτος (Anderson 1932: 39).

[43] [Callisth.] β 3.29 (Bergson p. 206); the somewhat abbreviated version γ 3.30a, Parthe p. 430; Anderson 1932: 39; Henkelman 2010: 334 n. 48; Herren 2011: 76).

[44] Ll. 5738–39, 5748–50 (Wagner 1881: 230; Anderson 1932: 39).

[45] Stoneman 2011: 8. For the editions of Syriac, Greek and Latin versions of the *Apocalypse* see Reinink 2003: 152 n. 14.

[46] Latin, ch. 8 (Sackur 1898: 74.1) (cf. ch. 11, Istrin 1897: 81.34; Stang 1996: 158); Greek, 1.5 (Istrin 1897: 19.8); Anderson 1932: 46–47; Herren 2011: 76. According to the Syriac version (8.6), Alexander transfers them from the east to the farthest north, beyond 'the gate of the world in the north' (Reinink 1993: ii.22; cf. Alexander 1985: 40–41; Martinez 1985: 133; Van Donzel & Schmidt 2010: 29, cf. 2009: 51).

[47] Ch. 8.7 (Reinink 1993: ii.22–23); cf. Alexander 1985: 41; Van Donzel & Schmidt 2010: 29 = 2009: 28, cf. 2009: 54. This is also the name of the mountains in the Syriac *Book of the Bee* 54, where Gog and Magog are placed beyond 'the gate of the world in the north' (Budge 1886: 128).

here also we should read 'Breasts' instead of 'Sons'.[48] Both variants, however, are present in Syriac texts.[49] Finally, the *Chronicle* of Michael the Syrian from the end of the 12th c. locates the original abode of the Turks/Gog and Magog 'inside' the mountains called the 'Breasts of the Earth', clearly a mistake for the 'Breasts of the North'.[50] The Armenian sources offer some further variants: thus in an extract of Pseudo Methodius' *Apocalypse* in the 12th c. *Book of Questions* two mountains named 'Northern C'oyk'' are mentioned together with 'Northern gates' and 'Northern extremes', which is probably a mistake for 'ends' (*cags*) or 'breasts' (*stink'*).[51] These *ubera aquilonis* were also mentioned in Pseudo Jerome's work,[52] and they are additionally represented in the form of two islands on the 13th c. Ebstorf map,[53] derived from his *Cosmography*.[54] Finally, they are mentioned in the *Epistola prudenti viro* (c. 1240), by the 13th c. Rudolf von Ems, who adds the toponym *Promontorium Boreum* to *Ubera Aquilonis*, and by the author of *The Travels of Sir John Mandeville*.[55]

Another similar toponym, the *Promunturium Boreum*, entered the Latin translations of the Alexander *Romance* from Orosius' work, but here it explicitly acquired a place usually occupied by the mythic mountains in the far north (this is another example of a paradigmatic shift that leaves the structure of a concept unchanged).[56] Thus in one Latin version of *Alexander Romance* two large *mountains* called 'Promuntorium' and 'Boreum' are used by Alexander to enclose the impure nations.[57] Finally, although Isidore of Seville does not mention *ubera aquilonis*, on a 12th-c. Isidorean world map immediately east of north a double island-mountain shaped in the form of breasts is depicted. On one of the 'peaks' there is an inscription that reads 'gog', on the other 'magog'.[58] The double-mountain is positioned in the same place as *ubera aqulonis* on the Ebstorf map and even the island nature of the mountain is preserved. Additionally, certain mediaeval Beatus maps depict *Mons Aquillonis* or *Mons Aquilo* in corresponding positions.[59]

Thus, these northern mountains are ubiquitous in various instantiations of the Alexander-romance, as well as in its resonances elsewhere. I will return to this cosmological feature in the closing section of this chapter, where I will offer an evaluation of their involvement in the models accounting for solar movement constructed at different points in the development of the Alexander-romance tradition.

14.5. Alexander in the far north in the Islamic tradition

As emphasised earlier, Islamic tradition—alongside the 'standard' localisation of the Land of Darkness episode in the 'solar' west (see above)—seems to have set this particular episode also in the north, in line with the accounts discussed immediately above, apparently with reference to the annual solar movement. The texts that follow this alternative localisation are few, but one in particular—Nizāmī's *Iskandar Nāma*—contains numerous explicit, if somewhat incoherent, references to the annual solar movement. What is more, this particular text details the localisation of the Land of Darkness in terms comparable to the description of the land of Laestrygonians in the *Odyssey*.

The sun's annual movement was thus treated in Islamic works that employed mythic discourse in recounting Alexander's exploits, in parallel to the treatment the concept received in the rich corpus of Islamic scientific literature. Both these types of sources belong to a common ethnographic context, and both are derived from earlier sources. But while the mythic tradition largely depended upon earlier Near Eastern traditions—occasionally mediated by Greek or Syriac sources, as when the *Alexander Legend* or Pseudo Callisthenes' *Romance* played the role of a conduit for Gilgameš-material—Islamic science owed most of its substance exclusively to its Greek predecessors (which perhaps also in this case occasionally mediated earlier Near Eastern materials). The concept of the northern limit of the habitable zone, intimately connected with the notion of the geographical arctic circle and the southern limit of a sunless region in the far north, was thus taken over from Ptolemy. Accordingly, some Islamic geographers placed the northern limit of the *oikoumenê* at Ptolemy's 63°.[60] Others somewhat modified this figure,

[48] Martinez 1985: 133, 174–75 n. 8; Schmidt 2008: 95; Van Donzel & Schmidt 2010: 30–31 n. 45 = 2009: 54–55.

[49] Martinez 1985: 174–75 n. 8; cf. Trumpf 1971: 327; Reinink 1993: ii.23 n. VIII,7.

[50] Dickens 2004: 47 with n. 296, after Chabot 1905: iii.152, 14.2.

[51] Van Donzel & Schmidt 2010: 43 = 2009: 40, cf. 2009: 55 n. 121. In another excerpt in Orbelian's Histoire *de la Siounie* 32 Alexander confined the impure nations in the far north beyond 'the gates of the north'(Brosset 1864: 93; Van Donzel & Schmidt 2010: 42 = 2009: 39).

[52] *Cosm.* 23, 32, 39–40, 41b, 60 (Herren 2011: 24–25, 32–33, 48–49, 52–53, 134–35).

[53] Woodward 1987: 310 Fig. 18.19.

[54] Anderson 1932: 42 n. 1, 43, 51 n. 6.

[55] *EPV* in Burnett 1984: 164; Rudolf's *Alexander* 5.17305, 17311 (Anderson 1932: 78); *The Travels of Sir John Mandeville* 29, where the Caspian mountains are called 'Uber', in which the Gog and Magog/the Lost Tribes are enclosed (Higgins 1997: 181–82; 2011: 157; Kohanski & Benson 2007: 82–83, ll. 2352–55; Bale 2012: 104; cf. Krappe 1942b: 335).

[56] Oros. 1.2.47; Anderson 1932: 50; Trumpf 1971: 327.

[57] *Historia de Preliis* J² 77 (Hilka 1920: 141; 1977: 4; cf. Anderson 1932: 50). This recension is the source of the accounts in the French prose *Alexander Romance* 77 (Hilka 1920: 141; cf. Anderson 1932: 50), the Middle Swedish poetic version of the *Romance* (ll. 4019–20; Ahlstrand 1862: 133; Anderson 1932: 50), and the Hebrew version of *Alexander Romance* according to MS Héb. 671.5, Bibliothèque Nationale, Paris (Van Bekkum 1994: 77; strangely, according to MS London, Jews' College no. 145, the mountains are located in the extreme *south*, Van Bekkum 1992: 119; cf. Schmidt 2008: 95 n. 17; Van Donzel & Schmidt 2010: 30 n. 45 = 2009: 55 n. 120).

[58] München, Staatsbibliothek, Clm 10058, f. 154v (Harvey 1996: 25).

[59] Branch I, Burgo de Osma, Archivo de la Catedral, Cod. 1, f. 34v–35r (Williams 1997: 8 Fig. 1); Branch IIa, New York, Pierpont Morgan Library, M. 644, f. 33v–34r (Williams 1997: 9 Fig. 2).

[60] Thus al-Khwārazmi, Suhrāb (Tibbetts 1992: 102, 105 n. 69), the anonymous author of *Hudūd al-ʿĀlam* (2.3, 5, 4.26; Minorsky 1970: 50–51, 59), al-Dimashqī (Nansen 1911: ii.212 n. 1) and al-Khāzini (Rosenthal 1967a: i.114–15).

but not significantly.[61] Indeed, al-Dimashqī described the area extending from the northern tropic (66°25') to the north pole (90°) as 'the climate of darkness',[62] perhaps as a nod to the mythic tradition. The notion of darkness in the far north, which is conditioned by the annual solar movement and intimately connected with these related geographical concepts, was also incorporated into Islamic mythic narratives, some of which involve Alexander.

In this respect one of the most telling treatments of Alexander's exploits is found in Nizāmī's *Iskandar Nāma* (late 12th c.).[63] Here, according to the first version given by the poet, Alexander searches for the Fountain of Life in the Land of Darkness located under the northern pole star.[64] As if guided by Khizr,[65] but apparently not escorted by him, he arrived at a place where the sunlight appeared, apparently from the direction of the north celestial pole, only for a brief moment, before it set again below the horizon.[66] Therefore, the sun was visible (on a certain day) only for a brief moment, before it disappeared below the horizon. In truth, it certainly would not appear on the day of the winter solstice—or on any other day, for that matter—on the northern horizon, but rather on the southern.[67] On the other hand, for an observer at the pole, the sun would not set at all during summer months but would only approach the northern horizon at 'sunset'. The following lines increase the confusion. First the poet mentions that the extremity of the equator is laid upon the horizon, and that it stops at or agrees with the north pole,[68] which probably means that the equator corresponds to the line of the horizon. This more or less clearly implies that Alexander had arrived at the north pole itself.[69] However, the following couplets, such as that describing the continuation of the voyage towards north,[70] which is hardly likely when one is at the very north pole, together with the notion of complete darkness,[71] as well as the description of the ends of the earth at the encircling sea,[72] show that the poet recognised the location of the Land of Darkness generally at the northern edge of the earth, and believed that an eternal darkness rules there.

However, the phenomena described by Nizāmī in couplet 59—with the qualifications adduced above—are applicable only to a region immediately south of the arctic circle.[73] Therefore, if we take the poet's words explicitly, he must have had in mind the latitude of 66° north or a region in the immediate vicinity. It follows that it is hard to reconcile this concept with the conditions characteristic for the very north pole mentioned in the immediately following couplet 60. Neither of these pieces of information are in agreement with the idea of an 'eternal darkness' expressed in couplets 61 and 67, nor with the concept of earth's boundary in the far north (couplet 64). The quality of geographical and astronomical information provided by Nizāmī is thus rather disappointing. However, he seems to have interpreted the itinerary of Alexander's voyage to the Land of Darkness in terms of the annual solar movement, more precisely, he located the regions traversed by the king in relation to the sun's behaviour on various latitudes at certain—recklessly defined—seasons or dates. In this he seems to have followed an earlier tradition.

Alexander's visit to the Land of Darkness was indeed associated with the area of the north pole by the authors earlier than Nizāmī. Thus al-Tabarī (early 10th c.) succinctly described how the king, after he conquered China, entered the dark area near the north pole, where the sun stood towards the south (?).[74] Similarly, al-Tha'labī (11th c.) mentions in a chapter heading Dhū l-Quarnayn's expedition to the Land of Darkness at the north pole.[75] However, in the text the Land of Darkness is located 'at the horn of the sun',[76] and Dhū l-Quarnayn attempts to reach it by travelling towards sunrise or sunset,[77] while the north pole mentioned in the heading does not appear in the narrative at all.[78] The actual narrative thus differs from the title in that it interprets Alexander's voyage in terms of the diurnal solar movement. Other Islamic authors, such as al-Gharnātī, associated Alexander's visit to the Land of Darkness with similar meteorological features characteristic for high northern latitudes.[79] Finally, an Indian *mappa mundi* depicts the Spring of Life, discovered by Moses, as a black rectangle at the very north pole.[80] Thus the tradition of the Land of Darkness at specifically the north pole certainly existed in Islamic mythic geography, and Nizāmī's rendering of the episode shows his acquaintance with the concept. At the same

[61] Al-Battānī, *De scientia astrorum* (60°) (Nallino 1903: i.17; Burnett 1982: 335); al-Mas'ūdī *Muruj* 1.180–81 (Sprenger 1841: 196–97, ch. 8) (60°); al-Khizin (60°45') (Rosenthal 1967a: i.115); Ibn Khaldūn (64°) (Rosenthal 1967b: 50, 55, 56); al-Idrīsī (somewhat to the north of 64°) (Ahmad 1992: 162–63; cf. Nansen 1911: ii.203); al-Farghānī (66°) (Tibbetts 1992: 102 n. 59); al-Dimashqī (63°, 66°10' or 66°25') (Nansen 1911: ii.212); Hāfiz-i Abrū (over 66°) (Ahmad 1992: 170). Larger deviations: 'others' (77°) (Rosenthal 1967a: i.115).

[62] Nansen 1911: ii.212.

[63] For Nizami's sources for the episode involving the search for the Fountain of Life, of which the most important was an Arabic version of Pseudo Callisthenes' novel, see Friedländer 1913: 210.

[64] *Iskand.* 68.26, 31 (Clarke 1881: 788; Friedländer 1913: 210–11; Casari 2011: 96; 2012: 179).

[65] *Iskand.* 68.40 (Clarke 1881: 789).

[66] *Iskand.* 68.59 (Clarke 1881: 791: 'a luminosity (of the sun) appeared from the (northern) pole of the sky; it (the luminosity) ascended (above the horizon) and descended (below) quickly in a moment'; Friedländer 1913: 211: 'Vom Pol des Himmels-kreises erschien (Sonnen)licht; es ging in einem Augenblick schnell auf und unter').

[67] Thus also Casari 2011: 96.

[68] *Iskand.* 68.60 (Clarke 1881: 791; Friedländer 1913: 212).

[69] Friedländer 1913: 212 n. 2; cf. Clarke 1881: 792.

[70] *Iskand.* 68.65 (Clarke 1881: 792; cf. Friedländer 1913: 212).

[71] *Iskand.* 68.61, 67 (Clarke 1881: 791, 792).

[72] *Iskand.* 68.64 (Clarke 1881: 791).

[73] Cf. Casari 2011: 96.

[74] *Ta'rīkh* 701 (thus Friedländer 1913: 232; Perlmann 1987: 94: 'the area of southern sun'; cf. Casari 2005: 237; 2006: 215; 2011: 97; 2012: 179).

[75] *'Arā'is, Dhū l-Quarnayn* (Friedländer 1913: 163; Brinner 2002: 616).

[76] Friedländer 1913: 165–66 (cf. Brinner 2002: 617: the land on 'the first ray of the sun'), var. on p. 166 n. 1: 'at both horns of the rising sun'.

[77] See Friedländer 1913: 166 for the former, Brinner 2002: 617 for the latter, and Friedländer 1913: 166 n. 6 for both variants.

[78] Cf. Friedländer 1913: 166 n. 6.

[79] *Murib* 9–10 (Bolshakov & Mongait 1971: 30), *Tuhfat* 117, 119 (Bolshakov & Mongait 1971: 58–59).

[80] Schwartzberg 1992b: 395–96 with Fig. 17.4; cf. Harley & Woodward 1992: Pl. 29.

time, he included in his version of Alexander's expedition into the Land of Darkness geographical and astronomical facts pertaining to, and meteorological phenomena obtaining at, the region below the fixed arctic circle. This region was associated in Greek thought, as we have seen in chs. 1 and 4, with the Laestrygonians, Syriê and Thule, but this particular model involving the Land of Darkness constructed by Islamic thinkers in order to engage with these phenomena was manifestly different from the Greek one. Furthermore, it was rather infrequently and only sporadically employed, and does not seem to have played an important role in their understanding and accounting for the phenomena of the far north.

It appears that already in the 10th c. Islamic authors were well aware of the meteorological phenomena characteristic for the far north either from their own experience or from second-hand reports. The region was in general located either at the limit or beyond the *oikoumenê*. Furthermore, it seems that by the same period these phenomena were associated with Alexander's exploits in the Land of Darkness, itself usually located with reference to the 'extremely northerly' land of the Bulghars.[81] Nizāmī's report, however chaotic, is thus an idiosyncratic version of a firmly established set of 'facts' of some interest for Islamic mythic geography. The notions of the north pole, eternal darkness, a 24-hour (or much longer) continuous daylight or night and the positions of sunrise and sunset on the horizon in these unusual circumstances were all incorporated in the narrative of Alexander's exploits, more precisely, in the episode concerned with the search for the Spring of Life in the Land of Darkness.

14.6. Conclusion

It is thus clear that according to one strand of evidence, perhaps primary (as exemplified by the *Epic of Gilgameš*), Alexander's exploits were associated with the diurnal solar movement, but the location of the Land of Darkness in the far north cannot be integrated into this model and should rather be associated with the sun's annual movement. The ambiguity could well date from an early period, as many of its facets are already present in Mesopotamian texts; moreover, it seems to be inherent in the very nature of the phenomena treated by the mythic models accounting for solar movement. One of the most prominent features of this tradition is indeed the interchangeability of motifs incorporated into the mythic models respectively accounting for the diurnal and annual solar movement. This interchangeability is conditioned specifically by the related nature of the decisive moments in both the diurnal and annual solar movement and by the interdependence of these two movements in general.

Among the most prominent geographical or cosmological features in the Alexander tradition are the omnipresent Breasts of the North. These northern mountains, regularly imagined as a two-peaked cosmological and/or geographical feature, preserved through the ages a vague connection with the sun in their capacity as the boundaries of the 'land of darkness', but lost in the process the particular association with the diurnal solar movement evidenced in the Mesopotamian tradition.[82] The loss of this association is already apparent in the Syriac *Alexander Legend* and some Greek recensions of Pseudo Callisthenes' novel, where this cosmological feature was instead associated with the north and the absence of sunlight. It could be argued that the association with (perpetual) darkness rather reflects the liminal position of the double mountain in relation to the annual movement of the sun (similar to the probably double-peaked Bear Mountain of the pre-Homeric Argonautic tradition discussed in chapter 3, one of whose names was indeed derived from 'twin breasts'), but this interpretation is perhaps too charitable.[83] If accepted, it would add this cosmological feature to an assemblage of motifs that were interchangeably used in building the models accounting for both aspects of solar movement.

It might appear that a 'solar' interpretation of the Land of Darkness from the Alexander romance tradition, i.e., a definition of this region in terms of the annual solar movement, based on its characteristic meteorological features, geographical position and probable derivation from the earlier tradition related to Gilgameš, epitomised in the presence of mount Māšu/Mûsās/*Mazzoi Borra* as a cosmological feature accounting for some aspects of solar movement, is perhaps a bit strained and does not measure up to the standard of rigorousness announced in the introduction and followed in the rest of this book. The main reason for this relaxation of criteria is the distance of the texts available for study from their sources (most notably the Gilgameš tradition), which conditioned that the references to solar movement in these texts are less explicit than one would desire. However, a clear presence of these references, supported by a comparison with

[81] The Land of Bulghars, placed in the vicinity of the Land of Darkness, represents another concept in Islamic tradition accounting for the meteorological phenomena of the far north (Ibn Fadlān *Risalat* 17, 206a.3–8 with Yāqūt's report (Nansen 1911: ii.144); al-Gharnātī *Murib* 14–15 (Bolshakov & Mongait 1971: 32), *Tuhfat* 119 (Bolshakov & Mongait 1971: 59; Stang 1996: 241); Nizāmī *Iskand.* 68.56, 58 (Clarke 1881: 790–91); Hamadānī (Casari 2006: 217–18); Ibn Battuta *Rihla* (Gibb 2005[1929]: 150–51); *Mesalek Alabsar fi memalek alamsar* or *Masalak-al-Absar* (Quatremère 1838: i.285, cited in Yule & Cordier 1903: ii.485 n. 1)), regularly experiencing similar meteorological phenomena (Ibn Fadlān *Risalat* 16–17, 205b11–206a8; al-Mas'ūdī *Muruj* 13 (Sprenger 1841: 415–16); al-Qazwīnī (Nansen 1911: ii.210); al-Gharnātī *Murib* 9–10 (Bolshakov & Mongait 1971: 30), *Tuhfat* 117 (Bolshakov & Mongait 1971: 58); Ibn Battuta *Rihla* (Gibb 2005[1929]: 150–51)). Independent accounts of the meteorological phenomena of the far north are given by al-Gūzgānī (Raverty 1881: i.268–69; cf. Minorsky 1970: 309), Marco Polo *Il Milione* 4.21 with Ramusio's version in Yule-Cordier 1903: ii.484–85 with n. 1 (for Polo's non-European sources for his account see Casari 2011: 98) and Sharaf ad-Din Ali Yazdi *Zafar-Nama* 3.13, Darby 1723: i.370 (Page 1973: 46).

[82] A two-peaked cosmological mountain (exemplified in the *akhet*-sign), sometimes conceived as a female goddess' breasts, was also involved in some Egyptian models accounting for the diurnal solar movement (see Appendix 2).

[83] Later texts that include the Breasts of the North in their narrative dropped even this minimal connection with solar movement, retaining only the association with the farthest north.

earlier sources from which these narratives were clearly derived (sometimes at several removes), justifies their inclusion in the class of mythic narratives referencing the solar movement.

The analysis of the Gilgameš and Alexander traditions has taken us back to the early periods of Mesopotamian history. The road leads back from the mediaeval Arabic and Persian legendary itineraries of Alexander, through the Qur'ān and more or less contemporary Syriac renderings of the Alexander legend, the somewhat earlier Armenian and Latin translations of the original Greek version of the *Alexander Romance* and the Greek original itself, to, finally, the cuneiform sources dating back to the first half of the 2nd millennium BC. The road, however, leads further back, to the second half of the 3rd millennium BC and the Sumerian literary renderings of the Gilgameš legend. In spite of this general conclusion, I have not attempted a thorough reconstruction of the transmission of motifs and narratives either diachronically or synchronically. This particularly applies to their appearance in the Greek ethnographic context, which is the main concern of this book. I am not in principle against direct or indirect genetic connections between the Greek and Mesopotamian tradition. This non-committal stance is motivated by my focus on recognising a class of myths referencing solar movement and interpreting the narratives belonging to this class in terms of solar phenomena, rather than on establishing their stemmata. I have occasionally commented on more obvious cases of correspondences between the two traditions, without discarding the possibility of similar interpretations in other cases. This apparent conservatism is thus a consequence of a different focus, rather than a definitive standpoint.

15

Conclusion

In the last decades the very concept of a conclusion of a scholarly book in the humanities has been questioned. West thus placed a conclusion as the *first* chapter of his study of the *Odyssey*; Graham offered a narrative instead of the hitherto used argument in his study of the development of Presocratic astronomy; and Lincoln in general criticised both the 'summarising' and 'theoretical' flavour of standard conclusions of scholarly books.[1]

Instead of following the tendency to use the concluding chapter as theoretical superstructure for the preceding main body of the text,[2] I have offered a rather extensive 'theoretical' introductory chapter, while the rest of the book often tacitly leaves the theory embedded in the exposition and analysis of specific problems.[3] Consequently, I will not try to rehearse this somewhat laborious exposition here. Instead, and in place of a synopsis of the book, which would be, due to its patently episodic or nonlinear overall structure, equally laborious and thus, I believe, unhelpful,[4] I will offer an outline of the main argument of the treatise together with a short narrative exposition of several 'practical' main points that can be extracted or abstracted from the main body of the text (where the actual arguments for these extractions/abstractions are provided).

The nonlinear structure of the book is primarily a consequence of the fact that it represents—to the best of my knowledge—the first attempt at reintroducing this particular subject-matter into a less marginal position of scholarly interest, which necessitated the establishment of criteria for, and the initial formation of, an acceptable body of evidence for analysis. This resulted in a book with a structure that does not follow a strict diachronical arrangement, i.e., it does not offer a systematic analysis of the development of the notions of solar movement through successive periods of Greek history. In addition, my study uses the concept of Greek ethnographic context as a methodological framework that allows a synoptic view of this body of evidence, focusing on particular subjects as treated within this cultural setting. Naturally, this resulted in an episodic or topical overall structure of the book.[5]

15.1. An outline of the main argument of the book

Before I discuss the 'practical' main points, I will summarily expound my position on the main subjects studied in this book.

I have attempted to analyse a number of motifs that appear in myths, as well as a smaller number of complete episodes or stories (sometimes these categories overlap significantly), that seem to be in some way derived from solar observations, i.e., that have solar movement as their referent. These observations were as a rule processed in terms of anthropomorphic causality, due to its significant cognitive potential and the absence of non-personalistic interpretative strategies. They were consequently communicated in the form of stories, whether through a creation of new sequential narratives or an adaptation of the already existing ones by the integration of information obtained through solar observations. I have recognised two main strategies by which this integration was achieved. The prevailing practice seems to have been a simple incorporation of observational or interpretative data, but, rarely, actual narratives were structured on an analogy with solar movement. These myths can be thus understood as narrative accounts of phenomena in terms of anthropomorphic agents. They are necessarily speculative, based upon both empirical observations and intuitive ideas, with the latter dominating the explanatory facet of myth. But modern philosophy and history of science nevertheless accept this cognitive strategy as somewhat similar to scientific approach, insofar as anthropomorphism, metaphor, analogy and model-building are also present up to a point in science in its attempts to access physical reality and with respect to the fact that mythic approach crucially differs from the scientific in terms of their application of different idioms, respectively personalistic and non-personalistic. Other differences between these two approaches appear to be in degree rather than kind. I have described symbolic modes of thought, including both myth and science, as a continuum with somewhat blurred boundaries and much overlapping between different approaches, rather than monolithic edifices clearly separated from each other. This continuum was implicitly observed by previous scholars and attempts were made for its nuanced division in the bodies of ancient knowledge of geography and astronomy. Crucially, some myths clearly shared their subject matter with 'meteorology', geography and astronomy, which means that their referents belong to the same physical reality as do the subjects of study of these particular scientific disciplines.

---

1 West 2014: vii, 1–4; Graham 2013: 227–28 and on to p. 237; Lincoln 1989: 171.

2 Lincoln 1989: 171.

3 Thus following Lincoln 1989: 172, but going against Csapo 2005: 1, who argued that providing a 'definition' at the beginning of one's study is strongly tendentious. I have stated my arguments for avoiding the pitfalls outlined by Csapo with respect to this practice in the introductory chapter.

4 Cf. Lincoln 1989: 171.

5 In principle, within this framework the testimonies coming from different periods are almost indiscriminately taken into account. However, I have opted for a more historical approach, fairly sensitive to the issues of chronology and diachronical relations. In the present case, the focus is thus heavily on the earliest tradition, with later testimonies only evoked as supporting evidence, and the Greek ethnographic context studied in this book could be qualified with the adjective 'early'.

It thus seems that the best strategy to assess a society's knowledge on any subject is to analyse all available sources for this knowledge, including that that would conventionally be labelled as non-scientific or even irrational. This is an evocation of a basically structuralist notion of ethnographic context, which intends to interpret myths of a society in conjunction with other sources that have identical or similar referents, but which do not belong to myth, ritual or religion. These sources belong to different spheres of human intellectual activities, which include technical and scientific texts. Even though mythic geography and astronomy (i.e., the body of knowledge that appears in myths but that would at present be classified as belonging to these respective scientific disciplines) often seem incommensurable with ancient *scientific* geography and astronomy, other branches or sections of these ancient disciplines (speculative astronomy, cosmology and descriptive astronomy in Couprie's classification or intuitive/naïve and scholarly/canonical geography in Dan, Geus & Guckelsberger's classification) are much more conductive to the comparison. Indeed, the notion of a continuum seems to be especially applicable to the ancient bodies of knowledge on particular subject matters. Thus, ancient literary tradition—including myths, poetry and fictional works—is accepted, with certain qualifications, as a relevant source for the study of ancient geography and astronomy, while at least the former shows a remarkably descriptive and narrative character, with the latter also not innocent of a similar portrayal. In this way, to gain full access to an ancient society's knowledge of some subject one must analyse both its scientific (if available) and non-scientific texts; if the former are lacking, which is more often the case than not, one is left with only the latter. But, at least in the Greek case, there is a significant continuity and overlapping between the two approaches, which allows a concurrent study of earlier mythic accounts with later non-personalistic ones.

The practice of creating narratives of the kind outlined above testifies to the importance these observations had in traditional societies, belonging to a body of knowledge important for the community in terms of calendar and cultic festivals, agricultural cycles, travel and trade, weather-forecast, seasonal migrations of game animals, etc. This body of knowledge is not particularly impressive from the point of view of modern science, but it was nevertheless shared by the proponents of the newly-formed Greek philosophical and scientific discourse—for an example of the former one could adduce a number of Presocratic, especially Milesian, cosmologies, for an example of the latter it is Pytheas who first comes to mind—despite the fact that it was precisely the proponents of the new non-personalistic paradigm who created the category of 'mythic' in the first place and who took an agonistic and polemical stance towards their predecessors. Thus, myths were relegated to a category of non-scientific, even irrational, as opposed to *logos*; those who wanted to exculpate them (and thus 'rationalise' them) had to read them as allegories, most often understood as intentional and with referents either in the realm of physics or ethics,[6] which is a practice still occasionally in use today, especially outside regular scholarship. But the crucial error was always in trying to fit the existing vast body of material within the framework of comprehensive theories of myth. Modern scholarship had little trouble in criticising and dismissing such comprehensive interpretative strategies. Unfortunately, in the process it also dismissed the key subject matter of the solar-myth paradigm, which is of crucial importance to my study: the observance and study of solar movement by the Greek and other civilisations of the ancient world outside of technical texts in which it was expounded in a non-personalistic idiom not unlike that of modern science.

In this book I have attempted to reintroduce this material from the unenviable margin it holds today and position it nearer to the centre of the polyparadigmatic studies of myth and classical religion. While the nature/solar-myth paradigm understood solar phenomena as the very pivot of its hermeneutic approach—often reading it into texts and traditions that had little or nothing to do with solar observations, sometimes stretching the evidence bizarrely far—its successors, especially those adhering to the socio-cultural framework of interpretation of religion who dominated the field in the 20th c. naturally withdrew from this subject matter and looked at it—and those who took it seriously—with a high level of suspicion and scepticism. This seems to be the way scholarship functions instinctively, but the return of the pendulum, which is what I am attempting here, is its another standard trait (although I prefer the 'spiral of scientific development' metaphor/analogy/model).

I have used simple criteria for the inclusion of motifs, episodes and complete stories in the category of myths susceptible for the analysis in terms of solar movement. The basic criterion is either an explicit reference to solar movement or one that does not require a strained interpretation. With that said, I must concede that at least some of my interpretations of references to solar movement in myths are open to discussion. Yet even if some of these less straightforward interpretations were to be refuted in the future, this will not affect my general position, since there are many clear and unambiguous references to solar movement in myths. The ones that first come to mind are, in Greek tradition, Helios' voyage in his golden cup and the threshold of the house of Night, referencing the diurnal, or Syriê and its 'turnings of the sun' and the Laestrygonian polar night, referencing the annual solar movement.[7] Any interpretation that would attempt to circumvent solar movement as one of hermeneutic keys for the elucidation of these mythic episodes would necessarily be much more

6 See Brisson 2004 and Hawes 2014.

7 As repeated several times in this book (paraphrasing R. Parker), it is the existence of a certain practice that needs to be accounted for, notwithstanding its apparent or real—however defined—marginality, so even if there were even less evidence for references to solar movement in Greek myths than there actually is, these would still have to be accounted for.

strained than the ones offered here, since it would have to explain away the clear solar references in these stories. With that said, I do not believe, nor have I claimed, that there exists a one-to-one correspondence between solar movement and these particular stories, or any other treated in this book, for that matter. The notable exception, up to a point, are some rare narratives or segments of narratives that were indeed explicitly built upon an analogy with the diurnal solar movement and/or account for it in a straightforward way—an example of the first category would be Gilgameš's voyage upon the sun's path, the example of the latter Helios' voyage in his golden cup. Thus, I readily accept that other interpretive approaches can cogently explain some facets of these traditions, as well as that in some cases these other approaches have a larger explanatory potential with respect to an entire episode or a complete narrative I have attempted to analyse. However, this does not diminish the relevance of solar movement as a useful hermeneutic tool in assessing them.

The most important geographical and astronomical concepts engaged by myth discussed in this book are all related to the diurnal and annual solar movement and their defining moments, especially the solstices. Here one could also add the notion of the antipodes defined in relation primarily to the diurnal solar movement and almost regularly having certain connections with one eschatological tenet or another and (in later terminology) the latitudinal zonal division of the earth. The question of the earth's sphericity was not discussed, since all these phenomena could in some form be as applicable to the flat earth as they are to the spherical. For example, one can hold the notion of the limit of the annual solar movement within the flat-earth framework, and interpret it in terms that do not contradict the principles conditioned by this framework or presuppose another, even though the introduction of the notion of a spherical earth dramatically changed the understanding, nature and definition of this particular limit. At the same time, the notion of latitude in terms of a flat earth is certainly distinct from that in terms of a spherical (to which we are accustomed to applying it), but it can nevertheless be heuristically applied retrospectively in order to refer to the concepts envisaged within the chronologically earlier framework.

Terms such as the Bear(s), Laestrygonians, Telepylus, Thule, Strophadas, Harpies, Syriê, Helixoia, Hyperboreans, (Hyperborean) Apollo, Ocean, Rhipaean Mountains, Pillars of Heracles, etc. (to constrain ourselves to the vocabulary of Greek mythic discourse), were used to designate certain concepts we would label geographical and astronomical (for the Greeks both would fall under the term 'meteorological') parallel to the more straightforward (from our perspective) terms such as arctic (always-visible) circle, fixed or geographical arctic circle, tropics, last circumpolar star(s), latitude, annual solar movement, solstices, horizon, etc. The first set of terms can conveniently or heuristically be called mythic, while the other certainly conforms to our notion of scientific terminology, whether they refer to a clearly defined concept, such as the fixed arctic circle, or to certain errant or, better, imprecise notions, such as the always-visible circle(s). Both sets of terms, however, refer to comparable phenomena easily observable in physical reality, one accounting for them in a narrative model-building mode, inherently speculative, the other in the form of a more systematic discourse, with an accent on empirical verifiability and logical consistency. These two modes result from the application of two different cognitive strategies, while the difference between them is idiomatic. Our unfamiliarity with the idiom, which is in the end comprehensible and in the main rational, should not prevent us from approaching it with a well-balanced application of the principle of charity.

The narrative model-building mode is characterised by anthropomorphic causality, which is probably the main reason for our dismissal of it as unscientific. It was probably used already in the Upper Palaeolithic, when the humans accounted for periodical processes in nature by recurring to analogues derived from their immediate environment. It is perhaps in this early period that the observations of solstices actually began, which can thenceforth be followed throughout the unrecorded and recorded history.[8] In the Neolithic there already appeared analog devices used for marking, measuring and predicting—and at the same time verifying—periodical temporal phenomena, among which the movements of the sun are certainly not the least important nor the ones that are the most difficult to observe. These phenomena and processes in physical reality were accounted for in the narrative model-building mode by recourse to cognitive strategy characterised by anthropomorphic causality and the use of metaphors and analogies. Some of these mythic models are available for study in the form of their later instantiations in Greek and other ancient traditions. This is not to claim that, for example, the hypothetical prehistoric myths referencing solar phenomena are identical to the Greek ones accounting for comparable events and processes, but rather that both were expressed in a similar idiom characterised by anthropomorphic causality and narrativisation, both creating models of solar movement out of a limited number of motifs extracted from human experience.

Crucially, I have attempted to interpret only one specific category of narratives within the larger corpus of what is usually styled myth. This category, formed by the application of a set of rigorous and unambiguous criteria, can be understood as consisting of narrative models accounting for some specific (astronomical and geographical) phenomena in physical reality. In the first part of the book, I have placed the strongest emphasis on the meteorological (in Greek terms) concept of solstices, but they are discussed together with other phenomena and processes related to the diurnal and annual solar movement, as well as other celestial phenomena in some

[8] Ruggles 2005: 318, 386; Hayden & Villeneuve 2011: 332, 347, 350–51.

way connected to these, in the first place, the fixed arctic circle.

The Presocratics, although continuing the work of their predecessors that operated within the mythic tradition, introduced a new idiom into the observance and study of nature, one in which personalistic causality was either expelled to the beginnings and confines of the *kosmos* or abandoned altogether (although never as completely as they themselves perhaps wished). At the same time, stories were no longer acceptable for those professedly scientific-minded authors for either describing or explaining the phenomena, and anybody who wanted to discuss such subjects had to comply with the new idiom. Even so, these myths continued to be used in various ways right to the end of antiquity and beyond even by the philosophers themselves. In a truly ironic twist, while once mythic stories accounted for the phenomena of physical reality, now these phenomena were introduced as explanantia of the ancient, sometimes apparently incomprehensible—or morally unacceptable—stories. These allegorical interpretations were often utterly unconvincing, adapted to their authors' tastes and intentions, as a rule blatantly ignoring the very text they were meant to elucidate.

15.2. The 'practical' main points of the book

As announced at the beginning of this conclusion, there are several important abstractions deriving from the main part of the book that were not mentioned in the introductory essay. I will here briefly outline these abstractions.

First, Nakassis' uni-/bi-polar model accounting for solar phenomena of sunrise and sunset was recognised as an indispensable hermeneutical tool in the understanding of various myths that have solar movement as their referent. This model, which I further developed, explains the frequent duplication of certain localisations in myths, most often in the far west and below the earth, through two alternative ways in which solar phenomena of sunrise and sunset were accounted for in mythic discourse: according to the 'uni-polar' model the alternation between night and day occurs at the *axis mundi*, represented by and located in the region beneath or below the inhabited world, whether in terms of a flat or spherical earth, while according to the 'bi-polar' model the sun rises and sets on the eastern and western horizon, respectively. In both cases, mythic discourse was employed in elaboration of these important natural phenomena, but two different models accounting for the diurnal solar movement were developed, both based on observations and both offering a—highly speculative—theory that coherently accounted for these observations, even though some important details were left unexplained by both models.

Second, the interchangeability of motifs used in building models accounting for respectively the diurnal and annual solar movement was recognised as a widespread practice, conditioned by the similarity and interdependence of these two movements. A related occurrence is the limited assemblage of motifs from which the mythic models accounting for both the diurnal and annual solar movement were built. This was conditioned by the fact that they were configured by typical action patterns extracted from human experience, in its turn conditioned by the process of anthropomorphisation of physical phenomena in general and solar behaviour in particular.

Third, a widespread interaction between the conceptual domains of solar movement and eschatology, encompassing various beliefs in posthumous fate of mankind, was documented in different traditions. Necessarily, this association could only have arisen—and could only have been maintained—in mythic traditions, where an argument from analogy (just as the sun as a (divine) being lives continually and resurrects both daily and annually, so do humans) could have been validly devised on the grounds of a personalistic account of causality of periodical natural phenomena and processes. This analogy was necessarily based on the observations of both the diurnal and annual solar movement, as well as on a cosmological framework that could accommodate the existence of 'another' world below or beneath the *oikoumenê*. A rather widespread cosmological understanding of the location of the Otherworld specifically related to the solar movement is thus a direct consequence of the construction of specific models of the universe and models of physical phenomena that are understandable in terms of such cosmological frameworks. Sometimes these understandings of the localisation of the Otherworld represented wholesome models (for example, in the New Kingdom Egyptian tradition) or, more frequently, appeared as components of models accounting for the solar movement and related cosmological features, such as the antipodes. Their existence is another testimony to the importance of solar phenomena and the mythic models designed to account for them in various branches of human intellectual activities, including eschatological speculations.

Finally, it appears that both the annual and diurnal solar movement were approached in different ethnographic contexts by an application of similar strategies. This resulted in, and is recognised through the existence of, shared cross-cultural models accounting for these phenomena, or rather two aspects of a single phenomenon. A prominent example of these shared or parallel traditions is precisely the omnipresent interaction of the conceptual domains of solar movement and eschatology noted immediately above. The study of these parallel traditions can thus offer a valuable outside perspective on the Greek understanding of the concept of solar movement.

15.3. A final word

The myths referencing solar movement represent a recognisable class of narratives if a clear set of criteria is applied in their recognition. They are not, however, a distinct group of stories in some way separated from other narratives current in a particular ethnographic context. The solar references embedded in these narratives—

or, less frequently, the narratives structured on an analogy with solar movement—should nevertheless be recognised as such and interpreted in terms of their referents or analogues. The fact that this subject matter was particularly unfortunately misused in one period of scholarship should not sentence it to a permanent exile to the margins of scholarly interest. The *nachleben* of mythic models accounting for the movement of the sun, or of the motifs used in their creation, in later more objective (i.e., less personalistic) traditions adds to the value of studying this particular class of myths or mythic models accounting for the solar movement in anthropomorphic terms.

At the end, I would like to offer a reconstruction of a 'standard' mythic model accounting for the solar movement. Naturally, this reconstructed 'standard' model is not an *Ur*-model, an original narrative from which all others are derived, but only an abstraction derived from the actual models themselves. One should not evaluate the existing models in terms of their correspondence with the reconstructed 'standard' model, but the latter can be understood as a heuristic tool in the analysis of myths suspected of either containing solar references or being built upon an analogy with solar movement.

The diurnal solar movement starts with sunrise, which is conceptualised as the sun-god passing through the horizon-gates or rising behind a (double) mountain or over an island. The sun-god most often travels over the sky in a chariot during daylight hours, while at sunset he (rarely, she) once again passes through the horizon-gates or sets behind a double mountain or over an island. These second gates (or, rarely, corresponding features) are often at the same time an entrance to the Otherworld, regularly conceptualised as a posthumous abode of humans; the cosmological location of this region thus stands in a direct relationship with the diurnal movement of the sun. In a uni-polar model accounting for the daylight/night exchange, the sunrise and sunset features coalesce into a single cosmological location set in the Otherworld. During the night the sun travels over the region of the antipodes, as a rule envisaged as having more or less pronounced eschatological traits, where the daylight and nighttime are reversed, or whose typical state is that of darkness, only illuminated by the sun during (our) night. The sun can travel in a boat (the most frequently used vehicle for its nocturnal journey) upon a circumambient Ocean by way of north, often with a high mountain or the elevation of the earth-plane blocking its light, or by way of south or at the underside of the earth (with these two concepts often coalescing), irrespective of the acceptance of a flat or spherical earth as the underlying cosmological framework in either alternative. This southern/underside passage often involves the concept of a cosmic nadir, a cosmological location equidistant from west and east, corresponding to the meridian midnight transit of the sun, whether at the underside of the earth or in the south upon its surface.

The models of the annual solar motion are constructed out of similar or identical motifs, with the sun god travelling north–south and back on the eastern and western horizon, as well as on the meridian. Using the imagery of chariot-race, he turns at the liminal points in his orbit/race at the solstices (sometimes the imagery of gates is also used in this context), whether on the horizon or in terms of his northern advance in latitude in general. The sun(-god) advances to its northern limit where its further movement is blocked by a high mountain-range, or it turns at a solstice island at the latitude of the geographical arctic circle. The annual movement of the sun causes some strange phenomena in high northern latitudes. Thus, when it 'stays' or 'sleeps' in the far north these regions experience an (almost) eternal day, but an (almost) eternal night when its southern/underside of the earth voyage takes a longer time to make. The respective daytime and nighttime lengths of the sun-god's diurnal routes thus vary depending on the season as well as on the latitude of the observer. Although these varying lengths are observable at all latitudes, the phenomena experienced by the observers at high northern latitudes are more pronounced and have a greater explanatory potential, which is why they captured the imagination of various authors irrespective of their literary, or descriptive or explanatory motivation.

List of citations

Abel, K. 1974. "Zone". *RE* Supp. 14: 989.1–1188.34.

Abrams, M. H. 1953. *The Mirror and the Lamp. Romantic Theory and the Critical Tradition.* London: Oxford University Press.

Abusch, T. 2001. "The Epic of Gilgamesh and the Homeric Epics". In Whiting R. M. (ed.), *Mythology and Mythologies. Methodological Approaches to Intercultural Influences.* Helsinki: The Neo-Assyrian Text Corpus Project, 1–6.

Ackerknecht, E. H. 1954. "On the Comparative Method in Anthropology". In Spencer, R. F. (ed.), *Method and Perspective in Anthropology*, Minneapolis: University of Minnesota Press, 117–25.

Ackerman, R. 1975. "Frazer on Myth and Ritual". *Journal of the History of Ideas* 36.1: 115–34.

Adkins, A. W. H. 1983. "Orality and Philosophy". In Robb, K. (ed.), *Language and Thought in Early Greek Philosophy.* La Salle: Hegeler Institute, 207–27.

Ahl, F. M. 1982. "Amber, Avallon, and Apollo's Singing Swan". *American Journal of Philology* 103.4: 373–411.

Ahlstrand, J. A. (ed.) 1862. *Konung Alexander.* Stockholm: Norstedt.

Ahmad, S. M. 1992. "Cartography of al-Sharīf al-Idrīsī". In Harley, J. B. & Woodward, D. (eds.), *The History of Cartography.* Vol. 2. Pt. 1. *Cartography in the Traditional Islamic and South Asian Societies.* Chicago: University of Chicago Press, 156–74.

Albinus, L. 2000. *The House of Hades.* Aarhus: Aarhus University Press.

Albright, W. F. 1919. "The Mouth of the Rivers". *American Journal of Semitic Languages and Literature* 35.4: 161–95.

Albright, W. F. 1920. "The Goddess of Life and Wisdom". *American Journal of Semitic Languages and Literature* 36.4: 258–94.

Aldred, C. 1987. *The Egyptians.* London: Thames & Hudson.

Alexander, P. J. 1985. *The Byzantine Apocalyptic Tradition.* Berkeley: University of California Press.

Alföldi, A. 1997. *Redeunt Saturnia regna.* Bonn: Habelt.

Alfonsi, L.1953. "Sui Papiri Schubart". *Aegyptus* 33.2: 297–314.

Algra, K. 1999. "The beginnings of cosmology". In Long, A. A. (ed.), *The Cambridge Companion to Early Greek Philosophy.* Cambridge: Cambridge University Press, 45–65.

Allen, A. 1993. *The Fragments of Mimnermus. Text and Commentary.* Stuttgart: Franz Steiner.

Allen, J. P. 1988. *Genesis in Egypt: The Philosophy of Ancient Egyptian Creation Accounts.* New Heaven: Yale University.

Allen, J. P. 2003. "The Egyptian Concept of the World". In O'Connor, D. & Quirke, S. (eds.), *Mysterious Lands.* London: UCL, 23–30.

Allen, T. G. 1974. *The Book of the Dead or Going Forth by Day.* Chicago: University of Chicago Press.

Allen W. E. D. 1960. "Ex Ponto V. Heniochi-Aea-Hayasa". *Bedi Karthlisa. Revue de Kartvelologie* 8–9.34–35: 79–92.

Alster, B. 1974. "On the Interpretation of the Sumerian Myth 'Inanna and Enki'". *Zeitschrift für Assyriologie und vorderasiatische Archäologie* 64: 20–34.

Alster, B. 1983. "Dilmun, Bahrain, and the alleged paradise in Sumerian Myth and Literature". In Potts, D. T. (ed.), *Dilmun: New studies in the archaeology and early history of Bahrain.* Berlin: D. Reimer, 39–74.

Alster, B. 1986. "Edin-na ú-sag-gá: Reconstruction, history and interpretation of a Sumerian cultic lament." In Hecker, K. & Sommerfeld, W. (eds.), *Keilschriftliche Literaturen: Ausgewählte Vorträge der XXXII Rencontre Assyriologique Internationale Münster, 8.–12.7.1985.* Berlin: Reimer, 19–31.

Anderson, A. R. 1927. "Alexander's Horns". *Transactions and Proceedings of the American Philological Association* 58: 100–22.

Anderson, A. R. 1932. *Alexander's Gate, Gog and Magog, and the Inclosed Nations.* Cambridge: Medieval Academy of America.

André, L.-N. 2010. "L'escale à Cyzique (Apollonios de Rhodes, Les Argonautiques, I, 922–1152): espaces phobiques, stéréotypes paysagers et transfiguration". *Rursus* 5, http://rursus.revues.org/422.

ANET = *Ancient Near Eastern Texts Relating to the Old Testament.* Pritchard, J. B. (ed.), Princeton: Princeton University Press, 1969³.

Anghelina, C. 2010. "The Homeric Gates of Horn and Ivory". *Museum Helveticum* 67: 65–72.

Anklesaria, B. T. 1956. *Zand-Akasih, Iranian or the Greater Bundahishn.* Bombay: Rahnumae Mazdayasnan Sabha.

Antonelli, L. 2009. "Auienus, Ora maritima(2009)". *Die Fragmente der Griechischen Historiker* Part V. Editor in Chief: H.-J. Gehrke. Brill Online, 2013. Reference. BNJ-contributors. 06 December 2013 <http://referenceworks.brillonline.com/entries/fragmente-der-griechischen-historiker-v/auienus-ora-maritima-2009-a2009>

ARE = Breasted, J. H. 1906–1907. *Ancient Records of Egypt*. 5 vols. Chicago: University of Chicago Press.

Arethusa ed. 1969 = *Porphyry: The Cave of the Nymphs in the Odyssey: A revised text with translation* by Seminar Classics 609, State University of New York at Buffalo. Arethusa Monograph 1. Buffalo: State University of New York.

Arnaud, P. 2014. "Mapping the Edges of the Earth: Approaches and Cartographical Problems". In Podosinov, A. V. (ed.), *The Periphery of the Classical World in Ancient Geography and Cartography*. Leuven: Peeters, 31–57.

Arnold, I. R. 1933. "Local Festivals at Delos". *American Journal of Archaeology* 37.3: 452–58.

Arvidsson, S. 2006. *Aryan Idols: Indo-European Mythology as Ideology and Science*. Chicago: University of Chicago Press.

Asheri, D., Lloyd, A. B. & Corcella, A. 2007. *A Commentary on Herodotus I–IV*. Oxford: Oxford University Press.

Asmis, E. 1991. "Crates on Poetic Criticism". *Phoenix* 46: 138–69.

Ataç, M.-A. 2004. "The 'Underworld Vision' of the Ninevite Intellectual Milieu". *Iraq* 66: 67–76.

Athanassaki, L., Martin, R. P. & Miller, J. F. (eds.) 2009. *Apolline Politics and Poetics: International Symposium*. Athens: European Cultural Centre of Delphi.

Attinger, P. 1984. "Enki et Nin[h]ursa[g]a". *Zeitschrift für Assyriologie und vorderasiatische Archäologie* 74: 1–52.

Aubert, J.-J. 2016. "Aristoteles (646)". *Brill's New Jacoby*. Worthington I. (ed.). Brill Online, http://referenceworks.brillonline.com/entries/brill-s-new-jacoby/aristoteles-646-a646.

Aujac, G. 1987a. "The Foundations of Theoretical Cartography in Archaic and Classical Greece". In Harley, J. B. & Woodward, D. (eds.), *The History of Cartography*. Vol.1. *Cartography in Prehistoric, Ancient, and Medieval Europe and the Mediterranean*. Chicago: University of Chicago Press, 130–47.

Aujac, G. 1987b. "The Growth of an Empirical Cartography in Hellenistic Greece". In Harley, J. B. & Woodward, D. (eds.), *The History of Cartography*. Vol.1. *Cartography in Prehistoric, Ancient, and Medieval Europe and the Mediterranean*. Chicago: University of Chicago Press, 148–60.

Aujac, G. 1987c. "Greek Cartography in the Early Roman World". In Harley, J. B. & Woodward, D. (eds.), *The History of Cartography*. Vol.1. *Cartography in Prehistoric, Ancient, and Medieval Europe and the Mediterranean*. Chicago: University of Chicago Press, 161–76.

Aujac, G. 1993. *La Sphère, instrument au service de la découverte du monde: D'Autolycos de Pitanè à Jean de Sacrobosco*. Caen: Paradigme.

Aujac, G. 2015. "The 'Revolution' of Ptolemy". In Bianchetti, S., Cataudella, M. R. & Gehrke, H.-J. (eds.), *Brill's Companion to Ancient Geography*. Leiden: Brill, 313–34.

Austin, C. & Olson, S. D. 2004. *Aristophanes'* Thesmophoriazusae. Oxford: Oxford University Press.

Bachvarova, M. R. 2005. "The Eastern Mediterranean Epic Tradition from *Bilgames and Akka* to the *Song of Release* to Homer's *Iliad*". *Greek, Roman, and Byzantine Studies* 45: 131–53.

Bachvarova, M. R. 2016. *From Hittite to Homer. The Anatolian Background of Early Greek Epic*. Cambridge: Cambridge University Press.

Bacon, J. R. 1925. *The Voyage of the Argonauts*. London: Methuen.

Bacon, J. R. 1931. "The Geography of the Orphic Argonautica". *The Classical Quarterly* 25.3–4: 172–83.

Bährens, G. (ed.) 1911. *XII Panegyrici Latini*. Leipzig: Teubner.

Bailer-Jones, D. M. 2002. "Models, Metaphors and Analogies". In Machamer, P. & Silberstein, M. (eds.), *Blackwell Guide to Philosophy of Science*. Oxford: Blackwell, 108–27.

Bailey, A. 1997/1998. *The Caves of the Sun: The Origin of Mythology*. London: Jonathan Cape/Pimlico.

Ballabriga, A. 1986. *Le Soleil et le Tartare: L'Image mythique du monde en Grèce archaïque*. Paris: École des hautes études en sciences sociales.

Ballabriga, A. 1998. *Les fictions d'Homère. L'invention mythologique et cosmographique dans l'Odyssée*. Paris: Presses universitaires de France.

Bale, A. 2012. *John Mandeville. The Book of Marvels and Travels*. Oxford: Oxford University Press.

Barbantani, S. 2018. "A Survey of Lyric Genres in Hellenistic Poetry: The Hymn. Transformation, Adaptation, Experimentation". *Erga-Logoi* 6.1: 61–135.

Barbour, I. 1976. *Myths, Models and Paradigms: A Comparative Study in Science and Religion*. New York: Harper & Row.

Barons, K. & Wissendorf, H. 1894–1915. *Latvju dainas*. 6 vols. Jelgava/Riga/St. Petersburg: Draviņš-Dravnieks/

Kalniņš & Deutschmans/Imperial Academy of Sciences.

Barrick, W. B. 2008. BMH *as Body Language: A Lexical and Iconographical Study of the Word* BMH *When Not a Reference to Cultic Phenomena in Biblical and Post-Biblical Hebrew*. New York: T &T Clark.

Beaulieu, M.-C. 2016. *The Sea in the Greek Imagination.* Philadelphia: University of Pennsylvania Press.

Beazley, C. R. 1897–1906. *The Dawn of Modern Geography*. 3 vols. Oxford: Clarendon Press.

Beck, R. 1977–78. "Interpreting the Ponza Zodiac, II". *Journal of Mithraic Studies* 2.2: 87–147.

Beck, R. 2006. *The Religion of the Mithras Cult in the Roman Empire*. Oxford: Oxford University Press.

Beekes, R. 2010. *Etymological Dictionary of Greek*. 2 vols. Leiden: Brill.

Behrens, H. 1998. *Die Ninegalla-Hymne. Die Wohnungsnahme Inannas in Nippur in altbabylonischer Zeit*. Stuttgart: Franz Steiner.

Bekker, I. (ed.) 1814–16. *Anecdota Graeca*. 2 vols. Berlin: G. C. Nauck.

Bekker, I. (ed.) 1833. *Apollonii Sophistae lexicon Homericum*. Berlin: Reimer.

Bekkum, W. J. Van (ed. and trans.) 1992. *A Hebrew Alexander Romance according to MS London, Jews' College no. 145*. Leuven: Peeters.

Bekkum, W. J. Van (ed. and trans.) 1994. *A Hebrew Alexander Romance According to MS Héb. 671.5 Paris, Bibliothèque Nationale*. Groningen: Styx.

Belmonte, J. A., Fekri, M., Abdel-Hadi, Y. A., Shaltout, M. & González García, A. C. 2010. "On the orientation of ancient Egyptian temples: (5) Testing the theory in Middle Egypt and Sudan". *Journal for the History of Astronomy* 41.1: 65–93.

Bérard, V. 1929. *Les Navigations d'Ulysse*. Volume 4. *Nausicaa et le Retour d'Ulysse*. Paris: Colin.

Berger, H. (ed.) 1880. *Die geographischen Fragmente des Eratosthenes*. Leipzig: Teubner.

Berger, H. 1903. *Geschichte der wissenschaftlichen Erdkunde der Griechen*. Leipzig: Veit.

Berger, H. 1904. *Mytische Kosmographie der Griechen*. Supplement Roscher, W. (ed.), *Lexikon der Griechischen und Römischen Mythologie*. Leipzig: Teubner.

Berggren, J. L. & Jones, A. 2000. *Ptolemy's Geography*. Princeton: Princeton University Press.

Bergson, L. 1965. *Der griechische Alexanderroman rezension Beta*. Stockholm: Almqvist & Wiksell.

Bernabé, A. (ed.) 2004–2005. *Poetae Epici Graeci. Testimonia et Fragmenta*. Pars II. Fasc. 1–2. *Orphicorum et Orphicis similium testimonia et fragmenta*. München: Saur.

Bernabé, A. (ed.) 2007. *Poetae Epici Graeci. Testimonia et Fragmenta*. Pars II. Fasc. 3. *Musaeus, Linus, Epimenides. Papyrus Derveni. Indices*. Berlin: De Gruyter.

Bernabé, A. & Jiménez San Cristóbal, A. I. 2011. "Are the 'Orphic' gold leaves Orphic?". In Edmonds, R. G. (ed.), *The 'Orphic' Gold Tablets and Greek Religion*. Cambridge: Cambridge University Press, 68–101.

Bertemes, F. 2008. "Die Kreisgrabenanlage von Goseck: Ein Beispiel für frühe Himmelsbeobachtungen". *Acta Praehistorica et Archaeologica* 40: 37–44.

Bertemes, F. & Schlosser, W. 2004. "Der Kreisgraben von Goseck und seine astronomischen Bezüge". In Meller, H. (ed.), *Der geschmiedete Himmel*. Stuttgart: Landesmuseum für Vorgeschichte, 48–51.

Betz, H. D. 1992[2]. *The Greek Magical Papyri in Translation*. Chicago: University of Chicago Press.

Bianchetti, S. 1998. *Pitea di Massalia: L'Oceano: Introduzione, testo, traduzione e commento*. Pisa: Istituto editoriale e poligrafici internazionali.

Bianchetti, S. 2014. "Ancient Perceptions and Representations of the Island Britannia". In Geus, K. & Thiering, M. (eds.), *Features of Common Sense Geography: Implicit Knowledge Structures in Ancient Geographical Texts*. Berlin: Lit, 115–30.

Bicknell, P. J. 1969. "Anaximenes' Astronomy". *Acta Classica* 12: 53–86.

Bidez, J. (ed.) 1913. *Peri agalmatôn*. In Bidez, J. (ed.), *Vie de Porphyre le philosophe néo-platonicien*. Leipzig: Teubner, 1–23.

Bietti Sestieri A. M. & Macnamara E. 2007. *Prehistoric Metal Artefacts from Italy (3500–720BC) in the British Museum*. London: British Museum.

Bilić, T. 2009. "Jadranska Ogigija i vučedolski bog metalurgije—dva primjera historijskog (dis) kontinuiteta". *Vjesnik Arheološkog muzeja u Zagrebu* 42: 73–84.

Bilić, T. 2012a. "Crates of Mallos and Pytheas of Massalia: Examples of Homeric Exegesis in Terms of Mathematical Geography". *Transactions of the American Philological Association* 142: 295–328.

Bilić, T. 2012b. "Apollo, Helios, and the Solstices in the Athenian, Delphian, and Delian Calendars". *Numen* 59: 509–32.

Bilić, T. 2013a. "Locations of Mythical Exile: Two Mythical Models Accounting for the Phenomenon of the Diurnal Solar Movement". *Mnemosyne* 66: 247–72.

Bilić, T. 2013b. "Solar Symbolism of 'Horns of Consecration'". In Šprajc, I. & Pehani, P. (eds.), *Ancient Cosmologies and Modern Prophets. Proceedings of the*

20th Conference of the European Society for Astronomy in Culture. Ljubljana: Slovene Anthropological Society, 105–20.

Bilić, T. 2016a. "The island of the sun: spatial aspect of solstices in early Greek thought". *Greek, Roman, and Byzantine Studies* 56.2: 195–224.

Bilić, T. 2016b. "The Swan Chariot of a Solar Deity: Greek Narratives and Prehistoric Iconography". *Documenta Praehistorica* 43: 445–65.

Bilić, T. 2016c. "*Orbis quadrifarius*: The transmission of Crates' theory of quadripartite earth in the Latin West". *Geographia Antiqua* 25: 129–45.

Bilić, T. 2017. "Bears, gates, and solstices. Myth and meteorology in Homer and Apollonius". *Mnemosyne* 70.1: 1–23.

Bilić, T. 2019. "Sailing to the Other World: Lower hemisphere and the antipodes from Crates to Columbus". *Orbis Terrarum* 17: 19–51.

Bilić, T. 2020a. "Pytheas and Hecataeus: Visions of the North in the late 4th century". *Greek, Roman, and Byzantine Studies* 60.4: 574–93.

Bilić, T. 2020b. "North vs. South: Alternative models for the diurnal solar movement in early Greece". *Symbolae Osloenses* 94 (in press).

Bilić, T. 2020c. "From Salona to the Milky Way—The epitaph of Constantina (?) from Manastirine". In Radman-Livaja, I. & Bilić, T. (eds.), *Monumenta marmore aereque perenniora. Zbornik radova u čast Anti Rendiću-Miočeviću / A volume dedicated to Ante Rendić-Miočević*. Zagreb: Arheološki muzej u Zagrebu / The Archaeological Museum in Zagreb, 82–91.

Bilić, T. 2021. "Calendric aspects of myths and cults involving Apollo's visit to Hyperborea". *Classical Journal* 116.3: 257–83.

Bilić, T. forthcoming. "Early identifications of Apollo with the physical Sun in Ancient Greece: Tradition and interpretation". *Mnemosyne*.

Black, J. & Green, A. 2000. *Gods, Demons and Symbols of Ancient Mesopotamia*. Austin: University of Texas Press.

Black, J., Cunningham, G., Robson, E. & Zólyomi, G. 2004. *The Literature of Ancient Sumer*. Oxford: Oxford University Press.

Black, M. 1962. *Models and Metaphors: Studies in Language and Philosophy*. Ithaca: Cornell University Press.

Blomqvist, J. 2002. "The geography of the Baltic in Greek eyes—From Ptolemy to Laskaris Kananos". In Amden, B., Flensted-Jensen, P., Nielsen, T. H., Schwartz, A. & Tortzen, C. G. (eds.), *Noctes Atticae. 34 Articles on Graeco-Roman Antiquity and its Nachleben*. Copenhagen: Museum Tusculanum Press, 36–51.

Blumenberg, H. 1985. *Work on Myth*. Cambridge: MIT Press.

Boardman, J. 1974. *Athenian Black Figure Vases*. London: Thames & Hudson.

Boardman, J. 1975. "Herakles, Peisistratos and Eleusis". *Journal of Hellenic Studies* 95: 1–12.

Boas, F. 1916. "The origin of totemism". *American Anthropologist* 18: 319–26.

Boas, F. 1921. *The Mind of Primitive Man*. New York: Macmillan.

Boas, F. 1938. "Mythology and Folklore". In Boas, F. (ed.), *General Anthropology*. Boston: Heath, 609–26.

Bober, H. 1956–57. "An illustrated medieval school-book of Bede's 'De Natura Rerum'". *Journal of the Walters Art Gallery* 19–20: 64–97.

Böckh, A. 1863. *Über die vierjährigen Sonnenkreise der Alten, vorzüglich den Eudoxischen*. Berlin: Reimer.

Böhl, F. M. T. 1936–37. "Die 50 Namen des Marduk". *Archiv für Orientforschung* 11: 191–218 = 1953. *Opera minora: studies en bijdragen op assyriologisch en oudtestamentisch terrain*. Groningen: Wolters, 282–312.

Böhmer, R. M. 1965. *Die Entwicklung der Glyptik während der Akkad-Zeit*. Berlin: De Gruyter.

Boll, F. 1903. *Sphaera. Neue griechische Texte und Untersuchungen zur Geschichte der Sternbilder*. Leipzig: Teubner.

Bolshakov, O. G. & Mongait, A. L. 1971. *Puteshestvie Abu Khamida al-Garnati v Vostochnuyu i Tsentral'nuyu Evropu (1131–1153)*. Moscow: Nauka.

Bolton, J. D. P. 1962. *Aristeas of Proconnesus*. Oxford: Clarendon Press.

Borg, B. E. 2013. *Crisis and Ambition: Tombs and Burial Customs in Third-Century CE Rome*. Oxford: Oxford University Press.

Borgeaud, P. 1988. *The Cult of Pan in Ancient Greece*. Chicago: University of Chicago Press.

Borger, R. 1967. "Das dritte 'Haus' der Serie *Bīt Rimki* (VR 50–51, Schollmeyer HGŠ nr. 1)". *Journal of Cuneiform Studies* 21: 1–17.

Bošković D. 1959. "Quelques observations sur le char cultuel de Dupljaja". *Archaeologia Iugoslavica* 3: 41–45, pls. 21–25.

des Bouvrie, S. 2002. "The definition of myth. Symbolic phenomena in ancient Greek culture". In des Bouvrie, S. (ed.), *Myth and Symbol I. Symbolic phenomena in ancient Greek culture*. Bergen: Norwegian Institute at Athens, 11–69.

Bouzek J. 1985. *The Aegean, Anatolia and Europe. Cultural Interrelations in the 2nd Millennium B.C.* Göteborg: Åströms.

Bowen, A. C. 2002. "Eudemus' History of Early Greek Astronomy: Two Hypotheses". In Bodnár, I. & Fortenbaugh, W. W. (eds.), *Eudemus of Rhodes.* New Brunswick: Transaction Publishers, 307–22.

Bowen, A. C. & Goldstein, B. R. 1988. "Meton of Athens and Astronomy in the Late Fifth Century B.C.". In Leichty, E., de J. Ellis, M. & Gerardi, P. (eds.), *A Scientific Humanist: Studies in Memory of Abraham Sachs*. Philadelphia: University Museum, 39–81.

Bowen, A. C. & Todd, R. B. 2004. *Cleomedes' Lectures on Astronomy. A Translation of The Heavens with an Introduction and Commentary*. Berkeley: University of California Press.

Boutsikas, E. 2007. *Astronomy and Ancient Greek Cult: An Application of Archaeoastronomy to Greek Religious Architecture, Cosmologies and Landscapes*. Diss. University of Leicester.

Boutsikas, E. 2020. *The Cosmos in Ancient Greek Religious Experience. Sacred Space, Memory, and Cognition*. Cambridge: Cambridge University Press.

Bowen, A. C. 2018. "Hellenistic Astronomy". In Keyser, P. T. & Scarborough, J. (eds.), *Oxford Handbook of Science and Medicine in the Classical World*. Oxford: Oxford University Press, 293–314.

Bowra, C. M. 1937: "The Proem of Parmenides". *Classical Philology* 32.2: 97–112.

Bradford, E. 1963. *Ulysses Found*. London: Hodder & Stoughton.

Bradley, R. 2006. "Danish razors and Swedish rocks: Cosmology and the Bronze Age landscape". *Antiquity* 80: 372–89.

Braund, S. 1992. *Lucan. Civil War.* Oxford: Oxford University Press.

Bredholt Christensen L. & Warburton, D. A. 2014. "Ian Hodder and the Neolithic." In Bredholt Christensen, L., Hammer, O. & Warburton, D. A. (eds.), *The Handbook of Religions in Ancient Europe*. London: Routledge, 45–52.

Bremmer, J. 1983. "Scapegoat rituals in Ancient Greece". *Harvard Studies in Classical Philology* 87: 299–320.

Bremmer, J. (ed.) 1988. *Interpretations of Greek Mythology*, London: Routledge.

Bremmer, J. 1988. "What is a Greek myth". In Bremmer, J. (ed.), *Interpretations of Greek Mythology*, London: Routledge, 1–9.

Bremmer J. 1994. *Greek Religion*. Oxford: Oxford University Press.

Bremmer, J. 1999. "Rationalization and Disenchantment in Ancient Greece: Max Weber among the Pythagoreans and Orphics?". In Buxton, R. (ed.), *From Myth to Reason? Studies in the Development of Greek Thought*. Oxford: Oxford University Press, 71–83.

Bremmer, J. 2002. *The Rise and Fall of the Afterlife*. London: Routledge.

Bremmer, J. 2008. *Greek Religion and Culture, the Bible and the Ancient Near East*. Leiden: Brill.

Bremmer, J. 2011. "A Brief History of the Study of Greek Mythology". In Dowden, K. & Livingstone, N. (eds.). *A Companion to Greek Mythology*. Chichester: Wiley-Blackwell, 527–47.

Bremmer, J. 2016. "Shamanism in Classical Scholarship: Where are We Now?" In Jackson, P. (ed.), *Horizons of Shamanism: A Triangular Approach to the History and Anthropology of Ecstatic Techniques*. Stockholm: Stockholm University Press, 52–78.

Bremmer, J. 2018. "Method and Madness in the Study of Greek Shamanism. The

Case of Peter Kingsley." *ASDIWAL. Revue genevoise d'anthropologie et d'histoire des*

religions 13: 93–109.

Bridgman, T. P. 2005. *Hyperboreans: Myth and History in Celtic-Hellenic Contacts*. London: Routledge.

Brinner, W. M. 2002. *'Arā'is al-majālis fī qiṣaṣ al-anbiyā' or 'Lives of the Prophets' as Recounted by Abū Isḥāq Aḥmad ibn Muḥammad ibn Ibrāhīm al-Tha'labī.* Leiden: Brill.

Brisson, L. 1998. *Plato the Myth Maker*. Chicago: University of Chicago Press.

Brisson, L. 2004. *How Philosophers Saved Myths*. Chicago: University of Chicago Press.

Broggiato, M. 2001. *Cratete di Mallo. I frammenti*. La Spezia: Agorà.

Brosset, M. 1864–66. *Histoire de la Siounie*. 2 vols. St. Petersburg: Imperial Academy of Sciences.

Brown, D. 2000. "The Cuneiform Conception of Celestial Space and Time". *Cambridge Archaeological Journal* 10.1: 103–22.

Brown, E. L. 1985. "Eumaeus' Native Isle". *Classical Journal* 80.4: 292–96.

Brown, J. P. 1968. "Cosmological Myth and the Tuna of Gibraltar". *Transactions and Proceedings of the American Philological Association* 99: 37–62.

Bruner, J. 1986. *Actual Minds, Possible Worlds.* Cambridge: Harvard University Press.

Bryce, T. R. 1986. *The Lycians in Literary and Epigraphic Sources*. Copenhagen: Museum Tusculanum Press.

Budge, E. A. W. 1886. *The Book of the Bee*. Oxford: Clarendon Press.

Budge, E. A. W. 1889. *The History of Alexander the Great Being the Syriac Version of the Pseudo-Callisthenes*. Cambridge: Cambridge University Press.

Budge, E. A. W. 1896. *The Life and Exploits of Alexander the Great*. London: C. J. Clay.

Budge, E.A.W. 1899. *The Book of the Dead. Facsimiles of the Papyri of Hunefer, Ȧnhai, Kerāsher and Netchemet*. London: British Museum.

Budge, E. A. W. 1905. *The Egyptian Heaven and Hell*. 3 vols. London: Paul, Trench, Trubner & Co.

Buffière, F. (ed.) 1962. *Héraclite: Allégories d'Homère*. Paris: Les Belles Lettres.

Bunbury, E. H. 1883². *A History of Ancient Geography*. 2 vols. London: John Murray.

Burch, G. B. 1954. "The Counter-Earth". *Osiris* 11: 267–94.

Burgess, J. 1999. "Gilgamesh and Odysseus in the Otherworld". *Echos du monde Classique / Classical Views* 18.2: 171–210.

Burgess, J. 2015. *Homer*. London: I. B. Tauris.

Burgess, J. 2016. "Localization of the *Odyssey*'s Underworld". *Cahiers des études anciennes* 53: 15–37, http://etudesanciennes.revues.org/906.

Burkert, W. 1972. *Lore and Science in Ancient Pythagoreanism*. Cambridge: Harvard University Press.

Burkert, W. 1975. "Apellai und Apollon". *Rheinisches Museum für Philologie* 118: 1–21.

Burkert, W. 1979a. *Structure and History in Greek Mythology and Ritual*. Berkeley: University of California Press.

Burkert, W. 1979b. "Mythisches Denken". In Poser, H. (ed.), *Philosophie und Mythos. Ein Kolloquium*. Berlin: De Gruyter, 16–39.

Burkert, W. 1982. "Literarische Texte und funktionaler Mythos: zu Istar and Atrahasis". In Assmann, J., Burkert, W. & Stolz, F. (eds.), *Funktionen und Leistungen des Mythos. Drei orientalische Beispiele*. Freiburg/Göttingen: Universitätsverlag/Vandenhoeck & Ruprecht, 63–82.

Burkert, W. 1983. *Homo necans*. Berkeley: University of California Press.

Burkert, W. 1985. *Greek Religion*. Cambridge: Harvard University Press.

Burkert, W. 1988. "Oriental and Greek Mythology: The Meeting of Parallels". In Bremmer, J. (ed.), *Interpretations of Greek Mythology*, London: Routledge, 10–40.

Burkert, W. 1992. *The Orientalizing Revolution: Near Eastern Influence on Greek Culture in the Early Archaic Age*. Cambridge: Harvard University Press.

Burkert, W. 2004. *Babylon Memphis Persepolis: Eastern Contexts of Greek Culture*. Cambridge: Harvard University Press.

Burnet, J. 1920³. *Early Greek Philosophy*. London: Black.

Burnett, C. 1982. *Hermann of Carinthia, De essentiis*. Leiden: Brill.

Burnett, C. 1984. "An Apocryphal Letter from the Arabic Philosopher Al-Kindi to Theodore, Frederick II's Astrologer, Concerning Gog and Magog, the Enclosed Nations, and the Scourge of the Mongols". *Viator* 15: 151–67.

Burnett, C. 1985. *Pseudo-Bede: De mundi celestis terrestrisque constitutione*. London: Warburg Institute, University of London.

Burton, R. F. 1897. *The Book of the Thousand Nights and a Night*. Vol. 4. London: Nichols.

Bury, J. B. 1906. "The Homeric and the historic Kimmerians". *Klio* 6: 79–88.

Butler, S. 1897. *The Authoress of the Odyssey*. London: Longmans, Green.

Butterworth, E. A. S. 1970. *The Tree at the Navel of the Earth*. Berlin: De Gruyter.

Buxton, R. 1981. "Introduction". In Gordon, R. L. (ed.), *Myth, Religion and Society. Structuralist Essays by M. Detienne, L. Gernet, J.-P. Vernant and P. Vidal-Naquet*. Cambridge: Cambridge University Press, ix–xvii.

Buxton, R. 1992. "Imaginary Greek Mountains". *Journal of Hellenic Studies* 112: 1–15.

Buxton, R. 1994. *Imaginary Greece*. Cambridge: Cambridge University Press.

Buxton, R. 1999. "Introduction". In Buxton, R. (ed.), *From Myth to Reason? Studies in the Development of Greek Thought*. Oxford: Oxford University Press, 1–21.

Buxton R.: 2009. *Forms of Astonishment Greek Myths of Metamorphosis*. Oxford: Oxford University Press.

Buxton R. 2017. "Landscapes of the Cyclopes". In Hawes, G. (ed.), *Myths on the Map. The Storied Landscapes of Ancient Greece*. Oxford: Oxford University Press, 52–64.

CAD = The Assyrian dictionary of the Oriental Institute of the University of Chicago. 1956–2006. 20 vols. Chicago: Oriental Institute.

Calame, C. 1996a. *Mythe et histoire dans l'Antiquité grecque*. Lausanne: Payot (= Calame 2003).

Calame, C. 1996b. "The rhetoric of muthos and logos. Forms of figurative discourse". In *Bristol colloquium on Greek myth Myth into logos? Programme abstracts (Bristol colloquium Myth into reason? July 24–28*), 23.

Calame, C. 1997. *Choruses of Young Women in Ancient Greece*. Lanham: Rowman & Littlefield.

Calame, C. 1999. "The Rhetoric of Muthos and Logos: Forms of Figurative Discourse". In Buxton, R. (ed.), *From Myth to Reason? Studies in the Development of Greek Thought*. Oxford: Oxford University Press, 119–43.

Calame, C. 2003. *Myth and History in Ancient Greece: The Symbolic Creation of a Colony*. Princeton: Princeton University Press (= Calame 1996a).

Calame, C. 2009a. *Greek Mythology: Poetics, Pragmatics and Fiction*. Cambridge: Cambridge University Press.

Calame, C. 2009b. *Poetic and Performative Memory in Ancient Greece: Heroic Reference and Ritual Gestures in Time and Space*. Washington: Center for Hellenic Studies.

Calame, C. 2011. "The Semiotics and Pragmatics of Myth". In Dowden, K. & Livingstone, N. (eds.). *A Companion to Greek Mythology*. Chichester: Wiley-Blackwell, 507–24.

Cameron, A. 1995. *Callimachus and his Critics*. Princeton: Princeton University Press.

Cameron, A. 2004. *Greek Mythography in the Roman World*. Oxford: Oxford University Press.

Campbell, D. A. 1982. *Greek Lyric Poetry: Volume I. Sappho and Alcaeus*. Cambridge: Harvard University Press.

Campbell, D. A. 1988. *Greek Lyric Poetry: Volume II. Anacreon, Anacreontea, Choral Lyric from Olympus to Alcman*. Cambridge: Harvard University Press.

Campbell, D. A. 1993. *Greek Lyric Poetry: Volume V. The New School of Poetry and Anonymous Songs and Hymns*. Cambridge: Harvard University Press.

Campbell, M.: *A Commentary on Apollonius Rhodius* Argonautica *III 1–471*. Leiden, 1994.

Caplice, R. 1973. "É.NUN in Mesopotamian Literature". *Orientalia* 42: 299–305.

Carey, J. 1994. "The Sun's Night Journey: A Pharaonic Image in Medieval Ireland". *Journal of the Warburg and Courtauld Institutes* 57: 14–34.

Carey, J. 2000. "Ireland and the Antipodes: The Heterodoxy of Virgil of Salzburg". In Wooding, J. M. (ed.), *The Otherworld Voyage in Early Irish Literature*. Dublin: Four Courts Press, 133–42.

Carey, J. 2009. *Apocrypha Hiberniae 2, Apocalyptica 1. In Tenga Bithnua: The Ever-new Tongue*. Corpus Christianorum Series Apocryphorum 16. Turnhout: Brepols.

Carpenter, R. 1946. *Folk Tale, Fiction and Saga in the Homeric Epics*. Berkeley: University of California Press.

Carpenter, R. 1973. *Beyond the Pillars of Hercules: The Classical World Seen Through the Eyes of Its Discoverers*. London: Tandem.

Cary, M. & Warmington, E. H. 1963. *The Ancient Explorers*. Harmondsworth: Penguin.

Casagrande-Kim, R. 2012. *The Journey to the Underworld: Topography, Landscape, and Divine Inhabitants of the Roman Hades*. Diss. Columbia University.

Casari, M. 2005. "Tramonto settentrionale: glossa a Cor. XVIII, 85–86". In Bernardini, M. & Tornesello, N. L. (eds.), *Scritti in onore di Giovanni M. D'Erme*. Naples: Università degli studi di Napoli L'orientale. 233–43.

Casari, M. 2006. "Il viaggio a settentrione: mitografia e geografia dall'età classica al medioevo arabo-persiano". In Carbonaro, G., Cassarino, M., Creazzo, E. & Lalomia, G. (eds.), *Medioevo Romanzo e Orientale. Il viaggio nelle letterature romanze e orientali*. Soveria Mannelli: Rubbettino, 213–28.

Casari, M. 2011. "Nizāmī's Cosmographic Vision and Alexander in Search of the Fountain of Life". In Bürgel, J.-C. & van Ruymbeke, C. (eds.), *A Key to the Treasure of the Hakīm: Artistic and Humanistic Aspects of Nizāmī Ganjavī's* Khamsa. Leiden: Brill, 95–105.

Casari, M. 2012. "The King Explorer: A Cosmographic Approach to the Persian Alexander". In Stoneman, R., Erickson, K. & Netton, I. (eds.), *The Alexander Romance in Persia and the East*. Groningen: Barkhuis, 175–203.

Caspers, E. C. D. & Govindankutty, A. 1978. "R. Thapar's Dravidian Hypothesis for the Locations of Meluhha, Dilmun and Makan: A Critical Reconsideration". *Journal of the Economic and Social History of the Orient* 21.2: 113–45.

Casson, L. 1989. *The Periplus Maris Erythraei*. Princeton: Princeton University Press.

Cavigneaux, A. & Al-Rawi, F. 2000. "La fin de Gilgameš, Enkidu et les Enfers d'apres les manuscrits d'Ur et de Meturan (Textes de Tell Haddad VIII)". *Iraq* 62: 1–19.

Celoria, F. 1992. *The Metamorphoses of Antoninus Liberalis*. London: Routledge.

Cerri, G. 2014. "L'Ade ad Oriente, viaggio quotidiano del carro del sole e direzione della corrente dell'oceano". In Breglia, L. & Moleti, A. (eds.), *Hespería. Tradizioni, Rotte, Paesaggi*. Paestum: Pandemos, 165–79.

Chabot, J.-B. 1899–1910. *Chronique de Michel le Syrien, Patriarche Jacobite d'Antioche (1166–1199)*. 4 vols. Paris: Ernest Leroux.

Champagne, R. A. 1992. *The Structuralists on Myth: An Introduction*. New York: Garland.

Chantraine, P. 1968–80. *Dictionnaire etymologique de la langue grecque: histoire des mots*. 4 vols. Paris: Klincksieck.

Cherniss, H. 1935. *Aristotle's Criticism of Presocratic Philosophy*. Baltimore: Johns Hopkins University Press.

Cherniss, H. & Helmbold, W. C. 1957. *Plutarch's Moralia*. Vol. 12. Cambridge: Harvard University Press.

Chicideanu-Sandor M. & Chicideanu I. 1990. "Contributions to the Study of the Gîrla Mare Anthropomorphic Statuettes". *Dacia* 34: 53–75.

Civil, M. 1967. "Šū-Sîn's Historical Inscriptions. Collection B". *Journal of Cuneiform Studies* 21: 24–38.

Civil, M. 1969. "The Sumerian Flood Story". In Lambert, W. G. & Millard, A. R. (eds.), *Atrahasis: The Babylonian Flood Story*. Oxford: Clarendon Press, 138–45.

Clagett, M. 1955/1957. *Greek Science in Antiquity*. New York/London: Abelard-Schuman.

Clagett, M. 1995. *Ancient Egyptian Science*. Vol. 2. Philadelphia: American Philosophical Society.

Clarke, H. W. (trans.) 1881. *Nizami: Sikandar Nama, e bara*. London: W. H. Allen.

Clarke, K. 1999. *Between Geography and History: Hellenistic Constructions of the Roman World*. Oxford: Oxford University Press.

Clarke, K. 2017. "Walking through history. Unlocking the mythical past". In Hawes, G. (ed.), *Myths on the Map. The Storied Landscapes of Ancient Greece*. Oxford: Oxford University Press, 14–31.

Clauss, J. J. 1993. *The Best of the Argonauts. The Redefinition of the Epic Hero in Book One of Apollonius' Argonautica*. Berkeley: University of California Press.

Clay, D. 2007. "Plato Philomythos". In Woodard, R. D. (ed.), *The Cambridge Companion to Greek Mythology*. Cambridge: Cambridge University Press, 210–36.

Cleaves, F. W. 1959. "An Early Mongolian Version of The Alexander Romance". *Harvard Journal of Asiatic Studies* 22: 1–99.

Clère, P. 1961. *La Porte d'Évergète à Karnak*. Mémoires publiés par les members de l'Institut français d'archéologie orientale du Cairo 84.2. Cairo: Institut français d'archéologie orientale du Caire.

Cohen, M. E. 1993. *The Cultic Calendars of the Ancient Near East*. Bethesda: CDL Press.

Coldstream J. N. 1984. "A Protogeometric Nature Goddess from Knossos". *British Institute of Classical Studies* 31: 93–104.

Coles J. M. & Harding, A. F. 1979. *The Bronze Age in Europe*. London: Methuen.

Collard, C. & Cropp, M. 2009. *Euripides: Oedipus-Chrysippus. Other Fragments*. Cambridge: Harvard University Press.

Cook, A. B. 1914–40. *Zeus: A Study in Ancient Religion*. 3 vols. Cambridge: Cambridge University Press.

Cook, E. 1992. "Ferrymen of Elysium and the Homeric Phaeacians" *Journal of Indo-European Studies* 20.3–4: 239–67.

Cornelius, F. 1925. "Zur Geographie der Odyssee". *Rheinisches Museum für Philologie* 74: 344–46.

Cornford, F. M. 1912. *From Religion to Philosophy*. New York: Cosimo Classics.

Cornford, F. M. 1952. *Principium Sapientiae*. Cambridge: Cambridge: Cambridge University Press.

Cornwall, P. B. 1946. "On the Location of Dilmun". *Bulletin of the American Schools of Oriental Research* 103: 3–11.

Couprie, D. L. 2009. "The Tilting of the Heavens in Presocratic Cosmology". *Apeiron* 42: 259–74.

Couprie, D. L. 2011. *Heaven and Earth in Ancient Greek Cosmology. From Thales to Heraclides Ponticus*. New York: Springer.

Couprie, D. L. 2015. "The paths of the celestial bodies according to Anaximenes". *Hyperboreus* 21: 5–32.

Couprie, D. L. 2018. *When the Earth was Flat*. Cham: Springer.

Couprie, D. L. 2020. "The Spiral Movement of the Sun on an Imaginary Cylinder According

to Empedocles and Anaximander". *Philologia Classica*, 15.1: 4–24.

Couprie, D. L., Hahn R. & Naddaf, G. 2003. *Anaximander in Context*. Albany: SUNY Press.

Coxon, A. H. 2009². *The Fragments of Parmenides*. Las Vegas: Parmenides.

Crane, G. 1988. *Calypso: Backgrounds and Conventions of the Odyssey*. Frankfurt am Main: Athenäum.

Croon, J. H. 1952. *The Herdsman of the Dead: Studies on Some Cults, Myths and Legends of the Ancient Greek Colonization-area*. Utrecht: de Vroede.

Cross, J. E. & Hill, T. D. 1982. *The* Prose Solomon and Saturn *and* Adrian and Ritheus. Toronto: University of Toronto Press.

Csapo, E. 2005. *Theories of Mythology*. Malden: Blackwell.

Cuenca, L. A. de & Brioso, M. 1980. *Calímaco. Himnos, epigramas y fragmentos*. Madrid: Gredos.

Cunliffe, B. 2002. *The Extraordinary Voyage of Pytheas the Greek*. New York: Penguin.

Cunningham, G. 2007. *Deliver Me from Evil: Mesopotamian incantations 2500–1500 BC*. Studia Pohl: Series Maior 17. Rome: Pontificio Istituto Biblico.

Currie, B. 2016. *Homer's Allusive Art*. Oxford: Oxford University Press.

Czeglédy, K. 1957. "The Syriac Legend Concerning Alexander the Great". *Acta Orientalia Academiae Scientiarum Hungaricae* 7: 231–49.

Dale, A. 2011. "Topics in Alcman's *Partheneion*". *Zeitschrift für Papyrologie und Epigraphik* 176: 24–38.

Dalley, S. 1991. "Gilgamesh in the Arabian Nights". *Journal of the Royal Asiatic Society* 1.1: 1–17.

Dalley, S. 1994. "The Tale of Bulūqiyā and the Alexander Romance in Jewish and Sufi Mystical Circles". In Reeves, J. C. (ed.), *Tracing the Threads. Studies in the Vitality of Jewish Pseudepigrapha*. Atlanta: Scholars Press, 239–69.

Dalley, S. (trans.) 2000. *Myths from Mesopotamia*. Oxford: Oxford University Press (revised edition).

Dan, A. 2016. "The rivers called Phasis". *Ancient West & East* 15: 245–77.

Dan, A. 2017. "The First of the *Bêta*: Notes on Eratosthenes' Invention of Geography". In Rico, C. & Dan, A. (eds.), *The Library of Alexandria—A cultural crossroads of the ancient World, Proceedings of the second Polis Institute Interdisciplinary Conference*. Jerusalem: Polis Institute Press, 165–222.

Dan, A., Geus, K. & Guckelsberger, K. 2014. "What is Common Sense Geography? Some Preliminary Thoughts from a Historical Perspective". In Geus, K. & Thiering, M. (eds.), *Features of Common Sense Geography: Implicit Knowledge Structures in Ancient Geographical Texts*. Berlin: Lit, 17–38.

Danoff, C. M. 1962. "Pontos Euxeinos". *RE* Suppl. 9: 866–1175.60, 1911.18–1920.40.

Darby, J. 1723. *The History of Timur-bec, Commonly call'd Tamerlain*. Vol. 1. London: J. Darby etc.

Dassow, E. von 1998. *The Egyptian Book of the Dead, The Book of Going forth by Day. The First Authentic Presentation of the Complete Papyrus of Ani*. San Francisco: Chronicle (revised edition).

Davidson, H. E. 1993. *The Lost Beliefs of Northern Europe*. London: Routledge.

Davies, M. 1988. "Stesichorus' Geryoneis and Its Folk-Tale Origins". *Classical Quarterly* 38.2: 277–90.

Davies, M. 1992. "Heracles in narrow straits". *Prometheus* 18.3: 217–26.

Davies, M. 2015. *The Theban Epics*. Washington: Center for Hellenic Studies.

Davies, M. 2016. *The Aethiopis: Neo-Neoanalysis Reanalyzed*. Washington: Center for Hellenic Studies.

Davies, M. & Finglass, P.J. 2014. *Stesichorus. The Poems*. Cambridge: Cambridge University Press.

Davis, D. L. 2009. *Commercial Navigation in the Greek and Roman World*. Diss. University of Texas at Austin.

Davis, W. M. 1977. "The Ascension-Myth in the Pyramid Texts". *Journal of Near Eastern Studies* 36.3: 161–79.

Deimel, A. 1912. *Enuma Eliš, sive Epos Babylonicum de Creatione Mundi*. Rome: Pontificio Istituto Biblico.

Deimel, A. 1914. *Pantheon Babylonicum*. Rome: Pontificio Istituto Biblico.

De Jong, I. J. F. 2001. *A Narratological Commentary on the Odyssey*. Cambridge: Cambridge University Press.

Dekker, E. 2013. *Illustrating the Phaenomena. Celestial Cartography in Antiquity and the Middle Ages*. Oxford: Oxford University Press.

Demakopolou, K., Eluere, C., Jensen, J., Jockenhövel, A. & Mohen, J.-P. 2000. *Gods and Heroes of Bronze Age Europe. The Roots of Odysseus*. Copenhagen: Nationalmuseet.

Derow, P. S. & Forrest, W. G. 1982. "An Inscription from Chios". *Annual of the British School at Athens* 77: 79–92.

De Solla Price, D. J. 1969. "Portable Sundials in Antiquity, including an Account of a New Example from Aphrodisias". *Centaurus* 14: 242–66.

Detienne, M. 1962. *Homère, Hésiode et Pythagore. Poésie et philosophie dans le pythagorisme ancien*. Bruxelles: Latomus.

Detienne, M. 1981. "The myth of 'Honeyed Orpheus'". In Gordon, R. L. (ed.), *Myth, Religion and Society. Structuralist Essays by M. Detienne, L. Gernet, J.-P. Vernant and P. Vidal-Naquet*. Cambridge: Cambridge University Press, 95–109.

Detienne, M. 1986. *The Creation of Mythology*. Chicago: University of Chicago Press.

Detienne, M. 1992. "Myth and Writing: The Mythographers". In Y. Bonnefoy (ed.), *Greek and Egyptian Mythologies*. Chicago: University of Chicago Press, 10–11.

Detienne, M. 1994. *The Gardens of Adonis. Spices in Greek Mythology*. Princeton University Press.

Detienne, M. 1997. "This is where I intend to build a glorious temple". *Arion* 4.3: 1–27.

Detienne, M. 1998. *Apollon le couteau à la main. Une approche expérimentale du polythéisme grec*. Paris: Gallimard.

Detienne, M. 1999. "Experimenting in the Field of Polytheisms" *Arion* 7.1: 127–49.

Detienne, M. 2009. *Comparative Anthropology of Ancient Greece*. Washington: Center for Hellenic Studies.

Detienne, M. & Vernant, J.-P. 1991. *Cunning Intelligence in Greek Culture and Society*. Chicago: University of Chicago Press.

Dickens, M. 2004. *Medieval Syriac Historian's Perceptions of the Turks*. MPhil. thesis University of Cambridge, www.oxuscom.com/Medieval_Syriac_Historians_on_the_Turks.pdf.

Dicks, D. R. 1960. *Geographical fragments of Hipparchus*. London: Athlone Press.

Dicks, D. R. 1966. "Solstices, Equinoxes, & the Presocratics". *Journal of Hellenic Studies* 86: 26–40.

Dicks, D. R. 1970. *Early Greek Astronomy to Aristotle*. Ithaca: Cornell University Press.

Dickson, K. 2007. "Enki and Ninhursag: The trickster in Paradise". *Journal of Near Eastern Studies* 66.1: 1–32.

Diederich, S. 2018. "Kartenkompetenz und Kartenbenutzung bei den römischen Eliten—Teil 1". *Orbis terrarum* 16: 55–136.

Diederich, S. 2019. "Kartenkompetenz und Kartenbenutzung bei den römischen Eliten—Teil 2". *Orbis terrarum* 17: 101–84.

Diels, H. 1920. *Antike Technik: Sieben Vorträge*[2]. Leipzig: Teubner.

Diels, H. & Kranz, W. 1952[6]. *Fragmente der Vorsokratiker*. 3 vols. Berlin: Weidmann.

Dietrich, A. 1908. "Mitteilungen und Hinweise". *Archiv für Religionswissenschaft* 11: 159–60.

Diggle, J. 1970. *Euripides: Phaethon*. Cambridge: Cambridge University Press.

Diggle, J. 1984. "On the Manuscripts and Text of Euripides, Medea: II. The Text". *Classical Quarterly* 34.1: 50–65.

Diggle, J. 1994. *Euripidea. Collected Essays*. Oxford: Clarendon Press.

Dignas, B. & Audley-Miller, L. 2018. "Preface". In Audley-Miller, L. & Dignas, B. (eds.), *Wandering Myths. Transcultural Uses of Myth in the Ancient World*. Berlin: De Gruyter, vii–xxxi.

Di Gregorio, L. (ed.) 1975. *Scholia vetera in Hesiodi Theogoniam*. Milano: Vita e pensiero.

Dijkstra, M. 1991. "The Weather-God on Two Mountains". *Ugarit Forschungen* 23: 127–40.

Diller, A. 1934. "Geographical Latitudes in Eratosthenes, Hipparchus and Posidonius". *Klio* 27: 258–69.

Diller, A. 1975. "Pytheas of Massalia". In Gillispie, C. (ed.), *Dictionary of Scientific Biography*. Vol. 11. New York: Charles Scribner's Sons, 225–26.

Dilke, O. A. W. 1985. *Greek and Roman Maps*. London: Thames & Hudson.

Dolin, E. F. 1962. "Parmenides and Hesiod". *Harvard Studies in Classical Philology* 66: 93–98.

Donald, M. 1991. *Origins of the Modern Mind*. Cambridge: Harvard University Press.

Doniger W. 2009. "Claude Lévi-Strauss's Theoretical and Actual Approaches to Myth". In Wiseman, B. (ed.), *The Cambridge Companion to Lévi-Strauss*. Cambridge: Cambridge University Press, 196–215.

Donzel, E. van & Schmidt, A. 2009/2010. *Gog and Magog in Early Eastern Christian and Islamic Sources*. Leiden: Brill.

Dörpfeld, W. & Rüter, H. 1925. *Homers Odyssee*. 2 vols. München: Buchenau & Reichert. Vol. 1: Dörpfeld, W. *Die Wiederherstellung des Ursprünglichen Epos von der Heimkehr des Odysseus nach dem Tageplan mit Beigaben über Homerische Geographie und Kultur* (Beigabe III: Homerische Geographie, pp. 205–69).

Dorson, R. 1955. "The Eclipse of Solar Mythology". *Journal of American Folklore*, 68.270: 393–416.

Doufikar-Aerts, F. 2010. *Alexander Magnus Arabicus. A Survey of the Alexander Tradition through Seven Centuries: from Pseudo-Callisthenes to Suri*. Louvain: Peeters.

Dowden, K. 1992. *The Uses of Greek Mythology*. London: Routledge.

Dowden, K. 2004. "The epic tradition in Greece". In Fowler, R. (ed.), *The Cambridge Companion to Homer*. Cambridge: Cambridge University Press, 188–205.

Dowden, K. 2011a. "Telling the Mythology: From Hesiod to the Fifth Century". In Dowden, K. & Livingstone, N. (eds.). *A Companion to Greek Mythology*. Chichester: Wiley-Blackwell, 47–72.

Dowden, K. 2011b. "Initiation: The Key to Myth". In Dowden, K. & Livingstone, N. (eds.). *A Companion to Greek Mythology*. Chichester: Wiley-Blackwell, 487–505.

Drachmann, A. B. 1903–27. *Scholia vetera in Pindari carmina*. 3 vols. Leipzig: Teubner.

Dräger, P. 2001. *Die Argonautika des Apollonios Rhodios: Das zweite Zorn-Epos der*

griechischen Literatur. Munich: K. G. Saur.

Dreyer, J. L. E. 1953[2]. *A History of Astronomy from Thales to Kepler*. New York: Dover.

Dubois de Montpe´reux, F. 1843. *Voyage autour du Caucase* 6. Paris: Gide.

Dueck, D. 2012. *Geography in Classical Antiquity*. Cambridge: Cambridge University Press.

Duff, J. D. 1928. *Lucan: The Civil War, Books I–X*. Cambridge: Harvard University Press.

Dunbabin, T. J. 1948. *The Western Greeks*. Oxford: Clarendon Press.

Dunbar, N. 1998. *Aristophanes: Birds*. Oxford: Oxford University Press.

Durkheim, E. 1995[1912]. *The Elementary Forms of Religious Life*. New York: Free Press.

Durrbach, F. 1911. "Fouilles de Délos, exécutées aux frais de M. le Duc de Loubat. Inscriptions financières (1906–1909) (1)". *Bulletin de Correspondence Hellénique* 35: 5–86, 243–87.

Edmonds, R. G. 2004. *Myths of the Underworld Journey*. Cambridge: Cambridge University Press.

Edmonds, R. G. 2011. "Underworld, Topography of". In Finkelberg, M. (ed.), *The Homer Encyclopedia*. Vol. 3. Chichester: Wiley-Blackwell.

Edson, E. 1997. *Mapping Time and Space*. London: British Library.

Edson, E. & Savage-Smith, E. 2000. "An Astrologer's Map: A Relic of Late Antiquity". *Imago Mundi* 52: 7–29.

Edmunds, L. (ed.) 1990. *Approaches to Greek Myth*. Baltimore: Johns Hopkins University Press.

Edmunds, L. 1990. "Introduction: The Practice of Greek Mythology". In Edmunds, L. (ed.), *Approaches to Greek Myth*. Baltimore: Johns Hopkins University Press, 1–20.

Edmunds, L. 2009. "A Hermeneutic Commentary on the Eschatological Passage in Pindar Olympian 2 (57–83)". In Walde, C. & Dill, U. (eds.), *Antike Mythen: Medien, Transformationen und Konstruktionen*. Berlin: De Gruyter, 662–77.

Egg M. 1991. "Ein neuer Kesselwagen aus Etrurien". *Jahrbuch des Römisch-Germanischen Zentralmuseums Mainz* 38: 191–220.

Egg M. & Kaul F. 2001. "Kultwagen." In Beck, H., Geuenich, D. & Steuer, H. (eds.), *Reallexikon der germanischen Altertumskunde*² 17. Berlin: De Gruyter, 463–78.

Eliade, M. 1958. *Patterns in Comparative Religion*. London: Sheed and Ward.

Elweskiöld, B. 2005. *John Philoponus against Cosmas Indicopleustes. A Christian Controversy on the Structure of the World in Sixth-Century Alexandria*. Diss. Lund.

Endsjø, D. Ø. 1997. "Placing the unplaceable: The making of Apollonius' Argonautic geography". *Greek, Roman, and Byzantine Studies* 38: 373–85.

Ernst, N. (ed.). 2006. *Die D-Scholien zur Odyssee*. Diss. Universität zu Köln, https://kups.ub.uni-koeln.de/1831/1/D-Scholien.pdf.

Espak, P. 2006. *Ancient Near Eastern gods Enki and Ea: Diachronical analysis of texts and images from the earliest sources to the Neo-Sumerian period*. MA thesis, Tartu University, www.utlib.ee/ekollekt/diss/mag/2006/b18272897/espakpeeter.pdf.

Evans, J. 1998. *The History and Practice of Ancient Astronomy*. New York: Oxford University Press.

Evans, J. & Berggren, J. L. 2006. *Geminos's Introduction to the Phenomena: A Translation and Study of a Hellenistic Survey of Astronomy*. Princeton: Princeton University Press.

van den Eynde, C. 1955. *Commentaire d'Išo'dad de Merv sur l'Ancien Testament*. I. *Genèse*. Corpus Scriptorum Christianorum Orientalium 156, Scriptores Syri 75. Louvain: Peeters.

Fantuzzi, M. 2006. Review of Jouan, F. 2004. *Euripide. Tragédies, tome VIII, 2e partie: Rhésos*. Paris: Les Belles Lettres. BMCR 2006.02.18, http://ccat.sas.upenn.edu/bmcr/2006/2006-02-18.html.

Faraone, C. A. 2004. "The Collapse of Celestial and Chthonic Realms in a Late Antique 'Apollonian Invocation' (PGM I 262–347)". in Boustan, R. S. & Reed, A. Y. (eds.), *Heavenly Realms and Earthly Realities in Late Antique Religions*. Cambridge: Cambridge University Press, 213–32.

Farnell, L. R. 1907. *The Cults of the Greek States*. Vol. 4. Oxford: Clarendon Press.

Faulkner, A. 2011. "Introduction. Modern Scholarship on the Homeric Hymns. Foundational Issues". In Faulkner, A. (ed.), *The Homeric Hymns. Interpretative Essays*. Oxford: Oxford University Press, 1–25.

Faulkner, R. O. 1962. *A Concise Dictionary of Middle Egyptian*. Oxford: Griffith Institute.

Faulkner, R. O. 1966. "The King and the Star-Religion in the Pyramid Texts". *Journal of Near Eastern Studies* 25.3: 153–61.

Faulkner, R. O. 1969. *The Ancient Egyptian Pyramid Texts*. Oxford: Clarendon Press.

Faulkner, R. O. & Andrews, C. 1972. *The Ancient Egyptian Book of the Dead*. Austin: University of Texas Press.

Faulkner, R. O. 1973–78. *The Ancient Egyptian Coffin Texts*. 3 vols. Warminster: Aris & Philips.

Fearn, D. 2007. *Bacchylides. Politics, Performance, Poetic Tradition*. Oxford: Oxford University Press.

Felli, C. 2006. "Lugal-usumagal: an Akkadian Governor and his Two Masters". In Taylor, P. (ed.), *The Iconography of Cylinder Seals*, London/Torino: Warburg Institute/ N. Agrano, 35–50.

Ferrari, G. 2008. Alcman and the Cosmos of Sparta. Chicago: University of Chicago Press.

Ferrari Pinney, G. & Sismondo Ridgway, B. 1981. "Herakles at the ends of the Earth". *Journal of Hellenic Studies* 101: 141–44.

Finkel, I. 1995. "A join to the Map of the World: A notable discovery". *British Museum Magazine* 23: 26–27.

Finkelberg, A. 1986. "On the Unity of Orphic and Milesian Thought". *Harvard Theological Review* 79.4: 321–35.

Finkelberg, M. 1998. "The Geography of the Prometheus Vinctus". *Rheinisches Museum für* Philologie 141: 119–41.

Fitch, E. 1912. "Apollonius Rhodius and Cyzicus". *American Journal of Philology* 33.1: 43–56.

Flygare, J. B. 2006. *The Babylonian Map of the World*. http://www.opgavebank.dk/opgaver/806.pdf.

Fontenrose, J. E. 1943. "Apollo and Sol in the Oaths of Aeneas and Latinus". *Classical Philology* 38.2: 137–38.

Fontenrose, J. E. 1971. *The Ritual Theory of Myth.* Berkeley: University of California Press.

Fontenrose, J. E. 1974. *Python: A Study of Delphic Myth and its Origins*. New York: Biblio & Tannen.

Fontenrose, J. E. 1978. *The Delphic Oracle.* Berkeley: University of California Press.

Fontenrose, J. E. 1981. *Orion: The Myth of the Hunter and the Huntress*. Berkeley: University of California Press.

Foster, B. 1995. *From Distant Days*. Bethesda: CDL Press.

Foster, B. 2001. *The Epic of Gilgamesh.* New York: Norton.

Foster, B. 2005[3]. *Before the Muses*. Bethesda: CDL Press

Fowler, R. L. 1993. "The Myth of Kephalos as an Aition of Rain-Magic (Pherekydes *FGrHist* 3 F 34)". *Zeitschrift für Papyrologie und Epigraphik* 97: 29–42.

Fowler, R. L. 1996. "Herodotos and his Contemporaries". *Journal of Hellenic Studies* 116: 62–87.

Fowler, R. L. 2001–13. *Early Greek Mythography.* Vol. 1: Text and Introduction. Vol. 2: *Commentary*. Oxford: Oxford University Press.

Fowler, R. L. 2004a. "The Homeric question". In Fowler, R. L. (ed.), *The Cambridge Companion to Homer.* Cambridge: Cambridge University Press, 220–32.

Fowler, R. L. 2004b. "Dateline". In Fowler, R. L. (ed.), *The Cambridge Companion to Homer.* Cambridge: Cambridge University Press, 376–77.

Fowler, R. L. 2011. "*Mythos* and *logos*". *Journal of Hellenic Studies* 131: 45–66.

Fowler, R. L. 2017. "Imaginary Itineraries in the Beyond". In Hawes, G. (ed.), *Myths on the Map. The Storied Landscapes of Ancient Greece.* Oxford: Oxford University Press, 243–60.

Frame, D. 1978. *The Myth of Return in Early Greek Epic.* New Heaven: Yale University Press.

Frame, D. 2009. *Hippota Nestor*. Washington: Center for Hellenic Studies.

Frankfort, H. 1933. *The Cenotaph of Seti I at Abydos*. Vol. 1. London: London Egypt Exploration Society.

Frankfort, H. 1934. "Gods and Myths on Sargonid Seals". *Iraq* 1.1: 2–29.

Frankfort, H. 1939. *Cylinder Seals*. London: Macmillan.

Frazer, J. G. 1913. *The Golden Bough: A Study in Magic and Religion.* Part 7. *Balder the Beautiful*. Vol. 2. London: Macmillan.

Friedlander, G. 1916. *Pirḳê de Rabbi Eliezer*. London: Kegan Paul, Trench, Trubner.

Friedländer, I. 1913. *Die Chadhirlegende und der Alexanderroman*. Leipzig: Teubner.

Furley, D. 1987. *The Greek Cosmologists*. Cambridge: Cambridge University Press.

Furley, W. D. & Bremer, J. M. 2001. *Greek Hymns. Volume I. The Texts in Translation.* Tübingen: Mohr Siebeck.

Gabolde, L. 1999. "Canope et les orientations nord-sud de Karnak étables par Thoutmosis III". *Revue d'Égyptologie* 50: 278–82.

Gagné, R. 2007. "Winds and Ancestors: The *Physika* of Orpheus". *Harvard Studies in Classical Philology* 103: 1–23.

Gagné, R. 2021. *Cosmography and the Idea of Hyperborea in Ancient Greece: A Philology of Worlds*. Cambridge: Cambridge University Press.

Gain, D. B. 1976. *The Aratus Ascribed to Germanicus Caesar*. London: Athlone.

Galinsky, G. K. 1969. "Aeneas' Invocation of Sol (Aeneid, XII, 176)". *American Journal of Philology* 90.4: 453–58.

Gantz, T. 1993. *Early Greek Myth. A Guide to Literary and Artistic Sources*. Baltimore: Johns Hopkins University Press.

Gardiner, A. H. 1957[3]. *Egyptian Grammar: Being an Introduction to the Study of Hieroglyphics*. London: Griffith Institute.

Garner, R. 1993. "Achilles In Locri: P. Oxy. 3876 Frr. 37–77". *Zeitschrift für Papyrologie und Epigraphik* 96: 153–65.

Gaster, T. H. 1977[2]. *Thespis: Ritual, Myth, and Drama in the Ancient Near East*. New York: Norton.

Gattei, S. 2009. *Karl Popper's Philosophy of Science.* New York: Routledge.

Gee, E. 2013. *Aratus and the Astronomical Tradition.* Oxford: Oxford University Press.

Gee, E. 2020. *Mapping the Afterlife: From Homer to Dante*. Oxford: Oxford University Press.

Gehrke, H.-J. 2015. "The 'Revolution' of Alexander the Great: Old and New in the World's View". In Bianchetti, S., Cataudella, M. R. & Gehrke, H.-J. (eds), *Brill's Companion to Ancient Geography.* Leiden: Brill, 78–97.

Geisthövel, W. 2008. *Homer's Mediterranean.* London: Haus.

Geldner, K. F. 1951. *Der Rig-Veda: aus dem Sanskrit ins Deutsche Übersetzt.* 3 vols. Cambridge/Wiesbaden: Harvard University Press/Harrassowitz.

Geller, M. J. 1980. "A Middle Assyrian Tablet of Utukkū Lemnūtu, Tablet 12". *Iraq* 42.1: 23–51.

Geller, M. J. 1985. *Forerunners to Udug-hul: Sumerian Exorcistic Incantations*. Stuttgart: Steiner.

Gelling, P. & Davidson, H. E. 1972. *The Chariot of the Sun and other Rites and Symbols of the Northern Bronze Age*. London: J. M. Dent.

George, A. 1992. *Babylonian Topographical Texts*. Leuven: Peeters.

George, A. 2003a. *The Babylonian Gilgamesh Epic*. 2 vols. Oxford: Oxford University Press.

George, A. 2003b. *The Epic of Gilgamesh*. London: Penguin.

Germain, G. 1954. *Genèse de l'Odyssée*. Paris: Presses Universitaires de France.

Gernet, L. 1981. *The Anthropology of Ancient Greece*. Baltimore: Johns Hopkins University Press.

Getty, R. J. 1948. "Some Astronomical Cruces in the Georgics". *Transactions and Proceedings of the American Philological Association* 79: 24–45.

Geus, K. 2000. "Utopie und Geographie. Zum Weltbild der Griechen in frühhellenistischer Zeit". *Orbis Terrarum* 6: 55–90.

Geus, K. 2004. "Measuring the earth and the oikoumene: zones, meridians, sphragides and some other geographical terms used by Eratosthenes of Kyrene". In Talbert, R. & Brodersen, K. (eds.), *Space in the Roman World: Its Perception and Presentation*. Münster: LIT, 11–26.

Geus, K. 2011[2002]. *Eratosthenes von Kyrene: Studien zur hellenistischen Kultur und Wissenschaftsgeschichte*. Oberhaid: Utopica.

Geus, K. 2013a. "Agennius Urbicus und die Antichthonen: Ein stoisches Weltbild im Corpus agrimensorum Romanorum". In Knobloch, E. & Möller, C. (eds), *In den Gefilden der römischen Feldmesser. Juristische, wissenschaftsgeschichtliche, historische und sprachliche Aspekte*: Berlin: De Gruyter, 113–30.

Geus, K. 2013b. "Claudius Ptolemy on Egypt and East Africa". In Buraselis, K., Stefanou, M. & Thompson, D. J. (eds.), *The Ptolemies, the Sea and the Nile. Studies in Waterborne Power*. Cambridge: Cambridge University Press, 218–31.

Geus, K. 2016. "'Er hat die Oikumene der römischen Herrschaft unterworfen'—Bemerkungen zu den Raumvorstellungen in der Zeit des Augustus". In Baltrusch, E. & Wendt, C. (eds.), *Der Erste: Augustus und der Beginn einer neuen Epoche*. Mainz: Philipp von Zabern, 76–85, 159–60, 163–64.

Geus, K. 2018. "Greek and Greco-Roman Geography". in Jones, A. & Taub, L. (eds.), *The Cambridge History of Science*. Vol. 1. *Ancient Science*. Cambridge: Cambridge University Press, 402–12.

Giampaglia, A. 1998. "Cratete di Mallo nel POxy 2888". *Rendiconti dello Reale Instituto Lombardo di Scienze e Lettere* 132: 503–18.

Giangiulio, M. 2001. "Constructing the Past: Colonial Traditions and the Writing of History. The Case of Cyrene". In Luraghi, N. (ed.), *The Historian's Craft in the Age of Herodotus*, Oxford: Oxford University Press, 116–37.

Gibb, H. A. R. 2005[1929]. *Ibn Battuta: Travels in Asia and Africa 1325–1354*. Abingdon: Routledge.

Giedion, S. 1962. *The Eternal Present: A Contribution on Constancy and Change*. 2 vols. New York: Pantheon Books.

Giere, R. N. 1999. *Science without Laws*. Chicago: Chicago: University of Chicago Press.

Gillies, M. M. 1928. *The* Argonautica *of Apollonius Rhodius. Book III. Edited with Introduction and Commentary*. Cambridge: Cambridge University Press.

Gisinger, F. 1924. "Geographie". *RE* Suppl. 4.521.43–685.64.

Gisinger, F. 1929. "Zur Geographie bei Hesiod". *Rheinisches Museum für Philologie* 78: 315–28.

Gisinger, F. 1937. "Oikoumene 1". *RE* 17.34.2123.39–2174.1.

Glerup, M. 2010. *Commentaries on Genesis 1–3. Severian of Gabala and Bede the Venerable*. Downers Grove: IVP Academic.

Goodison, L. 1989. *Death, Women and the Sun: Symbolism of Regeneration in Early Aegean Religion*. London: University of London, Institute of Classical Studies.

Goodison, L. 2004a. "What is Ancient Mediterranean Religion?" In Iles Johnston, S. (ed.), *Religions of the Ancient World. A Guide*. Cambridge: Harvard University Press, 3–16.

Goodison, L. 2004b. "Myth". In Iles Johnston, S. (ed.), *Religions of the Ancient World. A Guide*. Cambridge: Harvard University Press, 45–58.

Gössmann, P. F. 1950. *Planetarium Babylonicum, oder die sumerisch-babylonischen Stern-Namen*. Rome: Pontificio Istituto Biblico.

Gottschalk, H. B. 1980. *Heraclides of Pontus*. Oxford: Clarendon Press.

Gow, A. S. 1951. "Nicandrea: With Reference to Liddell and Scott, ed. 9". *Classical Quarterly* 1.3–4: 95–118.

Gow, A. S. 1952[2]. *Theocritus*. 2 vols. Cambridge: Cambridge University Press.

Gow, A. S. & Scholfield, A. F. 1953. *Nicander of Colophon: Poems and Poetical Fragments*. Cambridge: Cambridge University Press.

Graf, F. 1993. *Greek Mythology*. Baltimore: Johns Hopkins University Press.

Graf, F. 2003. "Initiation. A Concept with a Troubled History". In Dodd, D. B. & Faraone, C. A. (eds.), *Initiation in Ancient Greek Rituals and Narratives*. London: Routledge, 3–24.

Graf, F. 2009. *Apollo*. London: Routledge.

Grafton, A. T. & Swerdlow, N. M. 1986. "Greek Chronography in Roman Epic: The Calendrical Date of the Fall of Troy in the Aeneid". *Classical Quarterly* 36.1: 212–18.

Graham, D. W. 2006. *Explaining the Cosmos. Ionian Tradition of Scientific Philosophy*. Princeton: Princeton University Press.

Graham, D. W. 2010. *Texts of Early Greek Philosophy: The Complete Fragments and Selected Testimonies of the Major Presocratics*. 2 vols. Cambridge: Cambridge University Press.

Graham, D. W. 2013. *Science before Socrates, Parmenides, Anaxagoras, and the New Astronomy*. Oxford: Oxford University Press.

Graham, A. J. 1982. "The colonial expansion of Greece". In Boardman, J. & Hammond, N. G. L. (eds.), *The Cambridge Ancient History*[2]. Vol. 3. Part 3. Cambridge: Cambridge University Press, 83–162.

Graham, A. J. 2001. *Collected Papers on Greek Colonization*. Leiden: Brill.

Grant, E. 1994. *Planets, Stars, and Orbs. The Medieval Cosmos, 1200–1687*. Cambridge: Cambridge University Press.

Graves, R. 1956. *Lucan: Pharsalia. Dramatic Episodes of the Civil Wars*. Harmondsworth: Penguin.

Gray, C. D. 1901. *The Šamaš Religious Texts*. Chicago: University of Chicago Press.

Grayson, A. K. 1972. *Assyrian Royal Inscriptions*. Vol. 1. Wiesbaden: Harrassowitz.

Grayson, A. K. 1974–77. "The Empire of Sargon of Akkad". *Archiv für Orientforschung* 25: 56–64.

Green, A. 1985. "A Note on the 'Scorpion-Man' and Pazuzu". *Iraq* 47: 75–82.

Green, M. 1991. *The Sun Gods of Ancient Europe*. London: Batsford.

Green, P.: 2007. The Argonautika by Apollonios Rhodios. Berkeley: University of California Press.

Greene, W. C. (ed.) 1938. *Scholia Platonica*. Haverford: American Philological Association.

Griffith, R. 1897[2]. *The Hymns of the Rig-Veda*. Benares: E.J. Lazarus.

Griffith, R. D. 2001. "Sailing to Elysium: Menelaus' Afterlife (*Odyssey* 4.561–69) and Egyptian Religion". *Phoenix* 55.3–4: 213–43.

Gronewald, M. & Daniel, R. 2004a. "Ein neuer Sappho-Papyrus". *Zeitschrift für Papyrologie und Epigraphik* 147: 1–8.

Gronewald, M. & Daniel, R. 2004b. "Nachtrag zum neuen Sappho-Papyrus". *Zeitschrift für Papyrologie und Epigraphik* 149: 1–4.

Guggisberg, M. 1996. "Eine Reise von Knossos nach Strettweg". *Archäologischer Anzeiger* 2: 175–95.

Gundel, W. 1912. "Helike 3". *RE* 7.14: 2858.39–2862.32.

Gurney, O. R. 1955. "The Sultantepe Tablets (Continued). IV. The Cuthaean Legend of Naram-Sin". *Anatolian Studies* 5: 93–113.

Guštin, M. 1974. "Gomile starejše železne dobe iz okolice Boštanja". In Guštin, M. (ed.), *Varia Archaeologica*. Brežice: Posavski muzej Brežice, 87–119.

Güterbock, H. G. 1958. "The Composition of Hittite Prayers to the Sun". *Journal of the American Oriental Society* 78.4: 237–45.

Guthrie, W. K. C. 1962–81. *A History of Greek Philosophy*. 6 vols. Cambridge: Cambridge University Press.

Guthrie S. E. 1993. *Faces in the Clouds. A New Theory of Religion*. New York: Oxford University Press.

Hack, K. 1939. *God in Greek Philosophy to the Time of Socrates*. Princeton: Princeton University Press.

Hahn R. 2001. *Anaximander and the Architects*. Albany: SUNY Press.

Hainsworth, B. 1988. "Books V–VIII". In Heubeck, A., West, S. & Hainsworth, B. (eds.), *A Commentary on Homer's Odyssey*. Vol. 1: *Introduction and Books X–VIII*. Oxford: Clarendon Press, 247–485.

Hainsworth, B. 1993. *The Iliad: A Commentary. Vol. III: books 9–12*. Cambridge: Cambridge University Press.

Halbherr, F. 1890. "Iscrizioni Cretesi". In Comparetti, D. (ed.), *Museo italiano di antichità classica* 3. Firenze: E. Loescher, 559–748.

Halverson, J. 1992. "Havelock on Greek Orality and Literacy". *Journal of the History of Ideas* 53.1: 148–63.

Hämeen-Anttila, J. 2006. *The Last Pagans of Iraq. Ibn Waçshiyya and his* Nabatean Agriculture. Leiden: Brill.

Hammerstaedt, J. 2009. "The Cologne Sappho: Its Discovery and Textual Constitution". In Greene, E. & Skinner, M. B. (eds.), *The New Sappho on Old Age: Textual and Philosophical Issues*. Hellenic Studies Series 38. Washington, DC: Center for Hellenic Studies.

Hannah, R. 2009. *Time in Antiquity*. London: Routledge.

Hänsel, A. & Hänsel, B. 1997a. *Gaben an die Götter—Schätze der Bronzezeit Europas*. Berlin: Museum für Vor- Und Frühgeschichte.

Hänsel, A. & Hänsel, B. 1997b. "Herrscherinsignien der älteren Urnenfelderzeit. Ein Gefäßdepot aus dem

Saalegebiet Mitteldeutschlands". *Acta Praehistorica et Archaeologica* 29: 39–68.

Hänsel B. 1997. "Gaben an die Götter—Schätze der Bronzezeit Europas—Eine Einführung". In Hänsel, A. & Hänsel, B. (eds.), *Gaben an die Götter—Schätze der Bronzezeit Europas*. Berlin: Museum für Vor- Und Frühgeschichte, 11–22.

Hansen S. 2001. "Neolithic Sculpture. Some Remarks on an Old Problem". In Biehl, P. F. & Bertemes, F. (eds.), *The Archaeology of Cult and Religion*. Budapest: Archaeolingua, 37–52.

Hardie, P. R. 1983. "Atlas and Axis". *Classical Quarterly* 33.1: 220–28.

Hardie, P. R. 1986. *Virgil's* Aeneid. Cosmos *and* Imperium. Oxford: Clarendon Press.

Hardie, P. R. 1992. "Plutarch and the interpretation of myth". *Aufstieg und Niedergang der römischen Welt* 33: 4743–87.

Harding, A. F. 2000. *European Societies in the Bronze Age*. Cambridge: Cambridge University Press.

Harkin M. E. 2009. "Lévi-Strauss and History". In Wiseman, B. (ed.), *The Cambridge Companion to Lévi-Strauss*. Cambridge: Cambridge University Press, 39–58.

Harley, J. B. 1987. "The Map and the Development of the History of Cartography". In Harley, J. B. & Woodward, D. (eds.), *The History of Cartography*. Vol.1. *Cartography in Prehistoric, Ancient, and Medieval Europe and the Mediterranean*. Chicago: University of Chicago Press, 1–42.

Harley, J. B. & Woodward, D. (eds.) 1987. *The History of Cartography*. Vol.1. *Cartography in Prehistoric, Ancient, and Medieval Europe and the Mediterranean*. Chicago: University of Chicago Press.

Harley, J. B. & Woodward, D. (eds.) 1992. *The History of Cartography*. Vol. 2. Pt. 1. *Cartography in the Traditional Islamic and South Asian Societies*. Chicago: University of Chicago Press.

Harrison, J. E. 1882. *Myths of the Odyssey in Art and Literature*. London: Rivingtons.

Harrison, J. E. 1908a². *Prolegomena to the Study of Greek Religion*. Cambridge: Cambridge University Press.

Harrison, J. E. 1908b. "Helios-Hades". *Classical Review* 22.1: 12–16.

Harrison, J. E. 1912. *Themis*. Cambridge: Cambridge University Press.

Harvey, P. D. A. 1996. *Mappa Mundi: the Hereford World Map*. London: British Library.

Hasluck, F. W. 1910. *Cyzicus*. Cambridge: Cambridge University Press.

Haubold, J. 2013. *Greece and Mesopotamia. Dialogues in Literature*. Cambridge: Cambridge University Press.

Havelock, E. A. 1958. "Parmenides and Odysseus". *Harvard Studies in Classical Philology* 63: 133–43.

Havelock, E. A. 1963. *Preface to Plato*. Cambridge: Harvard University Press.

Havelock, E. A. 1966a. "Pre-Literacy and the Pre-Socratics". *Bulletin of the Institute of Classical Studies* 13: 44–67.

Havelock, E. A. 1966b. "Thoughtful Hesiod". *Yale Classical Studies* 20: 61–72.

Havelock, E. A. 1978. *The Greek Concept of Justice*. Cambridge: Harvard University Press.

Havelock, E. A. 1983. "The Linguistic Task of the Presocratics". In Robb, K. (ed.), *Language and Thought in Early Greek Philosophy*. La Salle: Hegeler Institute, 7–82.

Hawes, G. 2014. *Rationalizing Myth in Antiquity*. Oxford: Oxford University Press.

Hawes, G. 2017. "Of myths and maps". In Hawes, G. (ed.), *Myths on the Map. The Storied Landscapes of Ancient Greece*. Oxford: Oxford University Press, 1–13.

Hawkes, C. F. C. 1977. *Pytheas: Europe and the Greek Explorers*. Oxford: Blackwell.

Haycock, M. 2007. *Legendary Poems from the Book of Taliesin*. Aberystwyth: CMCS.

Hayden, B. & Villeneuve, S. 2011. "Astronomy in the Upper Palaeolithic?" *Cambridge Archaeological Journal* 21: 331–55.

Heath, T. L. 2004 [1913]. *Aristarchus of Samos, the Ancient Copernicus*. Mineola: Dover.

Heath, T. L. 1991[1932]. *Greek Astronomy*. Mineola: Dover.

Heidel, A. 1949². *Gilgamesh Epic and Old Testament Parallels*. Chicago: University of Chicago Press.

Heidel, A. 1951². *The Babylonian Genesis*. Chicago: University of Chicago Press.

Heidel, W. A. 1937. *The Frame of the Ancient Greek Maps*. New York: American Geographical Society.

Heimpel, W. 1986. "The Sun at Night and the Doors of Heaven". *Journal of Cuneiform Studies* 38.2: 127–51.

Henkelman, W. 2006. "The Birth of Gilgamesh (Ael. NA XII.21): A case-study in literary receptivity". In Rollinger, R. & Truschnegg, B. (eds.), *Altertum und Mittelmeerraum: Die Antike Welt diesseits und jenseits der Levante*. Stuttgart: Steiner, 807–56.

Henkelman, W. 2010. "Beware of Dim Cooks and Cunning Snakes: Gilgameš, Alexander, and the Loss of immortality". In Rollinger, R., Gufler, B., Lang, M. & Madreiter, I. (eds.), *Interkulturalität in der Alten*

Welt: Vorderasien, Hellas, Ägypten und die vielfältigen Ebenen des Kontakts. Wiesbaden: Harrassowitz, 323–59.

Henn, M. J. 2003. *Parmenides of Elea: A Verse Translation with Interpretative Essays and Commentary to the Text*. Westport: Praeger.

Henrichs, A. 1988. "Three Approaches to Greek Mythography." In Bremmer, J. (ed.), *Interpretations of Greek Mythology*, London: Routledge, 242–77.

Herren, M. 2011. *The Cosmography of Aethicus Ister*. Turnhout: Brepols.

Hesse, M. B. 1966. *Models and Analogies in Science*. Notre Dame: University of Notre Dame Press.

Hesse, M. B. 1988. "The Cognitive Claims of Metaphor". *Journal of Speculative Philosophy* 2.1: 1–16.

Heubeck, A. 1989. "Books IX–XII". In Heubeck, A. & Hoekstra, A. (eds.), *A Commentary on Homer's Odyssey*. Vol. 2. *Books IX–XVI*. Oxford: Clarendon Press, 1–143.

Hiatt, A. 2008. *Terra Incognita: Mapping the Antipodes before 1600*. London: British Library.

Higgins, I. M. 1997. *Writing East: The 'Travels' of Sir John Mandeville*. Philadelphia: University of Pennsylvania Press.

Higgins, I. M. 2011. *The Book of John Mandeville: With Related Texts*. Indianapolis: Hackett.

Highbarger, E. L. 1940. *The Gates* of *Dreams: An Archaeological Examination of Vergil, Aeneid VI, 893–899*. Baltimore: Johns Hopkins University Press.

Hilka, A. 1920. *Der altfranzösische Prosa-Alexanderroman*. Halle: Niemeyer.

Hilka, A. 1977. *Historia Alexandri Magni (Historia de Preliis): Rezension J^2 (Orosius-Rezension)*. Vol. 2. Meisenheim am Glan: Hain.

Hinke, W. J.: 1907. *A New Boundary Stone of Nebuchadrezzar I*. Philadelphia: University of Pennsylvania.

Hirschfeld, G. 1895. "Ἄρκτων ὄρος". *RE* 2.3.1172.49–54.

Hodder, I. 1987. "The Contextual Analysis of Symbolic Meanings". In Hodder, I. (ed.), *The Archaeology of Contextual Meanings*. Cambridge: Cambridge University Press, 1–10.

Hodder, I. 1990. *The Domestication of Europe*. Oxford: Blackwell.

Hodder, I. & Hutson S. 2003[3]. *Reading the Past. Current Approaches to Interpretation in Archaeology*. Cambridge: Cambridge University Press.

Hoekstra, A. 1957. "Hésiode et la tradition orale. Contribution à l'étude du style formulaire". *Mnemosyne* 10.3: 193–225.

Hoekstra, A.: Books XIII–XVI. In Heubeck, A. & Hoekstra, A. (eds.), *A Commentary on Homer's Odyssey*. Vol. 2. *Books IX–XVI*. Oxford: Clarendon Press, 145–287.

Höfer, O. 1897–1902. "Orthros". In Roscher, W. (ed.), *Lexikon der Griechischen und Römischen Mythologie*. Vol. 3/1. Leipzig: Teubner, 1215.12–1218.33.

Hoffner, H. 1998[2]. *Hittite Myths*. Atlanta: Society of Biblical Literature.

Holenweger, E. 2011. *Die anthropomorphe Tonplastik der Mittel- und Spätbronzezeit im mittel- bis unterdanubischen Gebiet. Eine Untersuchung zu ägäischen Traditionen und ihrer Verbreitung an der unteren Donau*. Diss. Universität des Saarlandes.

Hollis, A. S. 1991. "Callimachus *Aetia* Fr. 75.58–59 Pf." *Zeitschrift für Papyrologie und Epigraphik* 86: 11–13.

Holmes, T. R. 1907. *Ancient Britain and the Invasion of Julius Caesar*. Oxford: Clarendon Press.

Holzinger, C. von 1895: *Lykophron's* Alexandra, *Griechisch und Deutsch, mit erklärenden Anmerkungen*. Leipzig: Teubner.

Homolle, T. 1881. "Le calendrier délien". *Bulletin de correspondance hellénique* 5: 25–30.

Honko, L. 1984. "The problem of defining myth". In Dundes, A. (ed.), *Sacred Narrative. Readings in the Theory of Myth*. Berkeley: University of California Press, 41–52.

Hornung, E. 1999. *The Ancient Egyptian Books of the Afterlife*. Ithaca: Cornell University Press.

Horowitz, W. 1988. "The Babylonian Map of the World". *Iraq* 50, 147–65.

Horowitz, W. 1997. "The Great Wall of Sargon of Akkad" *N.A.B.U.* 48: 98.

Horowitz, W. 1998a. *Mesopotamian Cosmic Geography*. Winona Lake: Eisenbrauns.

Horowitz, W. 1998b. "The 364 Day Year in Mesopotamia, Again". *N.A.B.U.* 49: 49–51.

Horowitz, W. 2014. *The Three Stars Each: The Astrolabes and Related Texts*. Vienna: Institut für Orientalistik der Universität Wien.

Horton, R. 1960. "A Definition of Religion, and its Uses". *Journal of the Royal Anthropological Institute of Great Britain and Ireland* 90.2: 201–26.

Horton, R. 1962. "The Kalabari World-View: An Outline and Interpretation". *Africa* 32.3:197–220.

Horton, R. 1964. "Ritual Man in Africa". *Africa* 34.2: 85–104.

Horton, R. 1967a. "African Traditional Thought and Western Science. Part I. From Tradition to Science". *Africa: Journal of the International African Institute* 37.1: 50–71.

Horton, R. 1967b. "African Traditional Thought and Western Science. Part II. The 'Closed' and 'Open' Predicaments". *Africa: Journal of the International African Institute* 37.2: 155–87.

Horton, R. 1968. "Neo-Tylorianism: Sound Sense or Sinister Prejudice?" *Man* 3.4: 625–34.

Hoskin, M. 2001. *Tombs, Temples and Their Orientations: A New Perspective on Mediterranean Prehistory*. Bognor Regis: Ocarina.

Housman, A. E. 1912. *M. Manilii Astronomicon liber secundus*. London: G. Richards.

Howard-Carter, T. 1981. "The Tangible Evidence for the Earliest Dilmun". *Journal of Cuneiform Studies* 33.3–4: 210–23.

Howard-Carter, T. 1987. "Dilmun: At Sea or Not at Sea? A Review Article". *Journal of Cuneiform Studies* 39.1: 54–117.

Huffman, C. A. 1993. *Philolaus of Croton. Pythagorean and Presocratic*. Cambridge: Cambridge University Press.

Huld, M. E. 1999. "PIE 'bear' *Ursus arctos*, *Ursa Major*, and *Ursa Minor*". In Jones-Bley, K., Huld, M. E., Della Volpe, A. & Robbins Dexter, M. (eds.), *Proceedings of the tenth annual UCLA Indo-European conference*. Journal of Indo-European Studies Monograph Series 32. Washington: Institute for the Study of Man, 117–30.

Hume, R. E. (trans.) 1921. *The Thirteen Principal Upanishads*. London: Milford.

Hunger, H. 1992. *Astrological Reports to Assyrian Kings*. Helsinki: Helsinki University Press.

Hunger, H. & Pingree, D. 1989. *MUL.APIN: An Astronomical Compendium in Cuneiform*. Archiv für Orientforschung, Beiheft 24. Horn: F. Berger.

Hunger, H. & Pingree, D. 1999. *Astral Sciences in Mesopotamia*. Leiden: Brill.

Hunnius, C. 1904. *Das syrische Alexanderlied*. Göttingen: Dieterich.

Hunter, R. L. 1993. Jason and the Golden Fleece. Oxford: Clarendon Press.

Huxley, G. L. 1960. "Homerica I: Homeric Syrie; II: Eugamon; III: A Poem of the Homeridae". *Greek, Roman and Byzantine Studies* 3: 17–30.

Huxley, G. L. 1963a. "Studies in the Greek Astronomers I. Eudoxian Topics". *Greek, Roman and Byzantine Studies* 4: 83–96.

Huxley, G. L. 1963b. "Studies in the Greek Astronomers II. A fragment of Cleostratus of Tenedos". *Greek, Roman and Byzantine Studies* 4: 97–99.

Huxley, G. L. 1965. "Xenomedes of Keos". *Greek, Roman and Byzantine Studies* 6: 235–45.

Huxley, M. 1997. "The Shape of the Cosmos According to Cuneiform Sources". *Journal of the Royal Asiatic Society* 7.2: 189–98.

Huxley, M. 2000. "The Gates and Guardians in Sennacherib's Addition to the Temple of Assur". *Iraq* 62: 109–37.

Iaia, C. 2004. "Lo stile della 'barca solare ornitomorfa' nella toreutica italiana della prima età del ferro". In Negroni Catacchio, N. (ed.), *Preistoria e Protostoria in Etruria. Miti simboli decorazioni—ricerche e scavi. Atti del Sesto incontro di studi* 1. Milano: Centro studi di preistoria e archeologia, 307–18.

Iles Johnston, S. 2018. *The Story of Myth*. Cambridge: Harvard University Press.

Insoll, T. 2004. *Archaeology, Ritual, Religion*. London: Routledge.

Irby, G. L. 2012. "Mapping the World: Greek Initiatives from Homer to Eratosthenes". In Talbert, R. J. A. (ed.), *Ancient Perspectives. Maps and Their Place in Mesopotamia, Egypt, Greece & Rome*. Chicago: University of Chicago Press, 81–107

Irby-Massie, G. L. & Keyser, P. T. 2002. *Greek Science of the Hellenistic Era: A Sourcebook*. London: Routledge.

Isager, S. & Skydsgaard, J. E. 1992. *Ancient Greek Agriculture: An Introduction*. London: Routledge & Kegan Paul.

Istrin, V. M. 1897. *Otkrovenie Mefodija Patarskogo i apokrifičeskie videnija Daniila v vizantijskoj i slavjano-russkoj literaturah*. Moskva: Universitetskaja tipografija.

Iversen, P. A. 2017. "The Calendar on the Antikythera Mechanism and the Corinthian Family of Calendars". *Hesperia* 86.1: 129–203.

Jacob, C. 1988. "Ecriture, géométrie et dessin figuratif: essai de lecture d'une carte grecque". *Mappemonde* 1: 1–4.

Jacobsen, T. 1976. *The Treasures of Darkness: A History of Mesopotamian Religion*. New Heaven: Yale University Press.

Jacobsen, T. 1981. "The Eridu Genesis". *Journal of Biblical Literature* 100.4: 513–29.

Jacobsen, T. 1987. *The Harps that once... Sumerian poetry in translation*. New Haven: Yale University Press.

Jacobsen, T. & Alster, B. 2000. "Ningišzida's Boat-Ride to Hades". In George, A. R. & Finkel, I. L. (eds), *Wisdom, Gods and Literature: Studies in Assyriology in Honour of W. G. Lambert*. Winona Lake: Eisenbrauns, 315–44.

Jacoby, F. 1923/1957². *Fragmente der griechischen Historiker*. Teil 1. *Genealogie und Mythographie*. A. *Vorrede, Text, Addenda, Konkordanz [Nr. 1–63]*. Berlin/Leiden: Weidmann/Brill.

Jacoby, F. 1926. *Fragmente der griechischen Historiker. Teil 2- Zeitgeschichte. C. Kommentar zu Nr. 64–105.* Berlin: Weidmann.

Jacoby, F. 1955. *Fragmente der griechischen Historiker. Teil 3. Geschichte von Städten und Völkern. B. Kommentar zu Nr. 297–607. Text—Noten.* Leiden: Brill.

Jaeger, W. 1947. *The Theology of the Early Greek Philosophers*. Oxford: Clarendon Press.

Jaeger, W. 1946–47[3]. *Paideia: The Ideals of Greek Culture.* Vols. 1–2. Oxford: Blackwell.

Jamison, S. W. & Brereton, J. P. 2014. *The Rigveda. The Earliest Religious Poetry of India.* 3 vols. Oxford: Oxford University Press.

Janko, R. 1982. *Homer, Hesiod and the Hymns*. Cambridge: Cambridge University Press.

Janko, R. 1998. "The Homeric Poems as Oral Dictated Texts". *Classical Quarterly* 48.1: 1–13.

Janko, R. 2002. "The Derveni Papyrus: An Interim Text". *Zeitschrift für Papyrologie und Epigraphik* 141: 1–62.

Janni, P. 2015. "The Sea of the Greeks and Romans". In Bianchetti, S., Cataudella, M. R. & Gehrke, H.-J. (eds.), *Brill's Companion to Ancient Geography*. Leiden: Brill, 21–42.

Jastrow, M. & Clay, A. T. 1920. *An Old Babylonian version of the Gilgamesh Epic.* New Haven: Yale University Press.

Jebb, R. C. 1994[1905]. *Bacchylides: The Poems and Fragments*. Hildesheim: Olms.

Jensen, P. 1890. *Die Kosmologie der Babylonier.* Strassburg: Trübner.

Jensen, P. 1906. *Das Gilgamesch-Epos in der Weltliteratur*. Strassburg: Trübner.

Jeremias, A. 1887. *Die babylonisch-assyrischen Vorstellungen vom Leben nach dem Tode.* Leipzig: Hinrichs.

Jeremias, A. 1902. *The Babylonian Conception of Heaven and Hell.* London: Nutt.

Jeremias, A. 1911. *The Old Testament in the Light of the Ancient Near East.* 2 vols. London: Williams & Norgate.

Jeremias, A. 1913. *Handbuch der Altorientalischen Geisteskultur*. Leipzig: Hinrichs.

Jessen, O. 1895. "Argonautai2. *RE* 2.3.743.47–787.25.

Jockenhövel A. 1974. "Ein reich verziertes Protovillanova-Rasiermesser". *Prähistorische Bronzefunde* 20.1: 81–88, pls. 19–21.

Johnson, D. M. 1999. "Hesiod's Descriptions of Tartarus (*Theogony* 721–819)". *Phoenix* 53.1–2: 8–28.

Jones, A. 2017. *A Portable Cosmos. Revealing the Antikythera Mechanism, Scientific Wonder of the Ancient World.* Oxford: Oxford University Press.

Jouanna, J. 1999. *Hippocrates*. Baltimore: Johns Hopkins University Press.

Joyce, J. W. 1993. Lucan: Pharsalia. Ithaca: Cornell University Press.

Kahn, C. H. 1960. *Anaximander and the Origins of Greek Cosmology*. New York: Columbia University Press.

Kahn, C. H. 1970. "On Early Greek Astronomy". *Journal of Hellenic Studies* 90: 99–116.

Kahn, C. H. 1979. *The Art and Thought of Heraclitus.* Cambridge: Cambridge University Press.

Kahn, C. H. 2001. *Pythagoras and the Pythagoreans: A Brief History*. Indianapolis: Hackett.

Kalogeropoulos, K. 2007. "Miniature Clay Anthropomorphic Representations in Greece and Europe through the Late Mycenaean Period: Similarities and Differences". In Galanaki, I., Tomas, H., Galanakis, Y. & Laffineur, R. (eds.), *Between the Aegean and Baltic Seas. Prehistory across Borders.* Austin/Liège: University of Texas/Universitè de Liège, 257–66.

Kaplan, P. G. 2018. "Early Greek Geography". In Keyser, P. T. & Scarborough, J. (eds.), *Oxford Handbook of Science and Medicine in the Classical World.* Oxford: Oxford University Press, 195–213.

Käppel, L. 2001. "Bilder des Nordens im frühen antiken Griechenland". In Engel-Braunschmidt, A., Fouquet, G., von Hinden, W. & Schmidt, I. (eds.), *Ultima Thule: Bilder des Nordens von der Antike bis zur Gegenwart.* Frankfurt am Main: Lang, 11–27.

Katz, D. 2007a. "Sumerian Funerary Rituals in Context". In Laneri, N. (ed.), *Performing Death: Social Analyses of Funerary Traditions in the Ancient Near East and Mediterranean.* Chicago: University of Chicago Press, 167–88.

Katz, D. 2007b. "Enki and Ninhursaĝa, Part One: The Story of Dilmun". *Bibliotheca Orientalis* 64.5–6: 568–89.

Kaul F. 1998. *Ships on Bronzes. A Study in Bronze Age Religion and Iconography.* 2 vols. Copenhagen: Nationalmuseet.

Keaney, J. J. & Lamberton, R. 1996. *[Plutarch]: Essay on the Life and Poetry of Homer.* Atlanta: Scholars Press.

Kees, H. 1987. *Der Götterglaube im alten Ägypten*. Berlin: Akademie Verlag.

Keim, K. 2014. "Cosmology as Science or Cosmology as Theology? Reflections of the Astronomical Chapters of Pirke DeRabbi Eliezer". In Stern, S. & Burnett, C. (eds.), *Time, Astronomy, and Calendars in the Jewish Tradition*. Leiden: Brill, 41–63.

Keith, A. B. 1920. *Rigveda Brahmanas*. Cambridge: Harvard University Press.

Kessler, K. 1983. "Zu den keilschriftlichen Quellen des 2./1. Jahrtausends v. Chr. über Dilmun". In Potts, D. T. (ed.), *Dilmun: New studies in the archaeology and early history of Bahrain*. Berlin: D. Reimer, 147–60.

Kidd, D. 1997. *Aratus: Phaenomena*. Cambridge: Cambridge University Press.

Kidd, I. G. 1988. *Posidonius. Volume II: The Commentary*. 2 vols. Cambridge: Cambridge University Press.

Kiessling, E. 1914. "Ῥίπαια ὄρη". *RE* 12.1.846.47–916.11.

Kilian-Dirlmeier I. 1974. "Ein bronzener Kesselwagen aus Delphi". *Archäologisches Korrespondenzblatt* 4.4: 349–52.

King, L. W. 1902. *The Seven Tablets of Creation*. Vol. 1. London: Luzac.

King, L. W. 1912. *Babylonian Boundary Stones and Memorial-Tablets in the British Museum*. London: British Museum.

Kingsley, P. 1995. *Ancient Philosophy, Mystery and Magic. Empedocles and Pythagorean Tradition*. Oxford: Clarendon Press.

Kirk, G. S. 1962. *Heraclitus: The Cosmic Fragments*. Cambridge: Cambridge University Press.

Kirk, G. S. 1970. *Myth: Its Meaning and Functions in Ancient and Other Cultures*. Berkeley: University of California Press.

Kirk, G. S. 1972a. "Greek Mythology: Some New Perspectives". *Journal of Hellenic Studies* 92: 74–85.

Kirk, G. S. 1972b. "Aetiology, ritual, charter: three equivocal terms in the study of myths". *Yale Classical Studies* 22: 83–102.

Kirk, G. S. 1984[1973]. "On Defining Myths". In Dundes, A. (ed.), *Sacred Narrative. Readings in the Theory of Myth*. Berkeley: University of California Press, 53–61.

Kirk, G. S. 1974. *The Nature of Greek Myths*. London: Penguin.

Kirk, G. S. 1990. *The Iliad. A Commentary*. Volume 2. *Books 5–8*. Cambridge: Cambridge University Press.

Kirk, G. S., Raven, J. E. & Schofield, M. 1983². *The Presocratic Philosophers*. Cambridge: Cambridge University Press.

Kiss, V. 2007. "Contacts Along the Danube. A Boat Model from the Early Bronze Age". In Galanaki, I., Tomas, H., Galanakis, Y. & Laffineur, R. (eds.), *Between the Aegean and Baltic Seas. Prehistory across Borders*. Austin/Liège: University of Texas/Universitè de Liège, 119–29.

Klingbeil, M. G. 1999. *Yahweh Fighting from Heaven. God as a Warrior and as God of Heaven in the Hebrew Psalter and Ancient Near Eastern Iconography*. Fribourg/Göttingen: Universitätsverlag/Vandenhoeck & Ruprecht.

Klooster, J. J. H. 2012. "Apollonius of Rhodes". In de Jong, I. J. F. (ed.), *Space in Ancient Greek Literature: Studies in Ancient Greek Narrative*. Leiden: Brill, 55–76.

Knight, V. 1995. *The Renewal of Epic: Responses to Homer in the* Argonautica *of Apollonius*. Leiden: Brill.

Koch, J. 1991. "Der Mardukstern Nēberu". *Die Welt des Orients* 22: 48–72.

Koch-Westenholz, U. 1995. *Mesopotamian Astrology: An Introduction to Babylonian and Assyrian Celestial Divination*. Copenhagen: Museum Tusculanum Press.

Kohanski, T. & Benson, C. D. 2007. *The Book of John Mandeville*. Kalamazoo: Medieval Institute Publications.

Kominko, M. 2013a. *The World of Kosmas. Illustrated Byzantine Codices of the Christian Topography*. Cambridge: Cambridge University Press.

Kominko, M. 2013b. "The Science of the Flat Earth". In Anderson, J. C. *The Christian Topography of Kosmas Indikopleustes. Firenze, Biblioteca Medicea Laurenziana, Plut. 9.28*. Rome: Edizioni di Storia e Letteratura, 67–82.

Konaris, M. D. 2010. "The Greek Gods in Late Nineteenth- and Early Twentieth-Century German and British Scholarship". In Bremmer, J. N. & Erskine, A. (eds.), *The Gods of Ancient Greece. Identities and Transformations*. Edinburgh: Edinburgh University Press, 483–503.

Konaris, M. D. 2011. "Apollo in nineteenth-century scholarship. The case of Karl Otfried Müller". In Haysom, M. & Wallensten, J. (eds.), *Current Approaches to Religion in Ancient Greece. Papers Presented at a Symposium at the Swedish Institute at Athens, 17–19 April 2008*. Stockholm: Svenska Institutet i Athen, 13–22.

Konaris, M. D. 2016. *The Greek Gods in Modern Scholarship*. Oxford: Oxford University Press.

Kotar, P. C. 2011. "The Ethiopic Alexander Romance". In Zuwiyya, Z. D. (ed.), *A Companion to Alexander Literature in the Middle Ages*. Leiden: Brill, 157–75.

Kothe, H. 1970. "Apollons ethnokultureller Herkunft". *Klio* 52: 205–30.

Kouremenos, T., Parássoglou, G. M. & Tsantsanoglou, K. 2006. *The Derveni Papyrus*. Firenze: Olschki.

Kovacs, M. G. 1989. *The Epic of Gilgamesh*. Stanford: Stanford University Press.

Kozlowski, J. 2018. "Gilgamesh's quest for immortality (Gilg. IX–XI) as a narratological pattern for Odysseus's *nostos* (Od. V,1–XIII,187)". *Eirene. Studia Graeca et Latina* 54: 11–31.

Kramer, S. N. 1944. "Dilmun, the Land of the Living". *Bulletin of the American Schools of Oriental Research* 96: 18–28.

Kramer, S. N. 1956. *From the Tablets of Sumer: Twenty-Five Firsts in Man's Recorded History*. Indian Hills: Falcon's Wing Press.

Kramer, S. N. 1961². *Sumerian Mythology*. New York: Harper & Row.

Kramer, S. N. 1963a. *The Sumerians: Their History, Culture and Character*. Chicago: University of Chicago Press.

Kramer, S. N. 1963b. "Cuneiform Studies and the History of Literature: The Sumerian Sacred Marriage Texts". *Proceedings of the American Philosophical Society* 107: 485–527.

Kramer, S. N. 1983. "The Sumerian Deluge Myth: Reviewed and Revised". *Anatolian Studies* 33: 115–21.

Krappe, A. H. 1941. "The Subterraneous Voyage". *Philological Quarterly* 20: 119–30.

Krappe, A. H. 1942a. "ΑΠΟΛΛΩΝ ΚΥΚΝΟΣ". *Classical Philology* 37.4: 353–70.

Krappe, A. H. 1942b. "The Land of Darkness". *Philological Quarterly* 21: 334–46.

Krappe, A. H.: 1943. "Avallon". *Speculum* 18.3: 303–22.

Krappe, A. H. 1947: "Ἀπόλλων Ὄνος". *Classical Philology* 42.4: 223–34.

Kraus, H. J., Schmidt H. & Kranz, W. 1952. "Ein neues Hesiodfragment". *Rheinisches Museum für Philologie* 95: 217–28.

Krauskopf, I. 2013. "Gods and demons in the Etruscan pantheon". In Turfa, J. M. (ed.), *The Etruscan World*. Abingdon: Routledge, 513–38.

Krebernik, M. 1992. "Mesopotamian Myths at Ebla: *ARET* 5, 6 and *ARET* 5, 7". In Fronzaroli, P. (ed.), *Literature and Literary Language at Ebla*. Firenze: Università di Firenze, 63–149.

Kremer, R. L. 2020. "Experience and Observation in Hellenistic Astronomy". In Bowen, A. C. & Rochberg, F. (eds.), *Hellenistic Astronomy. The Science in Its Contexts*. Brill: Leiden, 190–218.

Kristiansen, K. 2010. "Rock art and religion". In Fredell, Å., Kristiansen, K. & Criado Boado, F. (eds.), *Representations and Communications: Creating an Archaeological Matrix of Late Prehistoric Rock Art*. Oxford: Oxbow, 93–115.

Kristiansen, K. 2014. "Religion and society in the Bronze Age". In Bredholt Christensen, L., Hammer, O. & Warburton, D. A. (eds.), *The Handbook of Religions in Ancient Europe*. London: Routledge, 77–92.

Kristiansen, K. & Larsson, T. B. 2005. *The Rise of Bronze Age Society*. Cambridge: Cambridge University Press.

Kroll, W. 1922. "Krates 16." *RE* 11.22.1634.29–1641.55.

Kubitschek, W. 1894. "Apellaios 2". *RE* 1.2685.26–2686.13.

Kuelzer, A. 2018. "Byzantine Geography,". In Keyser, P. T. & Scarborough, J. (eds.), *Oxford Handbook of Science and Medicine in the Classical World*. Oxford: Oxford University Press, 921–42.

Kuentz, C. 1920. "Autour d'une conception égyptienne méconnue: l'Akhit ou soi-disant horizon". *Bulletin de l'Institut français d'archéologie orientale* 17: 121–90.

Kuiper, F. B. J. 1983[1960]. *Ancient Indian Cosmogony*. New Delhi: Vikas.

Kurth, D. 2004. *The Temple of Edfu, A Guide by an Ancient Egyptian Priest*. Cairo: American University in Cairo Press.

LÄ = Helck, W. & Otto, E. (eds.) 1975–84. *Lexikon der Ägyptologie*. 5 vols. Wiesbaden: Harrassowitz.

Lakoff, G. & Johnson, M. 2003[1980]. *Metaphors We Live By*. Chicago: University of Chicago Press.

Lambert, S. 2002. "The Sacrificial Calendar of Athens". *The Annual of the British School at Athens* 97: 353–99.

Lambert, W. G. 1960. *Babylonian Wisdom Literature*. Oxford: Clarendon Press.

Lambert, W. G. 1989. "Notes on a Work of the Most Ancient Semitic Literature". *Journal of Cuneiform Studies* 41.1: 1–33.

Lambert, W. G. 2013. *Babylonian Creation Myths*. Winona Lake: Eisenbrauns.

Lambert, W. G. & Millard, A.R. (eds.) 1969. *Atrahasis: The Babylonian Flood Story*. Oxford: Clarendon Press.

Landsberger, B. & Jacobsen, T. 1955. "An Old Babylonian Charm against Merḫu" *Journal of Near Eastern Studies* 14.1: 14–21.

Landsberger, B. & Kinnier Wilson, J. V. 1961. "The Fifth Tablet of Enuma Eliš". *Journal of Near Eastern Studies* 20.3: 154–79.

Laroche, E. 1960. *Les hiéroglyphes hittites, Première partie, L'écriture*. Paris: Centre National de la Recherche Scientifique.

Larson, L. M. 1917. *The King's Mirror (Speculum regale—Konungs Skuggsjá)*. New York: The American-Scandinavian Foundation.

Lauer, J. F. 1851. *Geschichte der homerischen Poesie*. Berlin: Reimer.

Lebedev, A. V. 1989. "Heraclitus in P. Derveni". *Zeitschrift für Papyrologie und Epigraphik* 79: 39–47.

Lehmann-Haupt, C. F. 1921. "Kimmerier". *RE* 11.21.397.6–434.7.

Lehoux, D. 2007. *Astronomy, Weather, and Calendars in the Ancient World:* Parapegmata *and Related Texts in Classical and Near-Eastern Societies.* Cambridge: Cambridge University Press.

Lesko, L. H. 1971. "Some observations on the composition of the Book of two ways" *Journal of the American Oriental Society* 91.1: 30–43.

Lesko, L. H. 1972. *The Ancient Egyptian Book of Two Ways.* Berkeley: University of California Press.

Lesky, A. 1948. "Aia". *Wiener Studien* 63: 22–68.

Letica Z. 1973. *Antropomorfne figurine Bronzanog doba u Jugoslaviji.* Beograd: Filozofski fakultet.

Lettvin, J. Y. 1980. "The Gorgon's Eye". In Brecher, K. & Feirtag, M. (eds.), *Astronomy of the Ancients,* Cambridge: MIT Press, 133–51.

Levin, D. N. 1971. *Apollonius'* Argonautica *Re-examined.* Vol. 1. *The Neglected First and Second Books.* Leiden: Brill.

Levine, B. A. & de Tarragon, J.-M. 1984. "Dead Kings and Rephaim: The Patrons of the Ugaritic Dynasty". *Journal of the American Oriental Society* 104.4: 649–59.

Lévi-Strauss, C. 1976. *Structural Anthropology 2.* New York: Basic Books.

Lightfoot, J. L. 1999. *Parthenius of Nicaea: The Poetical Fragments and the Erotica Pathemata.* Oxford: Oxford University Press.

Lightfoot, J. L. 2003. *Lucian: On the Syrian Goddess,* Oxford: Oxford University Press.

Lightfoot, J. L. 2014. *Dionysius Periegetes: Description of the Known World.* Oxford: Oxford University Press.

LIMC = *Lexicon Iconographicum Mythologiae Classicae.* 1981–97. 8 vols. Zürich: Artemis.

Lieu, S. N. C. & Vermes, G. 1996. "Constantine's 'Pagan Vision'; The Anonymous Panegyric on Constantine (310), Pan. Lat. VII(6)". In Lieu, S. N. C. & Montserrat, D. (eds.), *From Constantine to Julian, Pagan and Byzantine Views: A Source History.* London: Routledge, 63–96.

Lincoln, B. 1989. *Discourse and the Construction of Society. Comparative Studies of Myth, Ritual, and Classification.* New York: Oxford University Press.

Lincoln, B. 1996. "Gendered Discourses: The Early History of 'Mythos' and 'Logos'". *History of Religions* 36.1: 1–12.

Lincoln, B. 1997. "Competing Discourses: Rethinking the Prehistory of *Mythos* and *Logos*". *Arethusa* 30: 341–67.

Lincoln, B. 1999. Theorizing Myth: Narrative Ideology and Scholarship. Chicago: University of Chicago Press.

Lincoln, B. 2002. "Retheorizing Myth". In des Bouvrie, S. (ed.), Myth and Symbol I. Symbolic Phenomena in Ancient Greek Culture. Bergen: Norwegian Institute at Athens, 215–32.

Lincoln, B. 2003. "The Initiatory Paradigm in Anthropology, Folklore, and History of Religions". In Dodd, D. B. & Faraone, C. A. (eds.), *Initiation in Ancient Greek Rituals and Narratives.* London: Routledge, 241–53.

Lincoln, B. 2009. "In Praise of the Chaotic". In Walde, C. & Dill, U. (eds.), *Antike Mythen: Medien, Transformationen und Konstruktionen.* Berlin: De Gruyter, 372–88.

Lindsay, A. 1965. *The Clashing Rocks: A Study of Early Greek Religion and Culture and the Origins of Drama.* London: Chapman & Hall.

Livingstone, A. 1986. *Mystical and Mythological Explanatory Works of Assyrian and Babylonian Scholars.* Oxford: Clarendon Press.

Lloyd, G. E. R. 1966. *Polarity and Analogy. Two Types of Argumentation in Early Greek Thought.* Cambridge: Cambridge University Press.

Lloyd, G. E. R. 1967. "Popper versus Kirk: A Controversy in the Interpretation of Greek Science". *British Journal for the Philosophy of Science* 18.1: 21–38.

Lloyd, G. E. R. 1970. *Early Greek Science: Thales to Aristotle.* New York: Norton.

Lloyd, G. E. R. 1972. "The social background of early Greek philosophy and science". In Daiches, D. & Thorlby, A. (eds.), *Literature and Western Civilization, 1. The Classical World,* London: Aldus Books, 381–95.

Lloyd, G. E. R. 1975. "Greek cosmologies". In Blacker, C. & Loewe, M. (eds.), *Ancient Cosmologies,* London: Allen & Unwin, 198–224.

Lloyd, G. E. R. 1989. *The Revolutions of Wisdom: Studies in the Claims and Practices of Ancient Greek Science.* Berkeley: University of California Press.

Lloyd, G. E. R. 1990. *Demystifying Mentalities.* Cambridge: Cambridge University Press.

Lloyd, G. E. R. 1991. *Methods and problems in Greek science.* Cambridge: Cambridge University Press.

Lloyd, G. E. R. 2003. "The Problem of Metaphor: Chinese Reflections". In Boys-Stones, G. R. (ed.), *Metaphor, Allegory, and the Classical Tradition. Ancient Thought and Modern Revisions.* Oxford: Oxford University Press, 101–14.

Lloyd-Jones, H. & Parsons, P. 1983: *Supplementum Hellenisticum.* Berlin: De Gruyter.

López-Ruiz, C. 2010. *When the Gods Were Born: Greek Cosmogonies and the Near East.* Cambridge: Harvard University Press.

López-Ruiz, C. 2014[2]. "Greek and Near Eastern Mythologies: A Story of Mediterranean Encounters".

In Edmunds, L. (ed.), *Approaches to Greek Myth*. Baltimore: Johns Hopkins University Press, 154–99.

Lorimer, H. L. 1950. *Homer and the Monuments*. London: Macmillan.

Lo Schiavo F. 2010. *Le fibule dell'Italia meridionale e della Sicilia: dall'eta del bronzo recente al VI secolo a. C.* 2. Prähistorische Bronzefunde 14.14. Stuttgart: Steiner.

Louden, B. 2011. *Homer's* Odyssey *and the Near East*. Cambridge: Cambridge University Press.

LSJ = Liddell, H. G., Scott, R. & Jones, H. S. 1996[9]. *A Greek-English Lexicon*. Oxford: Clarendon Press.

Lucas, D. W. 1968. *Aristotle: Poetics*. Oxford: Clarendon Press.

Lundström, V. 1902. *Laskaris Kananos' Reseanteckningar från de nordiska länderna*. Upsala/Leipzig: Lundequist/ Harrassowitz.

Lunz R. 1974. *Studien zur End-Bronzezeit und älteren Eisenzeit im Südalpenraum*. Firenze: Sansoni.

Maass, E. 1883. *Analecta Eratosthenica*. Philologische Untersuchungen 6. Berlin: Weidmann.

Maass, E. 1898. *Commentariorum in Aratum reliquiae*. Berlin: Weidmann.

Mabbett, I. W. 1983 "The Symbolism of Mount Meru". *History of Religions* 23.1: 64–83.

MacGillivray, J. A. 2004. "The Astronomical Labyrinth at Knossos". In Cadogan, G., Hatzaki, E. & Vassilakis, A. (eds.), *Knossos: Palace, City, State*. BSAS 12. London: British School at Athens, 329–38.

MacKenzie, D. N. 1964. "Zoroastrian Astrology in the 'Bundahišn'". *Bulletin of the School of Oriental and African Studies* 27.3: 511–29.

Macurdy, G. H. 1920. "The Hyperboreans Again, Abaris, and Helixoia". *Classical Review* 34.7–8: 137–41.

Mair, A. W. & Mair, G. R. 1921. *Callimachus: Hymns and Epigrams. Lycophron. Aratus*. Cambridge: Harvard University Press.

Majnarić-Pandžić, N. 1982. "O porijeklu srednjebrončanodobne antropomorfne plastike u jugoslavenskom Podunavlju". *Opvscvla Archaeologica* 7: 47–60.

Malkin, I. 1998. *The Returns of Odysseus: Colonization and Ethnicity*. Berkeley: University of California Press.

Mander, P. 1999. "*Jānua hominum et deorum* in the Sumerian Mythological Texts". *AION* 59.1–4: 93–108.

Manoledakis, M. 2015. "The *Odyssey*, the Black Sea, and an Endless Voyage to a Utopian Destination". *Ancient West & East* 14: 301–12

Mansfeld, J. 1990. *Studies in the Historiography of Greek Philosophy*. Assen: Van Gorcum.

Mansfeld, J. 2021. "An Early Greek Epic Narrative Structures in Parmenides' Poem and the Relation between Its Main Parts". *Mnemosyne*.

Mansfeld, J. & Runia, D. T. 2009. *Aëtiana. The Method and Intellectual Context of a Doxographer*. Volume 2. *The Compendium*. Leiden: Brill.

Mansfeld, J. & Runia, D. T. 2010. *Aëtiana. The Method and Intellectual Context of a Doxographer*. Volume 3. *Studies in the Doxographical Traditions of Ancient Philosophy*. Leiden: Brill.

Mansfeld, J. & Runia, D. T. 2020. *Aëtiana*. Volume 5. *An Edition of the Reconstructed Text of the Placita with a Commentary and a Collection of Related Texts*. Leiden: Brill.

Maraszek, R. 1997. "Kultgerät im mittleren Oderraum: Die Deichselwagen". In Hänsel, A. & Hänsel, B. (eds.), *Gaben an die Götter—Schätze der Bronzezeit Europas*. Berlin: Museum für Vor- Und Fruhgeschichte, 71–75.

Maravelia, A.-A. 2009. "In Quest of Time through the Ancient Egyptian Art: The *Ouroboros* Archetype in the Meta-physics of Eternity". *Parnassos* 51: 55–88.

Marconi, C. 2007. *Temple Decoration and Cultural Identity in the Archaic Greek World: The Metopes of Selinus*. Cambridge: Cambridge University Press.

Marcovich, M. 2001[2]. *Heraclitus: Greek Text with a Short Commentary*. Sankt Augustin: Academia.

Marinatos, N. 2000. *The Goddess and the Warrior. The Naked Goddess and Mistress of Animals in Early Greek Religion*. London: Routledge.

Marinatos, N. 2001. "The Cosmic Journey of Odysseus". *Numen* 48: 381–416.

Marinatos, N. & Wyatt, N. 2011. "Levantine, Egyptian and Greek Mythological Conceptions of the Beyond". In Dowden, K. & Livingstone, N. (eds.). *A Companion to Greek Mythology*. Chichester: Wiley-Blackwell, 383–410.

Marshack A. 1972. *The Roots of Civilization. The Cognitive Beginnings of Man's First Art, Symbol, and Notation*. New York: McGraw-Hill.

Martin, T. H. 1879. "Comment Homère s'orientait". *Mémoires de l'Académie des Inscriptions et Belles-Lettres* 19.2: 1–28.

Martin, T. H. 1892. "Astronomia". In Daremberg, Ch. & Saglio, E. (eds.), *Dictionnaire des Antiquités Grecques et Romaines*. Vol. 1. Part 1. Paris: Hachette, 476a–304a.

Martinez, F. J. 1985. *Eastern Christian Apocalyptic in the Early Muslim Period: Pseudo-Methodius and Pseudo-Athanasius*. Diss. Catholic University of America.

Marx, C. 2014. "Lokalisierung von Pytheas' und Ptolemaios' Thule". *Zeitschrift für Geodäsie, Geoinformation und Landmanagement* 139.3: 197–203.

Matthäus, H. 1980. "Mykenische Vogelbarken. Antithetische Tierprotomen in der Kunst des östlichen Mittelmeerraumes". *Archäologisches Korrespondenzblatt* 10.4: 319–30.

Matthews, V. J. 1974. *Panyassis of Halicarnassos: Text and Commentary*. Leiden: Brill.

Matthews, V. J. 1996. *Antimachus of Colophon: Text and Commentary*. Leiden: Brill.

Maxwell-Stuart, P. G. 1981. *Studies in Greek Colour Terminology*. Volume 1. *ΓΛΑΥΚΟΣ*. Leiden: Brill.

McCready, W. D. 1996. "Isidore, the Antipodeans, and the Shape of the Earth". *Isis* 87: 108–27.

McCrindle, J. W. 1897. *The Christian Topography of Cosmas, an Egyptian Monk*. London: Hakluyt Society.

McKirahan, R. D. 2010²: *Philosophy before Socrates*. Indianapolis: Hackett.

McPhail, C. 2014. "Pytheas of Massalia's Route of Travel". *Phoenix* 68.3–4: 247–57.

Meißner, B. 1894. *Alexander und Gilgamos*. Leipzig: Eduard Pfeiffer.

Meletinsky, E. M. 2000[1976]. *The Poetics of Myth*. London: Routledge.

von Merhart, G. 1952. "Studien über einige Gattungen von Bronzegefäßen". *Festschrift der Römisch-Germanischen Zentralmuseums in Mainz* 2: 1–71, pls. 1–26.

Merkelbach, R. 1953. "Zu Pap. Schubart 17". *Museum Helveticum* 10: 125–27.

Merkelbach, R. 1958. "Literarische Texte unter Ausschluß der christlichen". *Archiv für Papyrusforschung* 16: 82–129.

Merkelbach, R. 1977². *Die Quellen des Griechischen Alexanderromans*. München: Beck.

Merkelbach, R. & West, M. L. 1964. "Origin and Religious Meaning of Greek Tragedy and Comedy, According to the 'Erigone' of Eratosthenes". *History of Religions* 3.2: 175–90.

Merkelback, R. & West, M. L. (eds.) 1967. *Fragmenta Hesiodea*. Oxford: Clarendon Press.

Merkelbach, R. & Stauber, J. 2005. *Jenseits des Euphrat. Griechische Inschriften*. München: Saur.

Merrills, A. 2013. "Geography and memory in Isidore's Etymologies". In Lilley, K. D. (ed.), *Mapping Medieval Geographies: Geographical Encounters in the Latin West and Beyond, 300–1600*. Cambridge: Cambridge University Press, 45–64.

Merry, W. W., Riddell, J. & Monro, D. B. 1886–1901. *Homer's Odyssey*. 2 vols. Oxford: Clarendon Press.

Mette, H. J. 1936. *Sphairopoiia: Untersuchungen zur Kosmologie des Krates von Pergamon*. München: Beck.

Meuli, K. 1921. *Odyssee und Argonautika, untersuchungen zur griechischen Sagengeschichte und zum Epos*. Berlin: Weidmann.

Meuli, K. 1931. "Laistrygones". *RE* Suppl. 5.537.5–540.48.

Meyer, E. 1959. "Pylos 5". *RE* 23.46.2135.1–2161.57.

Meyer, D. 2001. "Apollonius Rhodius as a Hellenistic Geographer". In Papanghelis, T. D. & Rengakos, A. (eds.), *A Companion to Apollonius Rhodius*. Leiden: Brill, 217–35.

Milićević Bradač, M. 2005. "The Transfer of Symbols and Meanings: The Case of the 'Horns of Consecration'". *Documenta Praehistorica* 32: 187–96.

Miller, J. F. 2009. *Apollo, Augustus, and the Poets*. Cambridge: Cambridge University Press.

Miller, K. 1898. *Die ältesten Weltkarten*. Vol. 6. *Rekonstruierte Karten*. Stuttgart: Roth.

Minkowski, C. 2010. "Where the Black Antelope Roam: Dharma and Human Geography". In Raaflaub, K.A. & Talbert, R. J. A. (eds.), *Geography and Ethnography: Perception of the World in Pre-Modern Societies*. Chichester: Wiley-Blackwell, 9–31.

Minorsky, V. 1970². *Hudūd al-ʿĀlam: The Regions of the World: A Persian Geography*. London: Luzac.

Molina Moreno, F. 2001. "Bilder des heiligen Nordens in der Antike, Patristik und Mittelalter". In Engel Braunschmidt, A., Fouquet, G., von Hinden, W. & Schmidt, I. (eds.), *Ultima Thule: Bilder des Nordens von der Antike bis zur Gegenwart*. Frankfurt am Main: Lang, 47–65.

Mommsen, A. 1878. *Delphika*, Leipzig: Teubner.

Mommsen, T. H. (ed.) 1895. *C. Iulii Solini Collectanea rerum memorabilium*. Berlin: Weidmann.

Mooney, G. W. 1912. *The Argonautica*. London: Longmans, Green.

Moreau, A. 1996. "Quand Apollon devint Soleil". In Bakhouche, A., Moreau, A. & Turpin, J.C. (eds.), *Les astres. Actes du Colloque International de Montpellier, 23–25 mars 1995*. Montpellier: Université Paul Valéry, 11–35.

Moretti, G. 1994. "The Other World and the Antipodes. The Myth of the Unknown Countries between Antiquity and the Renaissance". In Haase, W. & Reinhold, M. (eds.), *The Classical Tradition and the Americas*. Vol. 1. *European Images of the Americas and the Classical Tradition*. Part 1. Berlin: De Gruyter, 241–84.

Morgan, J. D. 1985. "ΛΕΥΚΑΣ ΠΕΤΡΗ". *Classical Quarterly* 35.1: 229–32.

Morgan, J. R. 1985. "Lucian's True Histories and the Wonders beyond Thule of Antonius Diogenes". *Classical Quarterly* 35.2: 475–90.

Morgan, K. 2000. *Myth and Philosophy from the Presocratics to Plato.* Cambridge: Cambridge University Press.

Morgan, L. 1985. "Idea, Idiom and Iconography". In Darcque, P. & Poursat, J.-C. (eds.), *L'iconographie minoenne.* Paris: French School at Athens, 5–19.

Morris, S. 1997. "Homer and the Near East". In Morris, I. & Powell, B. (eds.), *A New Companion to Homer.* Leiden: Brill, 599–623.

Morrison, J. S. 1955. "Parmenides and Er". *Journal of Hellenic Studies* 75: 59–68.

Moscalu E. & Beda C. 1991. "Bujoru. Ein Grabhügel der Basarabi-Kultur mit Votivkesselwagen aus Rumänien". *Prähistorische Zeitschrift* 66.2: 197–218.

Most, G. W. 1999. "The poetics of early Greek philosophy". In Long, A. A. (ed.), *The Cambridge Companion to Early Greek Philosophy.* Cambridge: Cambridge University Press, 332–63.

Mourelatos, A. 2008². *The Route of Parmenides.* Las Vegas: Parmenides.

Müllenhoff, K. 1870–87. *Deutsche Altertumskunde.* 2 vols. Berlin: Weidmann.

Müller, K. 1839². *The History and Antiquities of the Doric Race.* 2 vols. London: J. Murray.

Müller-Graupa, E. 1942. "Orthros 1". *RE* 18.1.36.1495.26–1500.30.

Müller-Karpe, H. 1974. "Der Grab 871 von Veji, Grotta Gramiccia". *Prähistorische Bronzefunde* 20.1: 89–97, pls. 22–28.

Müller-Karpe, H. 1978–79. "Bronzezeitliche Heilszeichen". *Jahresbericht des Instituts für Vorgeschichte der Universität Frankfurt a.M.*: 9–28.

Müller-Karpe, H. 1980. *Handbuch der Vorgeschichte.* Band 4. *Bronzezeit.* München: Beck.

Munn, M. 2007. "From Science to Sophistry: The Path of the Sun, the Shape of the World, and the Place of Athens in the Cosmos". In Pierris, A. L. (ed.), *Physis and Nomos: Power, Justice and the Agonistical Ideal of Life in High Classicism.* Patras: Institute for Philosophical Research, 111–33.

Murphy, J. P. 1977. *Rufus Festus Avienus: Ora maritima.* Chicago: Ares.

Naddaf, G. 1998. "Translator's Introduction". In Naddaf, G. (ed. and trans.), Brisson, L. *Plato the Myth Maker.* Chicago: University of Chicago Press.

Naddaf, G. 2005. *The Greek Concept of Nature.* Albany: SUNY Press.

Naddaf, G. 2009. "Allegory and the Origins of Philosophy". In Wians, W. (ed), *Logos and Mythos: Philosophical Essays on Greek Literature.* Albany: SUNY Press, 99–131.

Nadeau, J. Y. 1970. "Ethiopians". *Classical Quarterly* 20.2: 339–49.

Nagel, G. 192.: "Un papyrus funéraire de la fin du Nouvel Empire [Louvre 3292 (inv.)]". *Bulletin de l'Institut français d'archéologie orientale* 29: 1–127.

Nagy, G. 1973. "Phaethon, Sappho's Phaon, and the White Rock of Leukas". *Harvard Studies in Classical Philology* 77: 137–77.

Nagy, G. 1990. *Greek Mythology and Poetics.* Ithaca: Cornell University Press.

Nagy, G. 2007. "Homer and Greek Myth". In Woodard, R. D. (ed.), *The Cambridge Companion to Greek Mythology.* Cambridge: Cambridge University Press, 52–82.

Nakassis, D. 2004. "Gemination at the Horizons: East and West in the Mythical Geography of Archaic Greek Epic". *Transactions of the American Philological Association* 134: 215–33.

Nakassis, D. 2020. "Homeric Geography". In Pache, C. O. (ed.), *The Cambridge Guide to Homer.* Cambridge: Cambridge University Press, 267–77.

Nallino, C. A. 1899–1907. *Al-Battānī sive Albattenii Opus astronomicum.* 3 vols. Milano: Hoepli.

Nansen, F. 1911. *In Northern Mists. Arctic Exploration in Early Times.* 2 vols. London: Heinemann.

Naville, E. 1886. *Ägyptische Todtenbuch der XVIII. bis XX. Dynastie.* 3 vols. Berlin: A. Ascher.

Nesselrath, H. G. 2005. "Where the Lord of the Sea Grants Passage to Sailors through the Deep-blue Mere no More: The Greeks and the Western Seas". *Greece & Rome* 52: 153–71.

Neugebauer, O. 1945. "The History of Ancient Astronomy: Problems and Methods". *Journal of Near Eastern Studies* 4.1: 1–38.

Neugebauer, O. 1964. "Notes on Ethiopic Astronomy". *Orientalia* 33.1: 49–71.

Neugebauer, O. 1972. "On Some Aspects of Early Greek Astronomy". *Proceedings of the American Philosophical Society* 116.3: 243–51.

Neugebauer, O. 1975. *A History of Ancient Mathematical Astronomy.* Vol. 2. Berlin: Springer.

Nicolet, C. 1991. *Space, Geography, and Politics in the Early Roman Empire.* Ann Arbor: University of Michigan Press.

Nielsen, M. 2002. "Greek Myth—Etruscan Symbol". In des Bouvrie, S. (ed.), *Myth and Symbol I. Symbolic*

Phenomena in Ancient Greek Culture. Bergen: Norwegian Institute at Athens, 171–98.

Nilsson, M. P. 1906. *Griechische Feste*. Leipzig: Teubner.

Nilsson, M. P. 1920. *Primitive Time-Reckoning*. Lund: Gleerup.

Nilsson, M. P. 1932. *The Mycenaean Origin of Greek Mythology*. Berkeley: University of California Press.

Nilsson, M. P. 1949². *A History of Greek Religion*. Oxford: Clarendon Press.

Nishimura-Jensen, J. 2000. "Unstable Geographies: The Moving Landscape in Apollonius' Argonautica and Callimachus' Hymn to Delos". *Transactions of the American Philological Association* 130: 287–317.

Nitzsch, G. W. 1840. *Erklärende Anmerkungen zu Homer's Odyssee* 3. Hannover: Hahn.

Noegel, S. 2004. "Apollonius' *Argonautika* and Egyptian Solar Mythology". *Classical World* 97.2: 123–36.

Noviomagus, J. (ed.) 1537. *Bedae Presbyteri Anglosaxonis, monachi benedicti, viri literatissimi, opuscula cumplura de temporum ratione diligenter castigata: atq[ue] illustrata veteribus quibusdam annotationibus vna cum scholijs in obscuriores aliquot locos*. Köln: Iohannes Prael.

Oatley, K. 1978. *Perceptions and Representations: The Theoretical Bases of Brain Research and Psychology*. New York: Free Press.

Obbink, D. 2009. "Sappho Fragments 58–59: Text, Apparatus Criticus, and Translation". In Greene, E. & Skinner, M. B. (eds.), *The New Sappho on Old Age: Textual and Philosophical Issues*. Hellenic Studies Series 38. Washington, DC: Center for Hellenic Studies.

Ober, J. 1989. "Models and Paradigms in Ancient History". *The Ancient History Bulletin* 3.6: 134–37.

Ogden, D. 2013. *Drakōn: Dragon Myth and Serpent Cult in the Greek and Roman Worlds*. Oxford: Oxford University Press.

O'Grady, P. F. 2002. *Thales of Miletus. The Beginnings of Western Science and Philosophy*. Aldershot: Ashgate.

Olbrycht, M. J. 2000. "The Cimmerian problem re-examined: the evidence of the Classical sources". in Pstrusińska, J. & Fear, A. (eds.), *Collectanea Celto-Asiatica Cracoviensia*. Cracow: Księgarnia Akademicka, 71–99.

Olmsted, G. 1994/2019. *The Gods of the Celts and the Indo-Europeans*. Budapest/Tazewell: Archaeolingua.

Onians, R. 1951. *The Origins of European Thought about the Body, the Mind, the Soul, the World, Time and Fate*. Cambridge: Cambridge University Press.

Ornan, T. 2005. *The Triumph of the Symbol. Pictorial Representation of Deities in Mesopotamia and the Biblical Image Ban*. Göttingen: Vandenhoeck & Ruprecht.

Ossendrijver, M. 2020. "Techniques of Measurement and Computation". In Bowen, A. C. & Rochberg, F. (eds.), *Hellenistic Astronomy. The Science in Its Contexts*. Brill: Leiden, 41–53.

Otten, H. 1961. "Eine Beschwörung der Unterirdischen aus Boğazköy". *Zeitschrift für Assyriologie und vorderasiatische Archäologie* 54: 114–57.

Packard, L. R. 1874. "On a Passage in Homer's Odyssey (x. 81–86)". *Transactions of the American Philological Association* 5: 31–41.

Page, D. 1955. *Sappho and Alcaeus*. Oxford: Clarendon Press.

Page, D. 1973. *Folktales in Homer's Odyssey*. Cambridge: Cambridge University Press.

Palincaş, N. 2012. "Investigating Bronze Age Social Organisation in the Lower Danube Region. The Case of the Žuto Brdo-Gârla Mare Area". *Istros* 18: 13–38.

Palincaş, N. 2013. "Notes on Sun Cult and Burials in the Romanian Middle and Late Bronze Age (ca 2000–1200 BC)". In Sîrbu, V. & Ştefănescu, R. (eds.), *The Thracians and their Neighbors in the Bronze and Iron Ages. Proceedings of the 12th International Congress of Thracology* 2. Braşov: Istros, 311–22.

Palmer, J. 2009. *Parmenides & Presocratic Philosophy*. Oxford: Oxford University Press.

Panchenko, D. 1999. "The shape of the earth in Archelaus, Democritus and Leucippus". *Hyperboreus* 5.1: 22–39.

Pannekoek, A. 1961. *A History of Astronomy*. London: Allen and Unwin.

Pare, C. 1987. "Der Zeremonialwagen der Urnenfelderzeit: seine Entstehung, Form und Verbreitung". In Barth, F. E., Biel, J., Egg, M., France-Lanord, A., Jaochim, H. E., Pare, C., Schauer, P. & Uenze, H. P. (eds.), *Vierrädrige Wagen der Hallstattzeit. Untersuchungen zu Geschichte und Technik*. Mainz: Römisch-Germanisches Zentralmuseum, 25–67.

Pare, C. 1989. "From Dupljaja to Delphi: The Ceremonial Use of Wagon in Later Prehistory". *Antiquity* 63: 80–100.

Pare, C. 2004. "Die Wagen der Bronzezeit in Mitteleuropa". In Fansa, M. & Burmeister, S. (eds.), *Rad und Wagen. Der Ursprung einer Innovation Wagen im Vorderen Orient und Europa*. Mainz: Von Zabern, 355–72.

Parker, R. 2005. *Polytheism and Society at Athens*. Oxford: Oxford University Press.

Parker, R. 2011. *On Greek Religion*. Ithaca: Cornell University Press.

Parker, S. B. 1997. *Ugaritic Narrative Poetry*. Atlanta: Society of Biblical Literature.

Parpola, A. 2004–2005. "The Nāsatyas, the Chariot and Proto-Aryan Religion". *Journal of Indological Studies* 16–17: 1–63.

Parpola, S. 1993. "The Assyrian Tree of Life: Tracing the Origins of Jewish Monotheism and Greek Philosophy". *Journal of Near Eastern Studies* 52.3: 161–208.

Parthe, F. 1969. *Der griechische Alexanderroman Rezension gamma, Buch III*. Meisenheim am Glan: Hain.

Paton, W. R. & Hicks, E. L. 1891. *The Inscriptions of Cos*. Oxford. Clarendon Press.

Patton, L. L. & Doniger, W. 1996. *Myth and Method*. Charlottesville: University Press of Virginia.

Pearson, A. C. (ed.) 1917. *The Fragments of Sophocles*. 3 vols. Cambridge: Cambridge University Press.

Pearson, M. P., Cleal, R., Marshall, P., Needham, S., Pollard, J., Richards, C., Ruggles, C., Sheridan, A., Thomas, J., Tilley, C., Welham, K., Chamberlain, A., Chenery, C., Evans, J., Knüsel, C., Linford, N., Martin, L., Montgomery, J., Payne, A. & Richards, M. 2007. "The age of Stonehenge". *Antiquity* 81: 617–39.

Pease, A. S. 1943. "The Son of Neptune". *Harvard Studies in Classical Philology* 54: 69–82.

Peek, W. 1979. "Zu den Gedichten des Marcellus von Side auf Regilla und das Triopion des Herodes Atticus". *Zeitschrift für Papyrologie und Epigraphik* 33: 76–84.

Pender, E. E. 2003. "Plato on Metaphor and Model". In Boys-Stones, G. R. (ed.), *Metaphor, Allegory, and the Classical Tradition. Ancient Thought and Modern Revisions*. Oxford: Oxford University Press, 55–81.

Penglase, C. 1994. *Greek Myths and Mesopotamia. Parallels and Influence in the Homeric Hymns and Hesiod*. London: Routledge.

Perlmann, M. (trans.) 1987. *The History of al-Tabari*. Vol. 4. *The Ancient Kingdoms*. Albany: SUNY Press.

Pescheck, C. 1972. "Ein Reicher Grabfund mit Kesselwagen aus Unterfranken". *Germania* 50: 29–56.

Petridou, G. 2015. *Divine Epiphany in Greek Literature and Culture*. Oxford: Oxford University Press.

Petrović, J. 1928–30. "Votivna kolica iz Dupljaje". *Starinar* 5: 21–28.

Pettersson, M. 1992. *Cults of Apollo at Sparta: The Hyakinthia, the Gymnopaidai and the Karneia*. Stockholm: Svenska Institutet i Athen.

Pfeiffer, R. 1922. *Kallimachosstudien*. München: Hueber.

Pfister, F. 1934. "Θεοφάνια". *RE* 15.10.2133.25–58.

Pfister, K. 1940. *Die Etrusker*. München: Bruckmann.

Piankoff, A. 1934. "The Sky-Goddess Nut and the Night Journey of the Sun". *Journal of Egyptian Archeology* 20: 57–61.

Piankoff, A. 1954. *The Tomb of Ramesses VI*. 2 vols. New York: Pantheon Books.

Piankoff, A. 1957. *Mythological Papyri*. 2 vols. New York: Pantheon Books.

Picard, C. 1940. "Trois urnes cinéraires sculptées au Musée Condé à Chantilly". *Monuments et mémoires de la Fondation Eugène Piot* 37: 73–103, pls. 6–8.

Pingree, D. 1978. "A History of Mathematical Astronomy in India". In Gillispie, C. (ed.), *Dictionary of Scientific Biography*. Vol. 15. Suppl. 1. New York: Charles Scribner's Sons, 533–633.

Pingree, D. 1981. *Jyotiçòāstra: Astral and Mathematical Literature*. Wiesbaden: Harrassowitz.

Pingree, D. & Walker, C. 1988. "A Babylonian Star-Catalogue: BM 78161". In Leichty, E., de J. Ellis, M. & Gerardi, P. (eds.), *A Scientific Humanist: Studies in Memory of Abraham Sachs*. Philadelphia: University Museum, 313–22.

Pirenne-Delforge, V. 2009. "Under Which Conditions Did the Greeks "Believe" in Their Myths? The Religious Criteria of Adherence". In Walde, C. & Dill, U. (eds.), *Antike Mythen: Medien, Transformationen und Konstruktionen*. Berlin: De Gruyter, 38–54.

Plofker, K. 2010. "Humans, Demons, Gods and Their Worlds: The Sacred and Scientific Cosmologies of India". In Raaflaub, K.A. & Talbert, R. J. A. (eds.), *Geography and Ethnography: Perception of the World in Pre-Modern Societies*. Chichester: Wiley-Blackwell, 32–41.

Pocock, L. G. 1958. "Gadflies and the Laestrygonians". *Classical Review* 8.2: 109–11.

Pocock, L. G. 1968. "A Note on *Odyssey* 10.86" *Classical Quarterly* 18.1: 1–3.

Podosinov, A. V. 2018. "Pagan and Christian elements in the early medieval cosmological and geographical concepts [in Russian]". In Guimon, T. V., Jackson, T. N., Melnikova, E. A. & Shchavelev, A. S. (eds.), *The Earliest States of Eastern Europe*, 2016: *Galina V. Glazyrina: in memoriam*. Moscow: Dmitry Pozharsky University, 196–210.

Podossinov, A. V. 2013. "Das Schwarze Meer in der geokartographischen Tradition der Antike und des frühen Mittelalters. IV. Odysseus im Schwarzen Meer? Okeanische Fahrten der griechischen Helden und archaisches Weltbild der Griechen". *Ancient West & East* 12: 205–36.

Poiss, T. 2014. "Looking for Bird's-Eye View in Ancient Greek Sources". In Geus, K. & Thiering, M. (eds.), *Features of Common Sense Geography: Implicit Knowledge Structures in Ancient Geographical Texts*. Berlin: Lit, 69–87.

Pomeroy, S. B. 2007. *The Murder of Regilla*. Cambridge: Harvard University Press.

Popper, K. 2005[1959[1]/1980[4]]. *The Logic of Scientific Discovery*. London: Routledge.

Popper, K. 1982. *Postscript to the Logic of Scientific Discovery*. Vol. 3. *Quantum Theory and the Schism in Physics*. London: Hutchinson.

Porada, E. 1981–82. "Cylinder Seals Found at Thebes in Boeotia". *Archiv für Orientforschung* 28: 1–70.

Porter, J. I. 1992. "Hermeneutic Lines and Circles: Aristarchus and Crates on the Exegesis of Homer." In Lamberton, R. & Keaney, J. J. (eds.), *Homer's Ancient Readers: The Hermeneutics of Greek Epic's Earliest Exegetes*. Princeton: Princeton University Press, 67–114.

Porter, J. I. 2001. "Review of Broggiato 2001". *BMCRev* 2003.11.33, http://bmcr.brynmawr.edu/2003/2003-11-33.html.

Potts, D. T. 1990. *The Arabian Gulf in the Antiquity*. Vol. 1. Oxford: Clarendon Press.

Potts, D. T. 2016[2]. *The Archaeology of Elam*. Cambridge: Cambridge University Press.

Powell, B. 1991. *Homer and the Origin of the Greek Alphabet*. Cambridge: Cambridge University Press.

Powell, B. 1997. "Homer and Writing". In Morris, I. & Powell, B. (eds.), *A New Companion to Homer*. Leiden: Brill, 3–32.

Powell, B. 2004. *Homer*. Malden: Blackwell.

Poynton, J. B. 1963. "Statius, Thebaid". *Classical Review* 13.3: 259–61.

Prechel, D. 1996. *Die Göttin Išhara: Ein Beitrag zur altorientalischen Religionsgeschichte*. Münster: Ugarit-Verlag.

Preisendanz, K. 1928. *Papyri Graecae Magicae*. Vol. 1. Leipzig: Teubner.

Pritchett, W. Kendrick. 1964. "Thucydides V 20". *Historia: Zeitschrift für Alte Geschichte* 13.1: 21–36.

Pritchett, W. Kendrick. 1999. "Postscript: The Athenian Calendars". *Zeitschrift für Papyrologie und Epigraphik* 128: 79–93.

Pritchett, W. Kendrick & Van der Waerden, B. L. 1961. "Thucydidean time-reckoning and Euctemon's seasonal calendar". *Bulletin de correspondance hellénique* 85: 17–52.

Pugliese Carratelli, G. (ed.) 1996. *The Western Greeks*. Milano: Bompiani.

Pulbrook, M. 1988. "The 'Aetia' of Callimachus" *The Maynooth Review/Revieú Mhá Nuad* 13: 44–64.

Purves, A. C. 2010. *Space and Time in Ancient Greek Narrative*. Cambridge: Cambridge University Press.

Quack, J. F. 2016. "On the Concomitancy of the Seemingly Incommensurable, or Why Egyptian Astral Tradition Needs to be Analyzed within Its Cultural Context". In Steele, J. M. (ed.), *The Circulation of Astronomical Knowledge in the Ancient World*. Brill: Leiden, 230–44.

Quatremère E.-M. 1838. *Notice de l'ouvrage qui pour titre: Mesalek alabsar fi memalek alamsar*. Notices et Extraits des Manuscrits de la Bibliothèque du roi 13. Paris: Imprimerie Royale.

Quinn, K. 1982. "The Poet and his Audience in the Augustan Age". *Aufstieg und Niedergang der römischen Welt* 30.1: 75–180.

Race, W. H. 2009. *Apollonius Rhodius*, Argonautica. Cambridge: Harvard University Press.

Ragavan, D. 2013. "Entering Other Worlds: Gates, Rituals and Cosmic Journeys in Sumerian Sources". In Ragavan, D. (ed.), *Heaven on Earth. Temples, Ritual and Cosmic Symbolism in the Ancient World*. Chicago: University of Chicago Press, 201–21.

Ramsey, I. 1964. *Models and Mystery*. London: Oxford University Press.

Randles, W. G. L. 1994. "Classical Models of World Geography and Their Transformation Following the Discovery of America". In Haase, W. & Reinhold, M. (eds.), *The Classical Tradition and the Americas*. Vol. 1. *European Images of the Americas and the Classical Tradition*. Part 1. Berlin: De Gruyter, 5–76.

Rathmann, M. 2015. "The *Tabula Peutingeriana* and Antique Cartography". In Bianchetti, S., Cataudella, M. R. & Gehrke, H.-J. (eds.), *Brill's Companion to Ancient Geography*. Leiden: Brill, 337–62.

Ratzon, E. 2015. "The Gates Cosmology of the Astronomical Book of Enoch". *Dead Sea Discoveries* 22: 93–111.

Raverty, H. G. 1881. *Tabakat-I-Nasiri: A General History of The Muhammadan Dynasties of Asia Including Hindustan from A.H. 194 (810 A.D.) to A.H. 658 (1260 A.D.) And the Irruption of the Infidel Mughals into Islam*. Vol. 1. London: Gilbert & Rivington.

Rawlinson, H. C., Norris, E., Smith, G. & Pinches, T. G. 1861–1909. *The Cuneiform Inscriptions of Western Asia*. Vols 1–5. London: R. E. Bowler.

Redford, D. B. 2001. *The Oxford Encyclopaedia of Ancient Egypt*. Vol. 3. Oxford/Cairo: Oxford University Press/ The American University in Cairo Press.

Rehm, A. 1941. *Parapegmastudien. Mit einem Anhang: Euktemon und das Buch De signis*. München: Beck.

Reiner, E. 1956. "Lipšur Litanies". *Journal of Near Eastern Studies* 15.3: 129–49.

Reiner, E. & Pingree, D. 1981. *Babylonian Planetary Omens*. Part 2. *Enuma Anu Enlil, Tablets 50–51*. Malibu: Undena Publications.

Reinink, G. J. 1993. *Die Syrische Apokalypse des Pseudo-Methodius*. 2 vols. Louvain: Peeters.

Reinink, G. J. 2003. "Alexander the Great in Seventh-Century Syriac 'Apocalyptic' Texts". *Bizantinorossika* 2: 150–78.

Renfrew, C. 1994. "The Archaeology of Religion". In Renfrew, C. & Zubow, E. B. W. (eds.), *The Ancient Mind: Elements of Cognitive Archaeology*. Cambridge: Cambridge University Press, 47–54.

Rhodes, P. J. & Osborne, T. 2003. *Greek Historical Inscriptions 404–323 BC*. Oxford: Oxford University Press.

Richardson, H. 1926- "The Myth of Er (Plato, Republic, 616b)". *Classical Quarterly* 20.3–4: 113–33.

Richter, G. M. A. 1946. "A Fourth-Century Bronze Hydria in New York". *American Journal of Archaeology* 50.3: 361–67.

Ridley, E. 1896. *The Pharsalia of Lucan*. London: Longmans, Green

Riedinger, R. 1989. *Pseudo-Kaisarios, Die Erotapokriseis*. Berlin: Akademie.

Riley, H. T. 1853. *The Pharsalia of Lucan*. London: Henry G. Bohn.

Robert, C. 1886. "Beiträge zum Griechischen Fest-kalender". *Hermes* 21.2: 161–78.

Robertson, N. 1983. "Greek Ritual Begging in Aid of Women's Fertility and Childbirth". *Transactions of the American Philological Association* 113: 143–69.

Robertson, N. 1996. "The Ancient Mother of the Gods: A Missing Chapter in the History of Greek Religion". In Lane, E. (ed.), *Cybele, Attis and Related Cults: Essays in Memory of M. J. Vermaseren*. Leiden: Brill, 239–304.

Robertson, N. 2010. *Religion and Reconciliation in Greek Cities: The Sacred Laws of Selinus and Cyrene*. Oxford: Oxford University Press.

Robinson, P. 2003. "'As for them who know them, they shall find their paths': Speculations on Ritual Landscapes in the 'Book of the Two Ways'". In O'Connor, D. & Quirke, S. (eds.), *Mysterious Lands*. London: UCL, 139–59.

Roby, C. A. 2014. "Seneca's Scientific Fictions: Models as Fictions in the *Natural Questions*". *Journal of Roman Studies* 104: 155–80.

Rochberg, F. 1992. "The Cultures of Ancient Science: Some Historical Reflections. Introduction". *Isis* 83.4: 547–53.

Rochberg, F. 2003. "Heaven and Earth: Divine-Human Relations in Mesopotamian Celestial Divination". In Noegel, S. B., Walker, J. L. & Wheeler, B. M. (eds.), *Prayer, Magic, and the Stars in the Ancient and Late Antique World*. University Park: Pennsylvania State University Press, 169–85.

Rochberg, F. 2004. *The Heavenly Writing: Divination, Horoscopy, and Astronomy in Mesopotamian Culture*. Cambridge: Cambridge University Press.

Rochberg, F. 2010. *In the Path of the Moon. Babylonian Celestial Divination and Its Legacy*. Leiden: Brill.

Rochberg, F. 2012. "The Expression of Terrestrial and Celestial Order in Ancient Mesopotamia". In Talbert, R. J. A. (ed.), *Ancient Perspectives. Maps and Their Place in Mesopotamia, Egypt, Greece & Rome*. Chicago: University of Chicago Press, 9–46.

Rochberg, F. 2020. "The Babylonian Contribution to Greco-Roman Astronomy". In Bowen, A. C. & Rochberg, F. (eds.), *Hellenistic Astronomy. The Science in Its Contexts*. Brill: Leiden, 147–59.

Rochberg-Halton, F. 1983. "Stellar Distances in Early Babylonian Astronomy: A New Perspective on the Hilprecht Text (HS 229)". *Journal of Near Eastern Studies* 42.3: 209–17.

Rogerson, J. W. 1974. *Myth in Old Testament Interpretation*. Berlin: De Gruyter.

Rohde, E. 1925. *Psyche*. London: Routledge and Kegan Paul.

Roller, D. W. 2003. *The World of Juba II and Kleopatra Selene: Royal Scholarship on Rome's African Frontier*. New York: Routledge.

Roller, D. W. 2006. *Through the Pillars of Herakles*. New York: Routledge.

Roller, D. W. 2010. *Eratosthenes'* Geography. Princeton: Princeton University Press.

Roller, D. W. 2015. *Ancient* Geography. *The Discovery of the World in. Classical Greece and Rome*. London: I. B. Tauris.

Roller, D. W. 2018. "Hellenistic Geography from Ephorus Through Strabo". In Keyser, P. T. & Scarborough, J. (eds.), *Oxford Handbook of Science and Medicine in the Classical World*. Oxford: Oxford University Press, 315–35.

Rollinger, R. 2006a. "The Terms 'Assyria' and 'Syria' again". *Journal of Near Eastern Studies* 65.4: 283–87.

Rollinger, R. 2006b. "Assyrios, Syrios, Syros und Leukosyros". *Die Welt des Orients* 36: 72–82.

Romm, J. 1989. "Herodotus and Mythic Geography: The Case of the Hyperboreans". *Transactions of the American Philological Association* 119: 97–113.

Romm, J. 1992. *The Edges of the Earth in Ancient Thought*. Princeton: Princeton University Press.

Romm, J. 1994: "New World and '*novos orbos*': Seneca in the Renaissance. Debate over Ancient Knowledge of the Americas". In Haase, W. & Reinhold, M. (eds.), *The Classical Tradition and the Americas*. Vol. 1. *European Images of the Americas and the Classical Tradition*. Part 1. Berlin: De Gruyter, 77–116.

Roseman, C. H. 1994. *Pytheas of Massalia: On the Ocean*. Chicago: Ares.

Rosenthal, F. 1967a². *Ibn Khaldun: The Muqaddimah. An Introduction to History*. 3 vols. Princeton: Princeton University Press.

Rosenthal, F. 1967b. *Ibn Khaldun: The Muqaddimah. An Introduction to History*. Edited and abridged by N. J. Dawood. Princeton: Princeton University Press.

Ross, G. 1971. "Neo-Tylorianism: A Reassessment". *Man* 6.1: 105–16.

Ross, M. T. 2020. "Hellenistic Egyptian Planetary Theory". In Bowen, A. C. & Rochberg, F. (eds.), *Hellenistic Astronomy. The Science in Its Contexts*. Brill: Leiden, 160–65.

Rossi, F. 1878–79. "Illustrazione di un papiro funerario del Museo egizio di Torino". *Atti della Reale Accademia delle scienze di Torino* 14: 1203–24.

Ruggles, C. 1999. *Astronomy in Prehistoric Britain and Ireland*. New Heaven: Yale University Press.

Ruggles, C. 2005. *Ancient Astronomy: An Encyclopedia of Cosmologies and Myth*. Santa Barbara: ABC-CLIO.

Rusten, J. S. 1982. *Dionysius Scytobrachion*. Opladen: Westdeutscher.

Rutherford, I. 1990. "Paeans by Simonides". *Harvard Studies in Classical Philology* 93: 169–209.

Rutherford, I. 2001. *Pindar's* Paeans. *A Reading of the Fragments with a Survey of the Genre*. Oxford: Oxford University Press.

Rutherford, I. 2009. "Hesiod and the Literary Traditions of the Near East". In Montanari, F., Rengakos, A. & Tsagalis, C. (eds.), *Brill's Companion to Hesiod*. Leiden: Brill, 9–35.

Rutherford, I. 2015. "Stesichorus the romantic". In Finglass, P. J. & Kelly, A. (eds.), *Stesichorus in Context*. Cambridge: Cambridge University Press, 98–108.

Rutherford, I. 2018. "Kingship in Heaven in Anatolia, Syria and Greece". In Audley-Miller, L. & Dignas, B. (eds.), *Wandering Myths. Transcultural Uses of Myth in the Ancient World*. Berlin: De Gruyter, 3–22.

Sackur, E. 1898. *Sibyllinische Texte und Forschungen*. Halle: Niemeyer.

Sale, W. 1961. "The Hyperborean Maidens on Delos". *Harvard Theological Review* 54.2: 75–89.

Sale, W. 1962. "The Story of Callisto in Hesiod". *Rheinisches Museum für Philologie* 105: 122–41.

Sale, W. 1965. "Callisto and the Virginity of Artemis". *Rheinisches Museum für Philologie* 108: 11–35.

Salt, A. & Boutsikas, E. 2005. "Knowing when to consult the oracle at Delphi". *Antiquity* 79: 564–72.

Sambursky, S. 1960². *The Physical World of the Greeks*. London: Routledge & Kegan Paul.

Samuel, A. E. 1972. *Greek and Roman Chronology: Calendars and Years in Classical Antiquity*. München: Beck.

Sandars, N. K. 1968. *Prehistoric Art in Europe*. Harmondsworth: Penguin.

Sandars, N. K. 1971. *Poems of Heaven and Hell from Ancient Mesopotamia*. Harmondsworth: Penguin.

Saura Sanjaume, M. 2006. *La Tomba de Sennedjem a Deir-El-Medina TT.1*. Diss. Universitat de Barcelona.

Scafi, A. 2006. *Mapping Paradise. A History of Heaven on Earth*. London: British Library.

Scafi, A. 2013. *Maps of Paradise*. London: British Library.

Schäfer, H. 1919. *Von ägyptischer Kunst besonders der Zeichenkunst*. 2 vols. Leipzig: Hinrichs.

Schäfer, H. 1928. *Ägyptische und heutige Kunst und Weltgebäude der alten Ägypter*. Berlin: De Gruyter.

Schäfer, H. 1935. "Altägyptische Bilder der auf- und untergehenden Sonne". *Zeitschrift für Ägyptische Sprache und Altertumskunde* 71: 15–38.

Schachter, A. 2016. *Boiotia in Antiquity. Selected papers*. Cambridge: Cambridge University Press.

Schaumberger, J. 1935. *Sternkunde und Sterndienst in Babel* 3. Münster: Aschendorff.

Schibli, H. S. 1990. *Pherekydes of Syros*. Oxford: Clarendon Press.

Schein, S. L. 1997. "The *Iliad*: Structure and Interpretation". In Morris, I. & Powell, B. (eds.), *A New Companion to Homer*. Leiden: Brill, 345–59.

Schier, W. 2008. "Zur astronomischen Orientierung der mittelneolitischen Kreisgrabenanlage von Ippesheim, Mittelfranken". *Acta Praehistorica et Archaeologica* 40: 45–55.

Schmidt, A. B. 1994. "Die 'Brüste des Nordens' und Alexanders Mauer gegen Gog und Magog". In Brandes, W. & Schmidt, B. B. (eds.), *Israel's Beneficent Dead: Ancestor Cult and Necromancy in Ancient Israelite Religion and Tradition*. Tübingen: Mohr.

Schnetz, J. 1925. "Rigilinus". *Philologus* 80: 109–12.

Schollmeyer, P. A. 1912. *Sumerisch-babylonische Hymnen und Gebete an Šamaš*. Paderborn: Schöningh.

Schönbeck, L. 1993. "Heraclitus revisited (Pap. Derveni col. I lines 7–11)". *Zeitschrift für Papyrologie und Epigraphik* 97: 7–22.

Schott, A. 1936. "Marduk und sein Stern". *Zeitschrift für Assyriologie und vorderasiatische Archäologie* 43: 124–45.

Schrader, H. (ed.) 1880–82. *Porphyrii Quaestionum Homericarum ad Iliadem pertinentium reliquiae*. Fasc. 1–2. Leipzig: Teubner.

Schultz, W. 1911. "Der Text und die unmittelbare Umgebung von Fragment 20 des Anaxagoras". *Archiv für Geschichte der Philosophie* 24.3: 323–42.

Schwartzberg, J. E. 1992a. "Cosmographical Mapping". In Harley, J. B. & Woodward, D. (eds.), *The History of Cartography*. Vol. 2. Pt. 1. *Cartography in the Traditional Islamic and South Asian Societies*. Chicago: University of Chicago Press, 332–87.

Schwartzberg, J. E. 1992b. "Geographical Mapping". In Harley, J. B. & Woodward, D. (eds.), *The History of Cartography*. Vol. 2. Pt. 1. *Cartography in the Traditional Islamic and South Asian Societies*. Chicago: University of Chicago Press, 388–493.

Scodel, R. 2003. "The Paths of Night and Day". *Ordia Prima* 2: 83–86.

Scurlock, J. A. 1988. "A Ghostly Light on bīt rimki?". *Journal of the American Oriental Society* 108.2: 203–209.

Scurlock, J. A. 2018- "Mesopotamian Beginnings for Greek Science?". In Keyser, P. T. & Scarborough, J. (eds.), *Oxford Handbook of Science and Medicine in the Classical World*. Oxford: Oxford University Press, 35–46.

Seaford, R. 1986. "Immortality, Salvation, and the Elements". *Harvard Studies in Classical Philology* 90: 1–26.

Seaford, R. 2004. *Money and the Early Greek Mind*. Cambridge: Cambridge University Press.

Segal, R. A. 1980a. "The Myth-Ritualist Theory of Religion". *Journal for the Scientific Study of Religion* 19.2: 173–85.

Segal, R. A. 1980b. "In Defense of Mythology: The History of Modern Theories of Myth". *Annals of Scholarship* 1: 3–49.

Segal, R. A. 1999. *Theorizing about Myth*. Amherst: University of Massachusetts Press.

Segal, R. A. 2001. "In Defense of the Comparative Method". *Numen* 48.3: 339–73.

Segal, R. A. 2004. *Myth: A Very Short Introduction*. Oxford: Oxford University Press.

Sethe, K. 1906. *Urkunden der 18. Dynastie*. Vol. 1. Leipzig: Hinrichs.

Sethe, K. 1914. *Urkunden der der 18. Dynastie I*. Leipzig: Hinrichs.

Seux, M.-J. 1976. *Hymnes et prières aux dieux de Babylonie et d'Assyrie*. Paris: Cerf.

Shaffer, A. 1963. *Sumerian Sources of the Tablet XII of the Epic of Gilgamesh*. Diss. University of Pennsylvania.

Shaltout, M. & Belmonte, J. 2005. "On the orientation of ancient Egyptian temples: (1) Upper Egypt and Lower Nubia". *Journal for the history of astronomy* 36.3: 273–98.

Shaltout, M. & Belmonte, J. 2006. "On the orientation of ancient Egyptian temples: (2) New experiments at the oases of the Western Desert". *Journal for the history of astronomy* 37.2: 173–92.

Shaltout, M., Belmonte, J. & Fekri, M. 2007. "On the Orientation of Ancient Egyptian Temples: (3) Key Points in Lower Egypt and Siwa Oasis, Part I". *Journal for the History of Astronomy* 38.2: 141–60. "Part II". *Journal for the History of Astronomy* 38.4: 413–42.

Shaltout, M., Belmonte, J. & Fekri, M. 2008. "On the Orientation of Ancient Egyptian Temples: (4) Epilogue in Serabit el Khadem and overview". *Journal for the History of Astronomy* 39.2: 181–211.

Shcheglov, D. A. 2005. "Hipparchus on the Latitude of Southern India". *Greek, Roman, and Byzantine Studies* 45: 359–80.

Shcheglov, D. A. 2003–2007. "Hipparchus' Table of Climata and Ptolemy's Geography". *Orbis terrarum* 9: 159–91.

Shewan, A. 1918. "Scheria-Corcyra". *Classical Philology* 13.4: 321–34.

Shipley, G. 2011. Pseudo-Skylax's Periplous—The Circumnavigation of the Inhabited World: Text, Translation and Commentary. Exeter: Bristol Phoenix Press.

Sick, D. H. 2004. "Mit(h)ra(s) and the Myths of the Sun". *Numen* 51.4: 432–67.

Silva Castillo, J. 1998. "*Nagbu*: totality or abyss in the first verse of Gilgamesh". *Iraq* 60: 229–31.

Simon, E. 1983. *Festivals of Attica*. Madison: University of Wisconsin Press.

Skinner, J. E. 2012. *The Invention of Greek Ethnography. From Homer to Herodotus*. Oxford: Oxford University Press.

Skinner, Q. 1969. "Meaning and Understanding in the History of Ideas". *History and Theory* 8.1: 3–53.

Smith, C. D. 1987. "Cartography in the Prehistoric Period in the Old World: Europe, the Middle East, and North Africa". In Harley, J. B. & Woodward, D. (eds.), *The History of Cartography*. Vol.1. *Cartography in Prehistoric, Ancient, and Medieval Europe and the Mediterranean*. Chicago: University of Chicago Press, 54–101.

Smith, M. S. 2003. "Astral Religion and the Representation of Divinity: The Cases of Ugarit and Judah". In Noegel, S. B., Walker, J. L. & Wheeler, B. M. (eds.), *Prayer, Magic, and the Stars in the Ancient and Late Antique World*. University Park: Pennsylvania State University Press, 187–206.

Smith, M. S. & Pitard, W. T. 2009. *The Ugaritic Baal Cycle*. Vol. 2. *Introduction with Text, Translation and Commentary of KTU/CAT 1.3–1.4*. Leiden: Brill.

Snell, B. 1964[3-4]. *Pindari Carmina cum Fragmentis*. 2 vols. Leipzig: Teubner.

Snell, B. & Kannicht, R. 2007[2]. *Tragicorum Graecorum Fragmenta* 2. Göttingen: Vandenhoeck & Ruprecht.

Soden, W. von 1942. "Neue Bruchstücke zur sechsten und siebenten Tafel des Weltschöpfungsepos Enuma elish". *Zeitschrift für Assyriologie und vorderasiatische Archäologie* 47: 1–26.

Solmsen, F. 1947. "Eratosthenes' Erigone: A Reconstruction". *Transactions and Proceedings of the American Philological Association* 78: 252–75.

Solomon, J. (ed.) 1994. *Apollo: Origins and Influences*. Tucson: University of Arizona Press.

Sørensen, P. M. 2002. "Þorr's Fishing Expedition (Hymiskviða)". In Acker, P. & Larrington, C. (eds.), *The Poetic Edda: Essays on Old Norse Mythology*. London: Routledge, 119–37.

Sourvinou-Inwood, C. 1995. *'Reading' Greek Death to the End of the Classical Period*. Oxford: Clarendon Press.

Spalinger, A. 2020. "Issues in Hellenistic Egyptian Astronomical Texts". In Bowen, A. C. & Rochberg, F. (eds.), *Hellenistic Astronomy. The Science in Its Contexts*. Brill: Leiden, 261–71.

Speleers, L. 1917. *Le papyrus de Nefer Renpet*. Bruxelle: Vromant.

Sprenger, A. 1841. *El-Mas'udi's Historical Encyclopaedia Entitled* Meadows of Gold and Mines of Gems. Vol 1. London: Oriental Translation Fund of Great Britain and Ireland.

Sprockhoff, E. 1954. "Nordische Bronzezeit und frühes Griechentum". *Jahrbuch des römisch-germanischen Zentralmuseums* 1: 28–110.

Spronk, K. 1999[2]. "Travellers". In van der Toorn, K., Becking, B. & van der Hoorst, P. W. (eds.), *Dictionary of Deities and Demons in the Bible*. Leiden: Brill, 876–77.

Stallard, A. J. 2010a. "Origins of the Idea of Antipodes: Errors, Assumptions and a Bare Few Facts". *Terrae incognitae* 42: 34–51 (reprinted in Stallard 2010b: 320–40).

Stallard, A. J. 2010b. *Antipodes to Terra Australis*. Diss. University of Queensland.

Stallard, A. J. 2016. *Antipodes, In Search of the Southern Continent*. Monash University Press.

Stang, H. 1996. *The Naming of Russia*. Oslo: Universitetet i Oslo, Slavisk-baltisk avdeling.

Starè, F. 1954. *Prazgodovinske Vače*. Ljubljana: Univerza.

Steele, L. D. 2002. "Mesopotamian Elements in the Proem of Parmenides? Correspondences between the Sun-Gods Helios and Shamash". *Classical Quarterly* 52.2: 583–88.

Steinkeller, P. 1992. "Early Semitic Literature and Third Millenium Seals with Mythological Motifs". In Fronzaroli, P. (ed.), *Literature and Literary Language at Ebla*. Florence: Università di Firenze, 243–75.

Stephens, S. A. 2003. *Seeing Double: Intercultural Poetics in Ptolemaic Alexandria*. Berkeley: University of California Press.

Stern, J. 2003. "Heraclitus the Paradoxographer: Περὶ Ἀπίστων, 'On Unbelievable Tales'". *Transactions of the American Philological Association* 133.1: 51–97.

Stevens, W. M. 1980. "The Figure of the Earth in Isidore's *De natura rerum*". Isis 71.2: 268–77.

Stiehle, R. 1856. "Der Geograph Artemidoros von Ephesos" *Philologus* 11: 193–244.

Stokes, M. C. 1962. "Hesiodic and Milesian Cosmogonies—I". *Phronesis* 7: 1–37.

Stoneman, R. 1991. *The Greek Alexander Romance*. London: Penguin.

Stoneman, R. 2003. "Alexander the Great in the Arabic Tradition". In Panayotokis, S., Zimmerman, M. & Keulen, W. (eds), *The Ancient Novel and Beyond*. Leiden: Brill, 3–21.

Stoneman, R. 2011. "Primary Sources from the Classical and Early Medieval Periods". In Zuwiyya Z. D. (ed.), *A Companion to Alexander Literature in the Middle Ages*. Leiden: Brill, 1–20.

Struck, P. T. 2004. *Birth of the Symbol. Ancient Readers at the Limits of Their Texts*. Princeton: Princeton University Press.

Stückelberger, A. 1990. "Sterngloben und Sternkarten: zur wissenschaftlichen Bedeutung des Leidener Aratus". *Museum Helveticum* 47.2: 70–81.

Suárez del la Torre, E. 2013. "Apollo and Dionysos: Intersections". In Bernabé, A., Herrero de Jáuregui, M., Jiménez San Cristóbal, A. I. & Martín Hernández, R. (eds.), *Redefining Dionysos*. Berlin: De Gruyter, 58–81.

Sulimani, I. 2017. "Imaginary Islands in the Hellenistic Era: Utopia on the Geographical Map". In Hawes, G. (ed.), *Myths on the Map. The Storied Landscapes of Ancient Greece*. Oxford: Oxford University Press, 221–42.

Summers, W. C. 1894. *A Study of the Argonautica of Valerius Flaccus*. Cambridge: Deighton, Bell.

Talbert, R.J.A. 2010. "The Roman Worldview: Beyond Recovery?". In Raaflaub, K.A. & Talbert, R. J. A. (eds.), *Geography and Ethnography: Perception of the World in Pre-Modern Societies*. Chichester: Wiley-Blackwell, 252–72.

Tallgren-Tuulio, O. J. 1936. *Du nouveau sur Idrīsī*. Studia Orientalia 6.3. Helsinki: Societas orientalis Fennica.

Tannery, P. 1893. *Recherches sur l'histoire de l'astronomie ancienne*. Paris: Gauthier-Villars.

Taran, L. 1965. *Parmenides, A Text with Translation, Commentary, and Critical Essays*. Princeton: Princeton University Press.

Tarn, W. W. 1932. "Alexander Helios and the Golden Age". *Journal of Roman Studies* 22: 135–60.

Tecco Hvala S. 2012. *Magdalenska gora: družbena struktura in grobni rituali železnodobne skupnosti. Magdalenska Gora: Social Structure and Burial Rites of the Iron Age Community*. Ljubljana: Založba ZRC.

Tecco Hvala, S., Dular, J. & Kocuvan, E. 2004. *Železnodobne gomile na Magdalenski gori. Eisenzeitliche Grabhügel auf der Magdalenska gora*. Ljubljana: Narodni muzej Slovenije.

Teržan, B. 1999. "An Outline of the Urnfield Culture Period in Slovenia". *Arheološki vestnik* 50: 97–143.

Thalmann, W. G. 2011. *Apollonius of Rhodes and the Spaces of Hellenism*. Oxford: Oxford University Press.

Theiler, W. 1982. *Posidonios: Die Fragmente*. 2 vols. Berlin: De Gruyter.

Thiel, H. van (ed.) 2014. *Scholia D in Iliadem*. Köln: Universitäts- und Stadtbibliothek Köln, https://kups.ub.uni-koeln.de/5586/1/vanthiel.pdf.

Thiele, G. 1898. *Antike Himmelsbilder*. Berlin: Weidmann.

Thomas, E. 1956. "Solar barks prow to prow". *Journal of Egyptian Archaeology* 42: 65–79.

Thompson, R. C. 1900. *The Reports of the Magicians and Astrologers of Nineveh and Babylon in the British Museum*. 2 vols. London: Luzac.

Thompson, R. C. 1940. "A Selection from the Cuneiform Historical Texts from Nineveh (1927–32)". *Iraq* 7.2: 85–131.

Thomson, J. O. 1948. *History of Ancient Geography*. Cambridge: Cambridge University Press.

Thorndike, L. 1949. *The* Sphere *of Sacrobosco and Its Commentators*. Chicago: University of Chicago Press.

Thorne, S. 2000. "Diktaian Zeus in Later Greek Tradition". In MacGillivray, J. A., Driessen, J. M. & Sackett, L. H. (eds.), *The Palaikastro Kouros. A Minoan Chryselephantine Statuette and its Aegean Bronze Age Context*. London: British School at Athens, 149–62.

Tibbetts, G. R. 1992. "The Beginnings of a Cartographic Tradition". In Harley, J. B. & Woodward, D. (eds.), *The History of Cartography*. Vol. 2. Pt. 1. *Cartography in the Traditional Islamic and South Asian Societies*. Chicago: University of Chicago Press, 90–107.

Tierney, J. J. 1959–60. "The Celtic Ethnography of Posidonius". *Proceedings of the Royal Irish Academy. Section C: Archaeology, Celtic Studies, History, Linguistics, Literature* 60: 189–275.

Tierney, J. J. 1967. *Dicuili Liber de Mensura Orbis Terrae*. Dublin: Dublin Institute for Advanced Studies.

Tigay, J. H. 1982. *The Evolution of the Gilgamesh Epic*. Philadelphia: University of Pennsylvania Press.

Tiverios, M. A. 1982. "Observations on the East Metopes of the Parthenon". *American Journal of Archaeology* 86.2: 227–29.

Tor, S. 2017. *Mortal and Divine in Early Greek Epistemology. A Study of Hesiod,*

Xenophanes and Parmenides. Cambridge: Cambridge University Press.

Tor, S. 2020. "Parmenides on the Soul". In Bartoš, H. & King, C. G. (eds.), *Heat, Pneuma*

and Soul in Greek Philosophy and Science. Cambridge: Cambridge University Press, 61–79.

Tozer, H. F. 1897. *A History of Ancient Geography*. Cambridge: Cambridge University Press.

Tracy, S. V. 1997. "The Structures in the Odyssey". In Morris, I. & Powell, B. (eds.), *A New Companion to Homer*. Leiden: Brill, 360–79.

Trépanier, S. 2010. "Early Greek Theology: God as Nature and Natural Gods". In Bremmer, J. N. & Erskine, A. (eds.), *The Gods of Ancient Greece. Identities and Transformations*. Edinburgh: Edinburgh University Press, 273–317.

Trumpf, J. 1971. "Alexander, die Bersiler und die Brüste des Nordens". *Byzantinische Zeitschrift* 64: 326–28.

Trumpf, J. 1974. *Anonymi Byzantini Vita Alexandri regis Macedonum*. Stuttgart: Teubner.

Trümpy, C. 1997. *Untersuchungen zu den altgriechischen Monatsnamen und Monatsfolgen*. Heidelberg: Universitätsverlag C. Winter.

Trypanis, C. A., Gelzer, T. & Whitman, C. 1958. *Callimachus: Aetia, Iambi, Hecale and other Fragments. Musaeus: Hero and Leander*. Cambridge: Harvard University Press.

Turner, F. M. 1993. *Contesting cultural authority. Essays in Victorian intellectual life*. Cambridge: Cambridge University Press.

Tzifopoulos, Y. Z. 2011. "Center, periphery, or peripheral center: A Cretan connection for the gold lamellae of Crete". In Edmonds, R. G. (ed.), *The 'Orphic' Gold Tablets and Greek Religion*. Cambridge: Cambridge University Press, 165–99.

Uhden, R. 1936. "Die Weltkarte des Martianus Capella". *Mnemosyne* 3: 97–124.

Unger, E. 1931. *Babylon. Die heilige Stadt nach der Beschreibung der Babylonier*. Berlin: De Gruyter.

Unger, E. 1937. "From the Cosmos Picture to the World Map". *Imago Mundi* 2: 1–7.

Ungnad, A. & Gressmann, H. 1911. *Das Gilgamesch-Epos. Neu übersetzt von Arthur Ungnad und gemeinverständlich erklärt von Hugo Gressmann.* Göttingen: Vandenhoeck & Ruprecht.

Van Bladel, K. 2008. "The *Alexander Legend* in the Qur'an 18:83–102". In Reynolds, G. S. (ed.), *The Qur'ān in Its Historical Context.* London: Routledge, 175–203.

Van Buren, E. D. 1955. "The Sun-God Rising". *Revue d'assyriologie et d'archéologie orientale* 49.1: 1–14.

Van der Valk, M. 1958. "On Apollodori Bibliotheca". *Revue des Études Grecques* 71.334–38: 100–68.

Van Gessel, B. H. L. 1998. *Onomasticon of the Hittite Pantheon* 1. Leiden: Brill.

Van Noorden, H. 2015. *Playing Hesiod. The 'Myth of the Races' in Classical Antiquity.* Cambridge: Cambridge University Press.

Van Paassen, C. 1957. *The Classical Tradition of Geography*, Groningen: J. B. Wolters.

Vasić, M. M. 1954. "Analiza dupljajskih kolica". *Starinar* 4.2: 376–80.

Vasić R. & Vasić V. 2003. "Bronzezeitliche und eisenzeitliche Vogeldarstellungen im Zentralbalkan". *Prähistorische Zeitschrift* 78.2: 156–89.

Vasić R. & Vasić V. 2003–2004. "Identifikacija ptičjih predstava u praistoriji". *Starinar* 53–54: 181–92.

Vergados, A. 2013. *The Homeric Hymn to Hermes. Introduction, Text and Commentary.* Berlin: De Gruyter.

Vermeule, E. 1981[1979]. *Aspects of Death in Early Greek Art and Poetry.* Berkeley: University of California Press.

Vernant, J.-P. 1982. *The Origins of Greek Thought.* Ithaca: Cornell University Press.

Vernant, J.-P. 1983. *Myth and Thought among the Greeks.* London: Routledge & Kegan Paul.

Vernant, J.-P. 1990. *Myth and Society in Ancient Greece.* New York: Zone Books.

Versnel, H. S. 1985–86. "Apollo and Mars One Hundred Years after Roscher". *Visible Religion* 4–5: 134–72.

Versnel, H. S. 1988. "Greek Myth and Ritual: The Case of Kronos." In Bremmer, J. (ed.), *Interpretations of Greek Mythology*, London: Routledge, 121–52.

Versnel, H. S. 1990a. "What's Sauce for the Goose Is Sauce for the Gander: Myth and Ritual, Old and New". In Edmunds, L. (ed.), *Approaches to Greek Myth.* Baltimore: Johns Hopkins University Press, 25–90.

Versnel, H. S. 1990b. *Inconsistencies in Greek and Roman Religion 1. Ter Unus. Isis, Dionysos, Hermes: Three Studies in Henotheism.* Leiden: Brill.

Versnel, H. S. 1994[2]. *Inconsistencies in Greek and Roman Religion 2. Transition and Reversal in Myth and Ritual.* Leiden: Brill.

Vidal-Naquet, P. 1981a. "The Black Hunter and the origin of the Athenian ephebeia". In Gordon, R. L. (ed.), *Myth, Religion and Society. Structuralist Essays by M. Detienne, L. Gernet, J.-P. Vernant and P. Vidal-Naquet.* Cambridge: Cambridge University Press, 147–62.

Vidal-Naquet, P. 1981b. "Recipes for Greek Adolescence". In Gordon, R. L. (ed.), *Myth, Religion and Society. Structuralist Essays by M. Detienne, L. Gernet, J.-P. Vernant and P. Vidal-Naquet.* Cambridge: Cambridge University Press, 163–85.

Vlastos, G. 1947. "Equality and Justice in Early Greek Cosmologies" *Classical Philology* 42.3: 156–78.

Vlastos, G. 1953. "Isonomia". *American Journal of Philology* 74.4: 337–66.

Vlastos, G. 1955. "Review of F.M. Cornford, *Principium Sapientiae*". *Gnomon* 27.2: 65–76.

Vlastos, G. 2005[1975]. *Plato's Universe. With a New Introduction by Luc Brisson.* Las Vegas: Parmenides.

Völcker, K. H. W. 1830. *Über homerische Geographie und Weltkunde.* Hannover: Hahn.

Von Hendy, A. 2001. *The Modern Construction of Myth.* Bloomington: Indiana University Press.

Von Reden, S. 1999. "Re-evaluating Gernet: Value and Greek Myth". In Buxton, R. (ed.), *From Myth to Reason? Studies in the Development of Greek Thought.* Oxford: Oxford University Press, 51–70.

Vos, H. 1963. "Die Bahnen von Nacht und Tag" *Mnemosyne* 16: 18–34.

de Vries, J. 1984[1967]. "Theories Concerning 'Nature Myths'". In Dundes, A. (ed.), *Sacred Narrative. Readings in the Theory of Myth.* Berkeley: University of California Press, 30–40.

Wachsmann S. 1998. *Seagoing Ships & Seamanship in the Bronze Age Levant.* College Station: Texas A & M University Press.

Wackernagel, J. 1916. *Sprachliche Untersuchungen zu Homer.* Göttingen: Vandenhoeck & Ruprecht.

Van der Waerden, B. L. 1949. "Babylonian Astronomy, II. The Thirty-six stars". *Journal of Near Eastern Studies* 8: 6–26.

Van der Waerden, B. L. 1974. *Science Awakening II. The Birth of Astronomy.* Leyden/New York: Noordhoff/ Oxford University Press.

Wagner, W. (ed.) 1881. *Trois poèmes grecs du Moyen-Age.* Berlin: Calvary.

Waitz, G. 1876. *Adami Gesta Hammaburgensis ecclesiae pontificum ex recensione Lappenbergii.* Hannover: Hahn.

Walbank, F. W. 1948. "The Geography of Polybius". *Classica et Mediaevalia* 9: 155–82.

Walbank, F. W. 1979. *A Historical Commentary on Polybius*. Vol. 3. Oxford: Clarendon Press.

Walbank, F. W. 2002. *Polybius, Rome and the Hellenistic World.* Cambridge: Cambridge University Press.

Walcot, P. 1979. "Cattle Raiding, Heroic Tradition, and Ritual: The Greek Evidence". *History of Religions* 18.4: 326–51.

Warner, A. G. & Warner, E. 1905–25. *The Shahnama of Firdausi*. 9 vols. London: Kegan Paul, Trench, Trubner.

WÄS = Erman, A. & Grapow, H. (eds.) 1971. *Wörterbuch der Ägyptischen Sprache*. Vols. 1–2, 4. Berlin: Akademie. *Die Belegstellen*. Vol. 1^2 (Berlin: Akademie, 1958). Grapow, H. & Erichsen, W. (eds.), Vol. 2 (Leipzig: Hinrich, 1940).

Watkins, C. 1995. *How to Kill a Dragon: Aspects of Indo-European Poetics*. Oxford: Oxford University Press.

Watkins, C. 2007. "The Golden Bowl: Thoughts on the New Sappho and its Asianic Background". *Classical Antiquity* 26.2: 305–25.

Weber, C. 2002. "The Dionysus in Aeneas". *Classical Philology* 97.4: 322–43.

Weidner, E. 1915. *Handbuch der Babylonischen Astronomie*. Vol. 1. Leipzig: Hinrich.

Weidner, E. 1957–71. "Fixsterne". In Ebeling, E., Meissner, B., Weidner, E. & von Soden, W. (eds.), *Reallexikon der Assyriologie* 3. Berlin: De Gruyter, 72–82.

Weinstock, S. 1946. "Martianus Capella and the Cosmic System of the Etruscans". *Journal of Roman Studies* 36.1–2: 101–29.

Welcker, F. G. 1832. "Die homerische Phäaken und die Insel der Seligen". *Rheinisches Museum für Philologie* 1: 219–83.

Wendel, C. (ed.) 1914. *Scholia in Theocritum Vetera*. Leipzig: Teubner.

Wensinck, A. J. 1921. "Tree and Bird as Cosmological Symbols in Western Asia". *Verhandelingen der Koninklijke Akademie van Wetenschappen te Amsterdam*, 2nd series, 22.1 (reprinted in *Studies of A. J. Wensinck*. 1978. New York: Arno Press).

Wenskus, O. 1990. *Astronomische Zeitangaben von Homer bis Theophrast*. Stuttgart: Steiner.

Wernicke, K. 1896. "Augeias 1". *RE* 2.4.2306.62–2310.4.

West, E. 2014. "Circe, Calypso, Hiḍimbā: The Odyssey and Graeco-Aryan Proto-Epic". *Journal of Indo-European Studies* 42.1: 144–74.

West, M. L. 1963. "Three Presocratic Cosmologies". *Classical Quarterly* 13.2: 154–76.

West, M. L. 1966. *Hesiod: Theogony*. Oxford: Clarendon Press.

West, M. L. 1971. *Early Greek Philosophy and the Orient*. Oxford: Clarendon Press.

West, M. L. 1978. *Hesiod: Works & Days*. Oxford: Clarendon Press.

West, M. L. 1979. "The Prometheus Trilogy". *Journal of Hellenic Studies* 99: 130–48.

West, M. L. 1983. *The Orphic Poems*. Oxford: Clarendon Press.

West, M. L. 1985. *The Hesiodic Catalogue of Women*. Oxford: Clarendon Press.

West, M. L. 1988. *Theogony and Works and Days*. Oxford: Oxford University Press.

West, M. L. 1994. "Ab ovo. Orpheus, Sanchuniathon, and the origins of the Ionian world model". *Classical Quarterly* 33: 289–307.

West, M. L. 1995. "The date of the Iliad". *Museum Helveticum* 52.6: 203–19.

West, M. L. 1997. *East Face of Helicon: West Asiatic Elements in Early Greek Poetry and Myth*. Oxford: Clarendon Press.

West, M. L. 2001. *Studies in the Text and Transmission of the Iliad*, München: Saur.

West, M. L. 2002. "'Eumelos': A Corinthian Epic Cycle?". *Journal of Hellenic Studies* 122: 109–33.

West, M. L. 2003. *Greek Epic Fragments*. Cambridge: Harvard University Press.

West, M. L. 2005a. "Odyssey and Argonautica". *Classical Quarterly* 55.1: 39–64.

West, M. L. 2005b. "The New Sappho". *Zeitschrift für Papyrologie und Epigraphik* 151: 1–9.

West, M. L. 2007a. *Indo-European poetry and myth*. Oxford: Oxford University Press.

West, M. L. 2007b. "Phasis and Aia". *Museum Helveticum* 64: 193–98.

West, M. L. 2013. *The Epic Cycle: A Commentary on the Lost Troy Epics*. Oxford: Oxford University Press.

West, M. L. 2014. *The Making of the Odyssey*. Oxford: Oxford University Press.

West, M. L. 2018. "Gilgāmeš and Homer: The Missing Link?". In Audley-Miller, L. & Dignas, B. (eds.), *Wandering Myths. Transcultural Uses of Myth in the Ancient World*. Berlin: De Gruyter, 265–80.

West, S. 2003. "'The Most Marvelous of All Seas'; The Greek Encounter with the Euxine". *Greece&Rome* 50.2: 151–67.

Westenholz, J. G. 1997. *Legends of the Kings of Akkade*. Winona Lake: Eisenbrauns.

Wheeler, B. M. 1998. "Moses or Alexander? Early Islamic Exegesis of Quran 18:60–56". *Journal of Near Eastern Studies* 57.3: 191–215.

Wheeler, B. M. 2003. "The Prophet Muçammad Dhu al-Qarnayn: His journey to the cities at the ends of the earth". *Bizantinorossika* 2: 179–219.

Whitaker, I. 1981–82. "The Problem of Pytheas' Thule". *Classical Journal* 77: 148–64.

Whitman, C. H. 1970. "Hera's Anvils". *Harvard Studies in Classical Philology* 74: 37–42.

Whitney, W. D. 1905. *Atharva-Veda Samhita*. Part 2. Cambridge: Harvard University Press.

Widdows, P. F. 1988. *Lucan's Civil War*. Bloomington: Indiana University Press.

Wiggermann, F. A. M. 1992. *Mesopotamian Protective Spirits. The Ritual Texts*. Groningen: Styx.

Wijsman H. J. W. 1998. "Thule Applied to Britain". *Latomus* 57: 318–23.

Wilamowitz-Moellendorff, U. von & Kiessling, A. 1884. *Homerische Untersuchungen*. Philologische Untersuchungen 7. Berlin: Weidmann.

Wilamowitz-Moellendorff, U. von 1903. "Apollon". *Hermes* 38: 575–86.

Wilamowitz-Moellendorff, U. von 1926. "Pherekydes". *Sitzungsberichte der Preußischen Akademie der Wissenschaften*, 125–46.

Wilkinson, R. H. 1992. *Reading Egyptian Art*. London: Thames & Hudson.

Wilkinson, R. H. 1994. *Symbol and Magic in Egyptian Art*. New York: Thames & Hudson.

Wilkinson, R. H. 2000. *The Complete Temples of Ancient Egypt*. London: Thames & Hudson.

Williams, J. 1997. "Isidore, Orosius and the Beatus Map". *Imago Mundi* 49: 7–32.

Wilson, H. H. (trans.) 1840. *The Vishńu Puráńa*. London: John Murray.

Winiarczyk, M. 2011. *Die hellenistischen Utopien*. Berlin: De Gruyter.

Winiarczyk, M. 2013. *The Sacred History of Euhemerus of Messene*, Berlin: De Gruyter.

Witzel, M. 2000. "The Home of the Aryans". In Hinze, A. & Tichy, E. (eds.), *Anusantatyi: Festschrift für Johanna Narten zum 70. Geburtstag*. Münchener Studien zur Sprachwissenschaft, Beihefte NF 19. Dettelbach: J. H. Roell, 283–338.

Wöhrle, G. 2012. *Die Milesier: Anaximander und Anaximenes*. Berlin: De Gruyter.

Wöhrle, G. 2014. *The Milesians: Thales*. Berlin: De Gruyter.

Wolf, A. 2009. *Homers Reise: Auf den Spuren des Odysseus*. Köln: Böhlau.

Wolohojian, A. M. 1969. *The Romance of Alexander the Great by Pseudo-Callisthenes*. New York: Columbia University Press.

Wolska-Conus, W. 1968. *Cosmas Indicopleustès. Topographie Chrétienne*. Vol. 1. Paris: Cerf.

Woodbury, L. 1966. "Equinox at Acragas: Pindar, Ol. 2.61–62". *Transactions and Proceedings of the American Philological Association* 97: 597–616.

Woods, C. E. 2004. "The Sun-God Tablet of Nabû-apla-iddina Revisited". *Journal of Cuneiform Studies* 56: 23–103.

Woods, C. E. 2009. "At the Edge of the World: Cosmological Conceptions of the Eastern Horizon in Mesopotamia". *Journal of Ancient Near Eastern Religions* 9.2: 183–239.

Woodward, D. 1987. "Medieval *Mappaemundi*". In Harley, J. B. & Woodward, D. (eds.), *The History of Cartography*. Vol.1. *Cartography in Prehistoric, Ancient, and Medieval Europe and the Mediterranean*. Chicago: University of Chicago Press, 286–370.

Woytowitsch, E. 1978. *Die Wagen der Bronze- und Frühen Eisenzeit in Italien*. München: Beck.

Wright, J. K. 1925. *The Geographical Lore of the Time of the Crusades*. New York: American Geographical Society.

Wünsche, A. 1905. *Die Sagen vom Lebensbaum und Lebenswasser*. Leipzig: E. Pfeiffer.

Wyatt, N. 2005. *The Mythic Mind. Essays on Cosmology and Religion in Ugaritic and Old Testament Literature*. London: Equinox.

Wyatt, W. F. Jr. 1966. "Short Accusative Plurals in Greek". *Transactions and Proceedings of the American Philological Association* 97: 617–43.

Yalouris, N. 1980. "Astral Representations in the Archaic and Classical Periods and Their Connection to Literary Sources". *American Journal of Archaeology* 84.3: 313–18.

Yates, F. 1964. *Giordano Bruno and the Hermetic Tradition*. London: Routledge & K. Paul.

Yule, H., Cordier, H. 1903[3]. *The Book of Ser Marco Polo*. Vol 2. London: J. Murray.

Zeller, E. 1881. *A History of Greek Philosophy from the Earliest Period to the Time of Socrates*. 2 vols. London: Longmans, Green.

Zhmud, L. 2006. *The Origin of the History of Science in Classical Antiquity*, Berlin: De Gruyter.

Zhmud, L. 2012. *Pythagoras and the Early Pythagorean.*, Oxford: Oxford University Press.

Zimmer, H. & Winckler, H. 1903[3]. *Die Keilinschriften und das Alte Testament von Eberhard Schrader*. Berlin: Reuther & Reichard.

Zuntz, G. 1971. *Persephone: Three Essays on Religion and Thought of Magna Graecia.* Oxford: Clarendon Press.

Zuwiyya, Z. D. 2011. "The Alexander Romance in the Arabic Tradition". In Zuwiyya, Z. D. (ed.), *A Companion to Alexander Literature in the Middle Ages*. Leiden: Brill, 73–112.

Appendix 1

Diurnal solar movement in Mesopotamian tradition

On a number of occasions, I have adduced Mesopotamian parallels for the concepts that were primarily found in Greek ethnographic setting. Rather than collating them into a single chapter, and in this way removing them from their topical context, it seemed more prudent to distribute the pertinent comments and shorter discussions at appropriate points in the text. The exception to this rule was chapter 13, where a large body of Mesopotamian material was adduced in the context of a discussion focused on a subject belonging to the Mesopotamian tradition itself; yet even here it was studied in relation to important Greek parallels. This approach, I submit, demonstrated the importance and appropriateness of Mesopotamian evidence with regard to the subjects analysed in this work and its significance for understanding the comparable Greek models accounting for solar phenomena. However, it seemed more appropriate to discuss the Mesopotamian notions of specifically the diurnal solar movement more thoroughly and systematically in an appendix, due to their sheer abundance and complexity, which would otherwise burden the main text. On the other hand, I will not repeat the particular discussions already elaborated in appropriate places throughout the book but will rather focus on those aspects of Mesopotamian ideas regarding the daily movement of the sun that were not discussed in detail in the main text.

A1.1. Solar mountains and gates

The model accounting for the diurnal solar movement involving mount Māšu and its gates was discussed in chapter 13, including both literary and complementary iconographic representations of the concept.[1] The prevailing Mesopotamian model of the diurnal solar movement involves the sun god Utu/Šamaš rising at dawn from the doors of the sky and/or mountain of sunrise and setting into the corresponding pair of doors/mountains at sunset; the gates are watched by two guardians.[2] This model is a manifestation of the bi-polar model accounting for the daylight/night exchange, but, with a slight change of emphasis, is not incompatible with the uni-polar model either. In the main part of the book, I have only summarily considered the testimonies for this model, but I will now attempt to assess them more thoroughly and to provide support for this general portrayal of the model.

According to the *Third House Prayer* Utu emerges from kur-gal, 'Great Mountain', kur-idim, '(subterranean?) Mountain of the Spring', and 'Holy Hill' (du_6-kù, ?= *apsû*), from the place where heaven and earth embrace at 'the base of heaven' (an-úr = *išid same*, i.e., the horizon).[3] This account strongly suggests a connection of the diurnal solar movement and especially its nightly segment with the underworld, as do ll. 9 and 68 of *Nungal Hymn*, where the mountain (kur) where Utu rises is identified with the netherworld, Urugal (on this topic see more below).[4] Furthermore, as shown in chapter 13, (the mountain of) the Ḫašur-tree and/or cedar were associated with sunrise and with Dilmun. It was also pointed out that Dilmun as 'the place where Utu rises' is further mentioned in the *Eridu Genesis*, while the tree that sometimes appears in the sunrise scene on Akkadian glyptic might represent the ḫašur (perhaps also cedar) tree from the texts.[5] In light of their respective functions, the proposed association of the mountain of Ḫašur with Māšu (in its capacity of a sunrise mountain, whether as an 'eastern' Māšu in the bi-polar or simply a nadir-situated Māšu in the uni-polar model accounting for the daylight/night exchange) seems like a plausible conclusion. In addition, mount Ḫašur is cited as the goal reached by the Euphrates and Tigris,[6] which again suggests its association with the confluence of the rivers, Eridu and Dilmun. Furthermore, a number of sources explicitly mention the mountains of Sunset and Sunrise, suggesting that the sun both rises from and sets into a cosmological mountain, most probably the double mountain of Māšu.[7]

This concept is accompanied by the notion of the gates of the sun through which it passes both at sunrise and sunset. Thus, in the *Enuma Eliš* we read how Marduk opened gates in both of Tiamat's ribcages, fastened bolts on both sides, and established zenith in her liver, which can only mean that the gates were on the opposite sides of heaven.[8] Elsewhere, the sun god is regularly described as passing

[1] Cf. Horowitz 1998a: 97 n. 3, 266, 331; Woods 2009: 192.

[2] Black & Green 2000: 184.

[3] Borger 1967: 3–4, ll. 1–4; Seux 1976: 223–24; Heimpel 1986: 143–44; Horowitz 1998a: 235, 315–16, 331; Woods 2009: 203–205; Lambert 2013: 304.

[4] Woods 2009: 211, 225 with n. 179; ETCSL 4.28.1. On urugal as the name for the underworld see Horowitz 1998a: 293–94.

[5] Woods 2009: 190–91 with figs. 6–12 on pp. 229–31.

[6] *KAR* 34: 14–15 (George 2003a: ii.864; Woods 2009: 202–203 n. 74).

[7] The Akkadian translation of the twelfth tablet of the *Utukkū Lemnūtu* (=*Udugḫul*) (Heimpel 1986: 145; Horowitz 1998a: 332; George 2003a: i.493 n. 169, with another similar passage in *KAR* 24.5–7; Woods 2009: 187); a Sumerian hymn to Enlil (Horowitz 1998a: 331; George 2003a: i.493 n. 169; Woods 2009: 186–87 n. 12); *Inanna and Sukalletuda* (Horowitz 1998a: 249; George 2003a: i.493 n. 169; Woods 2009: 186 n. 12); for the sources mentioning solely the mountain of sunrise (a hymn to Nungal; a hymn in which the mountain is called Nergal's) or sunset (*Udugḫul* = *Utukkū Lemnūtu* IV.61) see Horowitz 1998a: 361; George 2003a: ii.864; Woods 2009: 187 n. 15, 217 with n. 140. The former is sometimes named Buduḫudug (a *lipšur* litany, type I.1.4; Reiner 1956: 132–33; Horowitz 1998a: 331 n. 13; George 2003a: ii.863: a litany based on *SB Hh* XXII.5, cf. the Emar version, Msk 74115 obv. 28'; Woods 2009: 187).

[8] *EE* 5.9–11; Horowitz 1998a: 266.

through the door of heaven at sunrise and/or sunset,[9] or as having a gate of his own.[10] He is in a similar fashion associated with the keys or bolts of heaven.[11]

A1.2. Interacting conceptual domains: solar movement and eschatology

Unsurprisingly in the light of the evidence presented in chapter 12 (and sporadically elsewhere), where an intimate connection of the sun(-god) with the otherworld in various traditions was discussed, in cuneiform sources the gates through which the sun passes at sunset are closely associated with the underworld. As we have just seen, the corresponding cosmological feature, the solar mountain, also shares this association. Šamaš thus opens the gateway of the 'wide earth' (*ersetu rapaštu*, one of Akkadian names for the underworld) at sunset,[12] or alternatively the 'doors of the earth' (*dalat dadmē*, in the meaning of 'door of the (entire) earth', indicating that *dadmū* extends to the ends of the earth's surface where the sun passes through the gates of heaven),[13] and, as noted in chapter 12, in the Early Dynastic Šamaš myth the sun god actually arrives at the door of Abzu.[14] The place where the sun sets is on the way to—or actually is the same as—the gate of the underworld,[15] the 'gate of Utu' is explicitly located in the underworld,[16] while Utu's great doors in the west are mentioned in connection with the netherworld, grave, Nergal and the *ersetu rapaštu*.[17] It thus seems that the western gate of the sun corresponds in large measure to the entrance to the underworld.[18]

The notion of a solar deity passing (at sunset) through the gates of the underworld was known to the cultures bordering with and strongly influenced by Mesopotamia. Thus, the Hittite *Telepinus Myth* describes the path of the Sun goddess of the Dark Earth associated with the seven doors of the Dark Earth,[19] while an incantation text describes the Sun goddess of the Earth opening the gate, as well as the release of the Former Gods, the lords of the Earth.[20] Furthermore, the Hurrian goddess Allani ('the lady')/Hittite 'the Sun goddess of the Earth' is found at the Bolts of the Netherworld in the *Song of Release* (1, 34).[21]

A1.3. The Mesopotamian sun-god's 'house of Night'

Previously I have briefly discussed some Mesopotamian accounts of the sun's nightly path and its presence in the underworld and/or the Abzu (chapter 12). Besides the *Epic of Gilgameš*, where Gilgameš travels along the sun's underworld path (see chapter 13), other Mesopotamian texts attest to a similar tradition. Thus some mention the 'heaven's interior' or 'middle of heaven', apparently the lower celestial hemisphere, in a solar context.[22] Furthermore, in the *Third House Prayer* a number of locations which are almost certainly to be associated with the underworld are mentioned in the context of the sun's nightly path (see immediately above), while the sun is said to have set for Dumuzi toward the 'Earth of the Dead', suggesting its presence at this location following the

---

[9] *Sunset Prayer* 1–3 (Heimpel 1986: 129; Horowitz 1998a: 217 n. 19, 248; George 2003a: i.461); *Fourth House Prayer* 1–5 (Heimpel 1986: 130); *KAR* 7.2ff (Rochberg-Halton 1983: 214 n. 29; Heimpel 1986: 134); *Exaltation of Ištar* 26 = *TCL* 6.51 r. 1–2 (Rochberg-Halton 1983: 214 n. 29; Horowitz 1998a: 144–45, 267; 2014: 14–15; Woods 2009: 188 with n. 21; Lambert 2013: 173); 4R 17: 9–10 (Rochberg-Halton 1983: 214 n. 29). For the sun-god associated with the gate of heaven cf. '*Prayer to an Obscured God*' (Heimpel 1986: 138); *Great Hymn to Šamaš* 182–83 (Gray 1901: 20–23 (iv.14–15); Lambert 1960: 136–37; Rochberg-Halton 1983: 214 n. 28); *Bit Rimki* 4R 17:5–6 (*CAD* n1 256 s.v. *namzaqu*; Rochberg-Halton 1983: 214 n. 28; Horowitz 1998a: 267); *HSM* 7494 (*YOS* 11 23) and *AO* 7032 (Rochberg-Halton 1983: 214 n. 28; Heimpel 1986: 133; Horowitz 1998a: 66, 267); *Sixth House Prayer* 1–4 (Heimpel 1986: 133); 4R 20.2: 3–6 (Rochberg-Halton 1983: 214 n. 29; George 2003a: ii.814).

[10] The Akkadian introduction to the *Enūma Anu Enlil*, which might be more precisely associated with sunrise (l.13 in King 1902: i.126–27; Heidel 1951: 74; Koch-Westenholz 1995: 77; Rochberg 2003: 179; 2004: 70; Lambert 2013: 173); Etana myth A 5, 42 (Horowitz 1998a: 50–53; Ragavan 2013: 205).

[11] *ARET* 5 6 i.1 and *OIP* 99 326 (+) 342 i.1–2 (Lambert 1989: 5, 33); *Bit Mummu Prayer* (Seux 1976: 233; Heimpel 1986: 134; Woods 2009: 201 n. 67 (*UVB* 15, 36: 12)); the inscription on the Sin-Iddinam cone (Horowitz 1998a: 267); *Šu'ila Prayer OECT* 6.82ff + *AOAT* 34.23:8–12 (Heimpel 1986: 133); *PBS* 1/1 12: 11 (Rochberg-Halton 1983: 214 n. 28); *Bit Rimki* 4R 17:5–6 (*CAD* n1 256 s.v. *namzaqu*; Rochberg-Halton 1983: 214 n. 28; Horowitz 1998a: 267); *HSM* 7494 (*YOS* 11 23) and *AO* 7032 (Rochberg-Halton 1983: 214 n. 28; Heimpel 1986: 133; Horowitz 1998a: 66, 267); *Sixth House Prayer* 1–4 (Heimpel 1986: 133); 4R 20.2: 3–6 (Rochberg-Halton 1983: 214 n. 29; George 2003a: ii.814);

[12] Heimpel 1986: 146; Horowitz 1998a: 279.

[13] Gray 1901: 20–23 (iv.14–15); Lambert 1960: 136–37; Rochberg-Halton 1983: 214 n. 28; Horowitz 1998a: 300.

[14] *ARET* 5, 6: xii.4 (Steinkeller 1992: 258, n. 39; Krebernik 1992: 85; Espak 2006: 69–71).

[15] *KAR* 267 rev. 12–14 = *BMS* 53.18–22 (Scurlock 1988: 206–207; cf. George 2003a: i.500 n. 192; Woods 2009: 187 n. 18).

[16] *PDT* 1 563 obv. i 9–16 (Katz 2007a: 184 n. 80; Ragavan 2013: 212).

[17] *Utukkū Lemnūtu* (=*Udugḫul*) text *CT* 16.9.i.11–12, 10.iv.25–30 (= *Forerunners to Udug-hul* 252, 299–303, Geller 1985: 34–37) (Heimpel 1986: 148 n. 58; George 2003a: i.500 n. 192; Ragavan 2013: 203 with n. 12).

[18] It is possible that the narrative *How Grain Came to Sumer* (*TMH* NF 3, no. 5 = ETCSL 1.7.6), which includes a visit to kur (ll. 13–20), introduced by the visit to 'Utu of the seventy doors' (l. 31), indicates Utu's position at the gate of the netherworld (Katz 2007a: 179 n. 56; Ragavan 2013: 204). However, the kur seems to be described as the Cedar mountain situated to the north (?) (l. 6; alternatively: upwards), which either means that the netherworld was here imagined in the north or that the kur is here not meant as the netherworld at all; alternatively, one could think of a northern solar mountain, a cosmological feature involved in the models accounting for the annual solar movement. At the same time, in the *Underworld vision of an Assyrian prince* the gatekeeper of the underworld, Bidu (Nedu), is described as a Lion demon, *ugallu* (Wiggermann 1992: 170; Huxley 2000: 130; Ataç 2004: 71) (for *ugallu* as 'creatures of apsû' see Wiggermann 1992: 13, 15 (ll. 144, 185), 50); on the other hand, from the Middle Assyrian period onwards the *ugallu* are occasionally depicted propping the solar disk of Šamaš, and already on the Akkadian seals an *ugullu* appears next to Šamaš (Huxley 2000: 128, 129 Fig. 14, 133). Thus, a creature associated with the solar disk is at the same time a doorkeeper of the underworld.

[19] *KUB* xvii.10.iv.13–14 (Goetze in *ANET*[3] p. 128; Gaster 1977: 298, 312; Hoffner 1998: 17).

[20] Otten 1961: 120–21, i.46–48; West 1997: 153, 298; Bachvarova 2016: 96.

[21] Hoffner 1998: 67, 73; Bachvarova 2005: 134; 2016: 124–25.

[22] For the 'heaven's interior' in a solar context see *BA* 10.1, 11–14 (Kramer 1944: 19–20 n. 9; Heimpel 1986: 143; George 2003a: ii.864; Woods 2009: 190 with n. 29); *Sunset Prayer* 1–3 (Heimpel 1986: 129; Horowitz 1998a: 217 n. 19, 248; George 2003a: i.461); *Fourth House Prayer* 1–5 (Heimpel 1986: 130). For Utu/Šamaš rising from 'Heaven's Interior' see also Horowitz 1998a: 248–49; the Akkadian translation of AN.ŠÀ in a *Bit-Rimki* incantation is *išid šamê*, 'the horizon', rather than 'Heaven's Interior' (*qereb šamê*) (Horowitz 1998a: 249). For the interpretation of 'heaven's interior' as the lower celestial hemisphere see Heimpel 1986: 130–31, 140, 151.

sunset.[23] Additionally, in the Old Babylonian *Prayer to the Gods of the Night* a priest describes the night by evoking 'the closed gates that were open (during the day)'; he also mentions how Šamaš 'entered his *kummu* (chamber)'.[24] Does Šamaš' *kummu* represent his underworld path or simply his private chambers (which also had to be located somewhere)? I feel free to offer my conclusion in advance of the argument, and to claim that both propositions are correct, his chamber being certainly located in the underworld, somewhere on his path from sunset to sunrise, to which one could compare the discussion of the Greek localisations of the house of Night and its relation to the sun (chapter 9). In any case, the term deserves a more detailed discussion.

Thus, Utu emerges from his *kummu*/agrun at sunrise or enters it at sunset.[25] Several other texts bear testimony to a similar idea, although they do not explicitly mention Utu's chamber.[26] Sometimes Utu enters his house at sunset and exits it at sunrise;[27] elsewhere his house is located at 'level place', *ki-ùr*.[28] The latter suggests the idea of a cosmic nadir, a place opposite the sun's culmination at noon in the 'upper' hemisphere and an otherworld location *par excellence*. A similar concept is designated by the term *asurrakku*, which has the same meaning as the already mentioned *qereb šamê*, 'heaven's interior' or 'middle of heaven'.[29] But it undoubtedly designates Apsu, that is, *mû šaplītum*, 'lowlying waters, deep waters', which is practically a synonym for the expression *mû rūqūtu*, 'distant waters'.[30] Elsewhere Apsu is explicitly identified with *asurrakku*.[31] In addition, there are two possible etymologies for *asurrakku* that further persuasively connect it with the Apsu: '(place of/where) a.sur.ra (flowing waters, roiled waters, deep waters)'[32] or '(place of) distant/deep waters', with a.sud.ra = *asurrakku* according to MSL 17 150.[33] Thus the Apsu can through this term be associated with 'distant deep waters', but at the same time with the 'heaven's interior' or 'middle of heaven'. Furthermore, it can also be associated with the nocturnal solar voyage, and thus corresponds to the 'level place' in the cosmic nadir (or the latter is a part of the Apsu).

Now *kummu* in *EE* 1.75–76 is clearly identical with the *apsû*,[34] while the explicit *apsû* = AGRUN/*kummu* identification can be found in a number of texts.[35] It is thus possible to identify *kummu* (agrun) with Enki/Ea's Abzu/Apsu, but also with the region through which the sun passes between sunset and sunrise.[36] Similarly, the Early Dynastic Šamaš myth discussed in a similar context in chapter 12 probably describes an encounter between Šamaš and Enki/Ea in the latter's Abzu/Apsu, during the former's nocturnal journey.[37] In this myth Šamaš is described as a bringer of light to the Abzu,[38] probably as visiting the region of Enki/Ea during the night,[39] and as arriving to the fortress of Ea and the door of Apsu, also presumably during the night.[40] A similar concept is recognizable elsewhere, with Utu/Šamaš encountering Enki/Ea in the Abzu/Apsu,[41] himself—or his light—reaching the Apsu,[42] or in some other way being explicitly associated with Ea's realm.[43] The connection of Utu/Šamaš with the Abzu/Apsu is perhaps also present in the *Third House Prayer* (see above), where Utu apparently spends the night in du$_6$-kù, perhaps (in part) identical with *apsû*.[44]

Thus, the term agrun/*kummu* most probably describes a point midway between sunset and sunrise, with respect to the observer's horizon. In effect, this seems like another manifestation of the 'uni-polar' model of the diurnal solar

[23] Horowitz 1998a: 281.

[24] Heimpel 1986: 128.

[25] Sunrise: *Gilgameš, Enkidu, and the nether world* Version A 49, 93 (ETCSL after Shaffer 1963; cf. Foster 2001: 131–32; George 2003b: 180–81; Black *et. al.* 2004: 33–34); the Ur version of the Sumerian text (*UET* VI/1 60) that was used for the translation in the *EG* 12, rev. 7' (Cavigneaux & Al-Rawi 2000: 8; George 2003a: i.52–53, ii.769, 777; 2003b:190; cf. Heimpel 1986: 128; Foster 2001: 142); *Enmerkar and lord of Aratta* 132, 204 (Caplice 1973: 303 n. 20); *Lugalbanda in the mountain cave* 485 (Black *et. al.* 2004: 21); *CT* 16 36: 36–37 and *BIN* 2 22: 179–80 (Caplice 1973: 303 n. 20); *VS* xvii.14.17–18 = *YOS* xi.54.13–14 (Cavigneaux & Al-Rawi 2000: 8, comm. ad 7'); sunset: *Hymn to Inanna as Ninegala* (ETCSL *Inana D* 118–20 after Behrens 1998; cf. Caplice 1973: 303 n. 20; Heimpel 1986: 128; Horowitz 1998a: 251); cf. *RA* 32/1935: 180: 13, 181: 15 (Caplice 1973: 303 n. 20) (at night).

[26] *Lugalbanda in the mountain cave* 143 (Heimpel 1986: 128–29; Black *et. al.* 2004: 15) (Utu's house).

[27] *CT* XVII.19.i.34–39 (Horowitz 1998a: 251 n. 11).

[28] *Sunset Prayer* 1–3 (Heimpel 1986: 129–30).

[29] *Enuma Anu Enlil* ACh Ištar 29: 12–13 (Rochberg-Halton 1983: 214; Horowitz 1998a: 179); Horowitz 1998a: 238–39.

[30] Horowitz 1998a: 178–79, 306–307, 310.

[31] *STC* I 205 with 4R 26 no. 4: 51–52 (Horowitz 1998a: 310–11).

[32] In *Enki and Ninmah* 13 Enki is described in 'deep Apsu' (engur-bùru), 'where water flows' (a.sur.ra), while a bilingual incantation *CT* 17 34: 23–26 mentions 'Enki's flowing water' (a-su-ra ^{d}en-ki) (Horowitz 1998a: 138–39, 312–13; Lambert 2013: 334-35, 503 translates a.sur.ra as 'the abyss').

[33] Horowitz 1998a: 312–13.

[34] Marduk's *kummu* (*EE* 5.124; 6.52) is another case (Caplice 1973: 301–302).

[35] *ASKT* 12 23–24 (Caplice 1973: 302); *Utukkū Lemnūtu* (= *Udugḫul*) *CT* 16 15 v 34–35 (Caplice 1973: 303; Woods 2009: 205 n. 84); *CT* 42 13: 3–7 (ETCSL *Inanna G* 3–7 after Kramer 1963b: 503; cf. Caplice 1973: 303–304, who further quotes another parallel Sumerian text); *RA* 12/1915: 34: 12–13 (Caplice 1973: 304) (Eridu = agrun).

[36] Caplice 1973: 304, further cited by Heimpel 1986: 128–29 n. 12 and Horowitz 1998a: 251 n. 11. For É.NUN = *kummu* associated with Abzu see also Cunningham 2007: 12.

[37] *ARET* 5, 6 = *OIP* 99, 326+342; Steinkeller 1992: 257, 258, n. 39; Woods 2004: 71, n. 241; Espak 2006: 68–71; cf. Lambert 1989: 25–26, Krebernik 1992: 85–86.

[38] *ARET* 5, 6: i.9–ii.1; *OIP*, 99, 326: i.11–12; compare Lambert 1989: 6, 33.

[39] *ARET* 5, 6: ii.5–iii.2; *OIP* 99, 326: i.7–ii.7.

[40] *ARET* 5, 6: xii.2–xiv.2. My interpretation follows Espak 2006: 68–71.

[41] *Temple Hymns* 15–16; Sargonic seals (Steinkeller 1992: 258 n. 39; Woods 2004: 71, n. 241; 2009: 200 n. 67; cf. Frankfort 1934: 25 with Pl. IV.a, c = Böhmer 1965: nos. 488–89 (on seal no. 489 the doorman opens the gate of Apsu to Šamaš, Böhmer 1965: 84).

[42] Great Šamaš Hymn i.37, ii.1 (Gray 1901: 14–15; Schollmeyer 1912: 81, 87–88; Lambert 1960: 128–29; Heimpel 1986: 145 n. 52).

[43] Great Šamaš Hymn i.31 (Gray 1901: 12–13; Schollmeyer 1912: 81, 87; Lambert 1960: 126–27; Gaster 1977: 227–28); *Bit Mummu Prayer* (Seux 1976: 233; Woods 2009: 201 n. 67 (*UVB* 15, 36: 12). Cf. the Sippar Tablet where Šamaš, enthroned and inside some structure, together with his solar disk, is placed directly upon the Apsu (Ornan 2005: 63–65, 111–12, 241 Fig. 65), suggesting some sort of connection between them.

[44] On this identification see especially Seux 1976: 223 n. 3 and Lambert 2013: 302–305, 367–68, 524 (with du$_6$-kù also representing the underworld on p. 305); *contra* George 1992: 270 and Horowitz 1998a: 316 (more cautiously in Horowitz 2014: 81–82).

movement which accounts for the alternation between night and daytime by locating it at the *axis mundi*, in its turn represented by and located in the Abzu.

At the same time, it corresponds to the location of Hesiod's house of Night and Stesichorus' localisation of Helios' family, in terms of the sun's meridian passage at midnight below the underside of the earth or, alternatively, its imagined meridian passage upon the surface of the earth in the far south, in which case the location corresponds to a point upon a flat earth's surface where the farthest east meets the farthest west (chapter 9). In its capacity as an Eridu-Abzu-related location and considering that its situation was defined with respect to the daily path of the sun, Utnapištim's dwelling at the *pī nārāti*, 'the mouth of the two rivers', can also be brought into connection with this cosmological location and—at the same time—a feature involved in a model of the diurnal solar movement, whether in terms of the uni- or bi-polar models accounting for the daylight/night exchange (chapter 13). In addition, the *pī nārāti* can be brought into close association with the daily movement of the sun through its apparent identity with du_6-kù—out of which Utu rises—in its capacity as 'mountain of springs [of the waters of Apsû]',[45] as well as through the association of the 'solar' mount Ḫašur and the homonymous tree with Eridu-Dilmun at the confluence of rivers. Here, as elsewhere, 'solar' designates a feature involved in a model accounting for the solar movement.

All Mesopotamian testimonies discussed in this appendix refer to the diurnal solar movement. The presence of some familiar motifs (gates/(double) mountains of sunrise/sunset) and phenomena (the occurrence of both the uni- and bi-polar models of the diurnal solar movement accounting for the phenomena of sunrise and sunset), as well as the strong association of the nocturnal section of the daily path of the sun(-god) with the otherworld, supplement and support the conclusions reached in the second to last chapter of the book, where I focused on the aspect of Mesopotamian tradition interested in the diurnal solar movement.

[45] Woods 2009: 204–205.

Appendix 2

Diurnal solar movement in Egyptian tradition

The Egyptian models of solar movement are almost exclusively concerned with the diurnal motion of the sun in a solar barque over the sky and its nocturnal voyage through the underworld. Egyptian tradition adds several important new elements that clarify some implicit notions present in other ethnographic contexts, and certainly offers the most complete mythic and cosmological account of the diurnal solar movement, especially in the context of its interaction with the conceptual domain of eschatology. It contains several elements current in other traditions, such as a solar voyage in a boat, the concept of gates through which the sun passes at sunrise and sunset, but also during its nocturnal voyage, the (double) mountains of sunrise and sunset and an elaborate solar eschatology. In addition, here all these elements seem to be more explicit, pronounced and systematically arranged. As usual, the diurnal solar movement is inextricably merged with eschatological tenets, in Egypt even more pronouncedly than elsewhere.

A2.1. Books of the Netherworld

The 12-hour nocturnal journey of the sun(-god) is richly attested in the Egyptian New Kingdom Books of the Netherworld.[1] Sometimes the progress of the sun-god through the Netherworld (Duat) is portrayed systematically, describing a 12-hour voyage from sunset to sunrise, with gates between successive hours, at other times more arbitrarily (both in words and in pictures).[2] The concept is, however, undoubtedly older than the New Kingdom. The Middle Kingdom *Book of Two Ways* is apparently the earliest of the books of the Netherworld, their precursor of a kind.[3] It illustrates a solar voyage, with the deceased journeying in Re's barque, structured in a fashion similar to the later conceptions but without incorporating a systematic division into 12 hours or a similar scheme. Another difference from later Books is in that the voyage as presented in the *Book* most probably takes place with the diurnal passage of the sun, starting with its rising, which would mean that it takes place in the daytime sky, rather than underground.[4]

In addition, the sun-god is regularly depicted in his barque, thus clearly showing he was imagined sailing through the Netherworld, which is a concept present in other traditions discussed in chapter 12. The accepted date for these compositions, the 16th century BC at the earliest, is too late to be an inspiration for the *Epic of Gilgameš* and its comparable 12 double-hours nocturnal solar path, yet the appearance of a fully organized system as manifested in these Books, with at least several variants, suggests its somewhat earlier origin, which is indeed conclusively proven by the existence of the *Book of Two Ways*. With that said, there is no evidence that the Egyptian tradition in any way influenced the Mesopotamian understanding and conceptualisation of solar movement, and all similarities between the model as expounded in the *Epic of Gilgameš* and its Egyptian counterparts are most probably the results of parallel developments, conditioned by the use of similar approaches in addressing identical referents in the natural world.

A2.2. Gates and mountains

The analysis of these Egyptian New Kingdom 'texts', however, offers additional parallels to Gilgameš's voyage upon the sun-god's (nightly) path, and the models of solar movement they employ can be supplemented and corroborated by even earlier records, taking us back to the Pyramid Texts, thus in the mid-third millennium BC. This is evident, more than anywhere else, in the concept of a double mountain, gate or some other feature as a place of sunrise and sunset, conforming perfectly to the Mesopotamian (both textual and iconographical) evidence.

Double gates of either sunrise or sunset, often with strong eschatological resonances, with the gates of sunset sometimes corresponding to the gates of the netherworld, appear in Egyptian texts from the period of the Pyramid Texts onwards, with most of the late appearances in the

1 Hornung 1999: 26.

2 Systematic portrayals are found in the *Amduat* (Piankoff 1954: i.40, with pp. 230–31, 234–35, 238–39, 241, 245–46, 248, 255, 262, 270, 278, 286, 294, 300, 306, 312; translations of the book are found in Budge 1905, vol. 1 (long version), vol. 2.1–40 (short version) and Piankoff 1954: i.230–318; see also the description in Hornung 1999: 27–53), the *Book of Gates* (translations of the book are found in Budge 1905, vol. 2.43–306 and Piankoff 1954: i.140–224; see also the description in Hornung 1999: 55–77; this book is often compared with the BD spells 144–47 (Faulkner 1972: 133–37; Faulkner in Dassow 1998: Plate 11, 120–21), which could be used as substitutes for it (Hornung 1999: 58–59)), the *Book of Night* (Piankoff 1954: i.409; Hornung 1999: 123; translation of the book is found in Piankoff 1954: i.409–28; see also the description in Hornung 1999: 122–35 (cf. the version in Frankfort 1933: 68–71); for its compatibility with the Books of the Netherworld proper, from which it differs in that the sun here traverses the body of Nut, i.e., the sky, during its nocturnal voyage, see Hornung 1999: 112; with this *Book* compare further the part of the shadow clock text from the ceiling of Seti I's cenotaph, Clagett 1995: 387 n. 24; cf. Frankfort 1933: 76), unsystematic in the *Spell of the Twelve Caves* = ch. 168 of the BD (Hornung 1999: 54–55; Faulkner's translation of ch. 168 of the BD in Faulkner 1972: 162–66 and Dassow 1998: 126–27; illustrations in Naville 1886: i.CLXXXVII; Faulkner 1972: 168–69; Hornung 1999: 54 Fig. 26).

3 Known through two recensions, both found on the bottom of the sarcophagi found at Dar el Bersha: CT 1029–130 (long version), 1131–85 + 513, 587 (short version) (Hornung 1999: 9). Translations of the book are found in Lesko 1971 and Faulkner 1973–78 vol. 3.

4 Lesko 1971: 30; 1972: 43, 61–62, 135; Hornung 1999: 11; Robinson 2003: 142, 146; cf. CT 1029.vii.252. For the identical settings of the diurnal and nocturnal parts of the daily solar course acknowledged in the *Book* itself see CT 1089.vii.368.

New Kingdom Book of the Dead reflecting earlier forms already present in the Coffin Texts.[5] Other similar imagery was also employed in different but complementary renderings of this concept, representing paradigmatic variations within a clearly defined syntagmatic context: two turquoise sycamores of sunrise and two great rocks of sunset 'in the midst of the Western Sea'.[6] Rocks are suggestive of mountains, and indeed the mountain of sunrise (on solar mountains see more below) can be described as two or twin mountains,[7] or as two mounds or caverns.[8]

But the double form of the cosmological features involved in the models accounting for the daily solar movement is not obligatory. The notion of 'simple' gates of sunrise and sunset at the horizon, the latter more or less regularly associated with the otherworld, is indeed widespread in the Egyptian tradition. The 'doors (or gates) of the horizon' are among the most frequently mentioned features of this kind,[9] sometimes explicitly identified with the gate of the netherworld,[10] other times associated with sunrise.[11] Thus a close association can be discerned between the gates of sunrise and sunset at the horizon on one side and the netherworld on the other, representing another instantiation of the widespread interaction of solar movement and eschatology.

All these features—(double) doors/gates, two rocks or sycamores—naturally correspond to a number of toponyms mentioned in the New Kingdom books of the Underworld: the gate at the western horizon, which is sometimes referred to as 'the gate of the Great Netherworld', and its counterpart, the gate at the eastern horizon, opened for Re in the morning, who boards there his Morning Barge after traversing the Netherworld.[12]

The concept of solar gates is accompanied in Egyptian tradition by the notion of solar mountains—which can be portrayed as double, as we have just seen—behind which the sun rises at dawn and sets at sunset. Furthermore, the image of the solar mountain(s) was in the form of the telling design of the *akhet*-sign (Gardiner N27:) an inseparable component of the Egyptian models accounting for sunrise and sunset. Explicit testimonies on solar mountain(s) appear in Egyptian texts, although apparently not in the PT and only sporadically in the CT.

The mountain of sunset is sometimes called Manu, and it is regularly associated with the entrance to the Underworld.[13] Later Akhet Bakhu simply meant 'the East' and Akhet Manu 'the West'.[14] However, the Egyptian tradition does not localise the cosmological position of Mount Manu consistently; sometimes it represents the mountain of sunrise.[15] At the same time, the occasional ambiguity present in the localisation of Mount Manu is also apparent in the position of the 'eastern' mountain/region of Bakhu.[16]

The mountain of sunset is occasionally depicted in Theban tombs as a hill towards which the sun disk descends and out of which the prominent breasts of the goddess Nut emerge, together with her hands receiving the disk.[17]

[5] PT 275.416, 611.1726 (the solar connection is not explicit); CT 44.i.185, 50.i.223; CT 335a.iv.224 (with Faulkner CT i.267 n. 30), corresponding to BD ch. 17 (Faulkner 1972: 45; 1998: (34–35) (unpaginated); Allen 1974: 28 (only here the door is double)) with accompanying illustrations in Naville 1886: i.XXVIII A.g, D.a L.a (in the last case upon the *djew*-sign denoting mountain, Gardiner N26:); CT 1131.vii.473 (*Book of Two Ways*, Lesko 1972: 23); CT 1065.vii.324–25 (*Book of Two Ways*, Lesko 1972: 65–66), corresponding to BD ch. 130 (Faulkner 1972: 117; Allen 1974: 105 (only here the door is double)); BD ch. 68 (Faulkner 1972: 69; Allen 1974: 62 (only here the door is double)) with accompanying illustrations, also upon the *djew*-sign (Speleers 1917: Pl. 13 Fig. 27, with a description on p. 19; Piankoff 1957: i.33 with Fig. 16; Davis 1977: 172; Wilkinson 1992: 146 Fig. 3); BD ch. 15 (Allen 1974: 16, 19–20); perhaps also BD ch. 177 (Allen 1974: 186; in Faulkner in Dassow 1998: 131 the door is not double).

[6] CT 159.ii.363–64, 367–68 (= CT 161, BD 109, 149 2nd mound) (sycamores) (Faulkner 1972: 102, 139; Faulkner in Dassow 1998: 113, 121), with PT 470.916 and 568.1433, where one or two tall sycamore trees in the east of the sky are mentioned, suggesting an earlier origin of this concept; *Book of Night* 12.2 (Piankoff 1954: i.426–27) (rocks).

[7] CT 305.iv.59 (two); BD ch. 149 (twin), a variant of ch. 108 (where Bakhu is located in the east) (Faulkner 1972: 139; Faulkner in Dassow 1998: 121; 'twin' is missing in Allen 1974: 143).

[8] *Book of Caverns* 6th section and concluding representation (Budge 1899: Pl. 6; Piankoff 1954: i, Fig. 20 (between pp. 116–17), 124, 131, 134; Hornung 1999: 90, 95 Fig. 52). The text refers to the 'Mountains of the East' as 'two caverns of the West', which only means that the netherworld extended to the eastern horizon (Piankoff 1954: i.117, 119–20). In the papyrus of Khonsu-mes, it is the sunset mountain that is represented in a similar way (Piankoff 1957: i.27 Fig. 10, 214–15).

[9] PT 220.194, 246.255, 610.1720, 690.2095; CT 91.ii.62, 107.ii.120, 190.iii.98, 1060.vii.314 (*Book of Two Ways*, Lesko 1972: 71; Faulkner CT p. iii.140 n. 1), 1130.vii.462 (*Book of Two Ways*, Lesko 1972: 130); *ARE* ii.141, p. 61 (18th dynasty).

[10] Paheri's tomb (18th dynasty, Sethe 1906: i.iv.116.13; 1914: 57; Piankoff 1957: i.4); Cairo papyrus (Piankoff 1957: i.197); BD ch. 126 (Faulkner 1972: 115; Allen 1974: 102).

[11] PT 311.496; BD chs. 39, 44; cf. *ARE* ii.141, p. 61 (18th dynasty).

[12] *Book of Caverns* 1.1 (Piankoff 1954: i.49); *Book of Gates* 9.1–3, 10.2–3, 11.1, 12 (Piankoff 1954: i.198–201, 209–11, 217, 224); *Book of Amduat* Introductory text, 1.4, 2.1, 9.2, 10.2–3, 11.2, 12.1, 3 (Piankoff 1954: i.230, 239, 241, 296, 303–304, 308, 313, 316); *Book of Day* 1st hour (Piankoff 1954: i.389–91); *Book of Night* 12.3 (Piankoff 1954: i.428).

[13] BD chs. 15 (Faulkner 1972: 40; Faulkner in Dassow 1998: (55) (unpaginated); Allen 1974: 15, 20–23, 25–26) and 168 (Faulkner 1972: 165); the ceiling of the Sarcophagus Room in the cenotaph of Seti I in Abydos (Frankfort 1933: 70; Thomas 1956: 78); BM 826, Cairo 34051 (*ANET*3 p. 368); *ARE* ii.321, p. 134, ii.905, p. 369, iv.12, p. 8; the ceiling of Corridor B of the tomb of Ramesses IX, tomb 4 of Amenemophis in Tanis, the wall of Hall XIX of the Theban Tomb 33, a Persian-Ptolemaic sarcophagus lid (Cairo 29316), a Greco-Roman stele (Cairo 22141) (Thomas 1956: 78–79). For numerous other examples see further *WÄS* 2.29.11–16 (1971: 29; 1940: 45).

[14] Assmann in *LÄ* iii.4.6–14 s.v. *Horizont*.

[15] BD *The Introductory Hymn to the Sun-God Re* (Faulkner 1972: 27; Faulkner in Dassow 1998: (20–21) (unpaginated)); the papyrus of Gaut-Sushen A (Piankoff 1957: i.182); Papyrus Harris pl. 22 (*ARE* iv.246, p. 140); cf. Kurth in *LÄ* iii.1185.11, 1186.2–19 n. 7 s.v. *Manu*.

[16] Faulkner 1962: 78. Some examples are to be found in BD ch. 108 (Allen 1974: 85; Faulkner in Dassow 1998: 113), even though this chapter is largely derived from the CT 160 (on which see the continuation of this footnote); Papyrus Harris pl. 7 (*ARE* iv.209, p. 120); anonymous papyrus from Cairo (Piankoff 1957: i.196); for other examples see *WÄS* I.422.8–13 (1971: 422; 1958: 70). Examples of Bakhu as the mountain of sunset can be found in CT 160.ii.374–76, 386–87 (cf. Otto in *LÄ* i.594.1–4 s.v. *Bachu*) and the papyri of Khonsu-mes and Gaut-Sushen A (Piankoff 1957: i.27 Fig. 10, 182, 214).

[17] Piankoff 1957: i.41–42 Figs. 25–27 (Fig. 25 = Schäfer 1935: 27 Abb. 12, Fig. 26 = Schäfer 1935: 26 Abb. 10); Schäfer 1935: 27 Abb. 11, an illustration accompanying ch. 186 of the BD = Naville 1886: i.CCXII La. Variants in Schäfer 1935: 27 Abb. 13, 28, 34 Abb. 22–23; descriptions only in Schäfer 1935: 27 and in BD ch. 15 (Allen 1974: 15, 21, 26) (both without breasts).

A variant of this design incorporates the *akhet*-sign.[18] Clearly, the goddesses' breasts in this setting represent, somewhat redundantly, the *akhet*-sign, which suggests that they are also meant to represent it when the actual sign is absent.[19] Since both the arms and breasts of the goddess Nut either emerge out of the Western mountain or descend from the heavens, and as a rule are associated with sunset, it seems that her breasts represent the western *akhet*. This association of the goddess's breasts with solar movement—here diurnal—is interesting when compared with similar associations of female breasts and solar motion recognised in other, rather remote, traditions (even though there the association is with the annual solar movement, see chapter 13). It seems that we are witnessing here another example of the interchangeability of motifs used in the mythic models accounting respectively for the diurnal and annual solar movement, as well as the limited number of motifs used in the creation of both types of models.

Finally, the solar gates and mountains were sometimes, although infrequently, combined in the Egyptian tradition. Thus, there existed the 'gate of Manu',[20] and on the papyrus of Ta-udja-re the Lords of both Bakhu and Manu are invoked, with an utterance suggesting the 'opening' of the Western Mountain as well as the doors presumably located on, or identical with, the mountain.[21] Lastly, Manu is intimately connected with the solar gates in ch. 15 of BD.[22]

A2.3. The horizon-sign

The geminate nature of the mountain of sunrise certainly reflects and is reflected in the already mentioned New Kingdom *akhet*-sign.[23] During this period the sign depicted the sun between two mountains (Gardiner N27: 𓈌). This iconographical rendering of the concept of the horizon sun can, however, be compared with much earlier textual evidence. Thus, in PT 685.2064 the birth of the king (with Re in the afterlife?) is described as two mountains splitting apart; furthermore, we have seen that the mountain of sunrise was described as double in the CT. This might show that the concept evident in the later sign could have been conceived much earlier.[24] What is more, the textual evidence from the Old and Middle Kingdom period could be in its turn preceded by a prehistoric, that is, pre-unification, iconographic evidence. The *akhet*-concept can be recognized in an 'unmistakable design' of an ornament from the Amratian (Naquada I, c. 4000–3500 BC) pottery depicting a double mountain surrounded by waters from which a solar disk emerges and into which it descends.[25] The tradition of the double-peaked mountain associated with solar rising and setting was thus present in the Nile valley already in the 1st half of the 4th millennium.

Specific details of the *akhet*-concept, when something unambiguous can be adduced about them, are somewhat inconsistently rendered in its various iconographical instantiations. Sometimes the left tip of the mountain in the *akhet*-sign represented the 'western horizon', and the right tip the 'eastern horizon',[26] but normally only 'one' horizon was meant by a single sign. The *akhet*-sign is used in depicting the Eastern Mountain of sunrise in numerous illustrations,[27] occasionally also the Western Mountains of sunset,[28] and sometimes both these solar mountains.[29] Thus the *akhet*-sign—or, rather, concept—could represent at the same time either the western and eastern horizon singularly or both horizons simultaneously. Nevertheless, once again the double mountain appears both in the east and in the west. With respect to architecture, it is often claimed that temple pylons mimic the shape of the *akhet*-sign and/or the horizon mountains.[30] However, even though this assertion seems commonsensical, it does not appear to be supported by clear or overt evidence. At least in the Edfu temple, a pylon actually did represent the *akhet*, since here an inscription associates Isis and Nephthys—'who raise

[18] Sarcophagus of Khonsu, with the addition of the head of the cow goddess Hathor (Piankoff 1957: i.34–35 with Fig. 18); an illustration accompanying ch. 16A of the BD (Naville 1886: i.XXI Ba; Schäfer 1928: 108–109 with Abb. 32; 1935: 22–23 with Abb. 3); Sennedjem's tomb (Wilkinson 1992: 134 Fig. 4).

[19] Wilkinson 1992: 135.

[20] Clère 1961: Tab. 17, cited in Kurth in *LÄ* iii.1185.56–58 n. 5 s.v. *Manu*.

[21] Piankoff 1957: i.137–38.

[22] Allen 1974: 15–16, 20.

[23] On the concept of *akhet* see especially Kuentz 1920: 140 *passim*, esp. 189–90; also, *ARE* ii, p. 132 n. *c*; v, p. 132 s.v. *Horizon*; Piankoff 1934: 58, 59; Faulkner 1962: 4–5; Goelet in Dassow 1998: 143; Assmann in *LÄ* iii.3–7 s.v. *Horizont*; 2003: 58; Allen 1988: 6; 2003: 27.

[24] Cf. Piankoff 1957: i.31 n. 6.

[25] Piankoff 1957: i.30 (quote), 31 Fig. 11; cf. Giedion 1962: ii.129 Fig. 69; Smith 1987: 89 Fig. 4.38; Bilić 2013b: 110 with Fig. 2.

[26] The portal of the tomb of Ramesses X (Schäfer 1919: ii. Taf. 36.1; 1928: 89 Abb. 4, 100–101; 1935: 34 and Abb. 24 (further cited in Assmann in *LÄ* iii.6.62–7.1); Piankoff 1957: i.31–32 and Fig. 12); sarcophagus from the Rijks-Museum (Piankoff 1957: i.36 Fig. 19, right part of the illustration); cf. the entrance to the tomb of Ramesses VI (Piankoff 1954: i.9, Fig. 3 on p. 11) and the papyrus of Amon-em-Saf, (Piankoff 1957: i.42–43 with Fig. 28).

[27] Papyri: Rossi 1878–79: 1212–13, Tav. XV (cf.Piankoff 1957: i.170); Piankoff 1957: i.21 Fig. 38, 26 Figs. 7b (with p. 168) and 8, 39 Fig. 22, 62 Figs. 49–50, 95; coffin: Piankoff 1957: i.63 Fig. 51; the ceiling of the Roman-period so-called New Year chapel or temple in Dendera (Kees 1987: Tafel IIIb with a description on p. 84, where he argues that the tips of the sign represent the East and West, respectively; a tree placed on both tips of the sign perhaps represent the already mentioned two sycamores between which Re rises, cf. MacGillivray 2004: 331); cf. an illustration accompanying the 2nd mound of ch. 149 of the BD (Naville 1886: i.CLXVIII Aa; Quirke 2003: 166 Fig. 9.1), depicting a *djew*-sign with Re-Horakhty as the god inside the mound (the accompanying text describes the gates through which Re rises in the east of the sky and two turquoise sycamores between which he rises, Faulkner 1972: 140; Faulkner in Dassow 1998: 121; Allen 1974: 142).

[28] The figure of the Prologue to the *Book of Gates* (Piankoff 1954: i.137, 139, 140 Fig. 30; Hornung 1999: 66 Fig. 30); papyri (Piankoff 1957: i.26 Fig. 7a, 108, 194, 199); Sennedjem's tomb (Wilkinson 1992: 134 Fig. 4); cf. the depictions of double gates upon the *djew*-sign on an illustration accompanying ch. 17 of the BD (Naville 1886: i.XXVIII L.a), Sennedjem's tomb (Wilkinson 1992: 146 Fig. 3), and papyrus of Nefer Renpet (Speleers 1917: Pl. 13 Fig. 27; Piankoff 1957: i.33 Fig. 16), as well as the illustrations in the papyri of Hunefer (a standard designating the West resting upon a *djew*-sign, Budge 1899: 9, Pl. 8) and Anhay (the same standard resting upon a foreign land/hill-country sign (Gardiner N25: 𓈉), Budge 1899: 22, Pl. 1).

[29] The *Book of Nut* (Allen 1988: 2; 2003: 27); coffins (Piankoff 1957: i.34–36 with Figs. 18–19); an illustration accompanying ch. 16A of the BD (Naville 1886: i.XXI Ba; Schäfer 1928: 108–109 with Abb. 32; 1935: 22–23 with Abb. 3).

[30] Aldred 1987: 165; Wilkinson 1992: 135, 139; 1994: 159; 2000: 60, 77–78; Redford 2001: iii.372.

up the sun [god who shines on the horizon]'—with two towers of the Pylon K'.[31]

The iconography of the god Aker is another Egyptian variant of the model accounting for the diurnal solar movement, more precisely its defining points, i.e., sunrise and sunset. The god represents the East and the West and symbolizes the mountain of sunrise and sunset.[32] Already the PT mention 'the gate(s) of Aker', associated with the King's resurrection, but not in a positively solar context.[33] The god is regularly represented as a double sphinx, occasionally with a solar barque atop of him. He represents both the entrance and the exit of the sun from the underworld, that is, the east and the west.[34] The depictions of Aker on the illustrations accompanying the BD usually take the form of two lions sitting back to back, supporting the *akhet* sign.[35] In the Papyrus of Ani the lions are tellingly called Tomorrow and Yesterday, in the Louvre Papyrus 2292 Bakhu and Manu.[36] Finally, Aker and the *akhet* sign merge into a single representation of a ram-headed solar deity standing on a boat formed of an elongated human figure with a head at either end, probably representing either Aker or *akhet*-formed solar barque.[37]

Egyptian notions of (double) doors or gates at the horizon, often conceived as double horizon mountains, the *akhet*-sign, the double god Aker, or, sometimes, two sycamores, great rocks, mounds or caverns regularly associated with the sun's diurnal motion from the PT onwards (perhaps even earlier), are an excellent parallel to the twin solar Māšu mountain of the *Epic of Gilgameš*, showing that a model accounting for the diurnal solar movement involving similar features was known to the Egyptians from an early period. Furthermore, the elaboration of the sun's nightly voyage and its full-scale incorporation into Egyptian eschatological concepts was current as early as the Pyramid Texts, even if not in a systematised way with which we are familiar from later periods. Apparently, these various models were all intended to account for the sun's diurnal motion, since a reference to its annual motion is nowhere to be found.

The mythic models found in a developed form in the Egyptian tradition seem to throw some welcomed light on the comparable notions found in other traditions. The Egyptian systematisation of the diurnal and nocturnal solar voyage is rarely recognised in other ethnographic contexts in its complete form. But its elements (gates, (double) mountains, boat as the vehicle of the sun-god, strong associations with the otherworld, etc.) are found in them often enough. I do not argue for any genetic relationship between these traditions or an influence of the Egyptian tradition upon the others here. Rather, I submit that Egyptian evidence—although more often than not of a rather complicated nature—can sometimes provide heuristic tools for interpreting comparable occurrences in other traditions, being as it is an iteration of a shared cross-cultural phenomenon of accounting for the solar motion by recourse to anthropomorphic causality and by models devised upon this basic hermeneutic approach. Unfortunately, this only applies to the diurnal solar voyage, since the notions associated with the annual solar voyage in Egypt are almost non existent.[38]

31 Wilkinson 1992: 135; 1994: 159; Kurth 2004: 48.

32 Piankoff 1954: i.41, 344.

33 PT 437.796, 483.1014 (var. 'the gate of the Netherworld'), 610.1713.

34 *Book of Earth* part A, uppermost register (Piankoff 1954: i.330 Fig. 89, 345 Fig. 101 = 1957: i.27 Fig. 9; Hornung 1999: 100, 102, 104 Figs. 54–55); the papyrus of Khonsu-mes (Piankoff 1957: i.27 Fig. 10); cf. the *Book of Caverns* (Piankoff 1954: i, Fig. 12 between pp. 66–67). Probably also the so-called Dream Stele or Sphinx Stele of Thutmose IV (Wilkinson 1992: 135 with Fig. 3 on p. 134; 1994: 152–53, 158 with Ill. 106; cf. *ARE* ii.812–15, p. 321–24).

35 Naville 1886: i.XXVII; Budge 1899: Pl. 8 (Hunefer); Saura Sanjaume 2006: 106 Fig. 6.3 (Sennedjem); cf. in other contexts Piankoff 1957: i. 33 Fig. 15; Wilkinson 1992: 134 Fig. 2; Maravelia 2009: 62 Fig. 3b. Variants of this illustration are found in e.g., Piankoff 1957: i.41–42 with Fig. 27.

36 Ani: Nagel 1929: 54; Plate VII in Dassow 1998; Louvre: Nagel 1929: Pl. IV K with a description on pp. 53–54, cf. the papyrus of Gaut-sushen A (Piankoff 1957: i.182).

37 The tomb of Ramesses VI (Piankoff 1954: i Figs. 139 after p. 434, 140 on p. 436 (unpaginated)); the papyrus of Djed-Khonsu-iuf-ankh I (Piankoff 1957: i.24 Fig. 5).

38 For a few possible references to the annual solar movement in Egyptian tradition see Bilić 2013b: 110–13. Most of these involve or reflect the concept of *akhet*.

Index

www.ingramcontent.com/pod-product-compliance
Lightning Source LLC
LaVergne TN
LVHW080546210926
848481LV00025B/2746